Application of Artificial Intelligence in Industrial Process Modelling and Optimization (2nd Edition)

Application of Artificial Intelligence in Industrial Process Modelling and Optimization (2nd Edition)

Guest Editors

Sheng Du
Li Jin
Pan Yu
Hao Liu

MDPI

Basel • Beijing • Wuhan • Barcelona • Belgrade • Novi Sad • Cluj • Manchester

Guest Editors

Sheng Du
School of Automation
China University of Geosciences
Wuhan
China

Li Jin
School of Automation
China University of Geosciences
Wuhan
China

Pan Yu
School of Information Science and Technology
Beijing University of Technology
Beijing
China

Hao Liu
School of Automation
Hubei University of Science and Technology
Xianning
China

Editorial Office
MDPI AG
Grosspeteranlage 5
4052 Basel, Switzerland

This is a reprint of the Special Issue, published open access by the journal *Processes* (ISSN 2227-9717), freely accessible at: https://www.mdpi.com/journal/processes/special_issues/2169C2JH6R.

For citation purposes, cite each article independently as indicated on the article page online and as indicated below:

Lastname, A.A.; Lastname, B.B. Article Title. *Journal Name* **Year**, *Volume Number*, Page Range.

ISBN 978-3-7258-7679-2 (Hbk)
ISBN 978-3-7258-7680-8 (PDF)
https://doi.org/10.3390/books978-3-7258-7680-8

Contents

About the Editors

Sheng Du

Sheng Du received a B.S. degree in Measurement & Control Technology and Instruments and a Ph.D. in Control Science and Engineering from China University of Geosciences, Wuhan, China, in 2016 and 2021, respectively. He was a joint Ph.D. student with the Department of Electrical and Computer Engineering, University of Alberta, Edmonton, Canada, from 2019 to 2021 and a Postdoctoral Fellow in Control Science and Engineering, China University of Geosciences, from 2021 to 2023. He is currently a Professor with the School of Automation, China University of Geosciences, and his research interests include process control, intelligent control, intelligent optimization, computational intelligence, and artificial intelligence. Dr. Du is a Senior Member of the Chinese Association of Automation (CAA) and a winner of the Outstanding Doctoral Thesis Nomination Award of the CAA. He is also an Associate Editor of *Measurement and Control.*

Li Jin

Li Jin received a B.S. degree in Automation and a Ph.D. in Control Science and Engineering from China University of Geosciences, Wuhan, China, in 2016 and 2021, respectively. She was a joint Ph.D. student with the Department of Electrical Engineering and Electronics, University of Liverpool, Liverpool, U.K., from 2018 to 2020 and a Postdoctoral Fellow in Control Science and Engineering, China University of Geosciences, from 2022 to 2024. She is currently a Professor in the School of Automation, China University of Geosciences, and her current research interests include power system stability analysis and control, time-delay systems, and robust theory and its applications.

Pan Yu

Pan Yu received a B.S. degree in industrial automation and a Ph.D. in control engineering from Central South University, China, in 2014 and 2019, respectively. She was a research student with the Department of Electrical and Electronic Engineering, Chiba University, Japan, from 2017 to 2019. In 2019, she joined Beijing University of Technology, Beijing, China, where she is currently an Associate Professor with the Faculty of Information Technology. Her research interests include multi-agent systems, disturbance estimation and rejection, time-delay systems, and robust control.

Hao Liu

Hao Liu received a B.S. degree in aircraft engineering from the Shenyang Institute of Aeronautical Engineering in 2007, an M.S. degree in aircraft design from the Harbin Institute of Technology in 2009, and a Ph.D. in control science and engineering from the Harbin Institute of Technology in 2013. From February 2012 to August 2012, he was a Visiting Scholar with the IMT Institute for Advanced Studies Lucca, Italy. He is currently an Associate Professor with the Wuhan Institute of Technology, and his current research interests include cybersecurity, switched systems, multi-agent systems, set-membership estimation, and robotic control.

processes

MDPI

Review

A Review of Hybrid LSTM Models in Smart Cities

Bum-Jun Kim [1] and Il-Woo Nam [2,*]

[1] Spatial Design and Engineering, Handong Global University, Pohang-si 37554, Republic of Korea; ktbumjun@gmail.com

[2] School of Spatial Environment System Engineering, Handong Global University, Pohang-si 37554, Republic of Korea

* Correspondence: namiru@handong.edu

Abstract: Rapid global urbanization poses complex challenges that demand advanced data-driven forecasting solutions for smart cities. Traditional statistical and standalone Long Short-Term Memory (LSTM) models often struggle to capture non-linear dynamics and long-term dependencies in urban time-series data. This review critically examines hybrid LSTM models that integrate LSTM with complementary algorithms, including CNN, GRU, ARIMA, and SVM. These hybrid architectures aim to enhance prediction accuracy, integrate diverse data sources, and improve computational efficiency. This study systematically reviews principles, trends, and real-world applications, quantitatively evaluating hybrid LSTM models using performance metrics such as mean absolute error (MAE), root mean square error (RMSE), and the coefficient of determination (R^2), while identifying key study limitations. The case studies considered include traffic management, environmental monitoring, energy forecasting, public health, infrastructure assessment, and urban waste management. For example, hybrid models have achieved substantial accuracy improvements in traffic congestion forecasting, reducing their mean absolute error by up to 29%. Despite the inherent challenges related to structural complexity, interpretability, and data requirements, ongoing research on attention mechanisms, model compression, and explainable AI has significantly mitigated these limitations. Thus, hybrid LSTM models have emerged as vital analytical tools capable of robust spatiotemporal prediction, effectively supporting sustainable urban development and data-driven decision-making in evolving smart city environments.

Keywords: hybrid LSTM; smart city; time-series forecasting; spatiotemporal prediction; deep learning; traffic flow modeling; environmental monitoring; energy management

1. Introduction

As global urbanization accelerates, cities are increasingly facing complex challenges, such as traffic congestion, rising energy consumption, environmental pollution, and aging infrastructure. However, conventional administrative approaches have proven inadequate to address these multifaceted issues. Consequently, a paradigm shift toward data-driven decision-making and real-time forecasting has become essential. In this context, the concept of the "smart city" has emerged—defined as an intelligent urban system that integrates information and communication technology (ICT), the Internet of Things (IoT), and artificial intelligence (AI) to manage urban resources efficiently and provide improved public services [1]. To operate effectively, smart cities require the ability to collect, process, and analyze vast volumes of urban data and make optimal real-time decisions based on that analysis.

A key feature of smart city operations is their reliance on time-series data generated from sensors embedded in infrastructure, vehicles, and the environment. Examples include energy usage, traffic speed, waste levels, and air quality indices. Accurate forecasting of such data is critical for responsive and sustainable urban management. Traditional statistical models such as autoregressive integrated moving average (ARIMA) and machine learning algorithms such as Support Vector Machines (SVMs) and Random Forests have historically been used for time-series forecasting. However, these models often exhibit limited performance in capturing non-linear patterns or long-term dependencies. For example, while ARIMA models may yield high short-term prediction accuracy, they frequently fail to capture sudden fluctuations or non-linear patterns in traffic forecasting [2]. This results in significantly increased errors over extended time horizons.

To overcome these limitations, deep learning-based approaches have gained prominence, particularly Recurrent Neural Networks (RNNs) and, more specifically, the Long Short-Term Memory (LSTM) model. Originally proposed by Hochreiter and Schmidhuber in 1997, the LSTM architecture introduces cell states and gating mechanisms—input, forget, and output gates—that enable the model to effectively learn long-term dependencies [3]. Since then, LSTM has been successfully applied to various smart city domains, such as traffic forecasting, energy demand prediction, and air quality monitoring. Despite its strengths, the standard LSTM model remains constrained by its fixed structure and limited adaptability, which may hinder its prediction performance in highly dynamic environments.

To address these shortcomings, recent studies have focused on hybrid LSTM models [4]. Hybrid LSTM integrates LSTM with other machine learning or deep learning algorithms, such as CNNs, GRUs, ARIMA, and SVMs, to leverage the complementary strengths of each component. The concept of hybrid time-series modeling can be traced back to Zhang's [5] study in 2003 titled "Time series forecasting using a hybrid ARIMA and neural network model." The study introduced a new paradigm by combining ARIMA for linear trends and artificial neural networks (ANNs) for non-linear components. Zhang's approach has spurred extensive research in hybrid modeling, with its principles adapted to LSTM architecture, leading to various innovative hybrid designs. For instance, Bhanja and Das [6] proposed a hybrid deep learning model that effectively enhanced air quality time-series prediction performance by integrating convolutional layers with LSTM units.

Smart cities generate massive volumes of heterogeneous and dynamic data from various sources, including traffic sensors, energy meters, air quality monitors, and IoT devices. These data streams are often non-linear, highly volatile, and exhibit complex temporal and spatial dependencies. Managing such complexity requires predictive models capable of learning from multivariate time-series data while adapting to rapidly changing conditions.

Traditional models such as CNNs, SVMs, GRUs and others have made significant contributions to time-series forecasting but show inherent limitations when applied in smart city contexts. CNNs are effective in spatial pattern extraction but struggle with long-term dependencies. SVMs perform well on smaller datasets but lack the scalability and temporal learning capacity needed for real-time urban data. GRUs offer computational efficiency but are often less accurate in modeling deep temporal patterns. Hybrid LSTM models address these challenges by integrating the long-term dependency learning strength of LSTM with the complementary features of other algorithms. This integration enables more accurate, robust, and adaptive forecasting, making hybrid LSTM models essential tools for supporting complex decision-making in smart city systems.

The primary objectives of hybrid LSTM models are threefold. First, they enhance prediction accuracy and robustness across varying time horizons, as demonstrated in vehicle speed prediction tasks that integrate ARIMA's linear modeling with LSTM's non-linear pattern recognition [2]. Similarly, Yadav [7] applied a CNN-LSTM hybrid model for short-term load forecasting in smart grids, demonstrating notable improvements in accuracy and computational efficiency. Second, they enable the integration of heterogeneous data types, such as combining time-series data with spatial inputs or unstructured formats such as images and text. This capability is particularly valuable for air quality forecasting, in which sensor data and satellite imagery can be jointly analyzed. Third, to optimize computational efficiency, lightweight architectures such as CNN-GRU-LSTM allow for high-performance training even in resource-constrained environments. In smart cities, hybrid LSTM models serve as key analytical tools by synthesizing diverse data sources to support policymaking and real-time operations, thereby contributing to sustainable urban development [8].

The adoption of hybrid LSTM models is driven not only by technological advances, but also by societal demands. Accelerated urbanization, climate change, energy crises, and post-pandemic public safety concerns have highlighted the need for sophisticated real-time data analysis and forecasting systems. Therefore, hybrid LSTM models have been widely studied and deployed in both academic and industrial settings. The proliferation of IoT sensor networks across cities has enabled the continuous collection of vast time-series datasets. Hybrid LSTM models are particularly well suited to processing these data in real-time, using LSTM to capture long-term patterns and auxiliary algorithms to account for short-term variations and environmental fluctuations [9]. These capabilities position hybrid LSTM as a core technology in a smart city ecosystem.

Prediction model requirements vary significantly across different smart city applications. To meet these diverse and domain-specific demands, the architecture of hybrid LSTM models can be flexibly adapted and designed accordingly. For instance, traffic management systems often require real-time, low-latency predictions to support immediate operational decisions. In contrast, air quality monitoring and public health/safety applications demand high spatial resolution and the capacity to capture spatiotemporal variability. Furthermore, most application domains require models capable of handling seasonal trends, non-linear dynamics, and data irregularities.

To the best of the authors' knowledge, no prior review has specifically examined hybrid LSTM models within the comprehensive context of smart cities. While existing reviews have addressed AI applications or hybrid models in isolated domains—such as transportation or energy—these studies do not offer a unified framework centered on hybrid LSTM approaches across multiple smart city sectors. This paper addresses this gap by providing a comprehensive analysis of hybrid LSTM model applications across diverse smart city domains.

This review aims to analyze the application of hybrid LSTM models in smart city domains, such as transportation, environmental monitoring, energy management, public health, infrastructure maintenance, and waste disposal. It also provides a comprehensive perspective on the evolution, limitations, future expansion, and market potential of this emerging technology, underscoring the strategic significance of hybrid LSTM models in advancing smart city innovation.

2. Principles and Characteristics

This section analyzes the principles, development process, strengths, and weaknesses of hybrid LSTM models.

2.1. Mechanism

The operating principle of hybrid LSTM models leverages the strengths of each combined algorithm to analyze and predict time-series data from multiple perspectives.

Fundamentally, LSTM is a type of recurrent neural network with an exceptional capability to learn data patterns over time. The core of the LSTM architecture consists of cell states and gate mechanisms, as shown in Figure 1. This structure, also represented in Figure 2a, which comprises input, forget, and output gates, regulates the importance of previous information, selectively incorporates new data, and determines appropriate outputs [10]. This architecture enables effective learning even with time-series data with long-term dependencies.

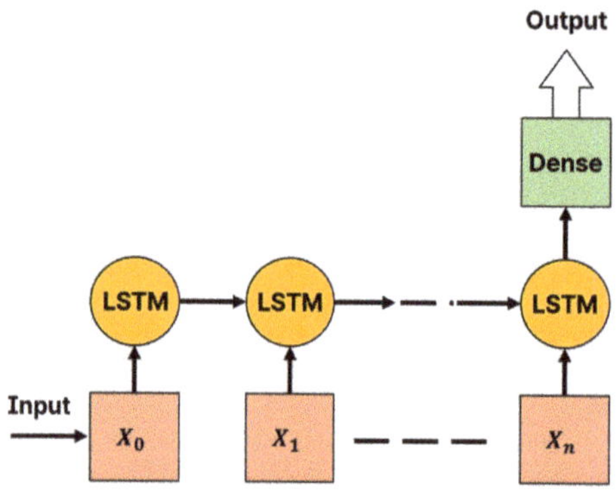

Figure 1. An example of the structure of an LSTM-based prediction model [10].

In hybrid LSTM models, the LSTM structure is combined with other algorithms. For example, gated recurrent units (GRUs) feature simplified gate mechanisms that enable rapid training. When integrated with LSTM, both computational efficiency and performance increase, allowing for the superior handling of problems requiring the simultaneous management of both long- and short-term dependencies.

Furthermore, the integration of ARIMA, a statistical model, into LSTM creates a structure in which ARIMA captures linear trends and seasonality while LSTM learns non-linear patterns. In this approach, data first passes through ARIMA to remove or extract linear components, and then the LSTM models the non-linearity of the residuals.

The data flows in the hybrid models vary according to integration method. In parallel integration, identical input data are fed simultaneously into multiple algorithms, generating outputs that are combined for final predictions. Conversely, in serial integration, data flow sequentially through algorithms. Examples include CNN outputs serving as LSTM inputs and the LSTM processing of ARIMA residuals.

Through these mechanisms, hybrid LSTM models can analyze and predict complex time-series patterns from multiple perspectives, delivering high performance across various smart city applications.

2.2. Technical Characteristics

Hybrid LSTM models enhance time-series prediction performance by combining LSTM's fundamental architecture with other algorithms. Their core attributes include complexity, flexibility, and scalability. Complexity refers to the ability to learn complex data patterns by exploiting the advantages of multiple algorithms. Flexibility denotes the capability of adjusting the model structure for various domains and data types. Scalability represents the potential for model expansion using additional modules or layers. These characteristics are crucial in the multidimensional and complex data environments of smart cities. Complexity captures non-linear urban data characteristics, flexibility enables adaptation to diverse urban challenges, and scalability addresses expanding data volumes as cities grow.

Hybrid integration methods can be categorized as either parallel or serial. LSTM-Prophet models represent a notable example of predicting energy consumption by eliminating seasonality and irregularities. CNN-LSTM models offer another important implementation. In these models, while the CNN processes spatial data, the LSTM handles temporal data. This architecture supports applications in network security and solar power generation forecasting. Particularly noteworthy are the GRU-based RNNs that efficiently handle missing values and noise using Soft GRU mechanisms. This capability makes them effective for traffic congestion prediction.

2.3. Development

Hybrid LSTM models have evolved continuously in terms of structure and function alongside technological advancements. Early hybrid models simply combined LSTM with other algorithms to simultaneously model linearity and non-linearity. With the increase in data complexity, diversification of data types, and emergence of various applications, hybrid LSTM models have developed into more sophisticated structures.

Recent models incorporate attention mechanisms, allowing them to highlight critical components of their input data. This approach complements LSTM's long-term dependency learning capabilities, improving prediction accuracy. Hybrid models combined with transformer architectures are also being developed. These strengthen parallel-processing capabilities and improve long-term dependency learning.

In smart city applications, models that consider spatial characteristics, such as CNN-LSTM and ConvLSTM, have received significant attention. ConvLSTM integrates convolutional operations into the LSTM structure and effectively learns spatiotemporal patterns. This integration exhibits excellent performance in traffic flow prediction and meteorological data analyses.

Hybrid models also incorporate ensemble techniques and improve prediction accuracy by combining predictions from various models. Recent studies have shown the active progression of model compression and optimization to increase computational efficiency. Examples include reducing the model parameters and training lightweight models through knowledge distillation.

In addition to the approaches already discussed, further extensions of LSTM have been exploited. Bidirectional LSTM (Bi-LSTM) processes input sequences in both forward and backward directions, improving context understanding in time-series prediction. LSTM with residual or skip connections enhances gradient flow and training stability, particularly in deeper architectures. Moreover, LSTM has been effectively combined with optimization algorithms (e.g., Particle Swarm Optimization, genetic algorithms) and signal processing techniques (e.g., wavelet transforms, empirical mode decomposition) to improve hyperparameter tuning, feature extraction, and noise reduction.

These developments have expanded the application scope of hybrid LSTM models and improved their performance across various domains.

2.4. Advantages and Disadvantages

Hybrid LSTM models have the following advantages:

1. High prediction accuracy and generalization ability: Compared to single models, hybrid LSTM models achieve higher prediction accuracy by combining the strengths of multiple algorithms. Considering both linearity and non-linearity simultaneously enables the effective learning of complex data patterns and ensures consistent performance across various scenarios.
2. Processing of complex non-linear data: These models handle complex non-linear data patterns by leveraging LSTM's long-term dependency learning capabilities alongside

the specialized functions of other algorithms. This is particularly useful for analyzing complex time-series data in smart cities, including traffic flow, energy consumption, and environmental changes.

3. Application potential across diverse fields: Hybrid LSTM models can be applied to various smart city domains, including transportation, energy, environment, health, and public safety. Their flexible structural adjustments based on data characteristics enable a broad application scope.
4. Real-time data-processing capability: These models effectively process and predict large volumes of real-time data, contributing to smart city real-time monitoring and rapid decision-making.

They also have the following disadvantages:

1. Complexity of model structure: Their complex structures combining multiple algorithms increase the training time and complicates model implementation and tuning. This results in increased computational resource requirements.
2. Requirement of large data and risk of overfitting: More complex models require larger training datasets. Insufficient data creates an overfitting risk, which degrades the generalization ability of the model.
3. Limitations in interpretability: The complex structures of hybrid models create difficulties in interpreting internal mechanisms and prediction results. This can affect the transparency and reliability of the models.
4. Importance of data quality and preprocessing: When combining various data sources, data quality and preprocessing significantly influence model performance. Handling data noise and missing values are crucial.

Several studies have addressed these limitations. Computational efficiency has been improved through model compression and optimization. Regularization techniques and data augmentation can prevent overfitting. Explainable AI (XAI) technology has been actively incorporated to enhance model interpretability. These efforts have aimed to increase the practicality and reliability of hybrid LSTM models.

3. Case Studies

This section examines the implementation of hybrid LSTM models across key smart city domains by analyzing real-world applications and their performances. This review analyzes use cases across six major areas: traffic management, air quality and environmental monitoring, energy management, health and public safety, urban infrastructure monitoring, and waste management.

3.1. *Traffic Management*

Traffic prediction for management purposes is challenged by the heterogeneous nature of data collected from IoT devices, sensors, and GPSs. These data are often noisy, incomplete, and exhibit complex temporal and spatial dependencies. Traditional models like ARIMA, SVMs, and CNNs struggle to capture both the linear and non-linear dynamics of urban traffic flow. Therefore, hybrid deep learning approaches are increasingly essential for achieving accurate and robust traffic forecasting.

IoT-based hybrid architectures have been widely explored for traffic management applications. Chahal et al. [11] proposed a hybrid model that combined seasonal autoregressive integrated moving average (SARIMA) with bidirectional long short-term memory (Bi-LSTM) for traffic congestion forecasting. SARIMA captures the linear components of IoT-based time-series data, whereas Bi-LSTM addresses non-linear patterns. The model further incorporates a Backpropagation Neural Network (BPNN) to optimize its combined output. By modeling the non-linear relationship between predictions, this multi-model approach improves forecasting accuracy and offers a more advanced alternative to basic model stacking.

These models exhibit statistically significant improvements over conventional methods by effectively capturing complex non-linear patterns in urban traffic data.

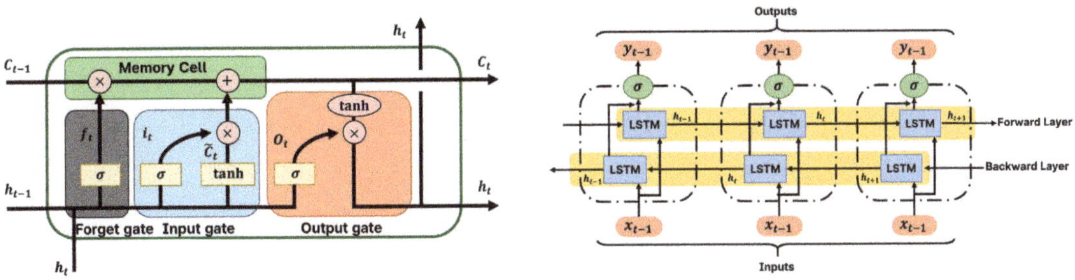

(a) (b)

Figure 2. (**a**) LSTM architecture; (**b**) Bi-LSTM architecture [12].

The dataset was sourced from the CityPulse EU FP7 project data, which included time-series data on vehicle speed and traffic volume. The proposed model was evaluated using mean absolute error (MAE), mean squared error (MSE), root mean square error (RMSE), and mean absolute percentage error (MAPE). The hybrid model consistently outperformed both the standalone LSTM and ARIMA models across all evaluation metrics.

The quantitative results demonstrated significant performance gains: the hybrid model achieved an MAE of 0.499 (29% improvement over standalone LSTM's 0.705), MSE of 0.337 (34% reduction compared to LSTM's 0.510), RMSE of 0.58 (compared to LSTM's 0.715), and MAPE of just 3% [11]. These metrics confirm the substantially enhanced predictive capabilities.

The results validate the effectiveness of the hybrid approach in addressing single-model limitations through the comprehensive modeling of both linear and non-linear data components. The proposed SARIMA–Bi-LSTM architecture represents a viable solution for smart city traffic prediction and management, with direct implications for urban mobility optimization and service quality enhancement.

In a follow-up study, Chahal et al. [12] introduced ARIMA-Bi-LSTM using an Aquila Optimizer (AB-AO) to improve traffic congestion forecasting in smart cities. As depicted in Figure 2b, the model combines ARIMA for capturing linear trends with Bi-LSTM for learning non-linear patterns. The performance was further enhanced through the hyper-parameter optimization of the Bi-LSTM component using an Aquila Optimizer (AO). The model was tested using traffic data from three cities provided by the CityPulse EU FP7 project.

The AB-AO model outperformed standalone ARIMA, Bi-LSTM, and the conventional ARIMA-LSTM combination across all evaluation metrics. It achieved the lowest MAE of 3.18, compared to 4.63 (ARIMA), 4.20 (Bi-LSTM), and 3.98 (ARIMA-LSTM) [12]. Similarly, it recorded the lowest MSE (18.74) and the lowest MAPE (0.21), significantly improving the forecast accuracy over competing models [12].

These results confirm the ability of the AB-AO model to effectively capture complex traffic patterns and highlight the value of AO-based optimization for enhancing predictive performance. This study positions the AB-AO approach as a high-performance and scalable solution for intelligent traffic management in smart cities.

Focusing on the practical application of IoT-based traffic data, Culita et al. [13] proposed the incorporation of Autoregressive Moving Average (ARMA) into LSTM to address congestion issues in Bucharest, Romania. By comparing the performance of the ARMA–LSTM model with that of a standard LSTM model, their study evaluated the benefits of the hybrid approach in traffic forecasting. Both models were assessed under short-term (25 min) and long-term (50 min) forecasting scenarios using RMSE and MAE as the primary evaluation metrics, as shown in Figure 3.

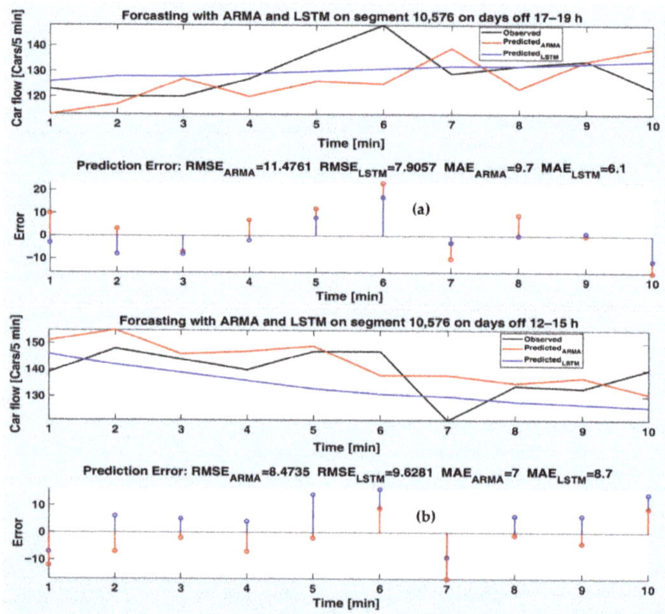

Figure 3. Measured and predicted traffic during 10 time steps for 10,576 segments: (**a**) during holidays, between 5:00 pm and 7:00 pm; (**b**) during holidays, between noon and 3:00 pm [13].

In short-term forecasting, the ARMA–LSTM model delivered RMSE values 8–12% lower than those of standalone LSTM [13]. For example, in one segment, LSTM recorded an RMSE of 13.22, while the hybrid model achieved an RMSE of 12.14 [13]. The MAE results followed a similar pattern, indicating greater precision in short-term predictions.

For long-term forecasting, the performance gap increases further. The ARMA–LSTM model recorded an RMSE of 16.42, compared to 18.77 from the LSTM, representing a 13% reduction in error [13]. The MAE was also 10–15% lower on average, demonstrating the effectiveness of the hybrid model in enhancing forecast accuracy across both time horizons [13].

The results demonstrate that the ARMA–LSTM model effectively addresses the structural limitations of standard LSTM models and delivers improved prediction accuracy. This enhancement is driven by the ARMA component, which enables the model to better capture the complex temporal patterns inherent in time-series data. Particularly for dynamic, non-linear datasets such as traffic flow, the ARMA–LSTM model offers more refined forecasts, underscoring its practical value in real-time traffic management applications.

Zafar et al. [14] proposed a hybrid LSTM-GRU model to predict traffic speed in smart cities by integrating heterogeneous data sources, including IoT sensor feeds, Google Maps traffic data, weather conditions, and holiday schedules, as shown in Figure 4. The model combines the temporal learning strength of LSTM with the computational efficiency of GRUs to capture complex spatiotemporal traffic patterns. To increase prediction accuracy, the model was trained and validated using large-scale traffic data collected at 15 min intervals in Islamabad, Pakistan.

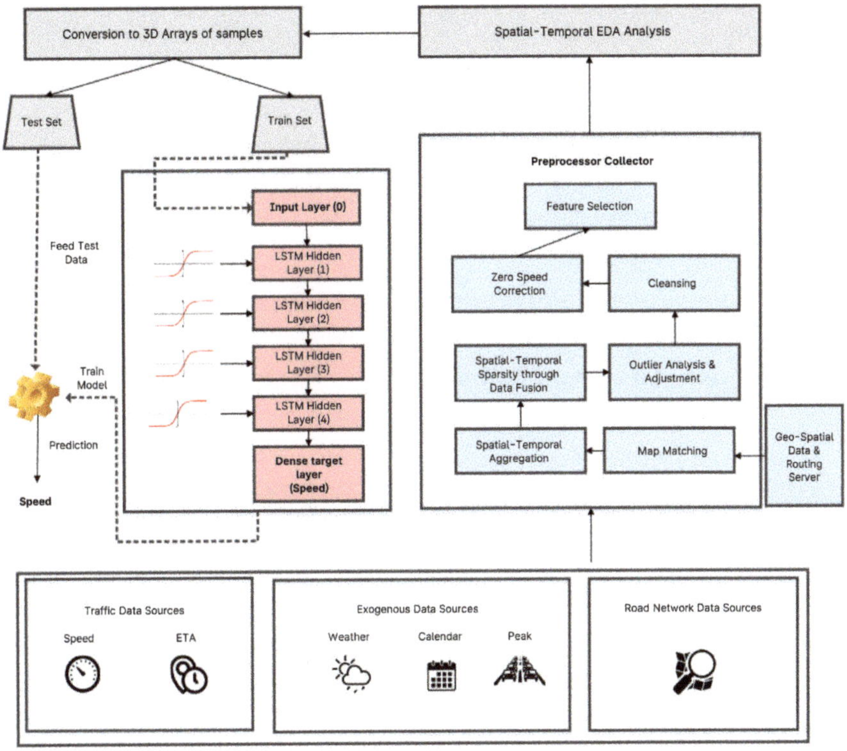

Figure 4. Predicted urban traffic vehicle speeds based on LSTM-GRU model [14].

According to Figure 5, the model demonstrated superior predictive accuracy over the standalone models. The hybrid LSTM-GRU model achieved an RMSE of 4.5, outperforming LSTM (4.86) and the GRU (5.05) [14]. It also recorded a lower MAE of 2.03, compared with 2.13 for LSTM and 2.29 for the GRU [14]. In terms of the MAPE, the hybrid model achieved the lowest error of 6.67%, whereas the LSTM and GRU recorded errors of 6.95% and 7.7%, respectively [14].

These results confirm the effectiveness of the hybrid LSTM-GRU model in processing complex urban traffic data. This approach offers practical benefits for real-time traffic management and system optimization in smart city applications.

Sengupta et al. [15] proposed a hybrid traffic flow prediction model that combined a Hidden Markov Model (HMM) with LSTM to improve short-term forecasting, ranging from 5 min to 1 h. Their objective was to enhance traffic management and alleviate congestion by accurately capturing short-term fluctuations.

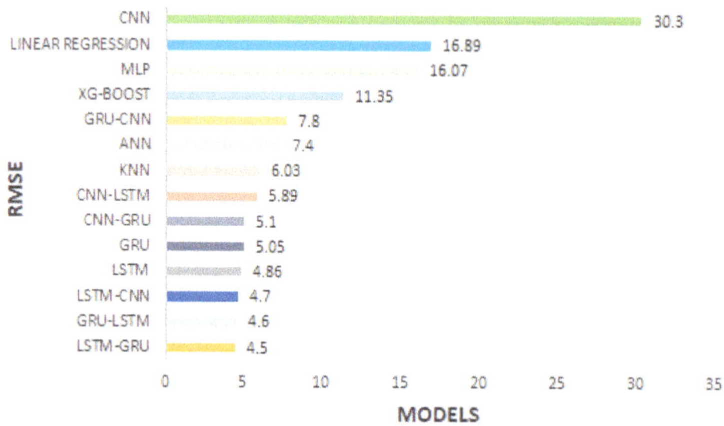

Figure 5. Comparison of root mean square error (RMSE) across predictive models [14].

The proposed model combines the ability of the HMM to identify shifts in the traffic state with LSTM's ability to effectively learn temporal dependencies. Among the two model architectures presented, the concatenated hybrid (C-Hybrid) structure, which combines the outputs from two LSTM networks, achieved the highest prediction accuracy.

Model performance was evaluated using real-world data from the California Department of Transportation's Performance Measurement System (PeMS). The C-Hybrid model recorded an RMSE of 0.4203 [15]. This is significantly lower than those of the stacked LSTM (0.8235) and AR-HMM (0.8766) models, reflecting an improvement of up to 52% [15]. These results demonstrate the superior ability of the hybrid model to predict sudden changes in traffic flow.

The study highlights the potential of combining an HMM and LSTM to improve prediction accuracy and reliability, offering practical value for advanced traffic management systems.

Chaoura et al. [16] suggested a hybrid LSTM model incorporating Particle Swarm Optimization (PSO) to enhance intersection traffic flow prediction. Their study aimed to mitigate intersection congestion and improve road safety, emergency response time, and overall traffic efficiency.

The hybrid architecture integrates LSTM's robust temporal learning capabilities with PSO's exceptional optimization functionality. PSO, which mimics collective intelligence in natural systems to identify optimal solutions, effectively optimizes the LSTM hyperparameters (learning rate and network architecture) to enhance predictive accuracy. This optimization process identifies the optimal model configurations, thereby enabling precise traffic predictions.

The study utilized 48,120 actual vehicle passage records collected from four distinct intersections in Nanjing, China, between November 2015 and June 2017 [16]. The data were preprocessed to optimize the characteristics of each intersection. The performance evaluation demonstrated that the proposed model outperformed the conventional LSTM, Random Forest (RF), k-neighbor (KN), and Decision Tree (DT) models across all intersections. For instance, at intersection 1, the model achieved an RMSE of 0.15258, indicating highly accurate predictions.

The performance assessment of the LSTM-PSO model showed its superior performance at all intersections compared with standalone LSTM and the widely used RF, KN, and DT models, as illustrated in Figures 6 and 7. At intersection 1, the model recorded an RMSE of 0.15258 and an MAE of 0.0898, demonstrating exceptional accuracy [16]. These results indicate the robust predictive capabilities of the LSTM-PSO model across complex and diverse intersecting environments.

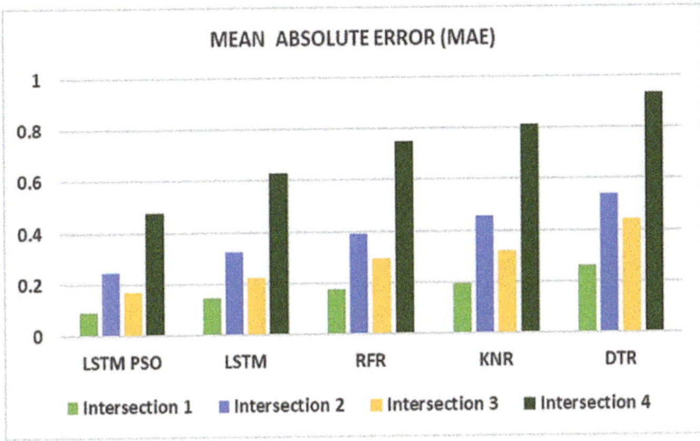

Figure 6. Model training MAE results across four intersections [16].

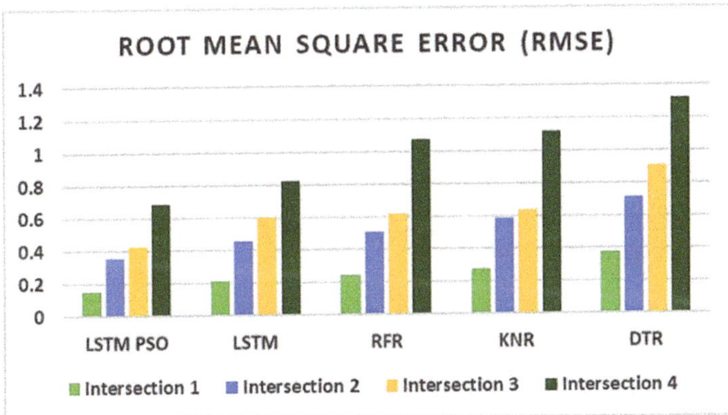

Figure 7. Model training RMSE results across four intersections [16].

In addition, the results validate the practical applicability of the LSTM-PSO hybrid approach for real-time traffic signal control and traffic management system implementation. Moreover, the results have significant potential contributions to the development and deployment of smart cities and intelligent transportation systems (ITSs).

In bicycle traffic flow forecasting, hybrid models have been proposed to address the non-linearity and complex distribution characteristics of time-series data. Myhrmann and Mabit [17] developed a Long Short-Term Memory Mixture Density Network (LSTMMDN) model for precise hourly bicycle flow prediction across urban areas. This architecture provides accurate exposure data for policy analyses such as bicycle traffic safety assessments.

The LSTMMDN integrates Mixture Density Network (MDN) components to predict various possible bicycle flow distributions probabilistically under identical conditions. This incorporation enables more realistic predictions than conventional models, which only forecast mean values, effectively capturing the traffic volume variability across different times of the day, days of the week, and weather conditions.

This study analyzed 64,664 hourly observations from bicycle counters in Copenhagen, Denmark, from 2017 to 2020, incorporating 17 weather and time variables [17]. Performance comparisons against the Danish Road Directorate's Seasonal Variation Factor (SVF) method and standard LSTM showed that the LSTMMDN achieved an up to 77% accuracy improvement compared with SVF methods, with an MSE of 0.102 [17]. Meanwhile, SVF recorded an MSE of 0.377 [17]. The model also consistently outperformed the standard LSTM implementations.

Visual analysis in Figure 8 confirmed the LSTMMDN's superior representation of actual traffic patterns. Using only 5% of the available data, the model maintained more than 50% higher accuracy than SVF, demonstrating strong performance with limited datasets [17]. In safety applications, LSTMMDN predictions improved the explanatory power of accident risk analysis by 5.5%, highlighting the importance of precise exposure data in safety assessments [17]. This research establishes the LSTMMDN as an effective tool for traffic management and policy development.

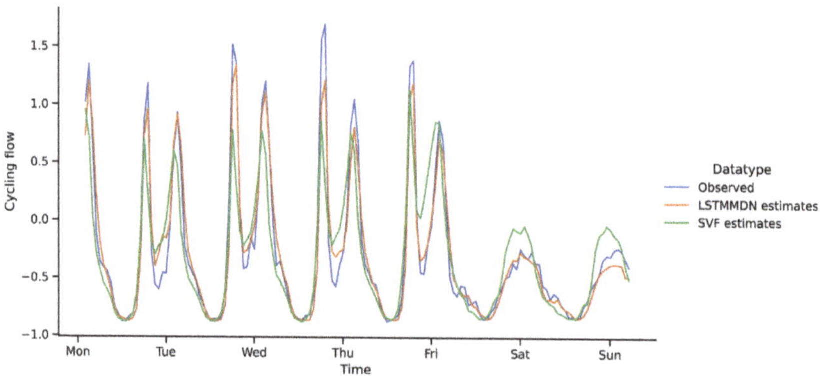

Figure 8. Bicycle flow visualization over 148 hours at a randomly selected monitoring site—actual counts (blue), LSTM-MDN predictions (orange), SVF-based predictions (green) [17].

Miao and Liao [18] proposed a hybrid CNN–LSTM-PSO model to improve traffic congestion forecasting in smart cities by leveraging real-time IoT-enabled sensor data. The goal was to enhance prediction accuracy and enable dynamic traffic signal control based on urban mobility patterns.

The model architecture integrates CNNs for spatial feature extraction, LSTM for capturing temporal dependencies in traffic patterns, and PSO for hyperparameter tuning. Sequential trends—such as vehicle flow, speed, and incidents—were modeled by embedding LSTM layers within the CNN framework. PSO effectively tuned key parameters, including learning rate and filter configurations, to accelerate convergence and enhance model stability.

Real-time traffic data were collected over one year from 50 intersections, generating approximately 1.2 million sensor records, with features such as vehicle count, speed, traffic density, incident indicators, and weather. Preprocessing included interpolation, normalization, and augmentation. The dataset was split into training (70%), validation (15%), and testing (15%).

The CNN–PSO model achieved 92% accuracy in congestion classification and reduced the MSE by 15% compared to baseline models [18]. It recorded an RMSE of 0.030 and an MAE of 0.025, outperforming SVM and RF approaches [18]. Its precision, recall, and F1-scores were 90%, 89%, and 89.5% [18]. Real-time deployment simulations showed a 20% reduction in average traffic delay and a 25% improvement in traffic flow efficiency [18].

A notable feature of the study was its real-time integration with traffic infrastructure, enabling dynamic signal control based on live LSTM-informed predictions. The model demonstrated low latency, high adaptability across urban conditions, and reduced computational overhead through PSO-based tuning compared to exhaustive search methods.

This work highlights the effectiveness of hybrid LSTM models in smart cities—fusing spatial and temporal learning with intelligent optimization to address complex, dynamic traffic systems. By embedding LSTM within a multimodal architecture, the model accurately captured non-linear and evolving patterns in urban mobility, offering a scalable, low-latency solution for intelligent transportation systems (ITSs) and real-time traffic management

3.2. Air Quality and Environmental Monitoring

Air quality and environmental factor prediction in smart cities is difficult due to noisy, incomplete, and varied sensor data, and rapidly changing conditions influenced by traffic, industry, and weather. Traditional models struggle with long-term and spatiotemporal patterns, while hybrid deep learning models improve accuracy and enable real-time decisions.

In a smart city, managing the environment through accurate predictions directly affects citizens' quality of life. Environmental monitoring primarily focuses on forecasting air quality and energy production.

Kataria and Puri [19] proposed a hybrid CNN-LSTM model for air quality prediction in smart cities. Their proposed architecture efficiently processes atmospheric pollution data to accurately forecast air quality index (AQI) values. They enhanced the prediction accuracy by combining the CNN architecture, which effectively extracts key features from data, with traditional LSTM, which learns temporal dependencies. Model performance was evaluated using two extensive public datasets from Beijing and Chennai.

The CNN-LSTM architecture processes data sequentially, beginning with the CNN learning spatial characteristics and transferring the extracted features to LSTM, followed by LSTM analysis of temporal dependency. This approach enables more sophisticated data interpretation and higher prediction accuracy compared with the operations performed by conventional LSTM models.

The results in Figure 9 demonstrated that the CNN-LSTM model significantly outperformed standalone LSTM models. Using a dataset from Beijing, China, the hybrid model recorded an RMSE of 23.14, which was approximately 3.5% lower than that of the LSTM model (23.97) [19]. For MAE, CNN-LSTM achieved 14.89, which was approximately 6.3% lower than LSTM's 15.89 [19]. The Chennai dataset showed similar improvements, with CNN-LSTM recording an RMSE of 30.85 and MAE of 26.69, both lower than LSTM's 31.52 and 29.89, respectively [19].

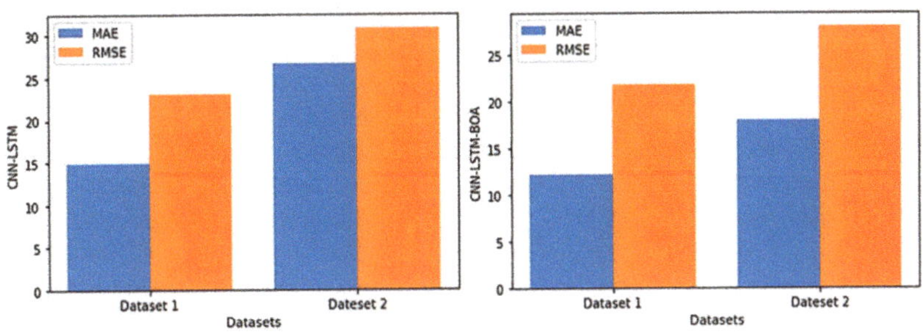

Figure 9. MAE and RMSE across various datasets [19].

These findings confirm that the CNN-LSTM model processes the spatiotemporal complexity of air quality data more effectively than standalone LSTM architectures. This approach is a valuable tool for predicting air pollution in smart cities, implementing early warning systems, supporting environmental policy development, and mitigating health risks.

To forecast the photovoltaic output in smart urban environments, Kim et al. [20] incorporated a seasonal autoregressive integrated moving average with exogenous factors (SARIMAX) model into LASTM in a stacked ensemble framework. Figure 10 shows the overall flow of Kim et al.'s study. Kim et al. attempted to incorporate SARIMAX's temporal processing capabilities into LSTM's advanced pattern recognition to improve prediction accuracy across diverse geographical locations.

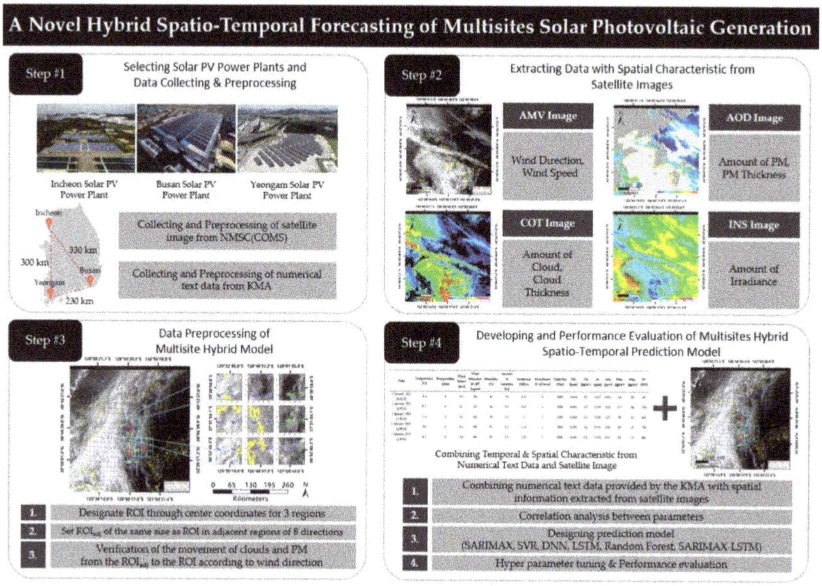

Figure 10. Overview of SARIMAX-LSTM hybrid model study [20].

The research team used predictions from the SARIMAX model, which excels in capturing seasonal characteristics and exogenous factors, as training data for the LSTM. It could handle both long- and short-term dependencies in time-series data to generate more refined forecasting results.

SARIMAX excels in analyzing data by considering seasonal characteristics and exogenous variables and effectively capturing temporal dependencies and seasonality. LSTM can simultaneously process both long- and short-term dependencies in time-series data using predictions from SARIMAX as training data to generate more refined results. The study evaluated the performance of the proposed SARIMAX-LSTM model using solar power plant data from three regions in South Korea: Incheon, Busan, and Yeongam.

As depicted in Figure 11, the empirical analysis revealed substantial improvements with the integrated approach: an MAE of 64.730, representing a 15.8% reduction compared to traditional LSTM's 76.913; an RMSE of 95.800, outperforming standard LSTM's 106.123 by 9.7%; and a symmetric mean absolute percentage error (SMAPE) of 19.891%, further confirming the superior performance compared to LSTM's 23.369% by 15% [20].

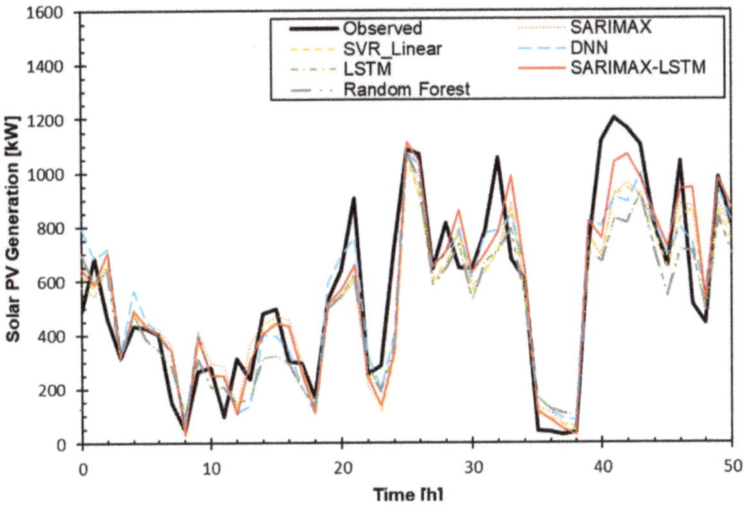

Figure 11. Comparison of predicted solar power generation values by each forecasting model [20].

These findings confirm the superior capability of the SARIMAX–LSTM model in reflecting seasonality and long-term dependencies in time-series data, providing enhanced precision for smart city solar power generation forecasting. This model is an important tool that can contribute to establishing efficient power supply plans and promoting the effective utilization of energy resources.

Mukhtar et al. [21] introduced a CNN-LSTM-ANN hybrid architecture for solar radiation prediction in smart cities. This framework focuses on learning spatiotemporal data patterns to more accurately forecast hourly solar radiation variations. As shown in Figure 12, the model employs a CNN to extract non-linear features, LSTM to learn temporal dependencies, and an ANN to perform the final prediction calculations, thereby enhancing the overall performance.

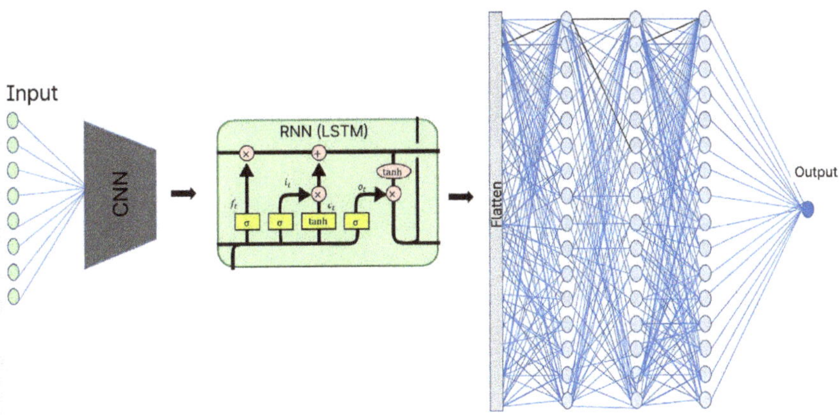

Figure 12. Structure of CNN-LSTM-ANN hybrid model [21].

The distinctive feature of the CNN-LSTM-ANN structure is its hierarchical organization: the CNN extracts spatial patterns from input data, LSTM processes long-term sequential dependencies, and the ANN computes the comprehensive prediction values.

This architecture was specifically engineered to effectively manage the spatiotemporal complexity of environmental data. The study evaluated the model's performance across varying climatic conditions using datasets from ten African countries.

The assessment results demonstrated that the CNN-LSTM-ANN model had superior predictive capabilities compared with the traditional ANN and CNN-ANN implementations. In Ethiopia, the CNN-LSTM-ANN recorded an MAE of 25.89 W/m^2, RMSE of 72.27 W/m^2, and MAPE of 9.31%, showing improvements over the ANN model's metrics (MAE: 28.13 W/m^2; RMSE: 73.81 W/m^2; MAPE: 10.28%) [21]. Similarly, in Somalia, the hybrid approach achieved an MAE of 16.60 W/m^2, RMSE of 51.54 W/m^2, and MAPE of 5.725%, again outperforming the ANN model (MAE: 22.21 W/m^2; RMSE: 52.80 W/m^2; MAPE: 7.78%) [21].

The study demonstrates how the CNN-LSTM-ANN hybrid architecture enhances prediction accuracy by capturing complex spatiotemporal patterns in environmental data. This model contributes to the efficient management and utilization of renewable energy resources in smart urban environments. Such advanced modeling approaches are critical in improving the reliability of energy management systems against climate variability and in strengthening sustainability initiatives in smart city development.

Zaini et al. [22] proposed a hybrid deep learning framework to improve the accuracy of urban PM2.5, aiming to maintain environmental sustainability and reduce health risks from air pollution. Their research addresses the deteriorating air quality in Kuala Lumpur, Malaysia, resulting from rapid industrialization, urbanization, traffic congestion, and increased energy consumption. The accurate prediction of PM2.5, which has a significant impact on public health and the environment, is essential.

The research team developed a hybrid architecture based on LSTM, integrating the Ensemble Empirical Mode Decomposition (EEMD) data decomposition technique with the metaheuristic optimization algorithms PSO and Sparrow Search Algorithm (SSA). Initially, they constructed PSO-LSTM and SSA-LSTM models to optimize the hyperparameters. Then, they developed EEMD-PSO-LSTM and EEMD-SSA-LSTM models based on the best-performing input configurations. In this architecture, LSTM learns long-term dependencies in time-series data, EEMD decomposes complex non-linear data into simpler components, and PSO/SSA handles LSTM hyperparameter optimization. The SSA demonstrated a more stable convergence than PSO and utilized the Levy Flight Trajectory to address local optimization issues.

This study utilized hourly time-series data collected from nine air quality monitoring stations and three meteorological stations in Malaysia from 1 January to 31 December 2019. The dataset included six air pollution variables (PM2.5, PM10, SO_2, NO_2, O_3, and CO) and five meteorological variables (temperature, relative humidity, wind speed, wind direction, and precipitation). The primary target areas were Batu Muda and Cheras, Kuala Lumpur. Missing values were addressed using linear interpolation, and the data were split into 70% training and 30% testing segments.

The performance evaluation used RMSE, MAE, MAPE, and the coefficient of determination (R^2). The initial models showed that the input variable combination M4 (air pollutants excluding PM2.5, SO_2, temperature, and relative humidity) delivered the best prediction performance. The SSA-LSTM model demonstrated a more stable performance than the PSO-LSTM model across multiple iterations. Models incorporating EEMD (EEMD-SSA-LSTM and EEMD-PSO-LSTM) generally outperformed the models without decomposition techniques. Notably, the EEMD-SSA-LSTM model achieved the best performance when PM2.5 was included from adjacent monitoring points (M5) as an additional input. In the Batu Muda area, this approach improved the RMSE by 2.65% and MAE by

9.31%, whereas in Cheras the improvements were 20.00% and 25.30%, respectively, as shown in Figure 13 [22].

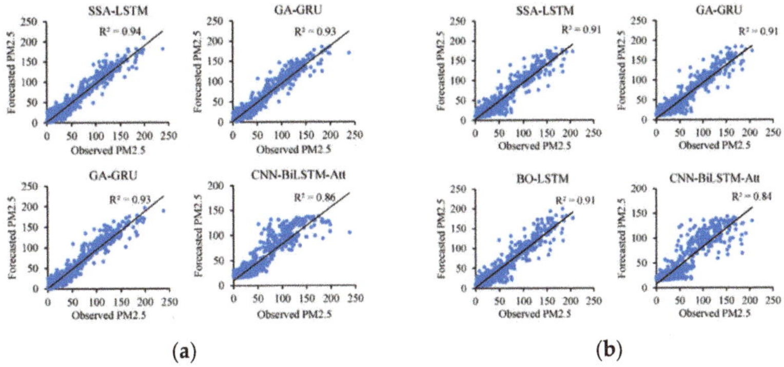

Figure 13. Performance comparison of various models: (**a**) Batu Muda and (**b**) Cheras [22].

The study demonstrates that a hybrid approach combining LSTM, data decomposition (EEMD), metaheuristic optimization (SSA), feature selection, and spatial data (neighboring PM2.5 information) effectively enhances the accuracy of air quality prediction. Such forecasting models can support practical response strategies, including environmental policy development, early warning systems, and emission control during pollution peaks, thereby contributing to environmental sustainability and public health protection. Future research could include incorporating meteorological variables from adjacent areas, improving computational efficiency, enhancing the day/night prediction performance, and applying the model to datasets from diverse regions.

Kaveh et al. [23] introduced a novel hybrid model called OA-LSTM for air pollution management, specifically for PM2.5 prediction. This architecture combines the Orchard Algorithm (OA) with Long Short-Term Memory networks to address the fundamental limitations of traditional LSTM implementations, including local minimum entrapment, high computational costs, and dependency on continuous objective functions, as illustrated in Figure 14. The OA employs a metaheuristic global search strategy inspired by orchard dynamics to efficiently perform optimization in high-dimensional parameter spaces. Additionally, the researchers implemented the Binary Orchard Algorithm (BOA) for sensitivity analysis and feature selection, optimizing the model by maintaining key predictive variables while minimizing prediction errors.

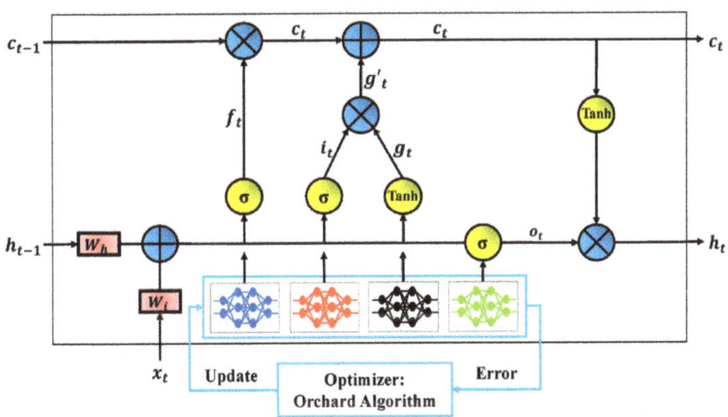

Figure 14. Structure of OA-LSTM [23].

The study utilized meteorological data, topographical features, PM2.5 pollution levels, and satellite imagery from Tehran, Iran, from 2014 to 2016. To prepare the model input dataset, data preprocessing involved noise reduction using Savitzky–Golay filtering, spatial interpolation through inverse distance weighting (IDW), and missing-value reconstruction via spline interpolation. The performance evaluation of the OA-LSTM framework employed multiple metrics, including RMSE, R^2, Standard Deviation Error (SDE), convergence trends, and computational complexity.

When compared against five machine learning models—traditional LSTM, an RNN, a Deep Neural Network (DNN), Random Forest (RF), and a Support Vector Machine (SVM)—the OA-LSTM architecture demonstrated superior performance, with an RMSE of 3.01 μg/m^3 and R^2 of 0.88 [23]. In contrast, conventional LSTM recorded an RMSE of 9.53 and R^2 of 0.65, indicating substantially lower performance [23]. OA-LSTM also exhibited greater stability with an SD Error of 0.05, outperforming all comparison models [23]. Computational efficiency testing revealed that OA-LSTM reached an RMSE below 10 μg/m^3 within 173 s, as shown by the convergence trend in Figure 15 [23]. Meanwhile, the RNN required 634 s, and the other models either failed to achieve this threshold or demanded significantly more processing time [23]. Through BOA-based feature selection, OA-LSTM achieved an RMSE of 5.12 μg/m^3 using only nine critical features, surpassing the performance of the RF and SVM models that utilized more extensive feature sets [23].

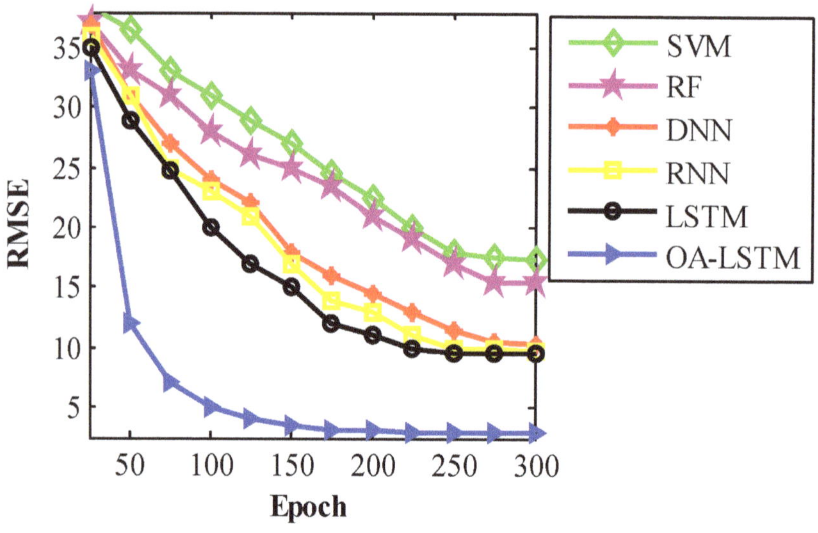

Figure 15. Convergence trend of proposed model based on RMSE [23].

This study demonstrates that integrating LSTM with an innovative OA optimization technique effectively addresses air pollution prediction challenges. The OA optimizes the LSTM weight and bias configurations, enhancing the convergence speed and model efficiency, whereas the BOA maximizes the prediction accuracy while minimizing variable redundancy. The proposed framework offers a scalable and efficient solution for urban air quality management and early warning systems for extreme pollution events.

Ratković et al. [24] conducted a comparative analysis of SVM and hybrid LSTM regression models for air pollution prediction in Niksic, Montenegro. Their study aims to forecast the Common Air Quality Index (CAQI) for upcoming periods and proposes a hybrid model that integrates a Genetic Algorithm (GA) with LSTM for hyperparameter optimization. The GA systematically searches for optimal combinations of time-step quantities and LSTM units per layer, thereby enhancing prediction accuracy and model reliability while preventing overfitting. Additionally, researchers have implemented bagging techniques to create an ensemble LSTM configuration, train multiple models on different data subsets, and combine their predictions to further improve performance.

This study used air pollution concentration data collected from monitoring stations in the Niksic region of Japan. After data preprocessing and feature engineering, the dataset was partitioned into training and testing segments and normalized. The performance evaluation employed multiple metrics, including the MSE, MAE, MAPE, and R^2, with an additional statistical analysis of the MAE values, summarized in Table 1.

Table 1. Evaluation results of SVM, optimized LSTM, and ensemble models for hourly CAQI prediction [24].

Performance Metric	SVM Model	Best LSTM Model	Ensemble Model
MSE	428.2885	58.8046	56.5810
MAE	13.8739	4.6657	4.5560
MAPE	0.2685	0.0923	0.0912
R^2	0.3820	0.9135	0.9168

The ensemble LSTM architecture demonstrated superior performance compared with the SVM model, with an MSE of 56.5810, MAE of 4.5560, MAPE of 0.0912, and R^2 of 0.9168 for CAQI prediction [24]. In contrast, the SVM model achieved a considerably lower performance, with an MSE of 428.2885, MAE of 13.8739, MAPE of 0.2685, and R^2 of 0.3820 [24]. Furthermore, the ensemble LSTM model showed improved metrics compared with the single LSTM implementation (MSE: 58.8046; MAE: 4.6657) and outperformed conventional LSTM architectures, as illustrated in Figure 16 [24].

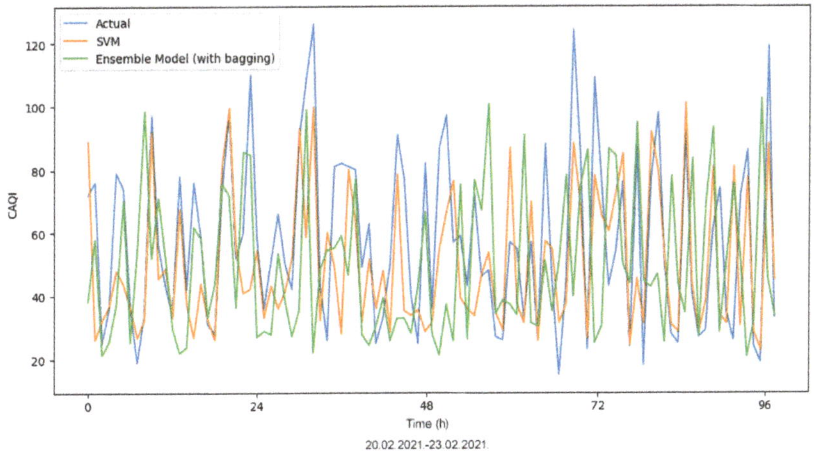

Figure 16. Actual values, SVM model predictions, and ensemble model predictions on subset of test dataset [24].

The study was the first demonstration of the effectiveness of a hybrid model combining LSTM, GA, and bagging techniques for air pollution prediction in Montenegro. The findings confirm that this integrated approach can significantly enhance the accuracy and reliability of hourly air pollution forecasts. The proposed modeling framework can be provided to relevant authorities to support pollution-level predictions and facilitate appropriate response measures.

Zeng et al. [25] proposed an innovative forecasting model combining the Extended Stationary Wavelet Transform (ESWT) with Nested Long Short-Term Memory (NLSTM). This was carried out in order to overcome the prediction limitations faced by conventional deep learning models, owing to time-series data volatility and the need for extensive historical records. ESWT extends the traditional Stationary Wavelet Transform (SWT) by decomposing the original data into high- and low-frequency components and then applying SWT again to generate six more-stable sub-signals, as illustrated in Figure 17. These components are individually processed through separate NLSTM networks for prediction, and the results are integrated using an Inverse Wavelet Transform (IWT). The NLSTM architecture enhances the long-term dependency learning capability by embedding additional LSTM cells within existing LSTM structures, thereby effectively expanding the memory capacity.

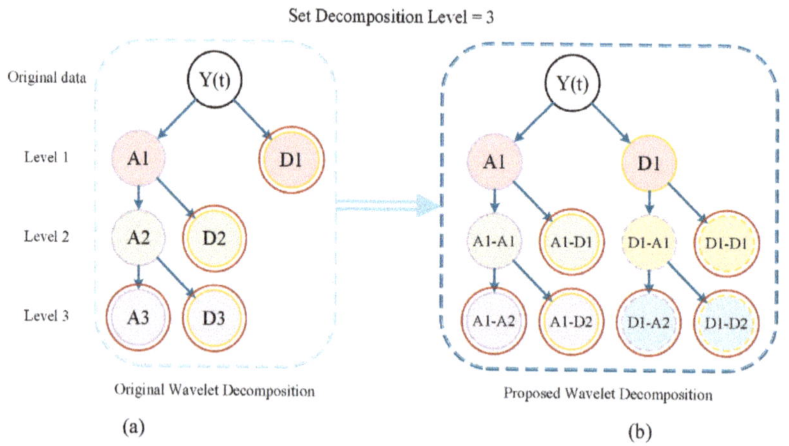

Figure 17. Decomposition of SWT and ESWT: (**a**) original SWT process; (**b**) proposed modified SWT [25].

This study utilized univariate PM2.5 time-series data collected from 12 monitoring stations around Beijing, China, from 2013 to 2017. The data underwent decomposition through ESWT into six sub-signals before being fed into the individual NLSTM models. The performance evaluation employed multiple metrics, including MAE, RMSE, MAPE, and R^2. The comparison of final prediction results and values for a subset of 72 data samples is shown in Figure 18.

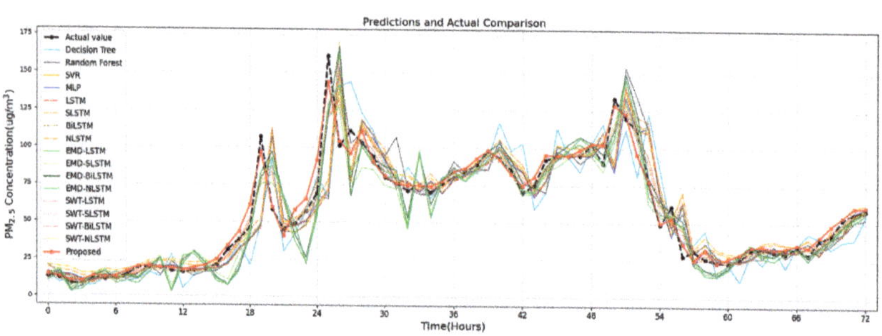

Figure 18. Comparison of final prediction results and actual values, displaying only 72 data samples for better visualization [25].

The proposed ESWT-NLSTM framework demonstrated superior performance across all evaluation metrics compared to various machine learning- and deep learning-based time-series prediction models. Specifically, ESWT-NLSTM recorded an MAE of 3.456, RMSE of 5.579, MAPE of 11.61%, and R^2 of 0.990, significantly outperforming both standard NLSTM (MAE = 9.290; RMSE = 17.968; MAPE = 31.60%; R^2 = 0.895) and EMD-NLSTM (MAE: 7.248; RMSE: 11.292; MAPE: 22.75%; R^2: 0.959) implementations [25]. The absolute error comparison between the proposed and existing models is illustrated in Figure 19. Notably, the model substantially reduced prediction lag effects, providing crucial advantages for peak value forecasting.

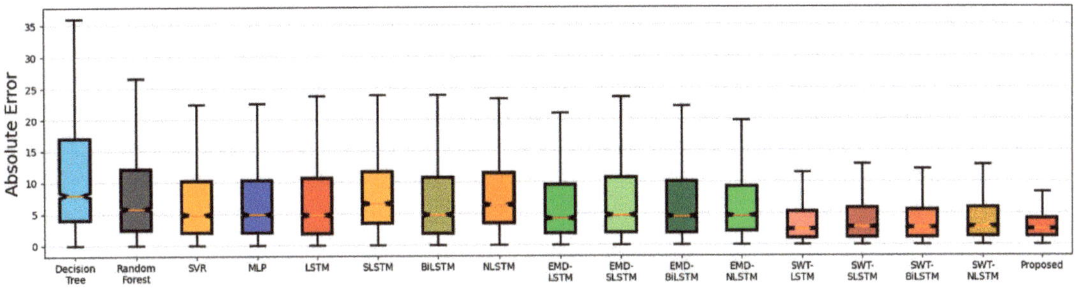

Figure 19. Absolute error of proposed model and existing models [25].

This study represents the first application of NLSTM for air pollution prediction. It demonstrates that the integration of ESWT with NLSTM creates an optimal hybrid architecture for highly volatile univariate time-series data. Improvements in prediction accuracy and a reduction in lag effects make this approach particularly suitable for operational air quality forecasting systems, especially pollution alert systems, which potentially play a vital role in public health protection.

Dalal et al. [26] developed a hybrid PSO–LSTM–RNN deep learning model for real-time air quality prediction in smart cities, targeting pollutants—such as SO_2, CO, O_3, and NO_2—to support urban environmental management and health risk mitigation.

The hybrid architecture integrated LSTM's ability to model long-term dependencies with RNNs' capability to capture sequential relationships in time-series data. PSO was employed specifically to optimize the LSTM neural network's weights, enhancing predictive accuracy by efficiently exploring high-dimensional parameter spaces. Additionally, a dropout regularization technique was integrated into the LSTM layer to address overfitting, improving the model's generalization capability in dynamic environments.

Compared to Gradient-Boosted Tree Regression, standalone LSTM, and SVM regression, the hybrid PSO–LSTM–RNN demonstrated superior performance—recording a significantly lower RMSE of 0.0184, MAE of 0.0082, and higher R^2 of 0.1227 [26]. This highlights its effectiveness in managing complex, non-linear air quality patterns common in urban environments.

The LSTM component was instrumental in leveraging gating mechanisms—input, forget, and output gates—to model complex time-series patterns, particularly in fluctuating pollutant concentrations from sensor networks. This contributed to enhanced long-term learning essential for real-time environmental monitoring.

This study demonstrates the practical applicability of hybrid LSTM-based models in smart cities for environmental monitoring and decision support. By incorporating PSO for precise weight optimization and combining multiple robust deep learning techniques, the proposed hybrid architecture offers an accurate and computationally efficient solution for real-time air quality forecasting in urban settings.

While the reviewed hybrid LSTM models demonstrate high accuracy in air quality and environmental monitoring, it is worth noting that most of the reviewed models were trained and validated using historical datasets. Consequently, their robustness against extreme data variability, long-term climate change, or rare events remains largely untested. This highlights a critical research gap and underscores the need for future studies to evaluate model performance under non-stationary or atypical conditions.

3.3. Energy Management

This section focuses on electrical energy consumption forecasting, covering various energy sources including natural gas, hydroelectric, wind, solar, and waste-to-energy conversion. The energy management approaches discussed primarily target smart grid infrastructure, smart homes, and smart buildings within urban environments.

Energy consumption forecasting in smart cities is challenging due to high-frequency, non-linear, and seasonal patterns driven by factors like population growth and weather. Traditional models often miss complex dependencies, while deep learning models may overfit or generalize poorly. Hybrid deep learning with decomposition techniques helps manage time-series variability, enabling more accurate predictions for planning and sustainability.

Consumption forecasting is a core research topic in the energy management domain. Arslan [27] proposed a hybrid model combining Prophet with Stacked Bidirectional LSTM to enhance energy consumption prediction in smart cities. Prophet offers strengths in decomposing time-series data into seasonality, trend, and irregularity components for separate analysis, whereas Stacked Bidirectional LSTM can simultaneously leverage both temporal sequences and future information, as illustrated in Figure 20. This integrated approach was designed to improve prediction accuracy by preserving seasonal patterns while reducing the impact of irregular fluctuations.

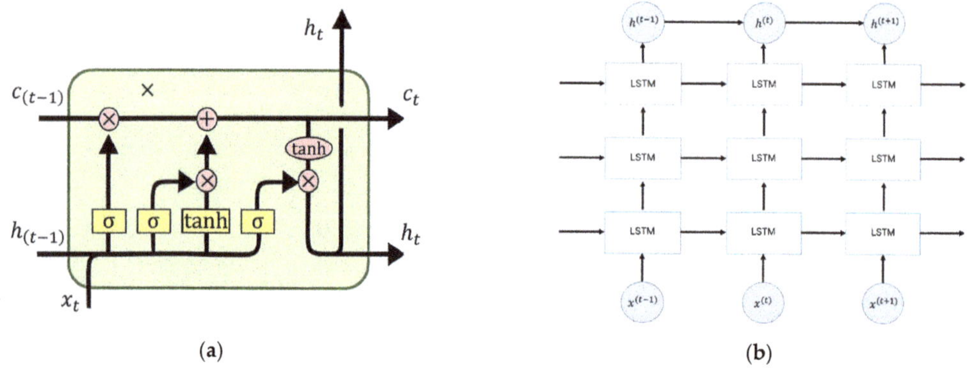

Figure 20. (**a**) Structure of LSTM cell; (**b**) overview of stacked LSTM network [27].

The key feature of the hybrid architecture lies in its complementary methodology: using Prophet to maintain seasonal characteristics while employing Stacked Bidirectional LSTM to learn non-linear and complex patterns in decomposed data. The framework implements Seasonal-Trend Decomposition using loess (STL) to separate data into seasonality, trend, and residual components. Prophet processes the original data to predict basic trends, including seasonality, whereas Stacked Bidirectional LSTM learns from the residual data to generate more sophisticated predictions. The final forecast combined the outputs of both models.

The performance evaluation demonstrated the hybrid model's superior results compared to standalone Prophet, Bi-LSTM, and deseasonalized LSTM (deBi-LSTM) implementations. For the Canadian dataset, the hybrid model achieved an RMSE of 2348.23, approximately 43% lower than Prophet's 4119.49 and 12% lower than Bi-LSTM's 2670.47 [27]. Similarly, for the French dataset, the hybrid model recorded an RMSE of 2147.37, showing improvements of 45% compared with Prophet (3931.96) and 1.3% compared with Bi-LSTM (2176.40) [27].

In conclusion, the hybrid model integrating Prophet with Stacked Bidirectional LSTM demonstrated enhanced accuracy in smart city energy consumption forecasting by simultaneously addressing seasonality and irregularity, compared with conventional approaches. This advancement has the potential to improve energy management efficiency in smart cities and enhance the precision of resource allocation and planning processes.

Ali et al. [28] introduced two hybrid models—ConvLSTM and CNN-LSTM—to predict energy consumption in smart homes. The primary objective was to improve the energy management efficiency in smart homes through accurate forecasting based on time-series data. ConvLSTM features an architecture capable of simultaneously processing two-dimensional spatial features and temporal dependencies. In contrast, CNN-LSTM is designed such that the CNN learns spatial features first, followed by LSTM learning temporal dependencies.

The key characteristic of the ConvLSTM model is its ability to learn spatiotemporal patterns through 2D ConvLSTM layers and then optimize temporal dependencies in subsequent LSTM layers to deliver high prediction accuracy. The CNN-LSTM operates by extracting key features through 1D Conv layers and learning time-series data flows via LSTM layers. Both architectures effectively combine spatial and temporal patterns to provide superior results compared with conventional approaches.

Performance evaluation demonstrated that ConvLSTM had the highest accuracy across all test scenarios. In the 1-day, 3-day, and 6-day prediction scenarios, ConvLSTM achieved MAE values of 3.69, 4.07, and 4.65, respectively, outperforming CNN-LSTM's MAE (3.80, 4.15, 4.69) and standalone LSTM (3.90, 4.24, 4.72) [28]. For MAPE, ConvLSTM recorded 18.48% for 1-day predictions, 21.27% for 3-day predictions, and 25.35% for 6-day predictions, showing improvements over CNN-LSTM (18.59%, 22.65%, and 25.20%, respectively) and LSTM (22.75%, 26.09%, and 28.03%, respectively), as illustrated in Figure 21 [28]. The RMSE values for ConvLSTM were also the lowest at 5.39, 5.88, and 6.54 for the 1-day, 3-day, and 6-day forecasts, respectively [28].

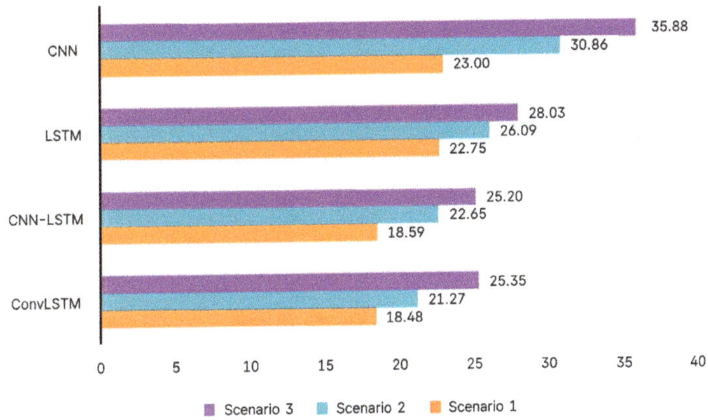

Figure 21. Comparison of MAPE values by predictive learning model [28].

In conclusion, both ConvLSTM and CNN-LSTM demonstrated superior performance in smart home energy consumption prediction compared to the conventional standalone

LSTM and CNN models. The ConvLSTM model maintained high accuracy even in long-term predictions and demonstrated excellent spatiotemporal pattern-learning capabilities. This research validates the idea that such hybrid architectures can serve as powerful tools for enhancing energy management efficiency in smart cities.

Alhussein et al. [29] proposed a hybrid CNN-LSTM model to enhance short-term electricity consumption forecasting for individual households in smart cities. Their study aimed to efficiently manage power supply and demand in smart grids while improving the accuracy of energy usage predictions. The model was evaluated using data from Australia's Smart Grid Smart City (SGSC) project, leveraging the complementary characteristics of CNNs and LSTM to significantly enhance the predictive performance, as shown in Figure 22.

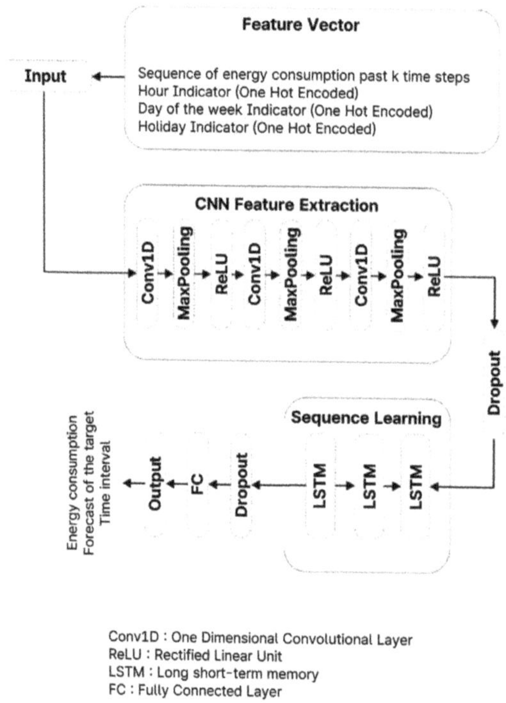

Figure 22. Overview of CNN-LSTM model [29].

The key feature of the CNN-LSTM architecture lies in its dual processing approach: the CNN extracts spatial patterns from data, whereas LSTM learns temporal dependencies, effectively capturing both short- and long-term patterns in time-series data. The CNN filters noise from the input data and extracts critical features, which are then passed to the LSTM layers to learn sequential dependencies. This hybrid approach combines the strengths of both techniques to deliver a higher prediction accuracy than conventional models.

The results demonstrated that the CNN-LSTM model outperformed traditional standalone LSTM models across all evaluation metrics. In single-step forecasting, the CNN-LSTM model achieved a MAPE of 40.38%, which was approximately 8.3% lower than that of the conventional LSTM model (44.06%) [29]. For multi-step forecasting, the CNN-LSTM model surpassed the LSTM model, with MAPE reductions of 4.01%, 4.76%, and 5.98% for its 1 h, 2 h, and 6 h predictions, respectively, as shown in Figure 23 [29].

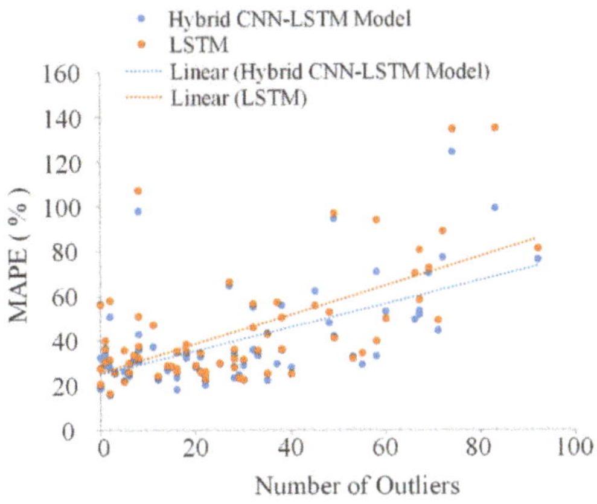

Figure 23. Comparison of MAPE values of proposed models [29].

In conclusion, the CNN-LSTM hybrid model demonstrated excellent performance in predicting individual household energy consumption in smart cities. This architecture is a powerful tool that can enhance the efficiency of power management systems and contribute to energy conservation and cost reduction strategies.

Karri et al. [30] proposed an IoT-based hybrid LSTM-MLP model aimed at improving waste-to-energy conversion efficiency in smart cities. Their study focused on effectively processing extensive IoT data to achieve accurate energy predictions. Within this architecture, LSTM handles the learning of temporal characteristics in IoT data, whereas a Multilayer Perceptron (MLP) manages complex feature extraction. This structural design was developed specifically to enhance the accuracy and reliability of massive IoT datasets.

The study utilized time-series data obtained from the Kaggle platform, incorporating waste generation volumes, energy conversion rates, environmental factors such as temperature and humidity, and IoT sensor data. Data preprocessing involved handling missing values, detecting and processing outliers, and applying normalization techniques (Min-Max Scaling, Z-score Normalization). Feature engineering implements temporal characteristics (day of the week and time), interaction features, correlation analysis, Gini importance, and Principal Component Analysis (PCA) for feature selection. During model training, the parameters were optimized through backpropagation using loss functions (MSE, Cross-Entropy Loss), with hyperparameter tuning conducted via cross-validation and grid search methodologies.

The performance evaluation employed metrics such as accuracy, precision, recall, and F1-score. The results demonstrated that the hybrid IoT-LSTM-MLP model achieved an excellent performance, with 97.6% accuracy, 98.2% precision, 96.8% recall, and 97.5% F1-score [30]. These metrics represent more than a 6% improvement in accuracy compared with the conventional ARIMA and SVM models [30]. These findings validate the effectiveness of temporal data processing using LSTM combined with feature extraction via the MLP. From a practical application perspective, this model can be integrated into dynamic energy management systems in smart cities, significantly contributing to waste-to-energy generation optimization and operational efficiency improvements.

Kaya et al. [31] conducted research utilizing a hybrid CNN-LSTM model to enhance consumption and production forecasting accuracy for multiple energy sources across Turkey. Their study aimed to overcome the limitations of existing prediction models for handling data complexity and to implement a more precise predictive performance. Within this architecture, the CNN extracts localized patterns from data, whereas LSTM learns long-term temporal dependencies.

The study employed hourly energy consumption and production data from 2018 to 2023 provided by the Energy Exchange Istanbul (EXIST). The dataset encompasses total consumption and production figures, as well as production data for various energy sources, including natural gas, hydroelectric dams, lignite, run-of-river hydroelectric, wind, and fuel oil. After transforming the data using a sliding window approach, they applied MinMaxScaler for normalization and partitioned the dataset into training (80%), validation (10% of training data), and testing (20%) segments.

The performance evaluation demonstrated that the hybrid CNN-LSTM model achieved superior results compared to conventional Linear Regression (LR), Random Forest (RF), Support Vector Regression (SVR), and standalone CNN and LSTM models [31]. Notably, the model recorded an R^2 value of 0.990 for total energy consumption prediction and exhibited the lowest RMSE and MAE values, with the highest R^2 scores across most energy source predictions [31]. These results confirm that the combination of a CNN and LSTM effectively captures both the spatial and temporal complexities in this data, significantly enhancing prediction accuracy. This research contributes to improving energy resource utilization efficiency and power system management precision while suggesting potential future extensions incorporating additional variables such as economic indicators and meteorological conditions.

Du [32] proposed a hybrid SVR-LSTM model with Random Forest optimization aimed at improving building energy consumption prediction accuracy to enable effective energy resource allocation and waste reduction. Through an SVM-LSTM-stacking architecture combining the strengths of Support Vector Regression and Long Short-Term Memory networks, the study sought to effectively analyze the complex and diverse energy consumption data characteristics that conventional models struggle to process. Within this framework, SVR handles complex, high-dimensional data patterns, whereas LSTM effectively learns the temporal characteristics of the energy consumption data. Random Forest was implemented using a stacking approach for feature selection and model optimization in order to enhance the overall prediction performance.

The study used data from 539 buildings collected between 2017 and 2020. The dataset comprised 12 variables: building size, construction year, total floor area, cooling area ratio, average monthly occupancy, chiller age, central cooling plant efficiency, LED usage status, and annual energy consumption. The data preprocessing primarily involved normalization and correlation analysis for appropriate variable selection because the dataset contained minimal missing values and outliers.

The performance evaluation employed metrics such as accuracy, mean squared error (MSE), and coefficient of determination (R^2). The results demonstrated that the hybrid model achieved over 90% accuracy on both the training and testing data, representing substantial performance improvements compared with standalone LSTM (R^2 of 0.542, approximately 54%) and SVR models (R^2 of 0.151, approximately 15%) [32]. Furthermore, the RF Stacking model exhibited a very low MSE (0.00047) with a high explanatory power ($R^2 = 0.898$) [32].

This hybrid architecture can be integrated into real-time monitoring and control systems for smart buildings to maximize energy efficiency while providing valuable data-driven guidelines for policy development and building design.

Sivarajan and Jebaseelan [33] proposed an innovative approach that combines a hybrid CNN-LSTM model with a Modified Sea Lion Algorithm (MSLA) for accurate energy demand prediction in IoT-based smart grids. To overcome the limitations of existing models in processing data complexity, they integrated a CNN's feature extraction capabilities with the LSTM's ability to learn time-series dependencies. Additionally, the MSLA was designed to further enhance model performance through hyperparameter tuning by implementing

Levy Flight Trajectory techniques to address local optimization issues in the conventional Sea Lion Algorithm.

The study utilized the actual energy consumption data from various household appliances in smart homes, including televisions, lighting, air conditioners, refrigerators, fans, and water heaters. The data underwent authentication, encryption, and normalization processes, as well as preprocessing, to improve prediction accuracy and processing efficiency.

The model evaluation employed metrics including MSE, RMSE, MAE, MAPE, and prediction and training times. The results demonstrated that the CNN+LSTM+MSLA hybrid model outperformed the conventional CNN-LSTM, standalone LSTM, and Linear Regression models across per minute, hourly, daily, and weekly data intervals. For instance, with per minute data, the hybrid model recorded an MSE of 0.034, an RMSE of 0.200, an MAE of 0.084, and a MAPE of 12.67%, representing the best performance across all metrics [33]. Similarly, the lowest MSE (0.198) and RMSE (0.438) were achieved for the hourly data [33]. The model also demonstrated improvements in reducing the training and prediction times.

This study validates that the combination of a hybrid CNN-LSTM architecture with MSLA optimization is highly effective for energy demand prediction in smart grids. The proposed model enables the efficient allocation of energy resources and minimization of waste, contributing to the enhanced stability and sustainability of smart grid systems.

Jin et al. [34] proposed a hybrid LSTM–BPNN-to-BPNN model to enhance long-term electricity peak load forecasting, a critical task for smart city energy planning and infrastructure resilience. The model addresses the non-linear and multivariate nature of peak demand, which is influenced by historical load patterns, economic indicators, and weather conditions.

The architecture consists of three components: an LSTM module to extract temporal features from historical peak load data, a BPNN module to capture the non-linear structure of projected future variables (e.g., temperature forecasts, economic development targets), and a final BPNN that fuses both outputs to predict monthly peak loads. The highest predicted monthly value is then selected as the forecasted annual peak load. This structure allows the model to leverage both temporal dependencies and static future information, offering a robust framework for forward-looking energy demand estimation.

The model was trained on a real-world dataset from a Chinese city (2012–2021), where it demonstrated strong performance. Compared to conventional models—ARIMA (MAPE: 4.63%), SVR (2.04%), LSTM (3.55%)—the hybrid model achieved the lowest MAPE of 1.47%, outperforming even standalone LSTM and BPNN approaches [34]. Its ability to incorporate exogenous and seasonal variables across multiple time resolutions proved essential to its accuracy.

In the context of smart cities, where energy systems must anticipate long-term load dynamics while adapting to policy and climate shifts, this hybrid model exemplifies the value of integrating LSTM's temporal learning with complementary machine learning layers. Its modularity and performance make it a viable candidate for energy infrastructure planning, peak load mitigation, and grid optimization strategies.

Although hybrid models for energy consumption prediction exhibit promising accuracy, the majority of evaluations are based on historical datasets and conducted under specific regional or contextual conditions. Their performance under extreme data variability, long-term climate trends, or atypical events remains largely unexplored. Furthermore, the reviewed studies utilize diverse datasets and evaluation metrics, hindering the direct comparability across models. This variability restricts the generalization and scalability of these current findings. Recognizing these limitations, future research should prioritize the development of standardized benchmarking frameworks to facilitate more interpretable and robust cross-context model evaluation.

3.4. Health and Public Safety

Health and public safety prediction in smart cities depends on diverse, often incomplete data from sources like IoT sensors and surveillance. Challenges include complex, non-linear, and spatiotemporal patterns in events like disease outbreaks or cyber threats. Traditional models fall short in handling such dynamics, making hybrid deep learning models vital for accurate, real-time crisis management.

Research has primarily focused on epidemic prediction and cybersecurity in the health and public safety domains. Khan et al. [35] proposed a hybrid CNN-LSTM model to predict COVID-19 hotspots in smart cities. The objective of this study was to develop a data-driven prediction model to facilitate the planning and implementation of efficient lockdown strategies. In this model, a CNN was employed to extract multiscale temporal features from the input time-series data. In contrast, LSTM was used to learn temporal dependencies based on those features, thereby enhancing the overall prediction performance.

The CNN-LSTM model integrates the strengths of both CNNs and LSTM, with a CNN being responsible for extracting spatial features from the data and LSTM handling temporal dependencies. Notably, the model was designed to utilize multiscale temporal features extracted from various CNN layers, enabling it to learn short-, medium-, and long-term dependencies simultaneously. This multiscale approach provides more accurate predictions than models that rely solely on single-scale features.

The performance evaluation demonstrated that the CNN-LSTM model significantly outperformed existing models, including ARIMA, standalone CNN, standalone LSTM, and FFNN. The CNN-LSTM model achieved a mean absolute error (MAE) of 74.27, compared to 120.76 for LSTM and 201.75 for ARIMA [35]. Regarding the root mean square error (RMSE), the model scored 106.90, which was considerably lower than the 179.41 for LSTM and 283.46 for ARIMA [35]. These results indicate that the hybrid model achieved an over 40% improvement in prediction accuracy compared with traditional models, as presented in Figure 24 [35]. This study confirmed that CNN-LSTM can be a powerful tool for managing public health emergencies by supporting proactive and precise containment strategies in smart cities.

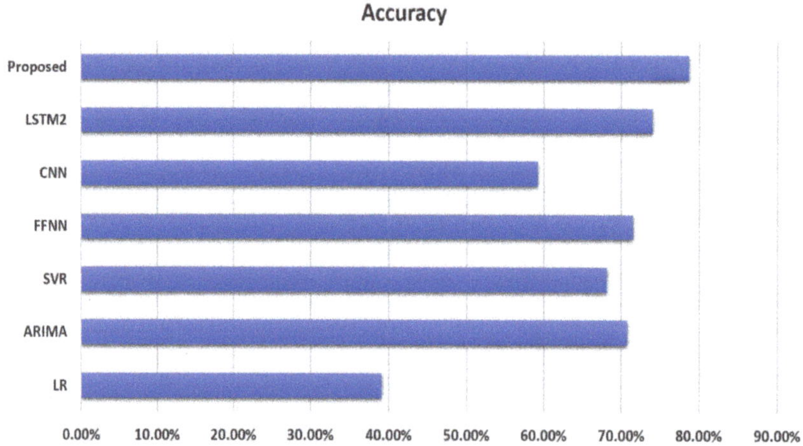

Figure 24. Accuracy comparison by predictive learning model used [35].

Kompunt et al. [36] proposed a hybrid MLP-LSTM model to predict the spread of COVID-19 across 77 provinces in Thailand. Figure 25 represents the overall architecture of the proposed system. The primary goal of this study was to forecast newly confirmed cases, total infections, and death counts that would support efficient disease control and resource allocation planning. In this model, LSTM processed time-series data by learning temporal dependencies. Meanwhile, an MLP handled static data, such as regional attributes and vaccination rates, to improve the model's predictive capability.

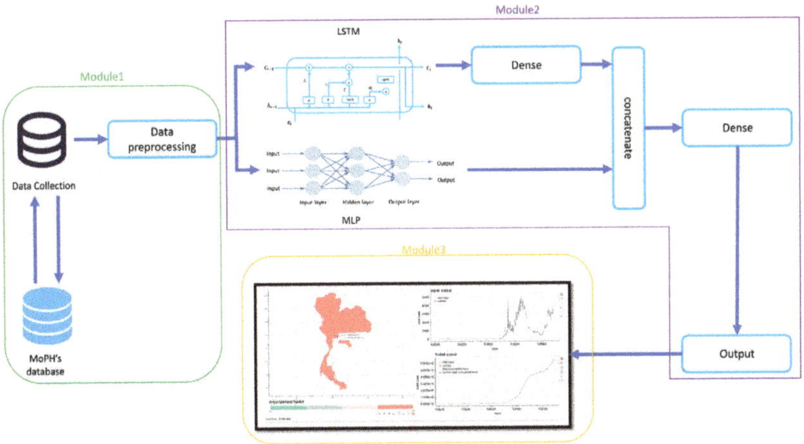

Figure 25. Three main modules of prediction system and their structures [36].

A key advantage of this MLP-LSTM model is its ability to learn dynamic temporal patterns and static features jointly. LSTM effectively handles long-term dependencies in time-series data, and the MLP increases prediction accuracy by incorporating diverse static variables. The model processes dynamic and static data streams separately and then integrate the results to generate the final prediction. The prediction outputs were visualized using a geographic information system (GIS) and made accessible via a web application.

In performance evaluations, the MLP-LSTM model consistently outperformed the standalone LSTM and unoptimized models. For example, when predicting the total infection cases, the LSTM model alone achieved an R^2 value of 0.9959 and an MAE of 300.97 [36]. In contrast, the MLP-LSTM model improved to 0.9980 and 100.60, respectively, indicating a 66.6% reduction in the error [36]. In death prediction, the LSTM model recorded an R^2 of 0.9944 and an MAE of 1.42, whereas MLP-LSTM achieved values of 0.9964 and 1.04, representing a 26.8% improvement in MAE [36]. These results validate the superior performance of the hybrid model in integrating heterogeneous data sources for pandemic management.

Sah et al. [37] developed a stacked hybrid LSTM-GRU model combining deep learning with statistical forecasting methods such as Prophet and ARIMA. Their study aimed to provide accurate predictions of COVID-19 confirmed and active cases across India to support healthcare resource planning. The Prophet and ARIMA models are effective at modeling linear trends and seasonality but are limited in handling non-linear, long-term dependencies. To overcome these shortcomings, a stacked LSTM-GRU architecture was implemented using an Adam optimizer and ReLU activation functions, enhancing its ability to learn complex time-series patterns.

The hybrid model significantly outperformed the individual statistical models. Specifically, in terms of prediction accuracy (R^2), the LSTM-GRU model achieved 0.74, whereas ARIMA and Prophet recorded 0.56 and 0.46, respectively [37]. For RMSE, the hybrid model scored 69.92, whereas ARIMA and Prophet scored 1260 and 568.58, respectively [37]. These results demonstrate that deep learning-based hybrid models are not only applicable to epidemic forecasting but can also be extended to other smart city domains such as traffic, energy, and environmental monitoring.

In the field of cybersecurity, Razib et al. [38] proposed a hybrid DNN-LSTM model to detect IoT-based security threats in smart environments. This model, as shown in Figure 26, was integrated with a Software-Defined Networking (SDN) framework to detect and mitigate various cyberattacks, such as DDoS, brute force, and infiltration threats. The DNN component learns non-linear patterns from the network traffic, whereas the LSTM component captures temporal dependencies. The fusion of these components allows the system to detect suspicious activities with greater precision in real time.

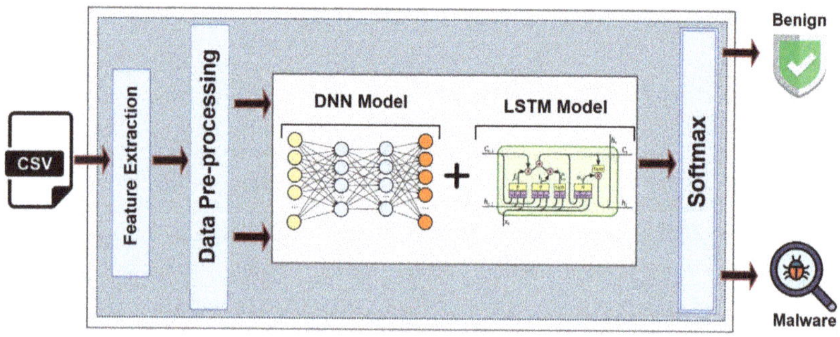

Figure 26. Proposed DNN-LSTM training model [38].

The DNN-LSTM model can analyze heterogeneous IoT data and monitor network traffic in real time. DNNs are well suited to learning complex patterns in high-dimensional data, whereas LSTM models have long- and short-term temporal dependencies. When integrated with SDN, the model enhances the network adaptability and scalability by dynamically controlling the data flow between IoT devices.

The performance evaluation showed that DNN-LSTM achieved 99.55% accuracy, significantly outperforming DNN-GRU (98.67%) and BLSTM (98.9%) [38]. Furthermore, the model recorded a precision of 99.36%, recall of 99.44%, and F1-score of 99.42%, all of which were 1–2% higher than those of the alternative models [38]. These findings demonstrate that DNN-LSTM is highly effective for the real-time detection and mitigation of cyber threats in IoT-driven smart city infrastructures.

Mohbey et al. [39] proposed a CNN-LSTM hybrid model to analyze sentiments related to monkeypox on social media. This study utilized 61,379 tweets collected between May and June 2022 to classify public sentiment into positive, negative, and neutral categories. The CNN component captured local text features, such as word position and contextual relationships, whereas the LSTM component modeled the sequential emotional flow within the tweets. The preprocessing steps included stop-word removal, punctuation, URL cleaning, text normalization, and stemming.

The performance metrics demonstrated that the CNN-LSTM model outperformed other classification algorithms, including standalone CNN, LSTM, SVM, Random Forest, and Decision Tree approaches. It achieved an accuracy of 94%, which was 6% higher than that of the LSTM model (88.1%) [39]. The ability of the hybrid architecture to capture long-term dependencies in lengthy texts proved advantageous for sentiment classification. This study validated CNN-LSTM as an effective tool for analyzing public sentiment during infectious disease outbreaks and guiding public health communication strategies in smart cities. Future studies should consider expanding the model to perform topic classification and sentiment analysis simultaneously.

Hassan et al. [40] proposed a hybrid deep learning model, CNN/weight-dropped LSTM (WDLSTM), for efficient intrusion detection in big data environments by leveraging the strengths of both convolutional neural networks and regularized recurrent units. The architecture includes two 1D convolutional layers for extracting spatial features, a max-pooling layer for dimensionality reduction, and a WDLSTM layer that maintains long-term temporal dependencies while preventing overfitting through drop-connect regularization.

The model was evaluated on the UNSW-NB15 dataset, comprising over 2.2 million instances across 10 attack categories. It achieved 97.17% accuracy in binary classification and 98.43% in multi-class classification, significantly outperforming baseline models such as the TSDL (two-stage deep learning) model, which showed only 89.13% accuracy on the same dataset [40]. Moreover, the CNN–WDLSTM model demonstrated a lower average execution time per instance (0.002383 ms) compared to TSDL (0.003372 ms), making it more suitable for real-time intrusion detection [40]. A Z-test confirmed the statistical significance of this performance improvement at a 95% confidence level.

While these studies demonstrate strong performance in epidemic prediction and cybersecurity, the generalizability of hybrid models to other smart city domains—characterized by different data structures, infrastructure configurations, and resource limitations—remains an open challenge. The cross-domain applicability of hybrid LSTM architectures has not been extensively examined, limiting insights into their transferability across diverse urban contexts. Future research that adapts these modeling strategies to structurally analogous domains may offer meaningful contributions to expanding the utility of hybrid approaches in smart city analytics.

3.5. Urban Infrastructure Monitoring

This section examines predictive models for key urban infrastructure types including seaports and maritime facilities, residential housing markets, coastal and tidal infrastructure, urban water supply systems, and structural health monitoring of bridges and reinforced concrete buildings.

Urban infrastructure monitoring in smart cities faces challenges due to heterogeneous, high-dimensional, and often incomplete sensor data. Prediction models must account for spatial–temporal dependencies and evolving infrastructure conditions. Traditional models struggle with adaptability to dynamic urban environments and real-time data processing. Advanced hybrid deep learning models are needed to enhance accuracy, resilience, and scalability in forecasting structural health and resilience.

Research on urban infrastructure monitoring has addressed logistics and real-estate forecasting. Cuong et al. [41] proposed a hybrid model combining Long Short-Term Memory (LSTM) and Random Forest (RF) to enhance the prediction of container throughput at the Busan Port, as shown in Figure 27. This model focuses on increasing forecast accuracy by addressing non-linear patterns and external disruptions in port operations. In the hybrid architecture, LSTM manages time-dependent patterns in the data, whereas RF processes the residuals produced by LSTM to correct prediction errors.

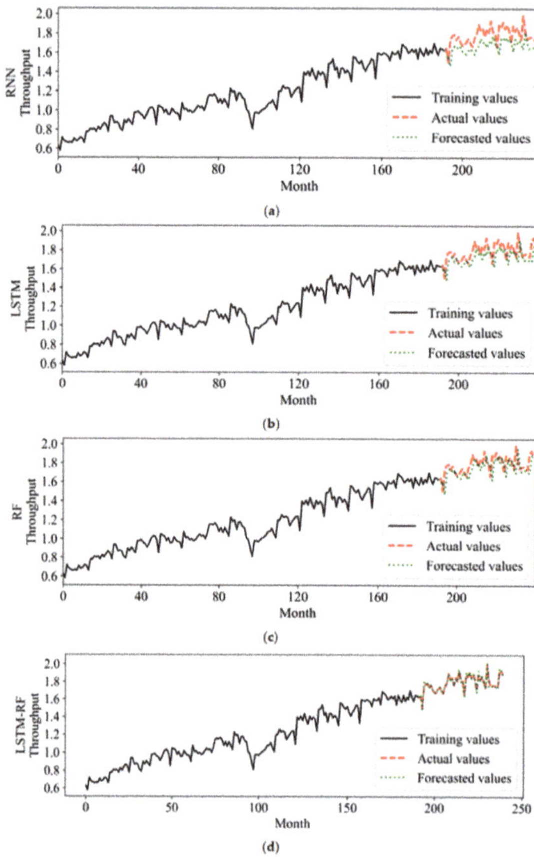

Figure 27. Comparison of actual and predicted container throughput: (**a**) RNN, (**b**) LSTM, (**c**) RF, (**d**) LSTM-RF [41].

The LSTM-RF model was designed to integrate the complementary strengths of the two algorithms: LSTM captured both long- and short-term dependencies to predict trends and patterns, whereas RF refined the remaining errors to improve the precision of the final predictions. Compared to the individual models, this hybrid approach demonstrated a stronger and more stable performance under variable conditions.

Performance evaluations demonstrated that the LSTM-RF model outperformed both the standalone LSTM and RF models. In the long-term prediction scenarios, LSTM-RF achieved a root mean square error (RMSE) of 0.019, compared with 0.1664 for LSTM and 0.0966 for RF [41]. Similarly, in terms of the mean absolute error (MAE), LSTM-RF achieved a value of 0.0004, which was significantly lower than 0.0277 (LSTM) and 0.0093 (RF) [41]. These findings highlight the ability of the hybrid model to handle non-linearity and external variability more effectively than traditional models.

Almaliki and Khattak [42] introduced a hybrid Temporal Convolutional Network (TCN) and LSTM model for tidal-level forecasting in coastal and port infrastructure. Their study aimed to construct a reliable forecasting framework capable of predicting both short-term (T+5 and T+10 days) and long-term (T+30 and T+60 days) tidal fluctuations. The proposed model was designed to combine TCN's ability to detect time-series patterns with LSTM's capacity to capture long-term dependencies, enabling the accurate modeling of both localized and periodic changes.

The study used high-precision hourly tidal data collected from Ras Tanura, Saudi Arabia, between 2016 and 2022 (±0.005 m accuracy). A sliding window of 30 d was applied to generate input sequences, and hyperparameters were optimized using the Covariance Matrix Adaptation Evolution Strategy (CMA-ES), thereby reducing human bias and improving learning efficiency.

In the performance assessments, the TCN-LSTM hybrid outperformed the standalone LSTM, GRU, and CNN models across all forecasting horizons. For example, in the T+5 d prediction, TCN-LSTM achieved an MAE of 0.073 and RMSE of 0.08 m, with a 25.5% lower MAE and 32.2% lower RMSE than LSTM [42]. For the T+30 and T+60 forecasts, the hybrid achieved MAEs of 0.050 and 0.054 and MAPEs of 5.39% and 4.93%, respectively [42]. The CNN exhibited the poorest performance, with MAE and MAPE values up to 52% higher than those of the hybrid model [42].

Uncertainty analysis further demonstrated the reliability of the hybrid model. For the T+60 predictions, TCN-LSTM recorded the lowest average uncertainty (0.301) and uncertainty ratio (0.376), substantially outperforming the GRU (0.545), LSTM (0.533), and the CNN (0.604) [42]. The researchers suggested expanding the model to include meteorological factors (e.g., storm surges) for integrated coastal forecasting and highlighted its applicability to urban planning, tidal energy management, and port design.

Temür et al. [43] used monthly housing sales data from Turkey (January 2008–April 2018) to propose a hybrid ARIMA-LSTM model. In the model, ARIMA models the linear components, whereas LSTM processes the residuals to capture non-linear relationships. This two-step process enables the independent modeling of both linear and non-linear patterns, with the final outputs generated through their combination, as illustrated in Figure 28.

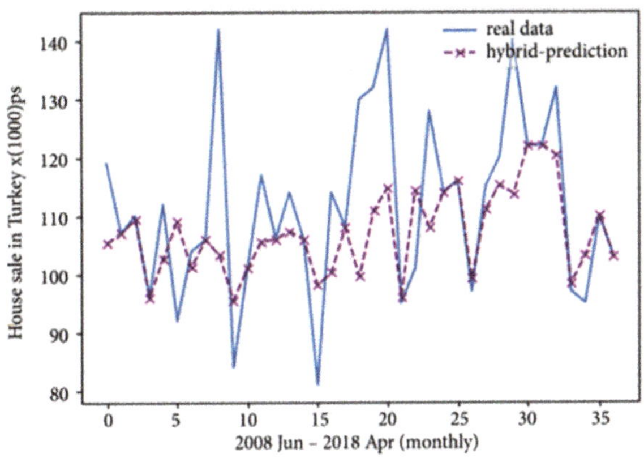

Figure 28. Comparison of actual real-estate transaction data and ARIMA-LSTM predictions [43].

The hybrid model outperformed the standalone ARIMA and LSTM models. It achieved a mean absolute percentage error (MAPE) of 0.072, compared with 0.121 for ARIMA and 0.15 LSTM, corresponding to error reductions of approximately 40% and 52%, respectively [43]. For the mean squared error (MSE), the hybrid model scored 175.6, compared to 280.4 (ARIMA) and 473.3 (LSTM), reflecting approximately 37% and 63% lower errors, respectively [43]. These error values are summarized in Table 2.

Table 2. Comparison of error values by predictive learning model [43].

Comparison of All Models				
Model	**MAE**	**MSE**	**RMSE**	**MAPE**
ARIMA (0,1,2)	12.801	280.400	16.745	0.121
LSTM (1000 epochs)	17.853	473.346	21.757	0.150
Hybrid (LSTM 1500 epochs–ARIMA (1,1,1))	8.847	175.611	13.252	0.072

Pu et al. [44] proposed a CNN-LSTM hybrid model for short-term urban water demand forecasting. The goal was to capture temporal and spatial patterns from complex water demand data influenced by variables such as temperature, rainfall, and hourly fluctuations. Accurate forecasting supports efficient water supply management and cost savings.

Performance comparisons showed that CNN-LSTM outperformed the ANN, Conv1D, GRUN, and LSTM models. For the 15 min forecasts, CNN-LSTM achieved a MAPE of 1.25% [44]. This was significantly lower than that of ANN (2.59%), Conv1D (3.05%), GRUN (2.39%), and LSTM (2.32%) [44]. In one-hour forecasts, the model achieved a MAPE of 2.97%, again outperforming ANN (4.08%), Conv1D (4.67%), GRUN (4.31%), and LSTM (4.49%) [44]. The RMSE values for CNN-LSTM were 79.23 m^3/h (15 min) and 201.13 m^3/h (1 h), reflecting substantial improvements in prediction accuracy [44].

Das and Guchhait [45] proposed a parallel hybrid GRU–LSTM(P) deep learning model for structural damage detection using time-series vibration data. Their objective was to identify and quantify the location and severity of structural damage based on experimentally obtained acceleration response data. The model was validated using real-world data from the Z24 bridge and achieved nearly 100% accuracy across 17 damage scenarios.

In the initial simulation phase, a two-span continuous beam was modeled as a simplified steel girder bridge, and 60 damage cases were created by simulating 10%, 30%, and 50% thickness reductions. Gaussian noise was added to simulate the real sensor environments. The data were input in raw form without preprocessing, thereby enhancing the practicality of the model.

Model comparisons showed that under high-noise conditions (10%), the standalone GRU achieved 91.77% accuracy and a 91.76 F1-score, and LSTM achieved a 93.04% accuracy and a 93.03 F1-score [45]. In contrast, the GRU–LSTM(P) model achieved superior results of 93.32% accuracy and a 93.30 F1-score [45]. The proposed model converged quickly (within 20–30 epochs) and was computationally efficient for real-time inference on platforms such as Google Colab.

The validation of the Z24 bridge confirmed the classification accuracy of the model across all damage scenarios. Compared to the CNN-LSTM model proposed by Yessoufou and Zhu [46], which achieved 86% accuracy, the GRU–LSTM(P) model demonstrated superior performance. The T-SNE visualizations revealed clear damage clusters, and the model accurately localized and quantified the damage using only raw acceleration data.

The study demonstrated the feasibility of real-time structural health monitoring using a high-performance deep learning model capable of processing raw vibration signals without preprocessing [46]. The model was resilient to noise and variations in the damage type, making it suitable for practical deployment in real-world infrastructure systems.

In 2022, Parida et al. [47] developed a hybrid CNN–LSTM deep learning model to monitor bond strength degradation in reinforced concrete structures using time-series data from both vibration and ultrasonic wave signals. Their objective was to evaluate bond quality degradation in rebar–concrete interfaces, a critical aspect of structural durability, through continuous, non-destructive monitoring.

The study employed experimental specimens with varying levels of artificially induced bond damage using accelerated corrosion, and data were collected using embedded sensors during cyclic loading.

Compared to standalone CNN and LSTM models, the hybrid CNN–LSTM achieved a higher predictive performance. Specifically, the CNN–LSTM model recorded an RMSD of 5.3% at a 10 mm displacement level, outperforming the standalone CNN (RMSD = 10.1%) and LSTM (RMSD = 7.0%) architectures [47]. These results underline the hybrid model's effectiveness in both regression and classification tasks.

A key strength of the study was its use of both ultrasonic and vibration data, enabling the model to fuse complementary signal characteristics. This multimodal input significantly improved the model's ability to detect early-stage degradation. The CNN–LSTM model also required fewer training epochs to converge and showed robustness under sensor noise and real-time fluctuations.

This work demonstrated the practical applicability of hybrid deep learning in structural health monitoring for concrete infrastructure. By combining spatial and temporal learning capabilities and leveraging multiple sensor modalities, the proposed approach offered a high-accuracy, low-latency solution for early warning systems in civil structures.

Although these studies demonstrate high predictive performance, most have not been tested under the practical constraints of real-world infrastructure, such as limited sensor coverage, data inconsistencies, and requirements for model interpretability. These implementation challenges, including the explainability of predictions and integration into operational decision systems, remain underexplored in the current literature. Future research should address these factors to support real-time adoption in smart city infrastructure monitoring.

3.6. *Urban Waste Management*

Predicting waste management in smart cities is challenging due to the dynamic influence of socioeconomic factors and seasonal trends. The data is often non-linear, multi-source, and exhibits temporal dependencies. Traditional models fail to capture complex relationships between input variables and waste output. Hybrid deep learning models have proven effective in improving prediction accuracy for sustainable waste management.

Hybrid LSTM models have also been applied in urban waste management, particularly to forecast waste generation, predict waste bin fill-levels, and develop smart waste classification systems.

Lin et al. [48] proposed a hybrid attention–LSTM-CNN model to accurately predict municipal solid waste (MSW) generation in smart cities. The primary goal of their study was to develop a forecasting system that integrated time-series data with socioeconomic variables (e.g., average household size, budget revenues, and public investment) to support waste policy and management decisions. The model leveraged the complementary strengths of the attention mechanism, 1D-CNN, and Bidirectional LSTM (Bi-LSTM) to enhance prediction accuracy, as shown in Figure 29.

The key feature of the Attention–LSTM-CNN model is its ability to process spatial and temporal patterns simultaneously. The attention mechanism selects important features from the input, which are then passed on to LSTM to learn the temporal dependencies. LSTM handles both long- and short-term data dependencies, whereas the CNN extracts spatial features, ultimately improving the prediction precision. This structure was specifically designed to effectively determine the non-linear relationships between complex socioeconomic variables and waste generation.

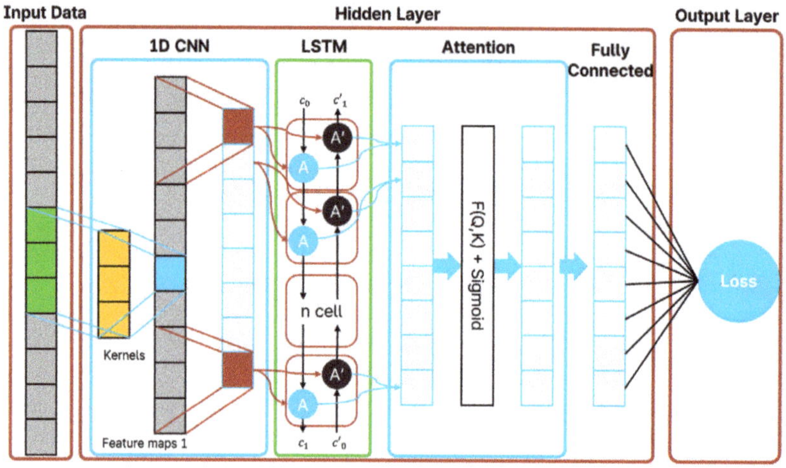

Figure 29. CNN-LSTM–Attention (C-L-A) integrated algorithm used for urban waste generation forecasting [48].

The experimental results demonstrated that the Attention–LSTM-CNN model outperformed the other hybrid and individual models. Based on the R^2 (coefficient of determination), the CNN-LSTM–Attention configuration achieved a value of 95.31%, compared with 89.45% for the Attention–CNN-LSTM (A-C-L) structure and 90.77% for the CNN–Attention–LSTM (C-A-L) structure [48]. The model also recorded the lowest mean squared error (MSE) of 0.032, confirming its superior predictive performance.

In conclusion, the Attention–LSTM-CNN model provided a promising solution for supporting data-driven decision-making in smart city waste management systems. Improving the accuracy of urban waste forecasting will contribute to more effective resource allocation, policy development, and the promotion of sustainable urban environments.

Lilhore et al. [49] introduced a smart waste classification framework that combined a CNN-LSTM hybrid model with transfer learning (TL) to address sustainable waste management in urban environments. This study aimed to automatically classify waste into organic and recyclable categories, thereby reducing the limitations of manual sorting, data imbalance, and model overfitting.

The proposed model was designed such that the CNN extracted spatial features from waste images, whereas the LSTM analyzed the temporal sequence relationships within those features. Transfer learning was applied using pre-trained weights from ImageNet to reduce training time and enhance generalization. The model was trained on the TrashNet image dataset from Kaggle, which included 27,025 images—17,005 of organic waste and 10,020 of recyclable waste.

Performance comparisons against models such as VGG-16, ResNet-34, ResNet-50, and AlexNet were conducted under two conditions (80:10:10 and 70:15:15 train/validation/test splits) and across 50 and 100 epochs. In all cases, the CNN-LSTM+TL hybrid model achieved the best performance, with 95.7% accuracy, 95.4% precision, 94.3% recall, and a 93.1% F1-score—improvements of 4–10% over conventional models [49]. The model also achieved the lowest error rate (0.0166%) and shortest training time (7586 s) in the 50-epoch setting [49].

To further enhance model performance, the study incorporated advanced data augmentation techniques, including background enhancement (BackMod), image rotation and flipping, and noise injection. Optimization was conducted using the Adaptive Moment

Estimator (AME) algorithm, which contributed to faster convergence and effective loss minimization.

Lilhore et al. [49] demonstrates the potential for deploying low-cost, high-accuracy, automated waste classification systems in smart cities. Their study makes important contributions to public sanitation, resource recovery, and environmental protection. Future work will explore the multi-class classification of waste images and the development of predictive models for waste generation using socioeconomic and service expenditure variables.

Fatovatikhah et al. [50] proposed a hybrid LSTM-SVM model to predict waste surges during flooding events. Their study focused on providing data for proactive responses and resource allocation by accurately forecasting various types of waste generated by floods.

The dataset was sourced from the U.S. Environmental Protection Agency's report "Advancing Sustainable Materials Management: Facts and Figures 2015," covering MSW generation and recycling data across nine time points between 1960 and 2015. To enhance the data volume, linear interpolation was applied between years, resulting in 56 data points that were converted to daily frequencies for model training.

In the hybrid architecture, the LSTM component learned sequential patterns, whereas the SVM was responsible for capturing non-linear relationships through classification or regression. The experiments compared the performances of standalone LSTM and hybrid LSTM-SVM models across different waste categories. The model was trained using the Adam optimizer (learning rate = 0.005) for up to 250 epochs.

Performance evaluation using RMSE and MAPE metrics showed that the LSTM-SVM hybrid model outperformed LSTM alone across all categories. On average, LSTM-SVM achieved an RMSE of 4.48 and a MAPE of 0.26, compared to 5.79 and 0.74 for LSTM, respectively [50]. For instance, in the Paper and Paperboard category, the RMSE decreased from 32.06 (LSTM) to 16.76 (LSTM-SVM), and in the Textile category, the MAPE dropped from 4.91 to 0.59 [50]. These results illustrate the model's ability to capture complex multivariate time-series patterns more effectively.

In addition to predictive modeling, the study proposes a waste trend analysis model based on the recycling potential across the U.S. regions (Northeast, South, Midwest, and West). An LSTM-based regression model was developed to rank regions according to their recycling capacities, offering practical insights into waste separation policies and regional resource recovery forecasting.

In 2023, Xie et al. [51] examined a hybrid modeling approach that integrated Temporal Convolutional Networks (TCNs) with Long Short-Term Memory (LSTM) networks to enhance the prediction accuracy of total nitrogen (TN) levels in wastewater treatment plant (WWTP) effluent, particularly in response to increasingly stringent discharge regulations. The model was developed using real-time operational data from a WWTP in Jiangsu Province, China.

By combining the strengths of both architectures, the TCN–LSTM model effectively captures temporal dynamics. The TCN component allows for modeling long-range dependencies with flexible receptive fields, while the LSTM layer further refines long-term temporal patterns, helping to preserve important features and mitigate overfitting, which often limits the performance of standalone deep learning models. This architecture proved resilient to large fluctuations in influent characteristics and the complexity of WWTP operations.

Evaluation using test data showed that the hybrid model outperformed standalone TCN and LSTM models, delivering a 33.1% improvement in real-time hourly prediction accuracy [51]. Compared to a traditional feedforward neural network (FFNN), the TCN–LSTM model demonstrated a 63.6% increase in predictive performance [51]. Additionally, it provided highly reliable early warnings of TN effluent concentrations up to eight hours in advance, with over 10% better reliability than baseline models.

Given its adaptability to varying influent and operational conditions, the proposed hybrid model holds promise for broader application across diverse WWTPs.

While these promising studies are based on controlled datasets, they do not address geographic variation, sensor infrastructure disparities, or real-time data constraints, especially in developing or disaster-prone areas. These limitations may affect the scalability and reliability of these models in real-world waste management systems. Further research is needed to assess their performance under varying conditions and capacities.

In conclusion, the hybrid LSTM-SVM model demonstrated robust and accurate forecasting capabilities for flood-induced waste management. The study offers a viable foundation for future applications in disaster preparedness and sustainable waste resource planning.

These developments have expanded the application scope of hybrid LSTM models and improved their performance across various domains.

3.7. Comparative Study

Comparisons and summaries regarding the studies in the six different domains are presented in Table 3. Based on this study comparison, analysis in relation to the different domains was carried out to identify domain-specific insights including trends, trade-offs, and the strengths/weaknesses of various hybrid LSTM architectures.

In traffic and energy management, models that integrate traditional time-series techniques—such as SARIMA, ARIMA, and Prophet—with LSTM have shown consistent improvements in capturing periodic trends and short-term fluctuations. These combinations leverage the interpretability of statistical models and the non-linear learning power of LSTM. For example, SARIMA–Bi-LSTM and Prophet–LSTM hybrids have demonstrated significant error reductions in vehicular flow and energy demand forecasting. The structure of these models is particularly suitable for domains where historical data patterns repeat with clear seasonal or cyclical behaviors.

Spatial–temporal domains, such as air quality monitoring and water demand prediction, benefit from hybrid architectures that incorporate CNNs alongside LSTM. These CNN-LSTM models effectively extract and preserve spatial correlations before passing sequences to the temporal model, thereby improving the forecasting of pollution levels or environmental variables that vary across both time and space. Furthermore, models that incorporate signal decomposition techniques—such as Ensemble Empirical Mode Decomposition (EEMD) or Extended Stationary Wavelet Transform (ESWT)—combined with LSTM have shown promise in reducing noise and isolating predictive components in environmental data.

In domains involving more heterogeneous and irregular data, such as public health and cyber-physical safety, ensemble models and parallel LSTM architectures have proven to be especially effective. For instance, GRU-LSTM or DNN-LSTM hybrids have achieved high levels of accuracy in tasks like COVID-19 hotspot prediction and IoT-based intrusion detection. These models improve generalization and robustness by mitigating individual model biases, though they often increase computational and interpretive complexity. The trade-off between performance and transparency becomes especially critical in domains where decision-making carries ethical or operational implications.

Table 3. Comparative summary of hybrid LSTM models across six application domains.

Domain	Study/ Authors	All Models Used	Best Performing Model	Evaluation Metrics	Best Reported Performance	Data Source/Notes
Traffic management	Chahal et al. [11]	Bi-LSTM, SARIMA, BPNN	SARIMA–Bi-LSTM combination	MAE, MSE, RMSE, MAPE	MAE: 0.499 (↓29%), RMSE: 0.58	CityPulse EU FP7, vehicle speed/volume
	Chahal et al. [12]	Bi-LSTM, ARIMA, Aquila Optimizer	ARIMA-Bi-LSTM with Aquila Optimizer	MAE, MSE, MAPE	MAE: 3.18 (↓32%), MAPE: 0.21	CityPulse EU FP7, 3 cities' traffic data
	Culita et al. [13]	LSTM, ARMA	ARMA–LSTM hybrid	RMSE, MAE	RMSE ↓8.2% (e.g., 13.22 → 12.14)	IoT traffic data, Bucharest (Romania)
	Zafar et al. [14]	LSTM, GRU	LSTM-GRU hybrid	RMSE, MAE, MAPE	MAE ↓10–15%	Islamabad traffic + weather + maps
	Sengupta et al. [15]	LSTM, HMM	C-Hybrid (HMM + LSTM)	RMSE	RMSE: 0.4203	California PeMS traffic data
	Chaoura et al. [16]	LSTM, PSO	LSTM optimized by PSO	RMSE, MAE	MAE: 0.0898 RMSE: 0.15258	Nanjing intersection data (48,120 records)
	Myhrmann & Mabit [17]	LSTM, MDN	LSTM–Mixture Density Network	MSE	MSE: 0.102	Copenhagen bicycle data (64,664 h)
	Miao & Liao [18]	LSTM, CNN, PSO	LSTM-CNN-PSO	MSE, RMSE, MAE,	RMSE: 0.03 MSE↓ 15%	City traffic data
Air quality and environmental monitoring	Kataria & Puri [19]	LSTM, CNN	CNN-LSTM hybrid	RMSE, MAE	Beijing RMSE ↓3.5%, MAE ↓6.3%	Air quality data (Beijing, Chennai)
	Kim et al. [20]	LSTM, SARIMAX	SARIMAX-LSTM stacked ensemble	MAE, RMSE, SMAPE	MAE ↓15.8%, RMSE↓ 9.7%	Solar power output (Incheon, Busan, Yeongam)
	Mukhtar et al. [21]	LSTM, CNN, ANN	CNN-LSTM-ANN hybrid	MAE, RMSE, MAPE	MAPE↓ 9.43%	Solar radiation, 10 African countries
	Zaini et al. [22]	LSTM, EEMD, PSO, SSA	EEMD-SSA-LSTM with neighbor data	RMSE, MAE, MAPE, R^2	MAE↓ 9.31% RMSE↓ 2.65%	PM2.5 and meteorology (Malaysia, 2019)
	Kaveh et al. [23]	LSTM, Orchard Algorithm (OA), BOA	OA-LSTM with BOA-based feature selection	RMSE, R^2, SD Error	R^2: 0.65 → 0.88	Tehran pollution, satellite + weather
	Ratković et al. [24]	LSTM, GA, Bagging, SVM	Ensemble LSTM optimized with GA	MSE, MAE, MAPE, R^2	MAPE↓ 66.0%	Niksic air quality, hourly CAQI data
	Zeng et al. [25]	NLSTM, ESWT	ESWT-NLSTM combination	MAE, RMSE, MAPE, R^2	RMSE ↓ up to 69%, MAE ↓62.8%	Beijing PM2.5, ESWT signal decomposition
	Dalal et al. [26]	LSTM, PSO, RNN	PSO–LSTM–RNN	MAE, RMSE, R^2	RMSE: 0.0184	City pollutants data
Energy management	Arslan [27]	Bi-LSTM, Prophet	Prophet + Bi-LSTM (stacked)	RMSE	RMSE ↓43% (Canada), ↓45% (France)	Smart city energy (Canada and France)
	Ali et al. [28]	LSTM, CNN, ConvLSTM	ConvLSTM (spatiotemporal forecasting)	MAE, MAPE, RMSE	MAE: 3.69–4.65, MAPE ↓18.5~18.8%	Smart home energy consumption
	Alhussein et al. [29]	LSTM, CNN	CNN-LSTM hybrid	MAPE (1-step and multi-step)	MAPE ↓6%	Australia SGSC smart grid data
	Karri et al. [30]	LSTM, MLP	LSTM-MLP hybrid for IoT data	Accuracy, precision, recall, F1-score	Accuracy↑ 6%	Kaggle IoT data on waste-to-energy
	Kaya et al. [31]	LSTM, CNN	CNN-LSTM for energy source forecasting	RMSE, MAE, R^2	R^2: 0.99	Turkey energy mix (2018–2023, EXIST)
	Du [32]	LSTM, SVR, RF	RF-stacked SVR-LSTM	Accuracy, MSE, R^2	Accuracy 90%	Building energy data (539 cases)
	Sivarajan & Jebaseelan [33]	LSTM, CNN, Modified Sea Lion Algorithm (MSLA)	CNN-LSTM + MSLA	MSE, RMSE, MAE, MAPE	RMSE: 0.438 → 0.2	IoT smart grid household appliances
	Jin et al. [34]	LSTM, BPNN	LSTM–BPNN-to-BPNN	MAPE	MAPE: 4.63%→1.47%	Copenhagen bicycle data (64,664 h)

Table 3. *Cont.*

Domain	Study/ Authors	All Models Used	Best Performing Model	Evaluation Metrics	Best Reported Performance	Data Source/Notes
Health and public safety	Khan et al. [35]	LSTM, CNN	CNN-LSTM (COVID-19 hotspot prediction)	MAE, RMSE	MAE: 74.27 (↓40%), RMSE: 106.90	COVID-19 time series (multiscale features)
	Kompunt et al. [36]	LSTM, MLP	MLP-LSTM integrated model	MAE, R^2	MAE ↓66.6%,	COVID-19 cases across 77 Thai provinces
	Sah et al. [37]	LSTM, GRU, Prophet, ARIMA	Stacked LSTM-GRU	R^2, RMSE	R^2: 0.74 RMSE: 69.92	India COVID data (confirmed, active cases)
	Razib et al. [38]	LSTM, DNN	DNN-LSTM for cyberattack detection	Accuracy, precision, recall, F1-score	Accuracy: 99.55%	IoT cyberattack detection in smart city SDN
	Mohbey et al. [39]	LSTM, CNN	CNN-LSTM for social sentiment analysis	Accuracy	Accuracy: 88.1%→ 94%	Monkeypox-related tweets (61,379 samples)
Urban infrastructure monitoring	Cuong et al. [41]	LSTM, Random Forest	LSTM-RF hybrid for port throughput	RMSE, MAE	MAE: 0.004	Busan port container throughput
	Almaliki & Khattak [42]	LSTM, TCN	TCN-LSTM for tidal forecasting	MAE, RMSE, MAPE, uncertainty	MAE ↓ 25.5% RMSE ↓ 32.3%	Ras Tanura (Saudi) tidal levels
	Temür et al. [43]	LSTM, ARIMA	ARIMA-LSTM for real estate	MAE, MSE, RMSE, MAPE	MAPE ↓ 52% MSE ↓ 63%	Turkey housing sales (2008–2018)
	Pu et al. [44]	LSTM, CNN	CNN-LSTM for water demand forecasting	MAPE, RMSE	MAPE ↓40–52%, RMSE ↓60.6%	Urban water usage (15 min and 1 h)
	Das & Guchhait [45]	LSTM, GRU	Parallel GRU–LSTM(P) for bridge damage	Accuracy, F1-score	Accuracy 93.32%, F1 93.3	Z24 bridge damage vibration data
	Parida et al. [47]	LSTM, CNN	Hybrid CNN–LSTM	RMSD	RMSD 5.3% at 10 mm displacement	Embedded sensor data (vibration & ultrasonic)
Urban waste management	Lin et al. [48]	CNN-LSTM-Attention	Attention–LSTM-CNN	R2, MSE	R^2: 95.31% MSE: 0.032	Municipal solid waste & socioeconomic data
	Lilhore et al. [49]	CNN, TL, CNN-LSTM	CNN-LSTM+TL	Accuracy, Precision	Accuracy 95.7% Precision 95.4%	TrashNet image
	Fatovatikhah et al. [50]	LSTM, SVM, Hybrid LSTM–SVM	LSTM–SVM	RMSE, MAPE	RMSE 4.48, MAPE 0.26	Effective for forecasting flood-related waste surges; strong multivariate pattern capture
	Xie et al. [51]	TCN, LSTM, TCN-LSTM	TCN-LSTM	Accuracy	Accuracy↑ 33.1%	Robust model for WWTP inflow prediction

Note: ↓ indicates a decrease in performance of the proposed model compared to the conventional model; ↑ indicates an increase; → indicates a change from A to B, meaning the value changed from A to B.

Urban infrastructure monitoring presents a unique challenge due to the need for real-time feedback and integration with physical sensor systems. In this context, hybrid LSTM models that include uncertainty quantification techniques or are optimized through evolutionary algorithms (e.g., GRU–LSTM with metaheuristic tuning) are well suited to handling the dynamic and localized nature of infrastructure data. However, their deployment often requires domain-specific customization and calibration with real-world systems. Across all domains, it is evident that the most effective hybrid configurations are those that align not only with data characteristics but also with the operational context in which predictions are applied.

4. Future Perspectives

This section presents a forward-looking perspective on hybrid LSTM models, focusing on technological advancement trajectories, application domain expansion, market growth projections, and implementation challenges.

The technological evolution of hybrid LSTM models is expected to advance primarily through architectural sophistication and improvements in computational efficiency. New hybrid structures incorporating transformers are likely to emerge, significantly enhancing long-term dependency learning and parallel-processing capabilities. Recent studies have demonstrated that adding self-attention layers to LSTM–CNN hybrids can further improve multi-step forecasting accuracy on urban traffic and energy datasets [52,53]. Recent research has identified limitations in RNN-based attention mechanisms for seasonal modeling, prompting the development of novel architectural solutions. A bibliometric survey of AI applications in smart cities, covering over 4400 publications between 2010 and 2023, shows exponential growth in neural network-based approaches, particularly in the transportation and energy domains [54–59]. While these bibliometric trends indicate rising academic interest, they do not yet confirm large-scale industrial adoption.

In particular, promising advancements are anticipated through CNN integration. Combinations with extended causal convolutions and Temporal Convolution Networks (TCNs) that effectively capture long-range dependencies in the temporal dimension show significant potential for time-series forecasting. Integration possibilities with quantum computing merit attention, whereas convergence with federated learning is expected to enhance model performance while maintaining data privacy protection.

The predictive capabilities expand beyond univariate time-series forecasting to probabilistic and multivariate predictions. Complex forecasting scenarios, such as those in retail, require sophisticated hybrid LSTM architectures capable of modeling interdependent relationships between different variables. Enhanced modeling capabilities for big data environments with multiple seasonality patterns are likely to gain prominence.

Innovative hybrid models that integrate global and local variables are designed to reflect specific time-series characteristics by combining hierarchical models with ensemble techniques. These advancements will enable more accurate predictions and decision-making across diverse smart city applications.

The utilization scope of hybrid LSTM models is projected to extend beyond the current transportation, environmental, and energy sectors to broader urban problem-solving domains. New applications are expected in disaster prediction and response, urban crime prevention, and social infrastructure life cycle forecasting. This technology can be applied to behavioral pattern analysis and demand forecasting for personalized citizen services.

Convergence with existing fields will accelerate. For instance, integration with digital twins could help develop tools that support optimal decision-making through various scenario simulations in virtual urban environments. Metaverse applications represent another emerging possibility.

The global smart city market was valued at USD 820.7 billion in 2023 and is projected to grow at a compound annual growth rate (CAGR) of 22.1% from 2024 to 2030 [60]. Other forecasts predict an even faster expansion—from USD 877.6 billion in 2024 to USD 3.76 trillion by 2030—reflecting a CAGR of approximately 29.4% [61,62] and indicating strong demand for AI-based technologies such as hybrid LSTM models. Patent landscape analyses also reveal a cumulative increase of 22,655 smart-city-related patents over the past 16 years, with annual filings peaking in recent years—underscoring heightened innovation and industry interest in this sector [63]. Amid this growth, smart city technologies increasingly rely on hybrid LSTM architectures to enhance real-time data processing and prediction accuracy. The Asia–Pacific region, driven by rapid urbanization and active government-led smart city initiatives, is expected to see explosive growth in demand for such technologies.

By industry, the smart mobility and smart energy sectors forecast the highest demand, reflecting the prioritization of urban traffic solutions and energy efficiency improvements.

The smart healthcare and smart building sectors will also experience rapid growth as IoT and AI technology adoption expands. Both government and private sector investments are expected to increase continuously, particularly aligned with ESG (Environmental, Social, Governance) management strengthening and sustainable urban operations.

Hybrid LSTM technologies provide a significant competitive advantage in terms of market growth. Real-time data processing and predictive capabilities are essential for maximizing the operational efficiency of smart cities and have become fundamental elements in various applications, including energy management, waste prediction, and traffic flow control. These technological advancements will further refine the smart city ecosystem and contribute to addressing the diverse challenges faced by global cities.

Several challenges must be addressed for the widespread implementation of H-LSTM models. Interpretability issues arising from model complexity require further improvement. Integrating explainable AI (XAI) technologies to create transparency in decision-making processes is crucial. Data quality management and standardization are urgent priorities.

Additionally, the development of unified meta-analytic frameworks and standardized benchmarking systems represents a critical future research priority to enhance the comparative validity of hybrid LSTM performance across different urban contexts. Such frameworks would address current limitations in generalizing model performance given the heterogeneous nature of datasets, temporal resolutions, and evaluation metrics used across various cities.

In response to these challenges, international standardization organizations have established data quality criteria and model evaluation metrics. Government initiatives are emerging to develop data governance frameworks and refine relevant regulations.

Hybrid LSTM models are expected to enhance the sustainability of smart cities significantly. They create environmental value through energy efficiency optimization and carbon emissions reduction while realizing social value by improving citizens' quality of life and urban service accessibility.

From an economic perspective, these models positively impact urban operational cost reductions and the creation of new business models. Predictive maintenance and optimized resource management can extend the lifespan of urban infrastructures and enhance operational efficiency. Public–private sector collaboration will maximize the potential of the hybrid LSTM model and promote sustainable smart city development.

This developmental trajectory faces several critical challenges. Therefore, the increasing computational time and cost requirements of more complex models must be addressed. Data quality issues when processing large volumes and difficulties in explaining model prediction processes also require resolution. Continuous research and innovation to overcome these technical limitations are necessary to provide opportunities to further enhance the practicality and reliability of hybrid LSTM models.

5. Conclusions

Hybrid LSTM models serve as powerful tools for effectively learning and predicting complex patterns in time-series data, thereby demonstrating their successful implementation across diverse smart city domains. These models contribute to urban problem-solving and quality-of-life improvements by delivering superior predictive performance compared with single models in various sectors, including traffic management, energy consumption forecasting, environmental monitoring, and public health and safety.

By combining the strengths of different algorithms, hybrid LSTM models simultaneously address data linearity and non-linearity while providing integrated spatiotemporal analysis. This capability is particularly significant in complex multidimensional data environments such as smart cities. Integrated analytical capacity is expected to play an increasingly critical role in sophisticated smart city decision-making.

Several challenges must be addressed for the successful implementation of these models. These include increased computational resource requirements and training time owing to model complexity, substantial data demands with overfitting risks, and limited model interpretability. In addition, data quality management, standardization, and privacy protection are important considerations.

Hybrid LSTM models have advanced significantly by integrating diverse techniques such as attention mechanisms, transformers, and optimization algorithms. These developments enhance the models' ability to capture long-term dependencies, spatial–temporal patterns, and complex data features. Techniques like ensemble learning, model compression, and Bi-LSTM further contribute to improved prediction accuracy and efficiency. As a result, hybrid LSTM models have broadened their applicability and demonstrated strong performance in sub-areas of the smart city.

This review has shown that hybrid LSTM models consistently outperform single models in specific smart city applications by leveraging complementary strengths across algorithms. However, their performance is context-dependent, often relying on historical datasets with limited attention to implementation feasibility, real-time constraints, or standard evaluation frameworks. Furthermore, a lack of model interpretability and insufficient validation under extreme or non-stationary conditions remain key methodological limitations across the literature.

Ongoing research and technological development are necessary to overcome these challenges. Examples include model compression and optimization, the implementation of explainable AI technologies, and the establishment of data governance frameworks. When governments, industries, and academia collaborate in these efforts, hybrid LSTM models will further evolve as core smart city technologies and contribute to sustainable urban development.

In conclusion, hybrid LSTM models will play a crucial role in addressing complex smart city challenges through technological advancements and application-domain expansion. They are expected to enhance the efficiency of urban management, deliver improved citizen services, and contribute to sustainable urban development. Concurrently, data-driven decision-making systems are becoming more sophisticated, significantly enhancing the operational efficiency of smart cities.

Author Contributions: Conceptualization, I.-W.N.; formal analysis, B.-J.K. and I.-W.N.; investigation, B.-J.K. and I.-W.N.; resources, B.-J.K. and I.-W.N.; writing—original draft preparation, B.-J.K. and I.-W.N.; writing—review and editing, B.-J.K. and I.-W.N.; visualization, I.-W.N.; supervision, I.-W.N.; project administration, I.-W.N.; funding acquisition, I.-W.N. All authors have read and agreed to the published version of the manuscript.

Funding: This work was supported by the Korea Institute of Energy Technology Evaluation and Planning (KETEP) and the Ministry of Trade, Industry, and Energy (MOTIE) of the Republic of Korea (Grant Number 20214000000010). This work was also supported by the Korea Hydro & Nuclear Power Co. (2025).

Data Availability Statement: No new data were created or analyzed in this study.

Acknowledgments: The authors appreciate the collaborative work of Manan Bhandari during the preparation of this manuscript.

Conflicts of Interest: The authors declare that this study received funding from the Korea Hydro & Nuclear Power Co. (2025). The funder was not involved in the study design, collection, analysis, interpretation of data, the writing of this article or the decision to submit it for publication.

References

1. Heidari, A.; Navimipour, N.J.; Unal, M. Applications of ML/DL in the Management of Smart Cities and Societies Based on New Trends in Information Technologies: A Systematic Literature Review. *Sustain. Cities Soc.* **2022**, *85*, 104089. [CrossRef]
2. Wang, W.; Ma, B.; Guo, X.; Chen, Y.; Xu, Y. A Hybrid ARIMA-LSTM Model for Short-Term Vehicle Speed Prediction. *Energies* **2024**, *17*, 3736. [CrossRef]
3. Hochreiter, S.; Schmidhuber, J. Long Short-Term Memory. *Neural Comput.* **1997**, *9*, 1735–1780. [CrossRef]
4. Sarker, I.H. Machine Learning: Algorithms, Real-World Applications and Research Directions. *SN Comput. Sci.* **2021**, *2*, 160. [CrossRef] [PubMed]
5. Zhang, G.P. Time Series Forecasting Using a Hybrid ARIMA and Neural Network Model. *Neurocomputing* **2003**, *50*, 159–175. [CrossRef]
6. Bhanja, S.; Das, A. A Hybrid Deep Learning Model for Air Quality Time Series Prediction. *Indones. J. Electr. Eng. Comput. Sci.* **2021**, *22*, 1611–1618. [CrossRef]
7. Yadav, S. A Comparative Study of ARIMA, Prophet and LSTM for Time Series Prediction. *J. Artif. Intell. Mach. Learn. Data Sci.* **2023**, *1*, 1813–1816. [CrossRef]
8. Schmidhuber, J. Deep Learning in Neural Networks: An Overview. *Neural Netw.* **2015**, *61*, 85–117. [CrossRef]
9. Goodfellow, I.; Bengio, Y.; Courville, A. *Deep Learning; Request Exam Copy View Preview Adaptive Computation and Machine Learning Series*; The MIT Press: San Francisco, CA, USA, 2016; ISBN 978-0-262-03561-3.
10. Buwaneswaran, M. *Implementation Differences in LSTM Layers—Tensorflow vs. Pytorch*; Towards Data Science: San Francisco, CA, USA, 2021.
11. Chahal, A.; Gulia, P.; Gill, N.S.; Priyadarshini, I. A Hybrid Univariate Traffic Congestion Prediction Model for IoT-Enabled Smart City. *Information* **2023**, *14*, 268. [CrossRef]
12. Chahal, A.; Gulia, P.; Gill, N.S.; Sultana, N. Aquila Optimizer-Based Hybrid Predictive Model for Traffic Congestion in an IoT-Enabled Smart City. *Int. J. Intell. Syst.* **2024**, *2024*, 5577278. [CrossRef]
13. Culita, J.; Caramihai, S.I.; Dumitrache, I.; Moisescu, M.A.; Sacala, I.S. An Hybrid Approach for Urban Traffic Prediction and Control in Smart Cities. *Sensors* **2020**, *20*, 7209. [CrossRef] [PubMed]
14. Zafar, N.; Haq, I.U.; Chughtai, J.-R.; Shafiq, O. Applying Hybrid Lstm-Gru Model Based on Heterogeneous Data Sources for Traffic Speed Prediction in Urban Areas. *Sensors* **2022**, *22*, 3348. [CrossRef]
15. Sengupta, A.; Das, A.; Guler, S.I. Hybrid Hidden Markov LSTM for Short-Term Traffic Flow Prediction. *arXiv* **2023**. [CrossRef]
16. Chaoura, C.; Lazar, H.; Jarir, Z. Traffic Flow Prediction at Intersections: Enhancing with a Hybrid LSTM-PSO Approach. *Int. J. Adv. Comput. Sci. Appl.* **2024**, *15*, 494. [CrossRef]
17. Myhrmann, M.S.; Mabit, S.E. Estimating City-Wide Hourly Bicycle Flow Using a Hybrid LSTM MDN. *Transp. Res. Part Policy Pract.* **2023**, *176*, 103783. [CrossRef]
18. Miao, Z.; Liao, Q. IoT-Based Traffic Prediction for Smart Cities. *IEEE Access* **2025**, *13*, 52369–52384. [CrossRef]
19. Kataria, A.; Puri, V. AI- and IoT-Based Hybrid Model for Air Quality Prediction in a Smart City with Network Assistance. *IET Netw.* **2022**, *11*, 221–233. [CrossRef]
20. Kim, B.; Suh, D.; Otto, M.-O.; Huh, J.-S. A Novel Hybrid Spatio-Temporal Forecasting of Multisite Solar Photovoltaic Generation. *Remote Sens.* **2021**, *13*, 2605. [CrossRef]
21. Mukhtar, M.; Oluwasanmi, A.; Yimen, N.; Qinxiu, Z.; Ukwuoma, C.C.; Ezurike, B.; Bamisile, O. Development and Comparison of Two Novel Hybrid Neural Network Models for Hourly Solar Radiation Prediction. *Appl. Sci.* **2022**, *12*, 1435. [CrossRef]
22. Zaini, N.; Ahmed, A.N.; Ean, L.W.; Chow, M.F.; Malek, M.A. Forecasting of Fine Particulate Matter Based on LSTM and Optimization Algorithm. *J. Clean. Prod.* **2023**, *427*, 139233. [CrossRef]
23. Kaveh, M.; Mesgari, M.S.; Kaveh, M. A Novel Evolutionary Deep Learning Approach for PM2.5 Prediction Using Remote Sensing and Spatial–Temporal Data: A Case Study of Tehran. *ISPRS Int. J. Geo Inf.* **2025**, *14*, 42. [CrossRef]
24. Ratković, K.; Kovač, N.; Simeunović, M. Hybrid LSTM Model to Predict the Level of Air Pollution in Montenegro. *Appl. Sci.* **2023**, *13*, 10152. [CrossRef]
25. Zeng, Y.; Chen, J.; Jin, N.; Jin, X.; Du, Y. Air Quality Forecasting with Hybrid LSTM and Extended Stationary Wavelet Transform. *Build. Environ.* **2022**, *213*, 108822. [CrossRef]
26. Dalal, S.; Lilhore, U.K.; Faujdar, N.; Samiya, S.; Jaglan, V.; Alroobaea, R.; Shaheen, M.; Ahmad, F. Optimising Air Quality Prediction in Smart Cities with Hybrid Particle Swarm Optimization-Long-Short Term Memory-Recurrent Neural Network Model. *IET Smart Cities* **2024**, *6*, 156–179. [CrossRef]
27. Arslan, S. A Hybrid Forecasting Model Using LSTM and Prophet for Energy Consumption with Decomposition of Time Series Data. *PeerJ Comput. Sci.* **2022**, *8*, e1001. [CrossRef]
28. Ou Ali, I.H.; Agga, A.; Ouassaid, M.; Maaroufi, M.; Elrashidi, A.; Kotb, H. Predicting Short-Term Energy Usage in a Smart Home Using Hybrid Deep Learning Models. *Front. Energy Res.* **2024**, *12*, 1323357. [CrossRef]

29. Alhussein, M.; Aurangzeb, K.; Haider, S.I. Hybrid CNN-LSTM Model for Short-Term Individual Household Load Forecasting. *IEEE Access* **2020**, *8*, 180544–180557. [CrossRef]
30. Karri, M.; KG, C.P.; Raha, S.; Kumar, P.; Gajbhare, B.P.; Bhatnagar, V. IoT and LSTM-MLP Integration for Efficient Waste-to-Energy Power Generation in Smart Cities. In Proceedings of the 2024 12th International Conference on Internet of Everything, Microwave, Embedded, Communication and Networks IEMECON 2024, Jaipur, India, 24–26 October 2024; pp. 1–6. [CrossRef]
31. Kaya, M.; Utku, A.; Canbay, Y. A Hybrid CNN-LSTM Model for Predicting Energy Consumption and Production Across Multiple Energy Sources. *J. Soft Comput. Artif. Intell.* **2024**, *5*, 63–73. [CrossRef]
32. Du, Z. Prediction of Building Energy Consumption Based on LSTM-SVR-Random Forest Hybrid Model. *Appl. Comput. Eng.* **2024**, *115*, 75–85. [CrossRef]
33. Sivarajan, S.; Sundar Singh Jebaseelan, S.D. Improving Energy Demand Prediction in IoT Based Smart Grids through Hybrid CNN-LSTM Modelling with Modified Sea Lion Algorithm. *Int. J. Electr. Electron. Eng.* **2023**, *10*, 221–231. [CrossRef]
34. Jin, B.; Zeng, G.; Lu, Z.; Peng, H.; Luo, S.; Yang, X.; Zhu, H.; Liu, M. Hybrid LSTM–BPNN-to-BPNN Model Considering Multi-Source Information for Forecasting Medium- and Long-Term Electricity Peak Load. *Energies* **2022**, *15*, 7584. [CrossRef]
35. Khan, S.D.; Alarabi, L.; Basalamah, S. Toward Smart Lockdown: A Novel Approach for COVID-19 Hotspots Prediction Using a Deep Hybrid Neural Network. *Computers* **2020**, *9*, 99. [CrossRef]
36. Kompunt, P.; Yongjoh, S.; Aimtongkham, P.; Muneesawang, P.; Faksri, K.; So-In, C. A Hybrid LSTM and MLP Scheme for COVID-19 Prediction: A Case Study in Thailand. *Trends Sci.* **2023**, *20*, 6884. [CrossRef]
37. Sah, S.; Surendiran, B.; Dhanalakshmi, R.; Mohanty, S.N.; Alenezi, F.; Polat, K. Forecasting COVID-19 Pandemic Using Prophet, ARIMA, and Hybrid Stacked LSTM-GRU Models in India. *Comput. Math. Methods Med.* **2022**, *2022*, 1556025. [CrossRef]
38. Razib, M.A.; Javeed, D.; Khan, M.T.; Alkanhel, R.; Muthanna, M.S.A. Cyber Threats Detection in Smart Environments Using SDN-Enabled DNN-LSTM Hybrid Framework. *IEEE Access* **2022**, *10*, 53015–53026. [CrossRef]
39. Mohbey, K.K.; Meena, G.; Kumar, S.; Lokesh, K. A CNN-LSTM-Based Hybrid Deep Learning Approach to Detect Sentiment Polarities on Monkeypox Tweets. *New Gener. Comput.* **2022**, *42*, 89–107. [CrossRef]
40. Hassan, M.M.; Gumaei, A.; Alsanad, A.; Alrubaian, M.; Fortino, G. A Hybrid Deep Learning Model for Efficient Intrusion Detection in Big Data Environment. *Inf. Sci.* **2020**, *513*, 386–396. [CrossRef]
41. Cuong, T.N.; You, S.-S.; Long, L.N.B.; Kim, H.-S. Seaport Resilience Analysis and Throughput Forecast Using a Deep Learning Approach: A Case Study of Busan Port. *Sustainability* **2022**, *14*, 13985. [CrossRef]
42. Almaliki, A.H.; Khattak, A. Short- and Long-Term Tidal Level Forecasting: A Novel Hybrid TCN + LSTM Framework. *J. Sea Res.* **2025**, *204*, 102577. [CrossRef]
43. Temür, A.S.; Akgün, M.; Temür, G. Predicting Housing Sales in Turkey Using ARIMA, LSTM and Hybrid Models. *J. Bus. Econ. Manag.* **2019**, *20*, 920–938. [CrossRef]
44. Pu, Z.; Yan, J.; Chen, L.; Li, Z.; Tian, W.; Tao, T.; Xin, K. A Hybrid Wavelet-CNN-LSTM Deep Learning Model for Short-Term Urban Water Demand Forecasting. *Front. Environ. Sci. Eng.* **2022**, *17*, 22. [CrossRef]
45. Das, T.; Guchhait, S. A Hybrid GRU and LSTM-Based Deep Learning Approach for Multiclass Structural Damage Identification Using Dynamic Acceleration Data. *Eng. Fail. Anal.* **2025**, *170*, 109259. [CrossRef]
46. Yessoufou, F.; Zhu, J. Classification and Regression-Based Convolutional Neural Network and Long Short-Term Memory Configuration for Bridge Damage Identification Using Long-Term Monitoring Vibration Data. *Struct. Health Monit.* **2023**, *22*, 4027–4054. [CrossRef]
47. Parida, L.; Moharana, S.; Ferreira, V.M.; Giri, S.K.; Ascensão, G. A Novel CNN-LSTM Hybrid Model for Prediction of Electro-Mechanical Impedance Signal Based Bond Strength Monitoring. *Sensors* **2022**, *22*, 9920. [CrossRef]
48. Lin, K.; Zhao, Y.; Kuo, J.-H. Deep Learning Hybrid Predictions for the Amount of Municipal Solid Waste: A Case Study in Shanghai. *Chemosphere* **2022**, *307*, 136119. [CrossRef]
49. Lilhore, U.K.; Simaiya, S.; Dalal, S.; Damaševičius, R. A Smart Waste Classification Model Using Hybrid CNN-LSTM with Transfer Learning for Sustainable Environment. *Multimed. Tools Appl.* **2023**, *83*, 29505–29529. [CrossRef]
50. Fatovatikhah, F.; Ahmedy, I.; Noor, R.M. Waste Prediction Approach Using Hybrid Long Short-Term Memory with Support Vector Machine. *Int. J. Comput. Intell. Syst.* **2024**, *17*, 103. [CrossRef]
51. Xie, Y.; Chen, Y.; Wei, Q.; Yin, H. A Hybrid Deep Learning Approach to Improve Real-Time Effluent Quality Prediction in Wastewater Treatment Plant. *Water Res.* **2024**, *250*, 121092. [CrossRef]
52. Vaswani, A.; Shazeer, N.; Parmar, N.; Uszkoreit, J.; Jones, L.; Gomez, A.N.; Kaiser, L.; Polosukhin, I. Attention Is All You Need. *arXiv* **2023**, arXiv:1706.03762.
53. Aslam, S.; Ullah, H.S. A Comprehensive Review of Smart Cities Components, Applications, and Technologies Based on Internet of Things. *arXiv* **2020**. [CrossRef]
54. Karger, E.; Rothweiler, A.; Brée, T.; Ahlemann, F. Building the Smart City of Tomorrow: A Bibliometric Analysis of Artificial Intelligence in Urbanization. *Urban Sci.* **2025**, *9*, 132. [CrossRef]

55. Ravisankar, A.; Ibraheem, I.K.; Raha, S.; Sakthi, T.; Ashok, J.; Tonk, A. Bibliometric Analysis of AI Research in Sustainable Smart Cities. In *Sustainable Development Goals*; CRC Press: Boca Raton, FL, USA, 2024; ISBN 978-1-003-46825-7.
56. Kousis, A.; Tjortjis, C. Data Mining Algorithms for Smart Cities: A Bibliometric Analysis. *Algorithms* **2021**, *14*, 242. [CrossRef]
57. Szpilko, D.; Naharro, F.J.; Lăzăroiu, G.; Nica, E.; de la Torre Gallegos, A. Artificial Intelligence in the Smart City—A Literature Review. *Eng. Manag. Prod. Serv.* **2023**, *15*, 53–75. [CrossRef]
58. Herath, H.M.K.K.M.B.; Mittal, M. Adoption of Artificial Intelligence in Smart Cities: A Comprehensive Review. *Int. J. Inf. Manag. Data Insights* **2022**, *2*, 100076. [CrossRef]
59. European Commission—Joint Research Centre. *Bibliometric Analysis of Scientific Publications and Patents on Smart Cities*; Publications Office: Luxembourg, 2023.
60. Grand View Research. *Smart Cities Market Size, Share & Trends Analysis Report by Solution, by End Use, by Region, and Segment Forecasts, 2024–2030*; Grand View Research: San Francisco, CA, USA, 2024.
61. Hassebo, A.; Tealab, M. Global Models of Smart Cities and Potential IoT Applications: A Review. *IoT* **2023**, *4*, 366–411. [CrossRef]
62. Xu, H.; Omitaomu, F.; Sabri, S.; Zlatanova, S.; Li, X.; Song, Y. Leveraging Generative AI for Urban Digital Twins: A Scoping Review on the Autonomous Generation of Urban Data, Scenarios, Designs, and 3D City Models for Smart City Advancement. *Urban Inform.* **2024**, *3*, 29. [CrossRef]
63. Kalleya, C.; Purnomo, A.; Madyatmadja, E.D.; Meiryani; Karmagatri, M. Smart City Applications: A Patent Landscape Exploration. *Procedia Comput. Sci.* **2023**, *227*, 981–989. [CrossRef]

processes

Article

A Low-Code Visual Framework for Deep Learning-Based Remaining Useful Life Prediction

Yuhan Lin [1,2,†], Jianhua Chen [2,†], Sijuan Chen [2,*], Yunfei Nie [2], Ming Wang [1], Bing Zhang [1], Ming Yang [2] and Jipu Wang [2]

[1] State Key Laboratory of Nuclear Power Safety Technology and Equipment, China Nuclear Power Engineering Co., Ltd., Shenzhen 518172, China; lin682423@outlook.com (Y.L.); wangming25436@163.com (M.W.); xiaohan1123@163.com (B.Z.)
[2] Shenzhen Key Laboratory of Nuclear and Radiation Safety, Institute for Advanced Study in Nuclear Energy & Safety, College of Physics and Optoelectronic Engineering, Shenzhen University, Shenzhen 518060, China; chenjianhua2023@email.szu.edu.cn (J.C.); nyf20020529@163.com (Y.N.); mingyang@szu.edu.cn (M.Y.); wangjipu@outlook.com (J.W.)
* Correspondence: chensijuan@szu.edu.cn
† These authors contributed equally to this work and should be considered as co-first authors.

Abstract: In the context of intelligent manufacturing, deep learning-based remaining useful life (RUL) prediction has become a research hotspot in the field of Prognostics and Health Management (PHM). The traditional approaches often require strong programming skills and repeated model building, posing a high entry barrier. To address this, in this study, we propose and implement a visualization tool that supports multiple model selections and result visualization and eliminates the need for complex coding and mathematical derivations, helping users to efficiently conduct RUL prediction with lower technical requirements. This study introduces and summarizes various novel neural network models for DL-based RUL prediction. The models are validated using the NASA and HNEI datasets, and among the validated models, the LSTM model best met the requirements for remaining useful life (RUL) prediction. In order to achieve the low-code usage of deep learning for RUL prediction, the following tasks were performed: (1) multiple models were developed using the Python (3.9.18) language and were implemented on the PyTorch (1.12.1) framework, providing users with the freedom to choose their desired model; (2) a user-friendly and low-code RUL prediction interface was built using Streamlit, enabling users to easily make predictions; (3) the visualization of prediction results was implemented using Matplotlib (3.8.2), allowing users to better understand and analyze the results. In addition, the tool offers functionalities such as automatic hyperparameter tuning to optimize the performance of the prediction model and reduce the complexity of operations.

Keywords: deep learning (DL); remaining useful life (RUL); user friendly and low code; visualization tool

1. Introduction

The inherent degradation of the health of long-running production equipment during its operation leads to an increasing risk of failure as the equipment continues to age. Additionally, the rapid advancement of technology has made production equipment structures more complex and intricate. As a result, when malfunctions occur, the costs of repairs and the losses from production interruptions increase significantly, adversely affecting the factory's economic efficiency. Statistical data indicate that repair costs typically range from 10 to 40% of a factory's production costs, representing a significant financial burden.

Consequently, factories must implement effective maintenance strategies to ensure the smooth and uninterrupted operation of their equipment. In light of these challenges, the field of Prognostics and Health Management (PHM) has gained considerable attention, offering promising solutions for addressing these issues.

The primary objective of PHM is the accurate prediction of the remaining useful life (RUL) of equipment [1,2]. In general, regardless of whether the equipment is in operational mode, the equipment's performance naturally degrades over time. However, accurately determining the extent of equipment degradation is challenging, making it difficult to ascertain whether the equipment can still perform its tasks and its remaining normal operational time. The value of RUL prediction lies in its ability to address this challenge.

Traditional remaining useful life (RUL) prediction methods largely rely on physical modeling and empirical formulas. While these methods have achieved some success in specific domains, they often require time-consuming modeling, struggle to adapt to environmental changes, and lack generalization capabilities when applied to complex systems. Data-driven RUL prediction has emerged as a research hotspot with the rapid development of big data and deep learning. Deep learning techniques, such as long short-term memory (LSTM) networks and convolutional neural networks (CNNs), can automatically extract features and capture complex temporal relationships, providing more accurate predictions. However, the practical application of deep learning still faces significant technical barriers—particularly in model construction, training, and tuning—which require deep expertise and considerable computational resources. Consequently, the challenge of lowering these technical barriers to enable non-experts to utilize deep learning for RUL prediction remains an urgent issue.

In this study, we propose a low-code deep learning platform that simplifies the application of deep learning models and reduces technical barriers. The platform provides an efficient and user-friendly tool for personnel in the field of equipment health management, allowing them to perform data importation, feature selection, model training, and prediction visualization through a graphical interface, all without the need for complex coding. This approach not only reduces the difficulty of applying deep learning but also promotes its widespread use in RUL prediction.

The main contributions of our work are as follows:

(1) Model Implementation: Multiple representative and novel neural network models suitable for time series RUL prediction tasks—such as LSTM, GRU, and 1D CNN models—were developed using the Python programming language and implemented on the PyTorch deep learning framework. This modular design allows users to freely select and switch between different models based on their task requirements or data characteristics.

(2) Interface Development: A low-code and interactive user interface was developed using the Streamlit framework. This interface supports operations such as data uploading, model selection, training initiation, and result downloading, enabling users without deep programming expertise to easily complete the RUL prediction process.

(3) Result Visualization: To improve the interpretability of model outputs, the system incorporates visualization modules built with Matplotlib (3.8.2). These modules provide intuitive plots such as training curves, prediction vs. actual RUL trends, and residual error distributions, helping users better evaluate model performance.

(4) Intelligent Parameter Optimization: To reduce the manual effort involved in model tuning, the system includes functions for automatic hyperparameter optimization. This feature leverages search algorithms (e.g., grid search or random search) to rec-

ommend the optimal training parameters, thereby enhancing the prediction accuracy and efficiency.

(5) Scalability and Extensibility: The architecture of the tool was designed with scalability in mind and supports the modular integration of additional models and datasets, making it suitable not only for academic research but also for industrial applications involving various types of equipment and operational scenarios.

The organization of the remainder of this paper is as follows. Section 2 provides a detailed literature review on the existing deep learning-based remaining useful life (RUL) prediction methods and other low-code platforms. Section 3 explains the theoretical background of deep learning and the low-code framework. Section 4 elaborates on the implementation of the proposed method and the system architecture. Section 5 discusses the experiments and result analysis. Section 6 presents the conclusions and future work.

2. Related Work

2.1. Overview of Deep Learning for RUL Prediction

Remaining useful life (RUL) prediction is a core issue in Prognostics and Health Management (PHM) that involves predicting the remaining lifespan of equipment by monitoring its operating condition, which helps with planning maintenance and preventing failures in advance. There are many RUL prediction methods, with the mainstream approaches involving model-based prediction, data-driven techniques, or a combination of both.

Gao Z. et al. [3] proposed a multiscale temporal memory Transformer framework for the remaining useful life (RUL) prediction of industrial robots, addressing challenges in detecting state changes and capturing multi-term temporal dependencies. Their method, validated on a self-built industrial robot platform, outperforms other advanced techniques by accurately locating state change points and achieving high-precision RUL prediction.

Wang L. et al. [4] proposed a novel predictive maintenance framework and developed a hybrid deep learning model for RUL prediction and two mixed-integer linear programming models to minimize/maximize the maintenance time, using a fast hybrid metaheuristic algorithm to solve large-scale problems effectively.

Zhou Y. et al. [5] proposed a deep ensemble RUL prediction method based on a multimodal interactive attention spatial–temporal network (MIASTN). The method integrates deep base learners, uses signal processing techniques to transform vibration data for improved interpretability, and employs a learning ensemble strategy.

Kim M. et al. [6] proposed a physics-informed deep learning framework for explainable remaining useful life (RUL) prediction. The framework enhances the RUL accuracy and interpretability by transforming raw measurements into five physics-informed features using a multiscale deep convolutional neural network to extract temporal patterns and applying layer-wise relevance propagation for failure mode identification.

Chen Z. et al. [7] proposed an attention-based deep learning framework for predicting the RUL of machines. The framework integrates both handcrafted features and automatically learned features, and the LSTM network learns sequential patterns from raw sensory data.

Cheng C. et al. [8] proposed a deep learning framework for predicting the RUL of rolling element bearings (REBs) that uses the Hilbert–Huang transform to extract a degradation indicator from raw vibrations, with a CNN to identify patterns and an ϵ-support vector regression model for RUL prediction, showing a superior performance across varying conditions.

Deep learning-based RUL prediction has gained significant attention for its robustness, wide applicability, and high precision due to the continuous advancement of sensor technology and the ongoing optimization of deep learning models. Deep learning has demonstrated significant performance advantages in RUL prediction but also has certain limitations, one of which is the high technical threshold required for its implementation. Practitioners must engage in data processing, model selection, and building the necessary computing framework, all of which require substantial programming skills and a foundational understanding of deep learning. This complexity can deter many potential users.

2.2. Low-Code Platform Applications

Low-code platforms (LCPs) enable users to quickly build applications with minimal coding through a visual interface. As data science and machine learning grow in popularity, LCPs are being adopted to speed up model development and data analysis. We explore their applications in these fields, focusing on open-source platforms and AutoML advancements.

The KNIME Analytics Platform [9] offers a wide range of extensions and plugins to support various data processing, analysis, modeling, and visualization functions, meeting the needs of complex data science tasks, and its modular design allows users to perform data analysis by dragging and dropping components, offering high flexibility. However, beginners may find the interface and functional modules of the KNIME Analytics Platform complex.

The Orange data mining toolbox [10] has a simple and user-friendly interface, making it suitable for beginners. Moreover, its modular design enables data analysis through drag-and-drop components, and it provides various machine learning algorithms. As an open-source and free project, users can freely download, modify, and extend its functionality. However, its performance is weaker when handling complex feature engineering or large-scale data.

AutoML [11,12] automates algorithm selection, feature engineering, and hyperparameter tuning, lowering the technical barrier for machine learning and helping non-experts quickly develop applications. However, while it saves time and effort, it has limited flexibility and may struggle with complex or specialized tasks, and due to its lack of transparency and high computational resource demands, it can face black-box issues and performance bottlenecks.

The table below (Table 1) provides a detailed comparison of the three different platform frameworks, covering their strengths and weaknesses in terms of their functionalities, performances, ease of use, and other aspects.

Table 1. A comparison of the three platform frameworks.

Platform/Tool	Advantages	Disadvantages
KNIME	Strong plugins and functionality, supports multiple data sources, AutoML integration	Steep learning curve, outdated interface, performance bottlenecks with large data
Orange	Simple to use, open-source, supports various algorithms	Less powerful than KNIME, limited data processing ability, fewer plugins
AutoML (General)	Automated workflows, quick prototyping, reduces human intervention	Low flexibility, black-box issue, high resource consumption

The low-code platform in this study is designed for beginners, allowing them to customize the number of layers in the model, and as a Python (3.9.18)-based software, it provides raw code to help users understand and build their own neural network models.

Unlike other platforms, our focus is on teaching users the model-building process and enabling the flexible application of deep learning techniques.

3. Theoretical Method

DL technology, as a forefront in the field of machine learning, derives its outstanding capabilities from its ability to automatically learn abstract features and complex patterns, thereby achieving highly accurate predictions and decision making across various domains [13]. At the technical level, deep learning models are typically constructed as multi-layer neural networks with a large number of parameters to handle high-dimensional data, such as images, text, and time series. Structures such as convolutional neural networks (CNN) and long short-term memory (LSTM) excel at processing spatial and temporal information [2,14]. The software developed in this project offers various foundational models, including CNN and LSTM models. For instance, with CNNs, users can adjust the model parameters in pursuit of improved predictive results, such as the input window size and the numbers of convolutional and normalization layers, among others.

Deep learning technology has shown significant potential in the RUL prediction field. Through training on sensor data from equipment or systems, deep learning models can predict their future lifespans and performances, aiding in more precise maintenance planning and resource management. Deep learning can also be used to identify subtle trends and potential anomalies in time series data, enhancing the RUL estimation accuracy.

3.1. *Basic Principles of the Model*

In this study, we comprehensively applied various deep learning models, including LSTM, GRU, CNN, and Transformer architectures. We provide a brief introduction to their core structures and fundamental principles below.

3.1.1. Long Short-Term Memory

Long short-term memory (LSTM) [15] was introduced by Hochreiter and Schmidhuber, and its core innovation lies in the introduction of memory cells and a triple-gating mechanism, which addresses the problem of long-term dependencies through a constant error carousel (CEC). The LSTM network structure consists of multiple memory and gate units. Each memory unit typically includes an input gate, an output gate, and an internal state, and these components work together as follows (Figure 1):

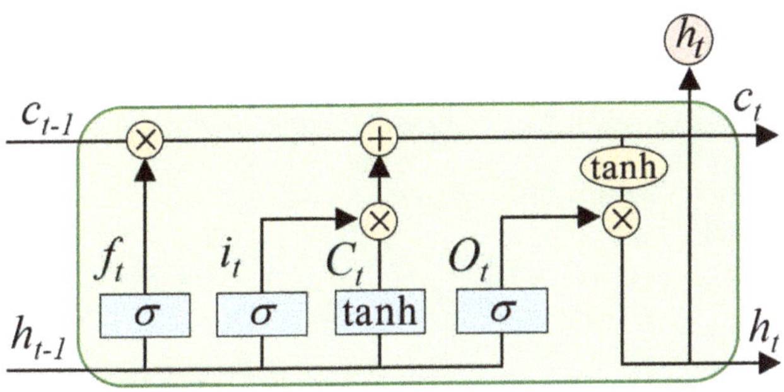

Figure 1. Diagram of the LSTM framework.

The input gate determines the degree to which new information enters the memory unit, and its activation value (i_t) is calculated via the sigmoid activation function using the current input (x_t) and the previous hidden state (h_{t-1}):

$$i_t = \sigma(W_i x_t + W_{hi} h_{t-1} + b_i) \tag{1}$$

where σ is the sigmoid activation function, W_i and W_h are the weight matrices, and b_i is the bias term.

The forget gate determines when the information in the memory unit is forgotten, and its activation value (f_t) is calculated via the sigmoid activation function using the current input (x_t) and the previous hidden state (h_{t-1}):

$$f_t = \sigma\left(W_{if}x_t + W_{hf}h_{t-1} + b_f\right) \tag{2}$$

where W_{if} and W_{hf} are the weight matrices, and b_f is the bias term. The internal state of the memory unit is updated by combining the outputs of the input and forget gates. First, the candidate state ($\tilde{c}_t$) is calculated as follows:

$$\tilde{c}_t = \tanh(W_{ic}x_t + W_{hc}h_{t-1} + b_c) \tag{3}$$

Then, the internal state (c_t) is updated as follows:

$$c_t = f_t \cdot c_{t-1} + i_t \cdot \tilde{c}_t \tag{4}$$

where $\tilde{c}_t$ represents the candidate state, c_t is the updated internal state, and tanh is the hyperbolic tangent activation function. W_{ic} and W_{hc} are the weight matrices, and b_c is the bias term. The output gate determines when the information in the memory unit is output. The activation value of the output gate (o_t) is calculated via the sigmoid activation function using the current input (x_t) and the previous hidden state (h_{t-1}):

$$o_t = \sigma(W_{io}x_t + W_{ho}h_{t-1} + b_o) \tag{5}$$

The output of the memory unit (h_t) is calculated as follows:

$$h_t = o_t \cdot \tanh(c_t) \tag{6}$$

where o_t is the activation value of the output gate, h_t is the output of the memory unit, W_{io} and W_{ho} are the weight matrices, and b_o is the bias term.

The LSTM employs a special gradient descent algorithm that combines truncated backpropagation to update the network weights. This method is as efficient as traditional backpropagation through time (BPTT) in terms of its computational performance, and by limiting the spread of error signals, it avoids gradient explosion and gradient disappearance.

3.1.2. Gated Recurrent Units

The GRU model was proposed by Cho et al., and its core innovation lies in the introduction of the update and reset gate mechanisms [16]. The GRU network structure consists of multiple cyclic units, each of which controls the flow of information through the update and reset gates (Figure 2).

The role of the update gate is to determine the degree to which the current unit state needs to be updated, and its activation value (z_t) is calculated via the sigmoid activation function using the current input (x_t) and the previous hidden state (h_{t-1}):

$$z_t = \sigma(W_z x_t + U_z h_{t-1} + b_z) \tag{7}$$

where σ is the sigmoid activation function, W_z and U_z are the weight matrices, and b_z is the bias term.

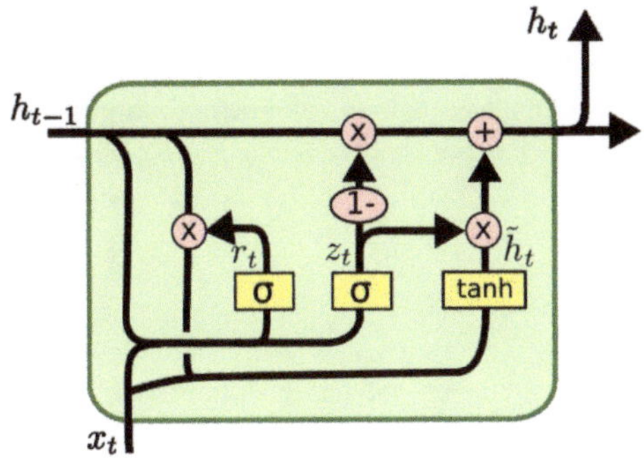

Figure 2. Diagram of the GRU framework.

The reset gate determines the extent to which the current unit state needs to be forgotten, and its activation value (r_t) is calculated via the sigmoid activation function using the current input (x_t) and the previous hidden state (h_{t-1}):

$$r_t = \sigma(W_r x_t + U_r h_{t-1} + b_r) \tag{8}$$

where W_r and U_r are the weight matrices, and b_r is the bias term. The candidate hidden state (Candidate Hidden State) combines the output of the reset gate to calculate the candidate hidden state ($\tilde{h}_t$):

$$\tilde{h}_t = \tanh(W x_t + U(r_t \odot h_{t-1}) + b_c) \tag{9}$$

where $\odot$ represents the element-wise multiplication, W and U are the weight matrices, b_c is the bias term, and tanh is the hyperbolic tangent activation function. The final hidden state (h_t) is obtained by updating the gate (z_t) to perform a weighted sum of the previous hidden state (h_{t-1}) and the candidate hidden state ($\tilde{h}_t$):

$$h_t = (1 - z_t) \cdot h_{t-1} + z_t \cdot \tilde{h}_t \tag{10}$$

where h_t is the hidden state of the current moment, and z_t is the activation value of the update gate.

Both the GRU and LSTM models are advanced variants of recurrent neural networks designed to address the issues of gradient disappearance and gradient explosion that traditional RNNs face when processing long sequences. The incorporated gating mechanisms control the information flow, thereby better capturing both short-term and long-term dependencies.

3.1.3. Convolutional Neural Networks

Convolutional neural networks (CNNs) are well-suited for image processing tasks due to their hierarchical feature extraction structure, inspired by biological visual systems [16]. CNNs excel at tasks such as image recognition and classification by automatically learning useful features from raw data.

The convolutional layer extracts features from an input image using small filters (kernels), sliding over the image and applying element-wise multiplication and summation to produce a feature map. The operation is as follows:

$$F(x, y) = (I * K)(x, y) = \sum_{i,j} I(x + i, y + j) K(i, j) \tag{11}$$

where I is the image, K is the kernel, and F is the feature map. This process reduces the number of parameters and helps capture spatial hierarchies.

Pooling reduces the spatial dimensions of the feature map, minimizing computation. Max pooling selects the maximum value from each region, while average pooling computes the average:

$$\mathrm{MaxPooling}(x, y) = \max(I(x, y)) \tag{12}$$

$$\mathrm{AvgPooling}(x, y) = \frac{1}{n}\sum_{i=1}^{n} I(x_i, y_i) \tag{13}$$

where I(x,y) represents the pixel values of the input image, and n is the number of elements in the pooling region. Pooling helps with downsampling the feature map, reducing the computational complexity, and making the network more robust to small translations or deformations of the input image.

Fully connected layers integrate the features to make final predictions. Each neuron connects to every neuron in the previous layer, and the output is computed as follows:

$$y = W \cdot x + b \tag{14}$$

where W is the weight matrix, x is the input, and b is the bias.

Activation functions introduce nonlinearity into the neural network, enabling it to learn complex patterns and relationships in the data. Without activation functions, the network would only be able to perform linear transformations, limiting its expressive power. One common activation function is the ReLU function:

$$\mathrm{ReLU}(x) = \max(0, x) \tag{15}$$

Backpropagation is the core algorithm for training convolutional neural networks, updating the parameters by calculating the gradients of the loss function with respect to the network parameters in order to minimize the loss. The weight update rule is as follows:

$$\theta = \theta - \eta\frac{\partial L}{\partial \theta} \tag{16}$$

where η is the learning rate, and L is the loss. Through repeated iterations, the network gradually optimizes its weights, improving its performance on the classification task.

3.1.4. Transformers

The Transformer model is a neural network architecture based on the attention mechanism and is designed to handle sequence-to-sequence tasks such as machine translation and text generation [13]. Moreover, the Transformer model completely abandons the traditional recurrent and convolutional network structures, relying solely on the attention mechanism to capture the global dependencies between inputs and outputs.

The Transformer processes sequence data through a multi-head self-attention mechanism and positional encoding, and it significantly improves the training efficiency by taking advantage of parallel computing.

Self-attention allows the model to consider all positions in the sequence when computing the representation of a particular position, calculating the correlations (typically via dot products) between elements and using them to weight and sum values, resulting in a new representation. Multi-head attention projects the input into multiple subspaces, performs self-attention independently in each, and concatenates the results before applying a linear transformation.

Because the Transformer does not rely on recurrent or convolutional structures, it cannot naturally capture positional information such as RNNs or CNNs. To address this, it adds positional encoding. The positional encoding is a vector with the same dimension as the input embedding and is used to represent the position of each element in the sequence. The positional encoding is calculated as follows:

The Transformer uses an encoder–decoder architecture, where the encoder converts the input sequence into a continuous representation, and the decoder generates the output sequence based on this representation.

3.2. DL-Based RUL Prediction

DL-based RUL prediction is a data-driven method that relies on deep neural network models to make precise predictions. These models use a vast amount of data collected from sensors and devices to achieve accurate predictions. Deep learning models can capture subtle variations in the equipment performance over time by automatically learning complex data patterns and features, thereby providing accurate estimates of the remaining life.

LSTM, a recurrent neural network (RNN) variant, is particularly suitable for time series data. LTSM possesses self-memory and the ability to handle long sequences of data, and its self-memory allows it to effectively manage long-term dependencies in time series data, which is crucial for RUL prediction. The equipment operating history often spans multiple time steps, where past operations and environmental conditions may significantly impact the current and future device status. LSTM can capture these intricate time dependencies, making it a powerful tool for remaining useful life prediction.

The GRU model simplifies the LSTM structure by including only the update and reset gates, and it directly stores information through the hidden state. In practical applications, whether to choose the GRU or LSTM model depends on the specific task requirements. The GRU model is used to predict when a device might fail at a certain point in the future, and its advantage lies in its efficient ability to handle time series data, making it particularly suitable for dynamic and changing industrial equipment prediction tasks.

CNNs, renowned for their automatic feature extraction capabilities, efficiently capture local and global features in time series data and are especially useful for analyzing the impacts of different operation and environmental condition time periods on the remaining life in the equipment's operational history. Utilizing CNNs for RUL prediction provides robust feature extraction abilities in time series data analysis, adeptly handling multi-dimensional data and thereby supporting an improved RUL prediction performance and maintenance efficiency. Additionally, CNNs have certain advantages in terms of their computational efficiency, reducing complexity and significantly improving data processing for large-scale datasets.

RUL prediction methods based on Transformer models make full use of their high expressive power and attention mechanism. The attention mechanism is a core component of the Transformer, mimicking how the human brain allocates resources [17], and it can automatically identify which parts of the time series data need more attention at different time points, which is crucial in the RUL prediction field. In the equipment's operational history, environmental variables cannot remain constant, and changes in the operations and environmental conditions during specific time periods may have a significant impact on the device lifespan. Through the attention mechanism, the Transformer can automatically identify and weigh these critical time periods and features. Furthermore, unlike traditional recurrent neural networks (RNNs) [18,19], the Transformer is not hindered by long-term dependency issues and can easily handle time series data, making it more suitable for handling long time series data. Additionally, the Transformer has the ability to extract contextual features (in RUL prediction, contextual features typically refer to factors such as the equipment operations, maintenance history, and environmental conditions). Traditional methods often require the manual or artificial generation of these features, but Transformer-based methods can automatically extract this context information from raw time series data [20], because the Transformer's self-attention mechanism allows it to capture the context relationships between different elements in the data, especially in time series data.

The software implemented in this study offers a range of deep learning models, including the four aforementioned models, enabling users to compare the predictive performances of different models for the same case, facilitating their selection based on specific use scenarios for improved prediction outcomes.

3.3. Low-Code Platform Design Concept

In this study, we designed a low-code, user-friendly visualization tool to lower the technical barriers for using deep learning in predicting the RUL of equipment. Many potential users find it difficult to directly use deep learning tools due to the high technical requirements. Therefore, simplifying the operational process was the core objective.

PyTorch (1.12.1) was chosen as the deep learning framework due to its dynamic computation graphs and flexibility, making it well-suited for handling complex tasks such as RUL prediction. The rich set of tools and libraries in PyTorch, such as Streamlit, further enhances the development efficiency. Streamlit, an innovative Python library, helps developers to quickly create interactive web applications due to its ease of use and high interactivity, eliminating the need for complex web development steps and further improving the accessibility of the technology.

To simplify the process, we developed a web-based interactive interface that allows users to upload data files and select features, labels, etc., without writing code for data preprocessing. Additionally, the platform provides a variety of predefined deep learning models (such as LSTM, GRU, CNN, and Transformer models), enabling users to easily select and adjust model parameters, such as the learning rate and training epochs, further reducing the technical barrier.

The platform also includes a visualization module, where users can view real-time charts to track the training progress and compare the predicted RUL with the actual RUL. This visualization helps users to better understand the model's behavior, enabling more effective decision making.

The platform uses a modular design, allowing users to flexibly combine different functional modules based on their needs. This flexibility ensures that both beginners and experienced users can efficiently complete RUL prediction tasks.

4. Software Design

Before designing and implementing the DL-based RUL prediction software, it is crucial to analyze the software's usage scenarios and user profiles. As an interactive platform, users can choose data files from the initial interface, and they can also select feature, label, time, and identifier columns from the data table. Moreover, users can actively participate in data preprocessing by setting parameters such as the data batch size and time window divisions, and during the model selection phase, they can choose the most suitable model based on the data provided. They can also configure model-specific parameters, such as the number of training iterations and the learning rate to optimize the performance. These settings are configured into the software's configuration files through the Flask framework. The software automatically reads this information to process the data and configure the model.

Once training is complete, the software outputs the evaluation metrics of the model training process and provides a visualization of these metrics. Users can select evaluation metric functions and examine the function graphs to gain insights into the model's training progress. The trained model is saved within the software for later use. During the testing and prediction phases, the software provides users access to the pretrained models. The prediction results are shown in the form of line graphs, allowing users to better understand the equipment's degradation trends.

4.1. Software Structure

4.1.1. Software Architecture Selection

The software architecture combines a Command-Line Interface (CLI) client with a web interface, providing users with multiple interactive pathways and enabling a distributed interactive application. Users initiate the interactive web interface hosted on a web server by inputting commands into the CLI client within their terminal. The web server receives requests from the CLI client and is responsible for executing the core logic of the application, data processing, and service calls. Users can communicate with the web server through their web browser, where the web interface provides visual data representation and functional operations.

To ensure the smooth operation of the CLI client, users need to maintain specific runtime environments and dependencies on their local terminals. This hybrid architecture allows users to interact with the software in a command-line manner in their terminals and view data and perform operations graphically through the web interface. This architecture is suitable for applications that require flexible interactions between users in both local and web interfaces, ensuring that the data and functionalities are presented visually while balancing the flexibility of the CLI and the user-friendliness of the web interface.

4.1.2. Software Architecture Implementation

The software architecture adheres to strict modular design principles with the primary goal of ensuring the system's high scalability and maintainability (The specific module development details can be found in Appendix A). The software is built in the form of a series of core modules, including the user interface, data processing, model selection, and visualization modules, among others. These modules collaborate with each other to provide a wide range of features and offer a comprehensive user experience (Figure 3).

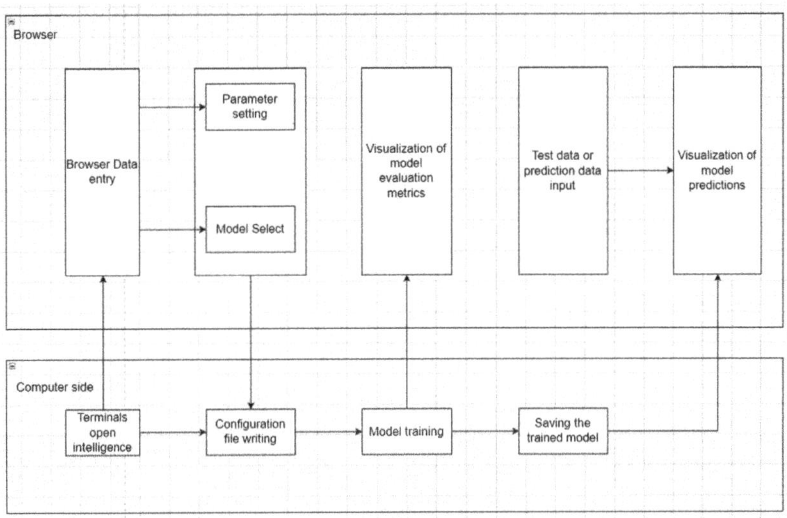

Figure 3. Overall software architecture.

The software features a modular architecture, and within this architecture, the user interface module allows users to import local data through a user-friendly interface. Users can also precisely specify parameters related to the data, including the feature, label, time, and sequence columns.

The data processing module undertakes critical data-related tasks, including but not limited to data partitioning and normalization and the creation of data windows. These tasks are essential to ensure that the data are prepared and cleaned before they are used by the model.

The model selection module is responsible for executing the training process of the machine learning model upon receiving the data prepared by the data processing module. During this process, users have the authority to choose the desired model architecture and adjust the model parameters as needed to meet their specific requirements. Once training is completed, the model is saved within the software for later testing and application (Figure 4).

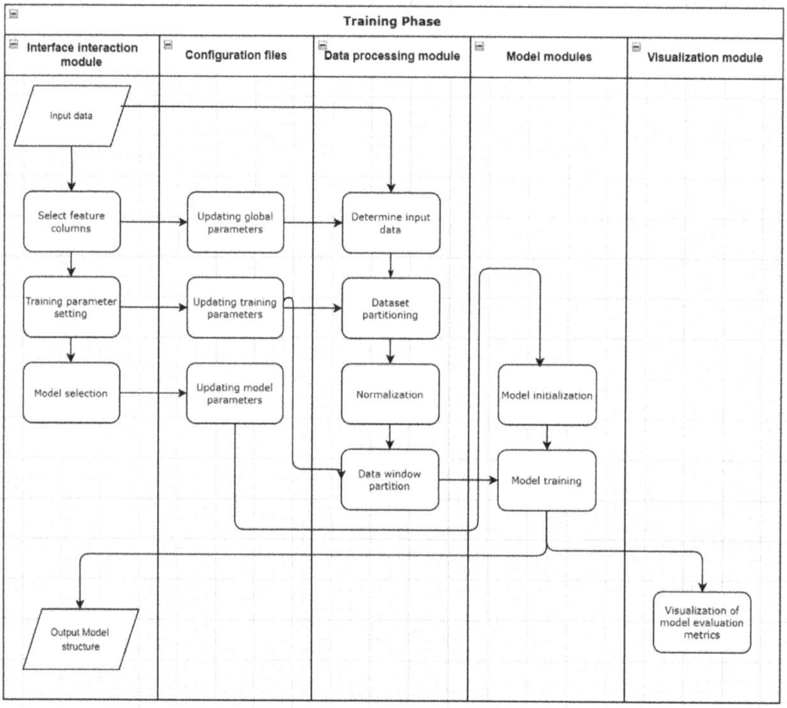

Figure 4. Architecture diagram of the software training phase.

Finally, the visualization module ensures that the model's evaluation metrics are presented in a highly visual manner, enabling users to better understand the model's performance and test results and gain profound insights into its behavior.

Users directly input data during the testing phase. At this point, the software automatically performs data integrity checks to verify whether the testing data include the selected feature columns from the training dataset. If the necessary feature columns are missing in the testing data, the software generates an error report and prevents users from proceeding with further actions. After ensuring the integrity of the testing data, the software normalizes these input data to ensure that they fall within an acceptable range for the subsequent model testing.

Users have the freedom to select one from multiple available testing models, and the software applies that model to the testing data provided by the user for predictions. The final test results are decoupled and presented to the user in the form of line graphs, allowing them to have a clear understanding of the model's performance and prediction results (Figure 5).

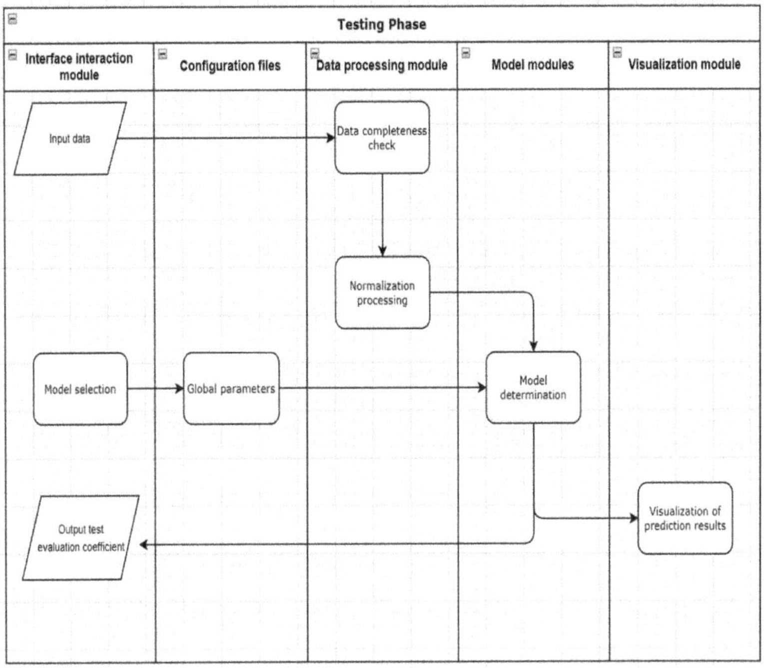

Figure 5. Architecture diagram of the software testing phase.

This modular design enhances the software's scalability and makes it easier to maintain. Each module in the system has well-defined tasks and responsibilities, helping to reduce the system complexity and improve maintainability, also providing a strong foundation for future feature expansion. The architecture offers users powerful data processing and model training capabilities, while ensuring a highly visual presentation of the data and model performance. As a result, the software's flexibility and user friendliness are improved. This modular design principle gives the software an excellent structure, allowing various tasks and components to work together effectively, providing users with outstanding data processing and model training capabilities.

4.1.3. Modular Design Principles

The software design adhered to a set of key modular design principles, including the Single Responsibility Principle (SRP) and Open–Closed Principle (OCP). These principles are foundational in the field of software engineering and aim to ensure a high level of maintainability, scalability, and low coupling in software systems. The Single Responsibility Principle mandates that each module has a well-defined responsibility, thereby increasing the code cohesion and clarity. The Open–Closed Principle emphasizes that systems should be open for extension but closed for modification, supporting seamless system functionality extension through the use of abstractions and interfaces without the need to modify existing code. The Interface Segregation Principle ensures that modules only depend on the minimal interfaces they need, reducing unnecessary dependencies and increasing module independence. The application of these principles ensures that each module in the system is responsible for a single, independent duty, and through appropriate abstraction and interface definitions, it allows for the extension of the system's functionality without modifying existing code, thereby guaranteeing maintainability and extensibility in the software.

4.2. Application Module Details

The software is composed of several major modules—the user interface, data processing, model selection, and visualization modules—which are all designed for deep learning-based equipment RUL prediction. Each module has distinct functions and responsibilities, collectively providing users with comprehensive data analysis and visualization tools.

4.2.1. Interface Interaction Module

The user interface module serves as the interface between the user and the software, offering a user-friendly interface for importing, processing, and displaying data (Figure 6). In terms of data, users input data files, select the feature columns to be used as input, and choose the label columns as output. If there is a device identifier column in the data table, users should select it as the sequence number column. Users can also select a model and decide on the data step size, model optimizer, training iterations, loss functions, and other model configurations within the interface.

Figure 6. Initial software interface.

We chose to use the Streamlit library to design the front-end interface, as Streamlit excels at transforming data science and machine learning results into interactive web applications. User input information is stored in the dataset.yaml configuration file through the Flask framework. This file provides the necessary parameters for the subsequent model training, such as the number of training epochs and the input and output dimensions.

4.2.2. Data Processing Module

The primary responsibility of this module is to process user input data, including dataset partitioning, normalization, and windowing. Users first define the categorization of the columns in the input data to facilitate the processing of data via this module (Figure 7).

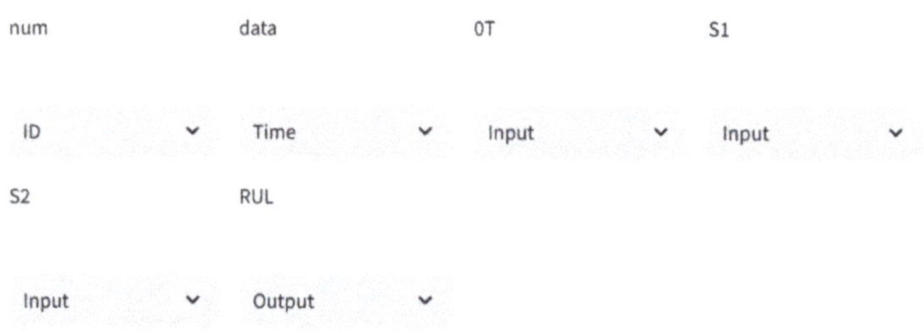

Figure 7. Determination of data columns.

Once the input data are determined, the module checks the existence of the files to ensure their accessibility. This is a crucial step, as data science applications often require loading data from external files or databases. By verifying that the file content is not empty, the module can prevent potential runtime errors and ensure smooth data loading.

The module proceeds to partition the dataset. Dataset partitioning is primarily performed using the train_test_split function from the sklearn library. The partition ratio is determined by the "Split" value in the dataset.yaml configuration file, and the partitioning is performed randomly. Data normalization is achieved using the MinMaxScaler function, which scales data values to the range [0, 1]. This normalization process aims to accelerate the model training, enhance the model stability, and improve the model's robustness to data with different scales.

Additionally, this module is responsible for data windowing. The module determines the data step size based on the "Step" variable in the dataset.yaml file and subsequently divides the data into windows to facilitate model training (Figure 8).

Figure 8. Data batch and time window size determination.

4.2.3. Model Selection Module

The model selection module (Figure 9) is one of the core components of the software, developed using Python and the PyTorch framework, and its primary objective is to provide users with a diverse range of model choices. To achieve this, we have developed various deep learning models, including convolutional neural networks (CNNs), long short-term memory (LSTM) networks, and Transformers, all centrally stored in the model.py file. This diversity of choices allows users to select the model structure that best suits their specific task requirements.

Figure 9. Model selection.

Moreover, shallow models may not meet the training requirements when dealing with complex tasks and large-scale datasets. Therefore, we provide options for setting the number of model layers, allowing users to personalize their models more effectively.

The selected model type, the number of model layers, and the other model parameters chosen by the user are written into the dataset.yaml file. During the model training phase, the get_model function retrieves essential model parameters from the dataset.yaml file, such as the model's hidden dimensions, the number of hidden layers, and the number of training epochs.

The software uses the Adam optimizer for model optimization. The Adam optimizer cleverly combines the gradient descent and momentum methods, adjusting the learning rate adaptively based on the moving averages of both the gradient and squared gradient for each parameter. This adaptation helps automatically adjust the learning rate during different training phases, improving the model stability and accelerating convergence for parameter optimization. Additionally, Adam introduces a momentum term to expedite convergence and reduce oscillations in gradient descent, and through a bias correction mechanism, it also corrects inaccuracies in gradient estimates during the early training stages.

The train_and_evaluate function within the module receives these parameters and the dataset, conducts the training, and outputs the trained model, along with the loss data generated during the training process.

4.2.4. Visualization Module

The visualization module was developed using the Matplotlib library, which is a widely used Python data visualization library in academic research and data analysis. The purpose of this module is to provide users with a convenient tool for data analysis and presentation, allowing for a clearer presentation of the loss values, root-mean-squared errors (RMSEs), and mean-squared errors (MSEs) during the training process, as well as the trends in the prediction results. The data generated during model training, as well as the prediction results from testing, are visualized through this module. These visualization tools assist users in better understanding the model performance and prediction trends, supporting their decision making and analysis tasks.

5. Software Operation and Practical Examples

5.1. Example Introduction

5.1.1. NASA Turbofan Engine Dataset

The software's testing data were obtained from NASA's Prognostics Center of Excellence (PCoE) Prognostics Data Repository [21]. The Turbofan Engine Degradation Simulation Dataset was provided by the NASA Ames Predictive Maintenance and Health Management Center of Excellence. This dataset, generated using the Commercial Modular Aero-Propulsion System Simulation (C-MAPSS) tool, simulates the degradation of turbofan engines under various operating conditions and fault modes.

The dataset consists of four different subsets (FD001, FD002, FD003, FD004), each of which operates under different operating conditions and failure modes (Table 2).

Table 2. Specific information for each subset.

Dataset	Number of Training Tracks	Number of Test Trajectories	Fault Pattern
FD001	100	100	1 (HPC degradation)
FD002	260	259	1 (HPC degradation)
FD003	100	100	2 (HPC degradation, fan degradation)
FD004	248	249	2 (HPC degradation, fan degradation)

Each engine has different initial wear and manufacturing variations when it starts, which users are unaware of (Table 3). These wear and manufacturing variations are considered normal and do not indicate a fault condition. The data are contaminated by sensor noise. Each time series begins with the engine running normally and then experiences a failure at a certain point. In the training set, the failure progresses until the system fails. In the test set, the time series ends at a certain point before the system fails. The goal is to predict the remaining operating cycle before the failure in the test set, especially the number of remaining cycles after the engine's last full operation. The actual remaining useful life (RUL) values of the test data are also provided.

Table 3. Detailed content of each dataset.

Unit number	Indicates the engine number.
Time (period)	Indicates the number of cycles in which the engine is running.
Operation settings	Includes three operational setting variables—the flight altitude, throttle analysis angle (TRA), and Mach number.
Sensor measurement	The measurements include 21 sensors, such as the total temperature at the fan inlet (T2) and the total temperature at the low-pressure compressor outlet (T24).

5.1.2. HNEI Battery RUL Dataset

The second dataset used in this study was the Hawaii Natural Energy Institute (HNEI) battery dataset [22], which supports research on battery remaining useful life (RUL) prediction. This dataset contains multiple cells subjected to repeated charging and discharging cycles under varying operating conditions (Table 4). Key features include the current, voltage, temperature, and capacity. Each time series represents the full lifecycle of a battery cell until the capacity drops below a specified threshold. Unlike the NASA turbofan engine dataset, the HNEI battery dataset represents electrochemical degradation processes. The goal is to predict how many cycles remain before the battery reaches its end of life, defined as 80% of the nominal capacity.

Table 4. Detailed content of each dataset.

cycle_index	The number of charge and discharge cycles of the battery is used to reflect its aging process.
discharge_time	The time (seconds) it takes for the battery to go from full charge to the set cut-off voltage in a complete discharge cycle is an indicator of health.
decrement_3.6–3.4 V	The time (sec) used to reduce the voltage from 3.6 V to 3.4 V is closely related to the voltage attenuation rate in this interval, which is a sensitive feature of the battery degradation process.
max_discharge_voltage	The highest voltage value during discharge (usually measured at the beginning of discharge) can be used to understand the decline in battery capacity.
min_charge_voltage	The minimum voltage during the charging process (usually measured at the beginning of the charging) reflects the change in the battery state to some extent.
time_at_4.15 V	The duration (seconds) that the battery voltage remains within the 4.15 V range reflects the characteristics of the constant voltage phase and can indirectly reveal the characteristics of the internal resistance and capacity of the battery.
constant_current_time (TCC)	The time (seconds) when the current is maintained constant during discharge corresponds to the change in the internal electrical characteristics and is often associated with the aging state of the battery.
charging_time	The discharge time variation reflects the remaining available capacity in seconds for a full charge cycle.
RUL (Remaining Useful Life)	The RUL is the predicted target value of the model.

Based on the original experimental records, the data are processed via feature engineering and normalized preprocessing to form structured time series samples. Each record represents the state characteristics of the battery in a specific cycle stage and corresponds to the corresponding RUL label.

5.2. Application Example

5.2.1. Introduction to Evaluation Metrics

To evaluate the performance of the deep learning model in predicting remaining useful life (RUL), we used several key metrics, including the mean-squared error (MSE), root-mean-squared error (RMSE), R^2, and weighted mean absolute percentage error (WMAPE).

The MSE (mean-squared error) is a commonly used metric to measure the average of the squared differences between predicted ($\hat{y}_i$) and actual (y_i) values. Specifically, for a dataset with real and predicted values, where there are data points in total, the formula for calculating the MSE is as follows:

$$\mathrm{MSE} = \frac{1}{n}\sum_{i=1}^{n}(y_i - \hat{y}_i)^2 \tag{17}$$

Because the MSE is the average of squared errors, it is always non-negative. The MSE is zero only when all the predicted values match the actual values exactly. Furthermore, because the MSE is the average of squared errors, larger errors are amplified, making it more sensitive to larger errors. For example, a difference of 2 between a predicted value and an actual value results in a squared error of 4, whereas a difference of 10 results in a squared error of 100, which significantly contributes to the MSE. Additionally, the MSE is measured in square units of the predicted and actual values. For instance, if both are measured in meters (m), the MSE is measured in square meters (m^2), which can make the MSE less intuitive in interpretation.

The RMSE (root-mean-squared error) is the squared root of the MSE. The RMSE performs the operation of squaring, averaging, and taking the square root of the error between the predicted and true values, and its calculation formula is as follows:

$$\mathrm{RMSE} = \sqrt{\frac{1}{n}\sum_{i=1}^{n}(y_i - \hat{y}_i)^2} \tag{18}$$

The root-mean-squared error (RMSE) is the square root of the mean-squared error (MSE), meaning that it retains the MSE's penalty for errors but, compared to the MSE, it has units that are consistent with those of the predicted and actual values. For example, if the units of the predicted and actual values are in meters (m), then the RMSE also has units of meters (m), making it more intuitive in interpretation. Similar to the MSE, the RMSE is non-negative, and it is only zero when all the predicted values exactly match the actual values.

Because the RMSE is the squared root of the squared error, it is still sensitive to large errors. However, due to the squared-root operation, the RMSE is less amplified by errors than the MSE.

The R^2 (determination coefficient) is an important index for evaluating the goodness of fit of a regression model. The R^2 reflects the explanatory power of the model for the fluctuations in the observed data, and its calculation formula is as follows:

$$R^2 = 1 - \frac{\sum_{i=1}^{n}(y_i - \hat{y}_i)^2}{\sum_{i=1}^{n}(y_i - \bar{y})^2} \tag{19}$$

where $\bar{y}$ represents the average of the true values. The R^2 value typically ranges from 0 to 1; the closer the value is to 1, the better the model explains the data changes; when the R^2 is 1, the model perfectly fits the data. The R^2 is sensitive to outliers and does not accurately reflect the actual size of prediction errors, so it should be used in conjunction with the MAE or MSE.

The WMAPE (weighted mean absolute percentage error) is a weighted version of the MAPE used to eliminate the sensitivity of the MAPE to zero values, and its calculation formula is as follows:

$$\text{WMAPE} = \frac{\sum_{i=1}^{n} |y_i - \hat{y}_i|}{\sum_{i=1}^{n} |y_i|} \tag{20}$$

Unlike the MAPE, the WMAPE does not average each sample but uses the weighted sum of the actual values as a normalization factor, giving greater weight to errors in larger samples. The value of this metric typically ranges from 0 to 1, with smaller values indicating lower error rates, often expressed as a percentage. The WMAPE is more stable, making it particularly suitable for practical applications where the actual values frequently contain many small or zero values.

5.2.2. NASA Dataset Application

The data input utilizes the FD002 dataset. In the initial interface, data are input, and the "oil temperature," "sensor measurement 1," and "sensor measurement 2" are selected as the data feature columns. The device run cycles are chosen as the time column. The remaining life is calculated by subtracting the current run cycle from the maximum run cycle for each device, and this value is set as the label column. In data processing, a time step of 5 is selected, and the data batch size is set to 32.

Regarding the model parameters, the training batch size is set to 64 rounds, and the Adam optimizer is selected. The model's learning rate can be adjusted through this optimizer to achieve the training goals faster. The hidden-layer dimension is set to 128, the number of layers is set to 3, and the FD004_68 dataset is used for prediction.

The training results and a sample prediction instance are as follows:

This case study demonstrates that in the task of predicting the remaining useful life (RUL) of equipment for the NASA dataset, the Transformer model outperformed the LSTM, GRU, and CNN models. The experimental results (Tables 5 and 6) show that the Transformer model achieved the highest R^2 score (0.9606) and the lowest WMAPE (0.0983), indicating its outstanding ability to capture the underlying patterns of time series data. These results highlight the advantages of the Transformer in complex time-dependent modeling for remaining useful life prediction. In contrast, the LSTM, GRU, and CNN models exhibited relatively lower prediction performances. Although the LSTM and CNN models achieved lower MSE and RMSE values compared with those of the Transformer model, their R^2 scores (0.7944 and 0.8269, respectively) and WMAPEs (both above 0.20) suggest that they were less effective at capturing the full variance and trend of the time series data. The GRU model performed the worst among the four, likely due to its simpler gating mechanism, which may limit its ability to model long-range dependencies in data. These shortcomings highlight the importance of basing the model selection on the task complexity and data characteristics.

In this specific case, users are able to use deep learning for equipment remaining life prediction without the need to delve into the intricacies of neural network construction. Instead, this interactive interface streamlines the workflow, allowing users to focus on key steps. First, users only need to provide training data without the need to construct complex neural network architectures themselves. Users can explicitly choose the training model that suits their problem and thereby shape the basic structure of the model effectively.

Furthermore, users are empowered to fine-tune the model parameters at appropriate times to meet their specific requirements. The model training process takes data into the user-defined model architecture and, through multiple iterations, automatically adjusts the model parameters to efficiently minimize prediction errors, enabling it to capture data features and patterns effectively to generate a robust model. After training, users can effectively test the model and obtain information on the predicted remaining life of the equipment in the testing phase. The key feature of this process is its simplicity, as in-depth knowledge of deep learning is not required to obtain high-quality prediction results from the system.

Table 5. Validation results of different algorithms on the NASA dataset.

	Numerical Visualization of Evaluation Indicators		
LSTM	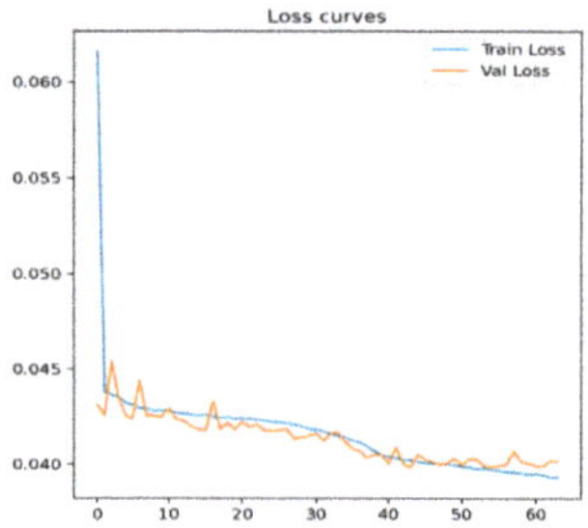	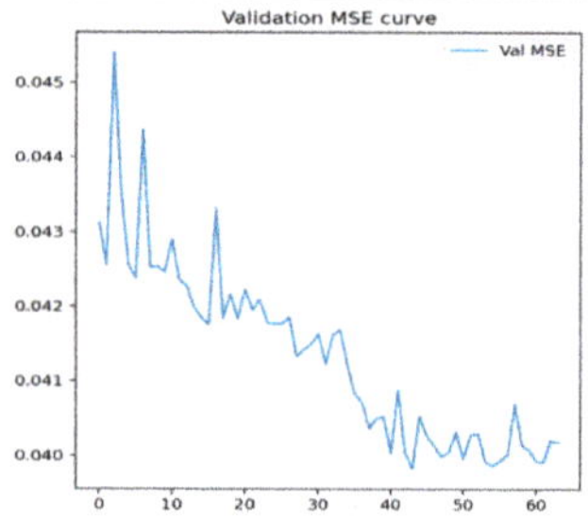	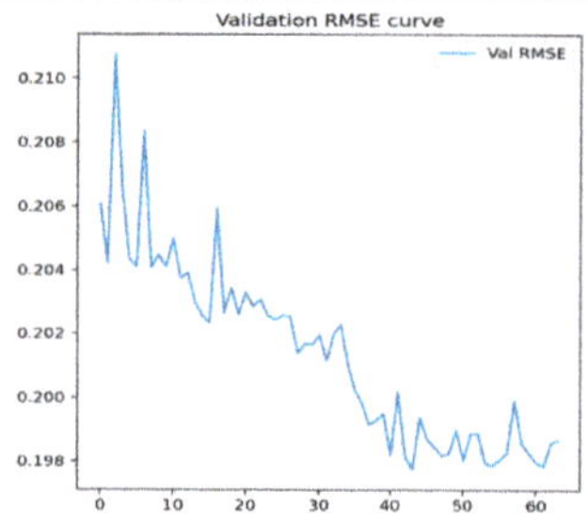
GRU	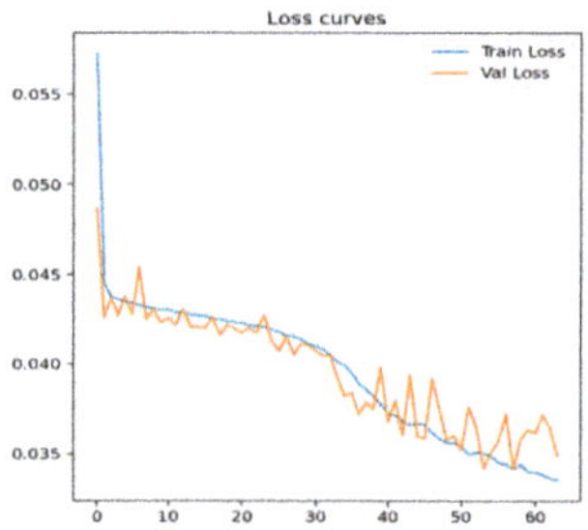	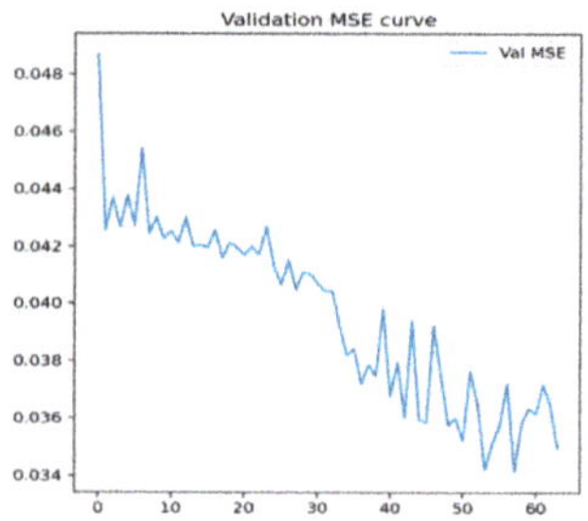	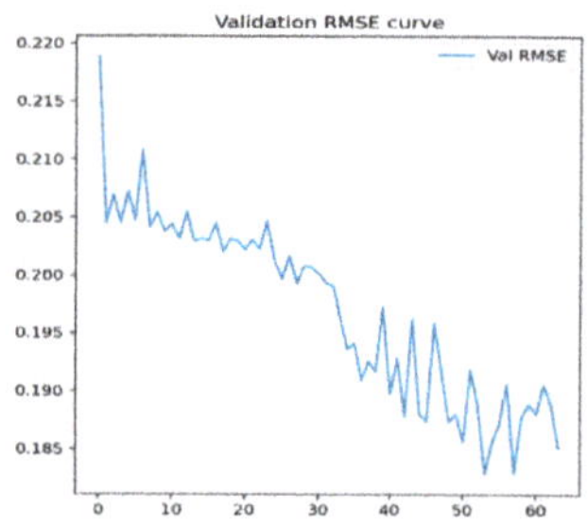
CNN	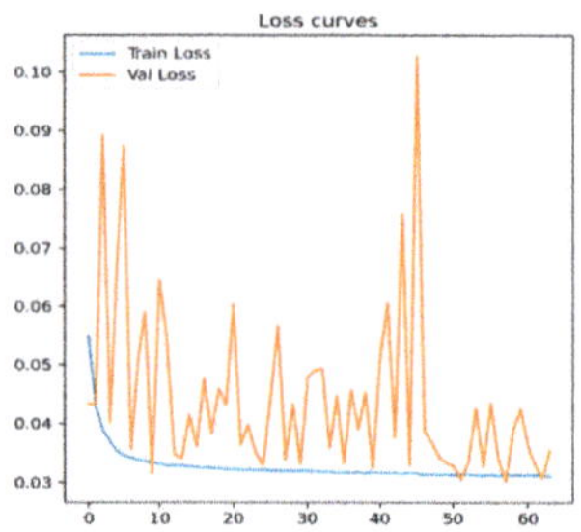	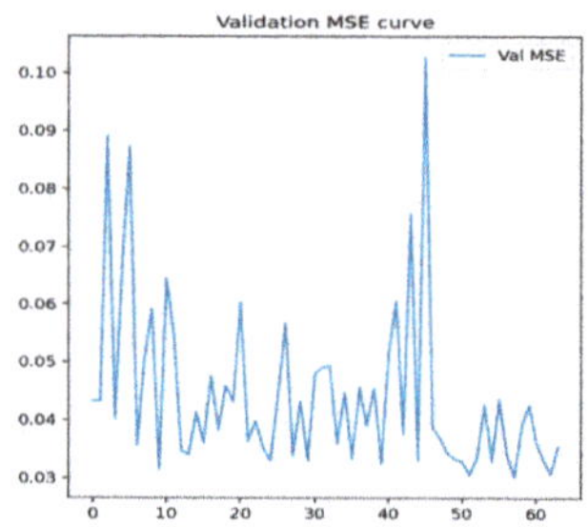	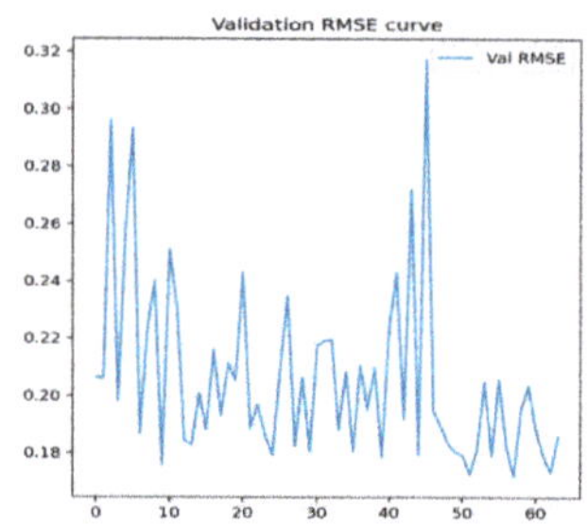
Transformer	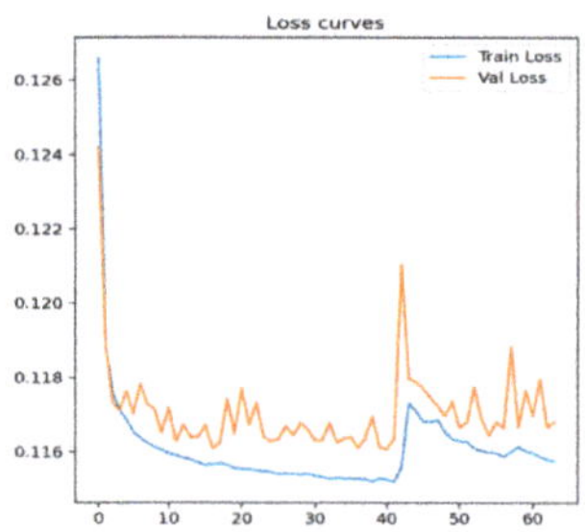	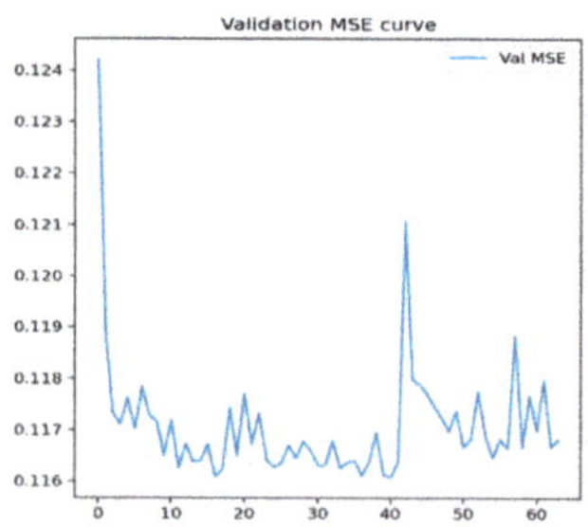	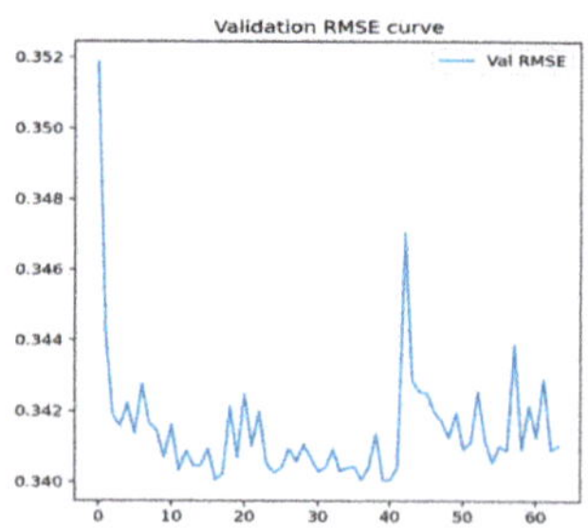

Table 6. Validation results of different algorithms on the NASA dataset.

	Results of Prediction	Test MSE	Test RMSE	R^2	WMAPE
LSTM	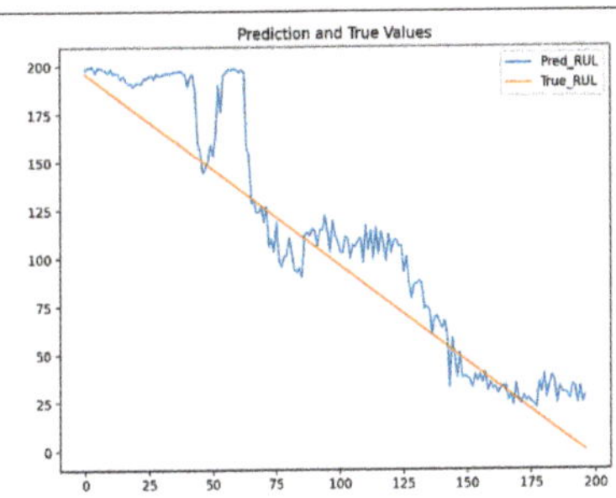	0.0136	0.1040	0.7944	0.2052
GRU	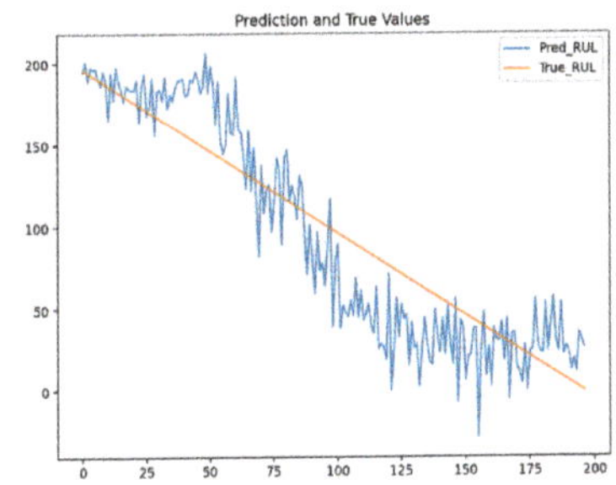	0.0299	0.1569	0.7596	0.2306
CNN	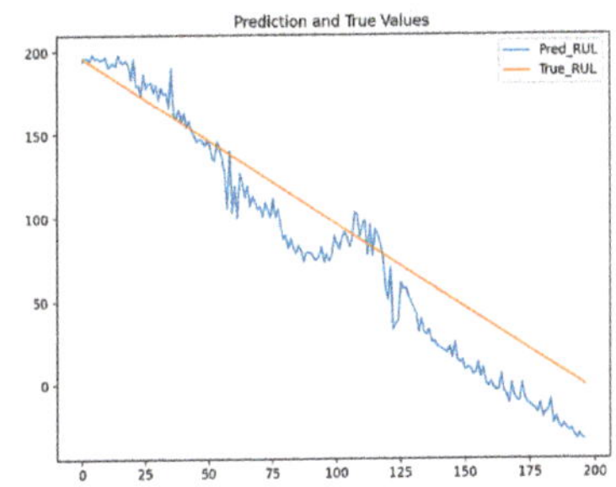	0.0209	0.1332	0.8269	0.2051
Transformer	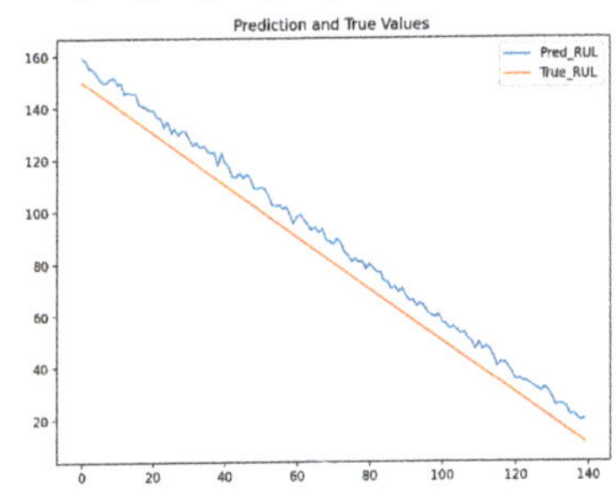	0.1485	0.3851	0.9606	0.098259

This approach has a positive impact on the popularization of deep learning applications by simplifying complex task workflows and improving the accessibility to deep learning technology for ordinary users. The user friendliness combined with the high-quality prediction results provided by the system gives it potential widespread applicability in the field of equipment remaining life prediction.

5.2.3. HNEI Battery Dataset Application

In this section, the application of the HNEI battery dataset is demonstrated using the developed software platform. The dataset comprises multiple lithium-ion battery cells subjected to repeated charge–discharge cycles under varying load conditions. The primary objective is to predict the remaining useful life (RUL) of the battery, defined as the number of cycles remaining before the battery capacity falls below a predefined threshold (typically 80% of the initial capacity).

In the data input interface, the HNEI dataset is loaded, and relevant features such as the "voltage," "current," and "temperature" are selected as the input variables. The "cycle" count is designated as the time column to represent the operational timeline of each battery unit. The target variable—i.e., the RUL label—is derived by subtracting the current cycle index from the total number of cycles recorded for that battery cell. This calculation yields a time-to-failure label consistent with the regression-based prediction framework used for the turbofan engine dataset.

For data preprocessing, a time window size of 20 is selected, which determines the length of the historical sequence used to predict the remaining life at each time step. The data batch size for training is set to 64, allowing for efficient mini-batch gradient updates. Prior to model training, the dataset is normalized to ensure the consistent scaling of the feature values and reduce the influence of unit differences among sensors.

Regarding the model configuration, the same deep learning architectures used in Section 5.2.1 are employed to facilitate the performance comparison. These include the GRU, LSTM, CNN, and Transformer models. The training process adopts the Adam optimizer, with a learning rate of 0.001 and a hidden-layer dimension of 128. Each model is trained for 64 epochs. This consistency in the model parameters ensures the comparability of performances across datasets and highlights the flexibility of the software in adapting to different data domains.

During training, the system automatically adjusts the model parameters through backpropagation and loss minimization to optimize the prediction performance. After training is completed, users can initiate the prediction phase on a test subset of the battery data. The predicted RUL values are visualized alongside ground-truth labels, enabling the intuitive evaluation of the model accuracy.

According to the table (Tables 7 and 8), the CNN model achieved the best prediction performance on the HNEI battery dataset. The test MSE was 0.0002, the RMSE was 0.015, the R^2 was as high as 0.9973, and the WMAPE was only 0.0244, showing the most superior performance. However, overall, the error gaps among the CNN, LSTM, and GRU models were relatively small, and all could fit the battery life data well, indicating that they all have strong stability and accuracy when dealing with such sequential data. The RMSE of the LSTM model was 0.0176 and the R^2 was 0.9963, while the RMSE of the GRU model was 0.0193, and the R^2 was 0.9955. Their performances were only slightly lower than that of the CNN.

In contrast, the performance of the Transformer model on this dataset was relatively weak. Although its R^2 still reached 0.9924, the MSE and RMSE are significantly higher than those of the other models. This might be because the battery sensor data have a strong linear correlation, and the Transformer is better at handling data with complex nonlinear relationships or long-range dependencies. In scenarios where the data features are relatively stable, and the structure is relatively simple, its advantages cannot be fully exerted, and instead, it may lead to a decline in the generalization ability.

Combining the predictive results of the turbofan engine dataset in Section 5.2.1, it can be seen that the different models performed differently on different types of data, further verifying the practicality and scalability of this software system. Users can flexibly choose the model according to specific data characteristics and their task requirements to achieve the optimal prediction effect.

Table 7. Validation results of different algorithms on the HNEI dataset.

Numerical Visualization of Evaluation Indicators

LSTM

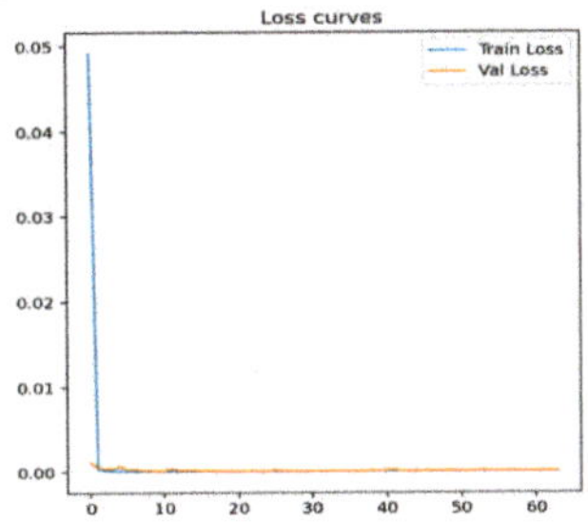

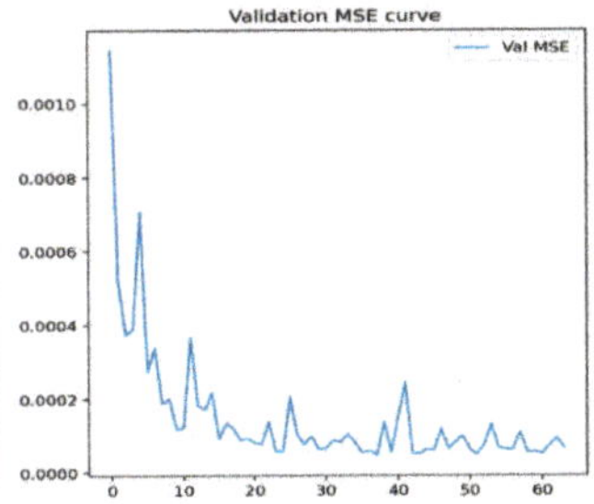

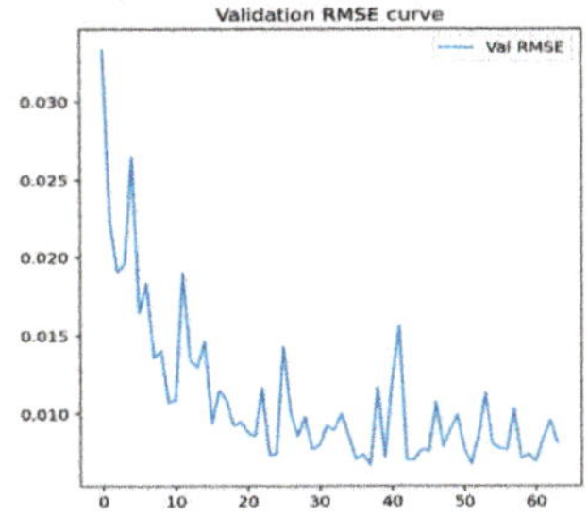

GRU

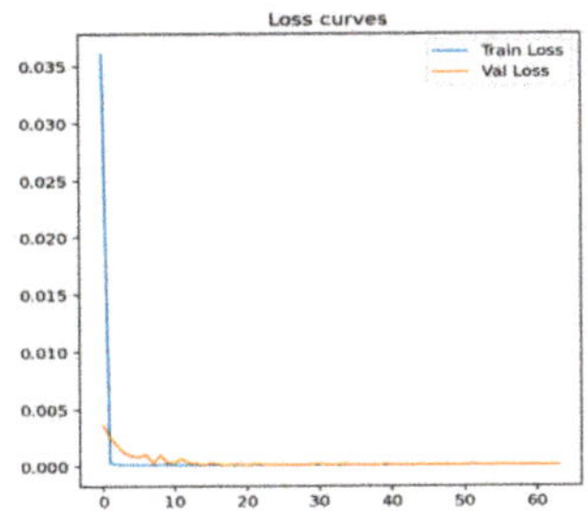

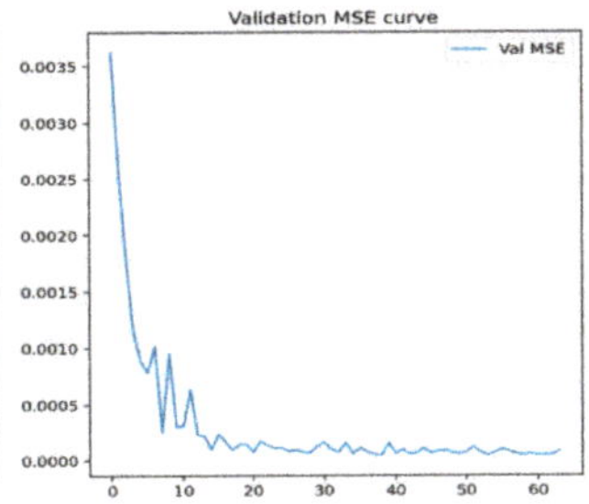

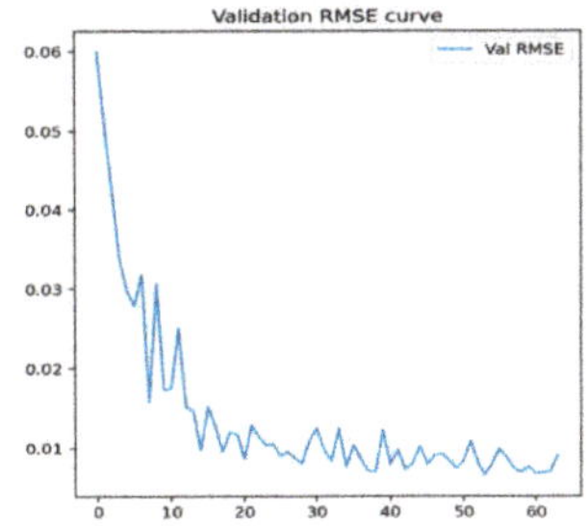

CNN

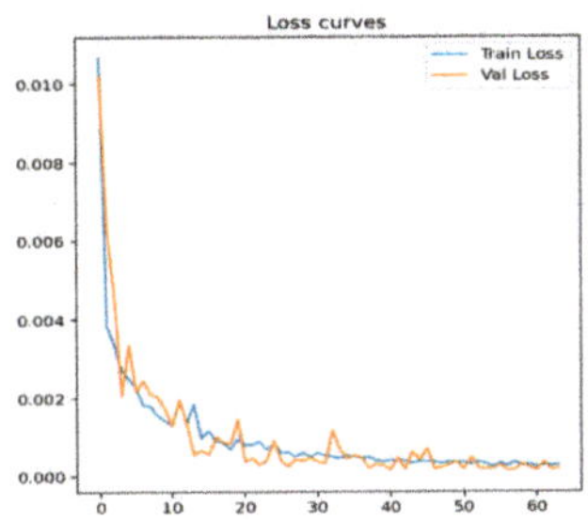

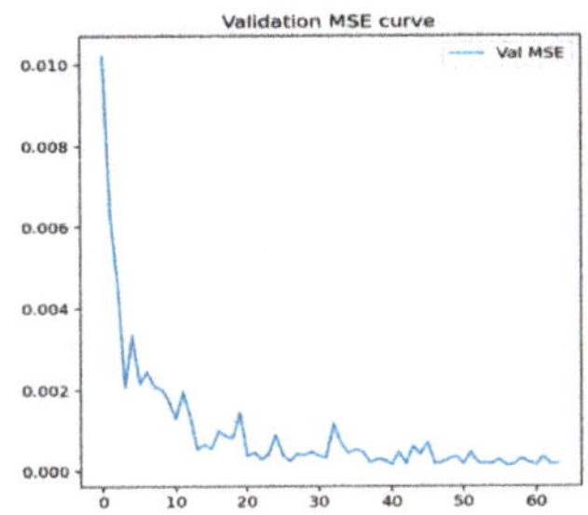

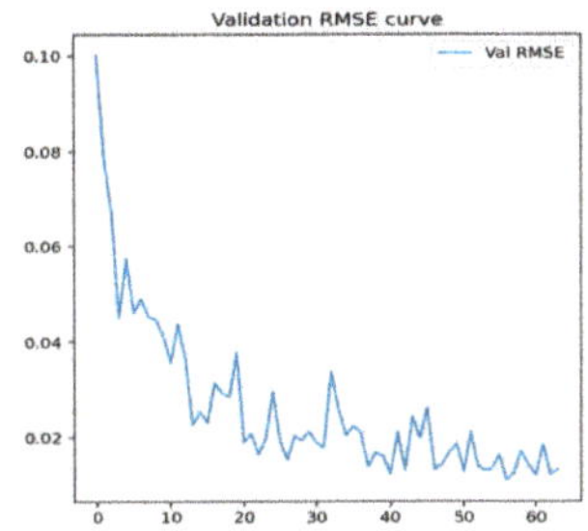

Transformer

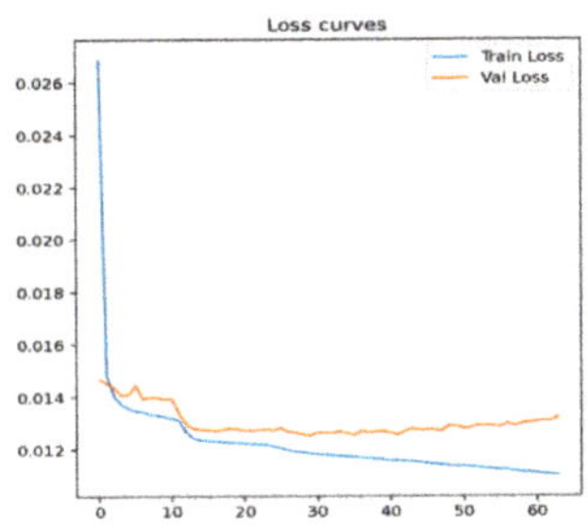

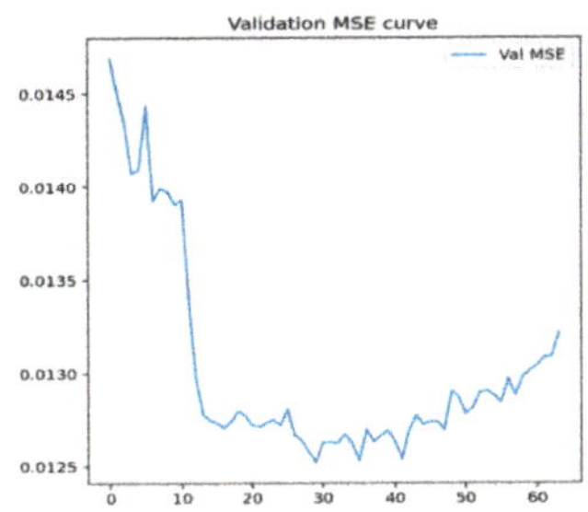

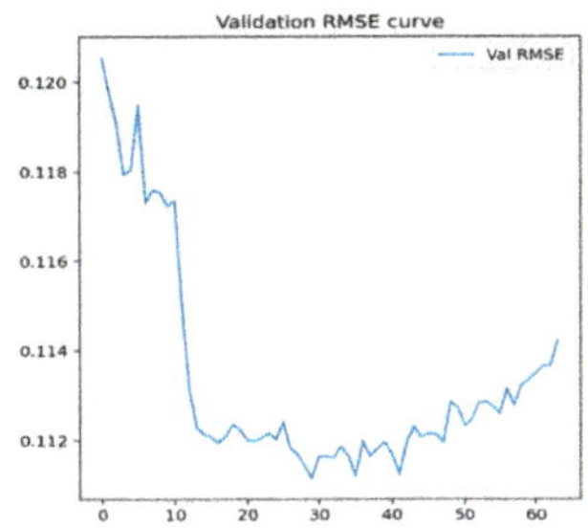

Table 8. Validation results of different algorithms on the HNEI dataset.

	Results of Prediction	Test MSE	Test RMSE	R^2	WMAPE
LSTM	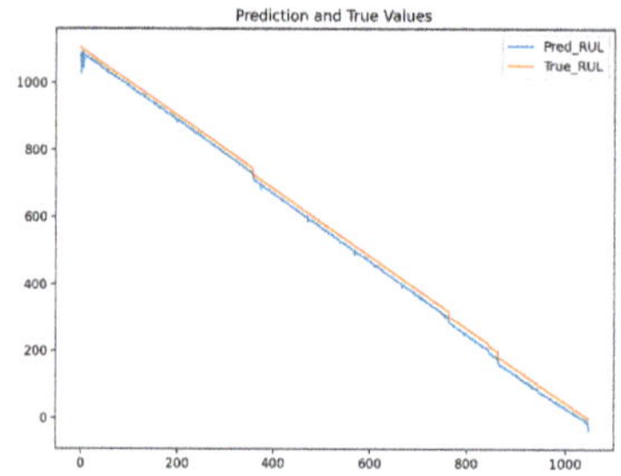	0.00029	0.0161	0.9963	0.0348
GRU	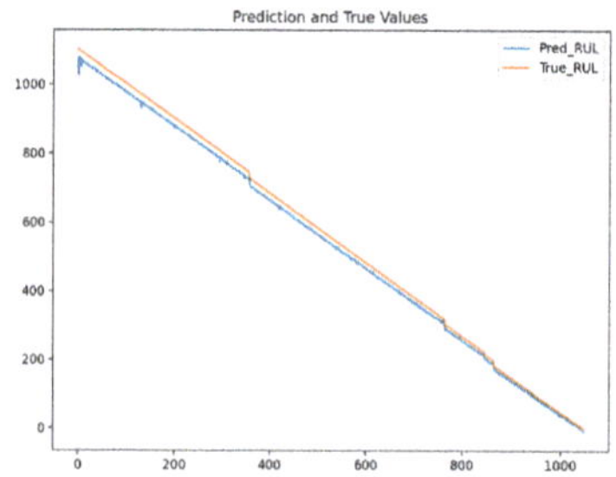	0.00022	0.0146	0.9973	0.0244
CNN	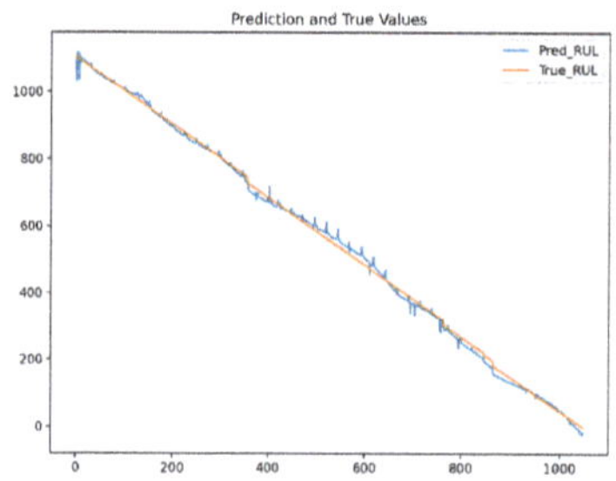	0.00014	0.0108	0.9955	0.0379
Transformer	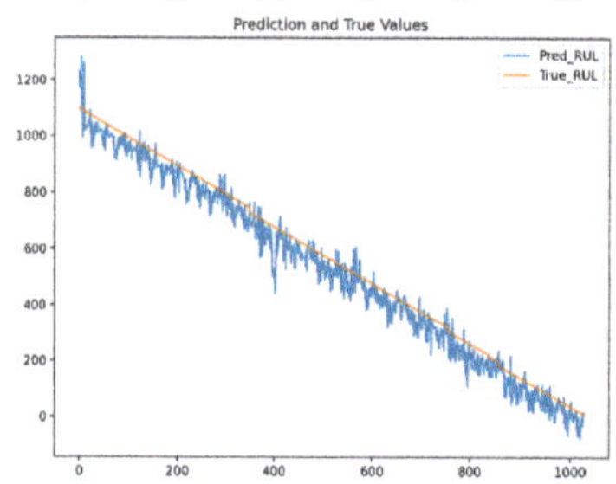	0.00167	0.0389	0.9924	0.030296

5.3. Model Training Efficiency and Resource Requirements

Considering that some users in practical applications may lack high-performance GPU devices, to lower the model configuration threshold and enhance the system's adaptability, we conducted experimental tests during the model training phase using an ordinary CPU environment to evaluate the training efficiency and resource consumption under common hardware conditions.

The experimental hardware platform was configured with an Intel Core i7-12700H processor and 16 GB of memory (Manufactured by Intel Corporation, headquartered in Santa Clara, CA, USA), and the operating system was Windows 11. Based on this, the training efficiencies of the four mainstream models (the LSTM, GRU, CNN, Transformer models) were evaluated for the NASA turbofan engine and HNEI battery datasets. The main statistical indicators included the total time (in seconds) required for the complete training process of a single model and the peak memory usage (in KB).

Table 9 presents the training time and memory usage of each model in the CPU environment for both datasets.

Table 9. The training time and memory usage of each model in the CPU environment for both datasets.

Model	Dataset	Training Time (s)	Memory Usage (kb)
LSTM	NASA	2325.80	1270
GRU	NASA	1023.85	983
CNN	NASA	2232.86	147
Transformer	NASA	9305.78	6690
LSTM	HNEI	363.83	1315
GRU	HNEI	549.52	987
CNN	HNEI	312.37	149
Transformer	HNEI	2060.61	2997

According to Table 9, the training time and memory usage of the Transformer model were significantly higher than those of the other models on both datasets, indicating its higher dependence on computing resources. In contrast, the GRU and LSTM models performed better in terms of the training time and memory consumption, possessing strong lightweight advantages and being suitable for resource-constrained environments. The CNN model is extremely efficient in terms of memory usage and is especially suitable for scenarios that require both prediction accuracy and deployment efficiency.

Furthermore, compared to the NASA dataset, the HNEI battery dataset generally shows a lower training time under the same model, which is related to its smaller sample size and time series length, as well as its relatively lower computational complexity.

Overall, this system can efficiently complete model training and inference without GPU acceleration, reducing the threshold of deep learning technology in engineering applications. Users can choose the appropriate model based on their hardware configuration and actual scenarios, taking into account the accuracy, efficiency, and resource consumption, providing diverse solutions for predicting the remaining life of equipment.

5.4. Comparative Analysis of Model Performance Across Datasets

To comprehensively evaluate the adaptability of the designed system to different types of data and the generalization ability of the models, in this section, we compare the model prediction results on the NASA turbofan engine and HNEI battery datasets (Table 10). These two datasets represent the typical degradation processes of mechanical systems and electrochemical systems, respectively, with different data patterns and degradation mechanisms, and possess good representativeness for comparison.

Table 10. The model predictions on the NASA turbofan engine and HNEI battery datasets.

Model	NASA Test MSE	NASA Test RMSE	HNEI Test MSE	HNEI Test RMSE
GRU	0.0299	0.1569	0.00029	0.0161
LSTM	0.0136	0.1040	0.00022	0.0146
CNN	0.0209	0.1332	0.00014	0.0108
Transformer	0.1485	0.3851	0.00167	0.0389

The LSTM and CNN models demonstrated excellent generalization capabilities on both datasets. Among them, the CNN model had the most outstanding performance on the HNEI battery dataset, with a test MSE of 0.00014 and an RMSE of only 0.0108, indicating its significant advantage in handling data with stationary and periodic characteristics. The LSTM model achieved the best performance on the NASA dataset (MSE: 0.0136; RMSE: 0.1040), demonstrating a strong modeling ability for complex mechanical degradation processes.

The GRU model achieved relatively stable prediction results on both datasets, performing slightly worse than the LSTM and CNN models, but with a significant advantage in terms of its training efficiency (refer to Section 5.3). Therefore, it has practical application value in lightweight deployment scenarios.

In contrast, the Transformer model performed unsatisfactorily on both datasets, with especially higher errors on the NASA dataset, indicating that it has difficulty fully learning the degradation patterns between sequences when dealing with small-scale time series data without rich contextual information, and it has a large resource overhead. In industrial applications, it is necessary to carefully consider the specific data scale and structure when deciding whether to use it.

The above analysis has verified the performance differences of different models on various types of datasets and has also demonstrated the multi-scenario adaptability of this system. The system enables users to flexibly select model architectures based on the specific data characteristics of their devices, which can not only adapt to the complex degradation trends of industrial equipment but also handle fine sequential data such as those for batteries, and it possesses excellent cross-domain prediction capabilities and scalability.

6. Conclusions

In this research, we discussed the crucial role of deep learning in the remaining life prediction of equipment, while also highlighting the considerable technical barriers and time and effort costs associated with multiple model constructions. In light of this, we developed deep learning-driven software for the remaining life prediction of equipment, which consists of the following four main modules: the interface interaction, data processing, model selection, and visualization modules. Users can input training and testing data on the interface, select data features, and set parameters, such as the time window size. The software provides various neural network models for the user's selection, such as the LSTM, CNN, GRU, and Transformer models, and also allows users to customize the training parameters. During the testing phase, users can input test data and choose a pretrained model, and the software will generate comparison plots between the predicted and actual values, as well as calculating the MSE evaluation metric.

Future work will focus on improving the user friendliness of the software and implementing automatic parameter tuning for models. Additionally, efforts will be directed towards further optimizing the model performance to enhance the model generalization and accuracy. Real users will be introduced for low-code platform validation when the implementation conditions are met.

Author Contributions: Methodology, S.C., M.Y. and J.W.; Software, Y.L.; Investigation, Y.N.; Writing—original draft, Y.L.; Writing—review & editing, J.C.; Supervision, S.C.; Project administration, M.W. and B.Z.; Funding acquisition, S.C. All authors have read and agreed to the published version of the manuscript.

Funding: This study was supported by the GuangDong Basic and Applied Basic Research Foundation (grant nos. 2022A1515110545 and 2024A1515012727), the Shenzhen Science and Technology Program (grant no. RCBS20231211090600001), the LingChuang Research Project of the China National Nuclear Corporation (grant no. CNNC-LCKY-202263), the Shenzhen Pengcheng Peacock Plan Talent Project (no. 827-000885), the Shenzhen Science and Technology Innovation Commission Key Technical Project (JSGG20210713091539014), and the Shenzhen Science and Technology Program (ZDSYS20230626091501002).

Data Availability Statement: The original contributions presented in this study are included in the article. Further inquiries can be directed to the corresponding author.

Conflicts of Interest: Authors Yuhan Lin, Ming Wang and Bing Zhang was employed by the company State Key Laboratory of Nuclear Power Safety Technology and Equipment, China Nuclear Power Engineering Co., Ltd. The remaining authors declare that the research was conducted in the absence of any commercial or financial relationships that could be construed as a potential conflict of interest.

Appendix A

Key Classes, Interfaces, and Functions

In the implementation of the low-code DL-based RUL prediction software, a series of well-defined Python classes and interface functions form the backbone of its functionality. These are distributed across several key modules, including the user interface (streamlit_version.py), training/testing logic (main.py), and model construction and evaluation (model.py, train.py, and predict.py). Below is a summary of the critical components.

For the model training, the primary entry point is the run_train(data) function. This function first reads the dataset.yaml configuration file to determine the input features, output labels, model architecture, and training parameters, and it then splits the input dataset into training and validation sets and uses the custom_data_loader() function to construct PyTorch-compatible DataLoader objects. The model structure is generated dynamically through the get_model() function, and the entire training process is carried out using the train_and_evaluate() function. Upon completion, the function returns the trained model object along with the evaluation metrics, including the loss, mean-squared error (MSE), and root-mean-squared error (RMSE), which are later used for visualization and analysis.

The testing phase is handled by the run_test (data, trained_model_path, has_labels=True) function, which loads a trained model and performs predictions on the user-provided test data. This function chooses between different test routines depending on the presence or absence of labels in the dataset; Transformer-based models are tested using test_model2(), while other models are evaluated using the standard test_model() function. The function also includes mechanisms to verify whether the required columns exist in the test data, improving the robustness and usability.

The model generation process is encapsulated in the get_model(config) function, which dynamically returns an appropriate model object based on the parameters defined in the configuration file. The supported models include LSTM, CNN, and GRU models and various Transformer variants. Users can freely select among these options through the front-end interface, and the system constructs the model instance accordingly.

For data loading, the custom_data_loader() and former_data_loader() functions are responsible for preprocessing raw data. These functions perform feature–label separation, normalization, and sliding-window segmentation to generate training samples based on the user's column selections. The former_data_loader() function is specifically designed to support Transformer-based models by formatting the input data to match the encoder's expected structure.

The training procedure itself is executed via the train_and_evaluate() function, which receives the model, data loaders, loss function, and optimizer and orchestrates the full training loop. This function tracks the loss values at each epoch and returns results for the subsequent visualization and evaluation.

The front-end interface is implemented using Streamlit, with the main logic contained in the streamlit_version.py file. Through this interface, users can upload datasets, select input, output, timestamp, and identifier columns and configure key parameters, such as the sequence length, batch size, model type, optimizer, loss function, and learning rate. All user inputs are automatically written to the dataset.yaml file for backend processing. Once training is complete, the model structure is displayed, and Matplotlib is used to render visualizations, such as training loss curves and prediction results, enabling users to better understand the model behavior and performance.

Configuration management is handled by the configs.read_configs() function, which ensures that all the training and testing parameters are consistently sourced from the same YAML file. This centralized parameter control improves maintainability and eliminates inconsistencies. Through the coordinated operation of these core classes and interfaces, the software achieves a complete closed-loop workflow from data ingestion to prediction, and it ensures the performance of deep learning models, while significantly lowering the operational complexity, offering users an efficient and user-friendly RUL prediction platform.

References

1. Ogunfowora, O.; Najjaran, H. A Transformer-based Framework For Multi-variate Time Series: A Remaining Useful Life Prediction Use Case. *arXiv* **2023**, arXiv:2308.09884.
2. Li, X.; Ding, Q.; Sun, J.Q. Remaining useful life estimation in prognostics using deep convolution neural networks. *Reliab. Eng. Syst. Saf.* **2018**, *172*, 1–11. [CrossRef]
3. Gao, Z.; Wang, C.; Wu, J.; Wang, Y.; Jiang, W.; Dai, T. Degradation-Aware Remaining Useful Life Prediction of Industrial Robot via Multiscale Temporal Memory Transformer Framework. *Reliab. Eng. Syst. Saf.* **2025**, *262*, 111176. [CrossRef]
4. Wang, L.; Zhao, X.; Pham, H. Novel formulations and metaheuristic algorithms for predictive maintenance of aircraft engines with remaining useful life prediction. *Reliab. Eng. Syst. Saf.* **2025**, *261*, 111064. [CrossRef]
5. Zhou, Y.; Wang, H.; Jin, H.; Liu, Y.; Liu, X.; Cao, Z. Remaining useful life prediction for machinery using multimodal interactive attention spatial–temporal networks with deep ensembles. *Expert Syst. Appl.* **2025**, *263*, 125808. [CrossRef]
6. Kim, M.; Yoo, S.; Son, S.; Chang, S.Y.; Oh, K.-Y. Physics-informed deep learning framework for explainable remaining useful life prediction. *Eng. Appl. Artif. Intell.* **2025**, *143*, 110072. [CrossRef]
7. Chen, Z.; Wu, M.; Zhao, R.; Guretno, F.; Yan, R.; Li, X. Machine Remaining Useful Life Prediction via an Attention Based Deep Learning Approach. *IEEE Trans. Ind. Electron.* **2020**, *68*, 2521–2531. [CrossRef]
8. Cheng, C.; Ma, G.; Zhang, Y.; Sun, M.; Teng, F.; Ding, H. A deep learning-based remaining useful life prediction approach for bearings. *IEEE/ASME Trans. Mechatron.* **2020**, *25*, 1243–1254. [CrossRef]
9. Francisco, O.V.; Rosaria, S. Machine learning for marketing on the KNIME Hub: The development of a live repository for marketing applications. *J. Bus. Res.* **2021**, *137*, 393–410. [CrossRef]
10. Demšar, J.; Zupan, B. Hands-on training about data clustering with orange data mining toolbox. *PLoS Comput. Biol.* **2024**, *20*, e1012574. [CrossRef]
11. Baratchi, M.; Wang, C.; Limmer, S.; van Rijn, J.N.; Hoos, H.; Bäck, T.; Olhofer, M. Automated machine learning: Past, present and future. *Artif. Intell. Rev.* **2024**, *57*, 122. [CrossRef]
12. Kok, C.L.; Tan, H.R.; Ho, C.K.; Lee, C.; Teo, T.H.; Tang, H. A Comparative Study of AI and Low-Code Platforms for SMEs: Insights into Microsoft Power Platform, Google AutoML and Amazon SageMaker. In Proceedings of the 2024 IEEE 17th International Symposium on Embedded Multicore/Many-core Systems-on-Chip (MCSoC), Kuala Lumpur, Malaysia, 16–19 December 2024; pp. 50–53. [CrossRef]
13. Vaswani, A.; Shazeer, N.; Parmar, N.; Uszkoreit, J.; Jones, L.; Gomez, A.N.; Kaiser, L.; Polosukhin, I. Attention is all you need. *arXiv* **2017**, arXiv:1706.03762.
14. Kim, Y. Convolutional neural networks for sentence classification. *arXiv* **2014**, arXiv:1408.5882.
15. Hochreiter, S.; Schmidhuber, J. Long short-term memory. *Neural Comput.* **1997**, *9*, 1735–1780. [CrossRef] [PubMed]
16. LeCun, Y.; Bottou, L.; Bengio, Y.; Haffner, P. Gradient-based learning applied to document recognition. *Proc. IEEE* **2002**, *86*, 2278–2324. [CrossRef]
17. Sutskever, I.; Vinyals, O.; Le, Q.V. Sequence to sequence learning with neural networks. *arXiv* **2014**, arXiv:1409.3215.
18. Graves, A. Generating sequences with recurrent neural networks. *arXiv* **2013**, arXiv:1308.0850.
19. Vinyals, O.; Toshev, A.; Bengio, S.; Erhan, D. Show and tell: A neural image caption generator. In Proceedings of the IEEE Conference on Computer Vision and Pattern Recognition, Boston, MA, USA, 7–12 June 2015; pp. 3156–3164.

20. Wu, H.; Xu, J.; Wang, J.; Long, M. Autoformer: Decomposition transformers with auto-correlation for long-term series forecasting. *arXiv* **2021**, arXiv:2106.13008.
21. Saxena, A.; Goebel, K. Turbofan Engine Degradation Simulation Data Set. In *NASA Prognostics Data Repository*; NASA Ames Research Center: Moffett Field, CA, USA, 2008.
22. Ignavinuales. Battery RUL Dataset [DB/OL]. 1 January 2023. Available online: https://github.com/ignavinuales/Battery_RUL_Prediction (accessed on 11 June 2024).

processes

MDPI

Article

Enhancing the Reliability and Durability of Micro-Sensors Using the Taguchi Method

Chi-Yuan Lee *, Jiann-Shing Shieh, Guan-Quan Huang, Chen-Kai Liu, Najsm Cox and Chia-Hao Chou

Department of Mechanical Engineering, Yuan Ze Fuel Cell Center, Yuan Ze University, Taoyuan 32003, Taiwan; jsshieh@saturn.yzu.edu.tw (J.-S.S.); s1135019@mail.yzu.edu.tw (G.-Q.H.); kevin88043027@gmail.com (C.-K.L.); coxnajsm@gmail.com (N.C.); hulkhower@gmail.com (C.-H.C.)

* Correspondence: cylee@saturn.yzu.edu.tw

Abstract: This study presents the development and optimization of a flexible integrated three-in-one micro-sensor using Micro-Electro-Mechanical Systems (MEMS) technology. To enhance its reliability and performance, the Taguchi Method was employed to analyze and optimize key fabrication parameters, including the electrode area, electrode thickness, and protective layer thickness. An L4 orthogonal array design enabled efficient experimentation with minimal runs. Experimental results demonstrate that optimized parameter combinations significantly improve sensor linearity, sensitivity, and reproducibility. Comparative analysis with commercial sensors shows the superior reliability of the self-fabricated sensor, particularly in airflow velocity detection. The findings validate the use of the Taguchi Method for robust MEMS sensor design and highlight its potential for industrial heating, ventilation, and air conditioning (HVAC) applications.

Keywords: Taguchi Method; MEMS; reliability; durability

1. Introduction

In the 1920s, Ronald Fisher introduced the concept of Design of Experiments (DOE) in agriculture. Today, DOE is not only used in agriculture but is also applied across various industries, such as medical, biotechnology, pharmaceuticals, and high-tech manufacturing. In the 1950s, Genichi Taguchi introduced the Taguchi Method, incorporating orthogonal arrays [1]. Although opinions in the scientific community are divided, the Taguchi design gained popularity due to its practical applicability, making it one of the most widely adopted experimental designs in both industry and scientific research. Its advantage comes from the use of effective orthogonal matrices, ensuring a balanced distribution of factor levels and minimizing the number of experimental runs needed. While most orthogonal arrays emphasize main effects, certain designs also enable the estimation of specific interactions [2].

DOE involves designing experiments by selecting experimental conditions and minimizing external factors to mitigate their influence on the experiment. Its key feature is the ability to efficiently collect data or identify correlations and issues between factors through fewer experiments, with reduced cost and time. DOE is widely applicable and is used to create conditions that can prove whether a hypothesis is correct or incorrect. Proper experimental design requires controlling all variables that may influence the results and allows for measuring variables that change due to controlled factors. The former is usually referred to as the independent variable, while the latter is called the dependent or response variable. If the experiment is well-controlled, the results will allow for a reliable and effective distinction between the relationships of independent and dependent variables [3].

This article differs from the previously submitted article in that the Taguchi Method is used to improve the design of a three-in-one micro-sensor and compared with commercial sensors. The results show that sensors of different designs have different accuracies and reliabilities, which proves that the Taguchi Method can improve the performance of the sensor.

1.1. Micro-Velocity Sensor

Wang et al. [4] developed an analytical model and evaluated a velocity sensor, built on a ceramic substrate and enclosed in a plastic casing, within a wind tunnel. The experimental results revealed that the sensor's output exhibited high sensitivity and exceptional linearity. Agrawal et al. [5] designed a velocity sensor based on a coiled cantilever structure made of silicon nitride, utilizing platinum resistors as the resistive element, which demonstrated remarkably high sensitivity. Xu et al. [6] proposed a novel velocity sensor system based on a self-powered triboelectric sensor and optoelectronic technology for simultaneous velocity and direction sensing. Yi et al. [7] examined the effect of dust on MEMS-based hot-wire velocity sensors. The measurement results indicated that at the same wind speed, the sensor's output voltage with dust on its surface was slightly lower than that of a clean, dust-free sensor, leading to reduced measurement sensitivity and accuracy. Yi et al. [8] introduced a newly designed velocity sensor featuring a dual-layer inductor, marking the first instance of enhanced sensitivity. Two identical inductors were constructed on the front and back of a flexible substrate. When wind flows over the substrate, both inductors bend in response to the wind pressure. Consequently, the mutual inductance effect significantly amplifies the overall change in inductance, serving as a measurement of velocity. Mu et al. [9] established a velocity model for a damper torque sensor through analysis of resistive characteristics and investigated the sensor's coefficient using Computational Fluid Dynamics (CFD) simulations and experiments. The results demonstrated that under consistent intake static pressure, as the damper opening angle increases, wind velocity escalates, while the torque variation range remains limited. Ligeza et al. [10] examined the static and dynamic characteristics of hot-wire sensors with different filament diameters. Their findings revealed that the sensitivity, represented by the slope of the characteristic curve (dI/dV), increased significantly as the filament diameter grew. Additionally, it was observed that the dI/dV slope decreased with increasing velocity.

1.2. Micro-Temperature Sensor

Hao et al. [11], by utilizing optimized electron beam parameters, demonstrated that the temperature coefficient of resistance in the range of 30 °C to 80 °C was 1.18%/°C, with a response time of 0.17 s. Their investigation into the effect of electron beam dosage on sensor performance revealed that the temperature sensors produced through this technique exhibited high sensitivity, rapid response, excellent linearity, and minimal hysteresis. Miao et al. [12] presented a wireless, chipless radio frequency identification sensor that incorporates a composite material of Fe_2O_3-Co_3O_4, SnO_2, and rGO for accurate temperature measurement. The sensor exhibited sensitivities of −0.293 MHz/°C and −2.12 dB/°C (temperature variation from 10 °C to 85 °C), with a temperature response duration of 15.3 s and a recovery duration of 21.7 s. Lin et al. [13] constructed and validated an ultra-high-sensitivity optical fiber temperature sensor using polydimethylsiloxane (PDMS) and the vinyl effect. The results revealed an extraordinarily high sensitivity of −4.08658 nm/°C and outstanding linearity of 0.99972, with experiments confirming its stability and reliability in micro-vibrational environments. Xu et al. [14] presented a relative humidity (RH) and temperature sensor that employs forward Brillouin scattering (FBS) in a single-mode optical fiber coated with polyimide (PI). The measured temperature values showed a strong correlation with RH fluctuations within the 35–85% range. Pan et al. [15] introduced a flexible temperature sensor that utilizes a combination of polyaniline/graphene (GPANI) and polyvinyl butyral (PVB) film, supported by a substrate made of Indium tin oxide (ITO) and Polyethylene Terephthalate (PET). The GPANI particles embedded within the PVB film not only facilitate temperature measurement but also respond to external pressure due to their slight deformation. Tursunniyaz et al. [16] designed a resistive temperature sensor using aerosol jet printing, employing nano-material inks to form simple, cost-effective, and environmentally friendly patterns on the film. Their findings demonstrated that the sensor's resistance varies linearly with temperature between 22 °C and 150 °C, throughout both heating and cooling cycles. Tursunniyaz et al. [17] also introduced a resistive temperature sensor fabricated via aerosol jet technology, utilizing a mixture of nickel–copper–silver nanoparticle ink. The results showed that the sensitivity improved by 300% with a 40% silver and copper/nickel toluene ink blend.

1.3. Micro-Humidity Sensor

Liu et al. [18] presented a novel high-performance Film Bulk Acoustic Resonator (FBAR) design that incorporated a polyimide (PI) film. The PI film provided structural support while also enabling humidity sensing functions, marking its first application in back-slot FBARs. Mu et al. [19] enhanced a MEMS capacitive relative humidity sensor by integrating HNTs-NH_2 into the sensing layer using polyimide (PI). The findings demonstrated that the modified nanotubes substantially improved the sensor's humidity sensing capabilities. Additionally, the introduction of hydrophilic additives within the nanotube structure significantly increased its sensitivity. Li et al. [20] introduced a highly sensitive optical fiber grating humidity sensor featuring coatings of polyimide (PI) and graphene layers. Experimental results revealed that this sensor had a humidity sensitivity 1.8 times greater than that of the sensor coated solely with PI film. Zhu et al. [21] developed a resistive humidity sensor utilizing an ultrathin graphene oxide (GO) film, composed of stacked monolayers of GO. The sensor demonstrated response and recovery times of under 1 s across the relative humidity (RH) range of 10% to 95%. Che et al. [22] designed a flexible device for moisture detection based on a precisely engineered Thermoplastic Polyamide Elastomer (TPAE). The resistance of the TPAE moisture sensor decreased as the relative humidity (RH) increased, demonstrating negative moisture sensitivity. Zhang et al. [23] developed modified MXene composites through PEI/$AgNO_3$ crosslinking, showcasing remarkable physicochemical stability and moisture sensitivity, which makes them ideal for real-time humidity monitoring applications. The optimized PEI/$AgNO_3$ crosslinking improved the interlayer structure of the MXene nanosheets, strengthening the composite's structural stability and promoting the movement of water molecules.

2. Research Methods

2.1. Integration of a Three-in-One Micro-Sensor with the Taguchi Method

This study employs Design of Experiments (DOE) to develop an orthogonal array (OA), dividing it into three factors—electrode area, electrode thickness, and protective layer thickness—each with two levels (parameter values). The two parameter levels for the three control factors were determined based on laboratory experience and existing literature, with size selection guided by past research findings from our laboratory [24]. Following the Taguchi Method, an L4 orthogonal array was constructed, as shown in Table 1, featuring four different designs of the integrated three-in-one micro-sensor. By analyzing the process parameters, we aim to optimize a high-accuracy integrated three-in-one micro-sensor, as illustrated in Figure 1.

Table 1. L4 orthogonal array.

	Electrode Area (μm)	Electrode Thickness (μm)	Protective Layer Thickness (μm)
Sensor 1	510 × 400 (temperature) 590 × 500 (humidity) 350 × 350 (velocity)	0.2	20
Sensor 2	510 × 400 (temperature) 590 × 500 (humidity) 350 × 350 (velocity)	0.15	15
Sensor 3	590 × 450 (temperature) 1065 × 1050 (humidity) 400 × 400 (velocity)	0.2	15
Sensor 4	590 × 450 (temperature) 1065 × 1050 (humidity) 400 × 400 (velocity)	0.15	20

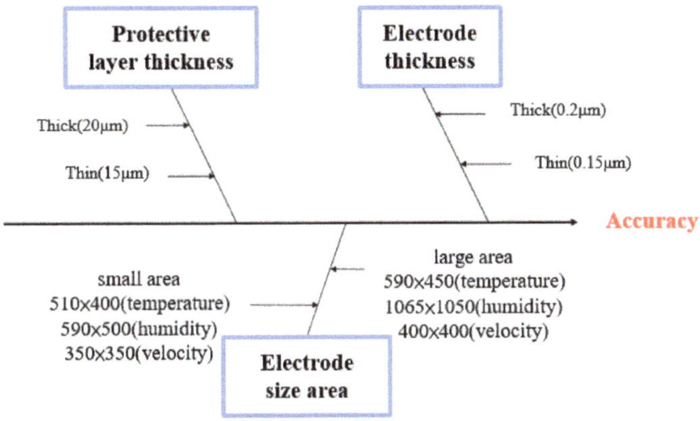

Figure 1. Process parameter factor analysis chart.

2.2. Design and Principle of the Integrated 3-in-1 Micro-Sensor

This study developed a compact three-in-one micro-sensor utilizing MEMS technology, with gold (Au) selected as the sensing element due to its excellent physical and chemical properties, as well as with its straightforward fabrication process. To minimize thermal interference between sensing elements, a spacing of 400 μm was maintained between the hot-wire anemometer and the RTD, and a PI protective layer (LTC® 9320) (Fujifilm, Tokyo, Japan) was applied to provide thermal insulation. As the applied voltage was very small, localized heating was negligible.

2.3. Micro-Velocity Sensor

Currently, several types of micro-velocity sensors have been created using MEMS technology. These sensors are generally classified into two primary categories: hot-wire type and calorimetric type. Among these, hot-wire micro-velocity sensors are particularly suited for this research due to their compact size, low energy consumption, high sensitivity, and accuracy, as well as the simplicity of fabrication, driving, and signal output circuit design. Velocity sensing methods can be further categorized into three types: constant temperature (CT), constant voltage (CV), and constant current (CC). Given that the Arduino microprocessor used in this study can only output and read voltage, incorporating a current-to-voltage module would increase costs. Therefore, this study adopts the Constant Voltage method and employs a custom-designed circuit to monitor voltage changes. The circuit includes a Wheatstone bridge and a voltage signal amplifier. The micro-velocity sensor's electrode dimensions are 350 μm × 350 μm and 400 μm × 400 μm, with a minimum line width and spacing of 10 μm.

2.4. Micro-Temperature Sensor

In this study the micro-temperature sensor falls under the category of Resistance Temperature Detectors (RTD), using gold (Au) as the electrode material. Metals exhibit thermal expansion and contraction with temperature variations, causing their electrical resistivity to change with temperature. The electrodes are designed in a serpentine structure to enhance resistance, with electrode areas of 510 μm × 400 μm and 590 μm × 450 μm and a minimum line width and spacing of 10 μm.

2.5. Micro-Humidity Sensor

The micro-humidity sensor employed in this research is a resistive micro-humidity sensor that incorporates a novel type of negative photoresist (Fujifilm LTC® PI 9305, Fujifilm, Tokyo, Japan) as the dielectric layer. This material is noted for its high durability, resistance to acids and bases, corrosion resistance, and moisture-absorbing characteristics, and it must generally remain non-conductive. As the moisture absorbed by the sensing material increases, its volume also expands, resulting in an increase in circuit resistance. The electrode dimensions are 590 μm × 500 μm and 1065 μm × 1050 μm, with a minimum line width and spacing of 10 μm.

2.6. Integration of the Three-in-One Micro-Sensor

Figure 2 illustrates the integrated design diagram of the three-in-one micro-sensor, and Table 2 outlines the principle and design of the three types of sensors. The sizes of the micro-velocity sensors are 350 μm × 350 μm and 400 μm × 400 μm; the micro-temperature sensors measure 510 μm × 400 μm and 590 μm × 450 μm; and the micro-humidity sensors have dimensions of 590 μm × 500 μm and 1065 μm × 1050 μm. Both the minimum line width and spacing are 10 μm, and the separation between the micro-sensors is 400 μm. Given the close arrangement of the micro-sensors, they can be treated as a single point for measuring wind speed, temperature, and humidity on a macroscopic scale, facilitating the simultaneous real-time monitoring of these three physical parameters. Potential electrical and structural interference effects were mitigated during mask design by optimizing sensor placement and spacing, while independent Wheatstone bridge circuits were employed to isolate electrical signals.

Table 2. Sensing principle of flexible three-in-one micro-sensor.

Micro-temperature sensor		This study uses a Resistance Temperature Detector (RTD) with a snake-shaped electrode to increase resistance. The temperature-sensitive resistance material is gold (Au), which has stable chemical properties, a simple manufacturing process, and high linearity.

Table 2. *Cont.*

Micro-humidity sensor		This study uses a resistive humidity sensor. Its electrode type is an interdigitated electrode structure. There is a moisture-sensing material film (PI 9305, Fujifilm, Tokyo, Japan) above the electrode. When the moisture absorbed by the moisture-sensing film increases, its dielectric constant It will also increase as the ambient humidity increases. The moisture-sensitive film material is PI 9305, which has stable chemical properties and high linearity in moisture-sensing characteristics.
Micro-velocity sensor		This study uses a snake-shaped hot-wire micro-velocity sensor. The main measurement structure of the hot-wire micro-velocity sensor is a resistance heater. The power supply is used to provide a constant voltage input to generate a heat source so that the electrode becomes a resistance heater, generating stable temperature field. When the gas increases with the wind speed and the heat is taken away, the resistance value of the heater will decrease accordingly. The resistance material is gold (Au), which has stable chemical properties, a simple manufacturing process, and high linearity.

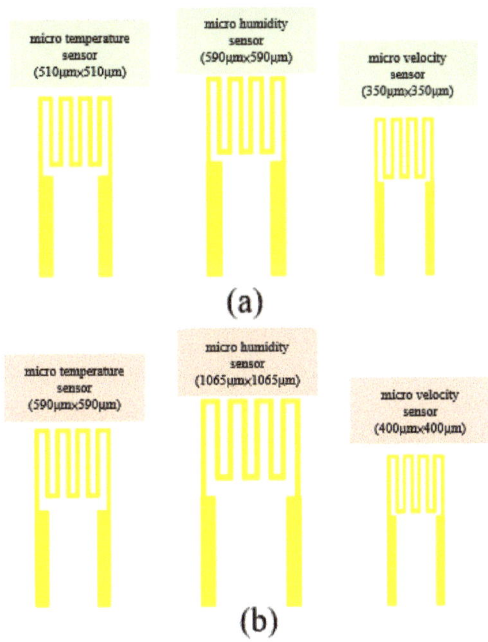

Figure 2. Two design diagrams of the integrated three-in-one micro-sensor: (**a**) first type; (**b**) second type.

2.7. Fabrication Process of the Integrated Three-in-One Micro-Sensor

This study employs MEMS technology to combine three sensing structures: velocity, temperature, and humidity. To ensure that the integrated three-in-one micro-sensor can provide flexibility, long-lasting durability, and sustained measurement within the HVAC cold air duct, this study selects a polyimide (PI) film as the substrate for the integrated sensor. The chosen PI film has a thickness of 50 μm, and the protective layer as well as the dielectric layer utilize liquid polyimides LTC® 9320 and 9305 both from Fujifilm,

Tokyo, Japan. To prevent any damage, surface micromachining techniques [25] from the MEMS process are employed. The fabrication process mainly includes deposition, lithography, and lift-off methods. In previous experiments, our laboratory employed wet etching to create micro-sensors, which frequently led to excessive etching of line widths, complicating adherence to specifications for back-end circuit integration. Consequently, this study adopts the lift-off technique to improve the stability of micro-sensor quality, with the overall fabrication workflow illustrated in Figure 3.

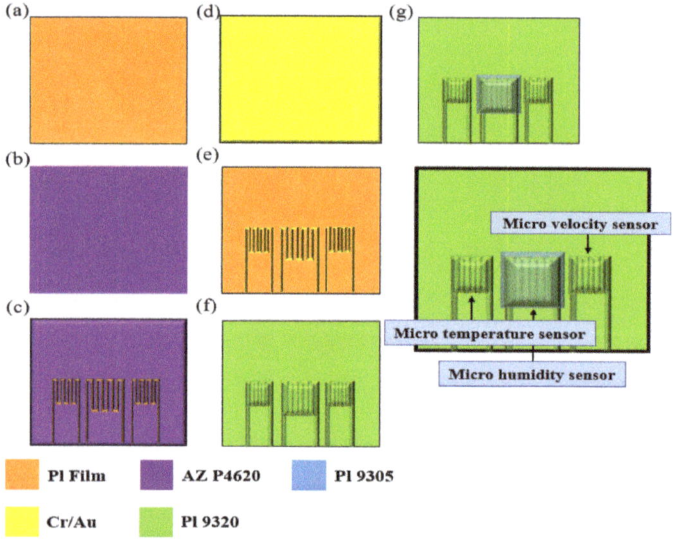

Figure 3. Flowchart for the fabrication process of an integrated three-in-one micro-sensor: (**a**) cleaning and fixing of polyimide film; (**b**) coating; (**c**) exposure; (**d**) evaporation; (**e**) lift off; (**f**) protective layer fabrication; (**g**) dielectric layer fabrication.

2.8. Calibration of the Integrated Three-in-One Micro-Sensor

Four sets of micro-velocity sensors were positioned in a programmable constant temperature and humidity testing apparatus. The calibration curve demonstrates that the sensor and calibration instrument are highly accurate. The calibration range for the custom micro-velocity sensor spanned from 0 m/s to 10 m/s, with data collected at intervals of 1 m/s. Since temperature affects the micro-velocity sensor, the environmental temperature during calibration varied from 5 °C to 40 °C, with increments of 5 °C. This enabled the collection of the relationship curve between velocity and the corresponding raw voltage readings from the Arduino microprocessor, leading to the calibration curve, which was subsequently normalized. As illustrated in Figures 4–7, V_f denotes the voltage reading at the current velocity, while V_0 signifies the voltage reading at 0 m/s. The y-axis unit $((V_f - V_0)/V_0)$ indicates the dimensionless change in the voltage reading.

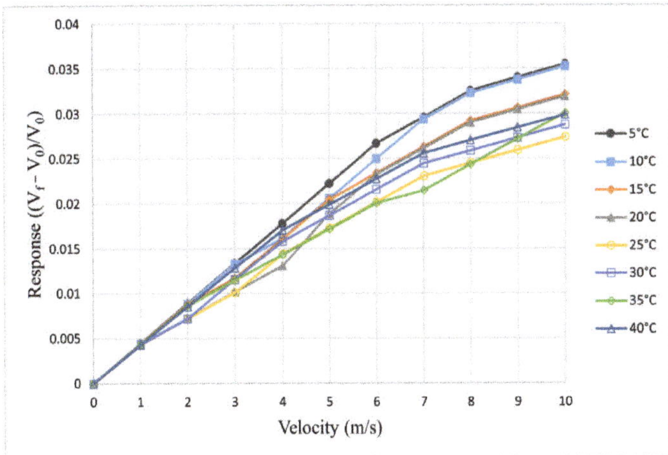

Figure 4. Calibration curve of the micro-velocity sensor (dimensionless) (sensor 1).

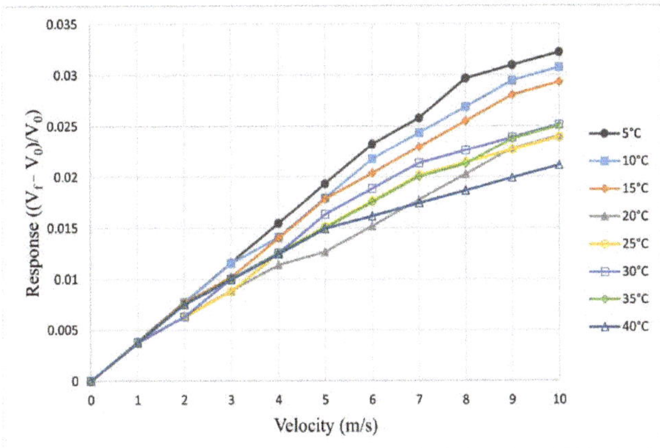

Figure 5. Calibration curve of the micro-velocity sensor (dimensionless) (sensor 2).

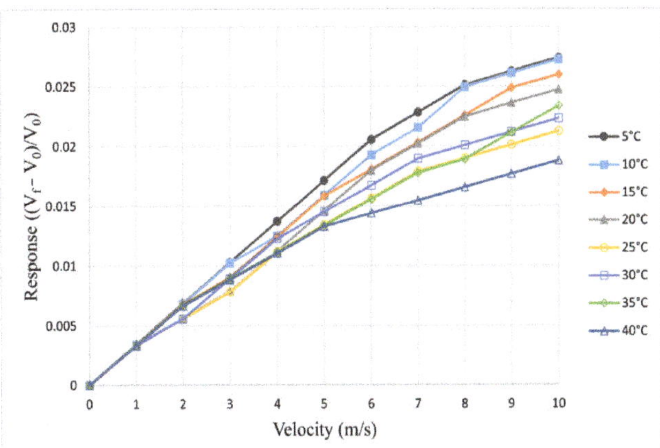

Figure 6. Calibration curve of the micro-velocity sensor (dimensionless) (sensor 3).

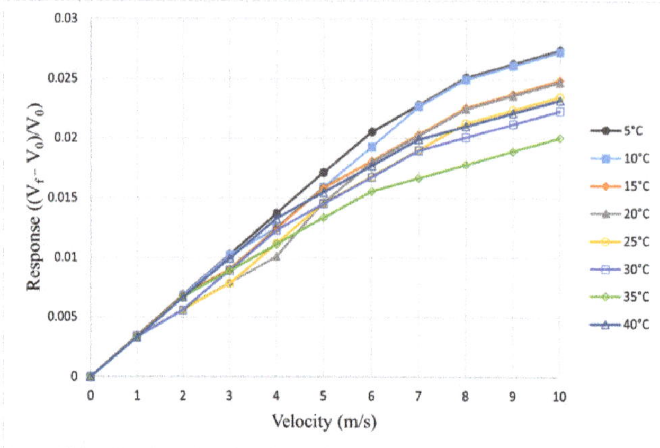

Figure 7. Calibration curve of the micro-velocity sensor (dimensionless) (sensor 4).

Four sets of micro-temperature sensors were positioned in a programmable constant temperature and humidity testing apparatus. The calibration curve demonstrates that both the sensor and the calibration instrument exhibit high accuracy. The calibration range for the micro-temperature sensors spanned from 5 °C to 40 °C, with data collected at 5 °C intervals. This allowed the establishment of the relationship between temperature and the corresponding raw voltage readings from the Arduino microprocessor, resulting in the calibration curve. The data were then normalized for analysis. As shown in Figure 8, V_f represents the voltage reading at the current temperature, and V_0 represents the voltage reading at 5 °C. The y-axis, expressed as $((V_f - V_0)/V_0)$, illustrates the dimensionless variation in the voltage reading, indicating how the sensor's output changes with temperature.

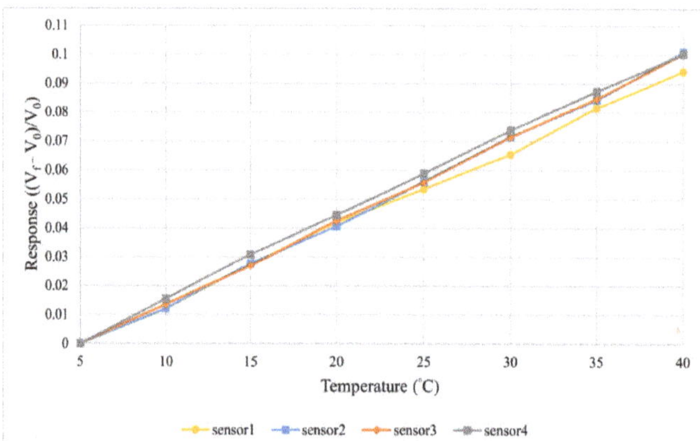

Figure 8. Calibration curve of the micro-temperature sensor (dimensionless).

Four sets of micro-humidity sensors were placed within a programmable constant temperature and humidity testing system. The calibration curve demonstrates that the sensor and calibration instrument are highly accurate. The calibration range for the micro-humidity sensors spanned from 70% RH to 95% RH, with data gathered at intervals of 5% RH. Since temperature affects the resistance values of the micro-humidity sensors, this experiment calibrated the sensors at temperatures from 5 °C to 40 °C, with increments of 5 °C. Moreover, due to the hysteresis effect of hygroscopic materials, the resistance values at the same humidity level may exhibit slight variations during the humidification and dehumidification processes. Consequently, this experiment calibrated the micro-humidity sensors by transitioning from low humidity to high humidity and vice versa. After stabilizing for 30 min each time, the relationship curve between humidity and the corresponding raw voltage readings from the Arduino was documented, resulting in the calibration curve, which was subsequently normalized. As illustrated in Figures 9–12, V_f denotes the voltage reading at the current humidity, while V_0 indicates the voltage reading at 70% RH. The y-axis unit $((V_f - V_0)/V_0)$ signifies the dimensionless change in the voltage reading.

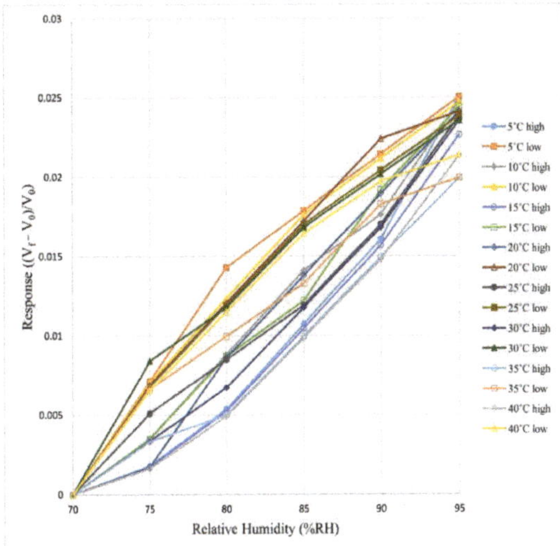

Figure 9. Calibration curve of the micro-humidity sensor (dimensionless) (sensor 1).

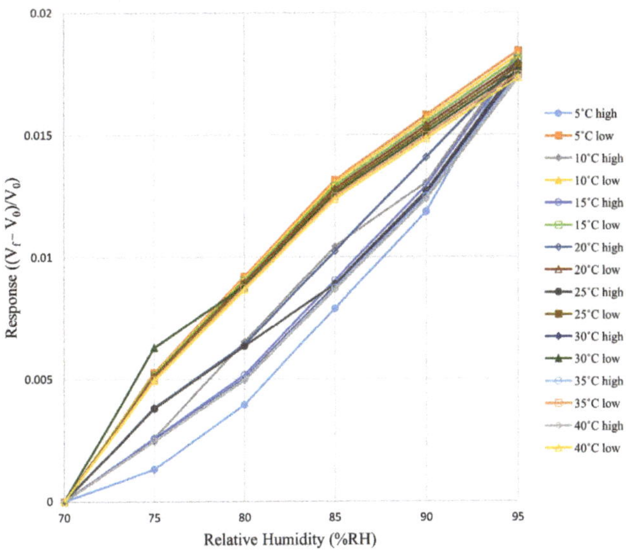

Figure 10. Calibration curve of the micro-humidity sensor (dimensionless) (sensor 2).

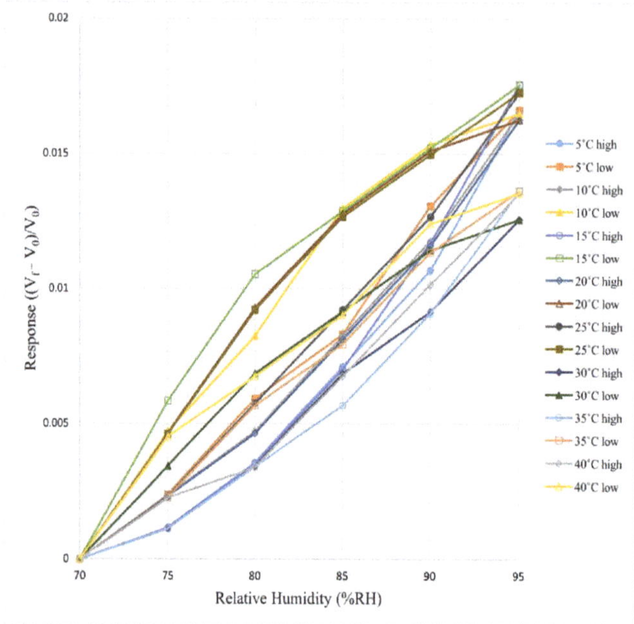

Figure 11. Calibration curve of the micro-humidity sensor (dimensionless) (sensor 3).

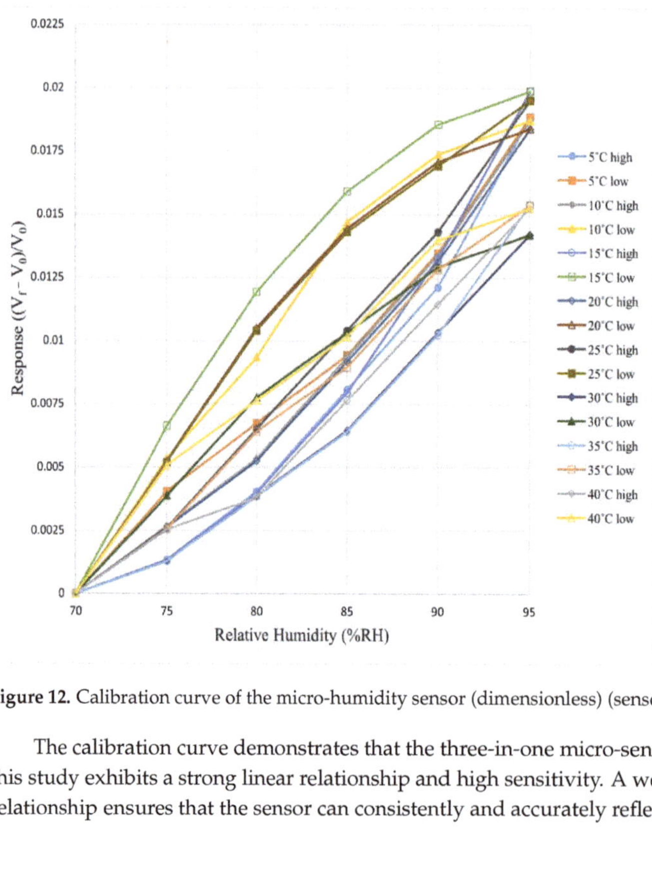

Figure 12. Calibration curve of the micro-humidity sensor (dimensionless) (sensor 4).

The calibration curve demonstrates that the three-in-one micro-sensor developed in this study exhibits a strong linear relationship and high sensitivity. A well-defined linear relationship ensures that the sensor can consistently and accurately reflect the correlation between input and output, minimizing the impact of nonlinear errors on data accuracy. High sensitivity enables the sensor to effectively detect small changes, improving measurement precision and reducing signal acquisition limitations, thereby showcasing its excellent performance. By analyzing the linear variations in the calibration plots, we can determine which of the four different-sized sensors performs best in terms of accuracy and overall performance. The results indicate that sensor 1 achieves the best performance in velocity measurements, while all four sensors exhibit excellent accuracy in temperature measurements. For humidity sensing, sensor 2 demonstrates superior performance. The exact reasons for these differences remain unclear. However, we speculate that variations in sensor dimensions, electrode thickness, or protective layer thickness during the manufacturing process may influence the sensing of physical quantities. This observation provides valuable insight for future research directions.

Experimental comparisons indicate that even with a ±5% deviation in electrode thickness, the calibration curves retain high linearity and stable sensitivity (within ±5%), and the overall response trend remains unaffected. This robustness is consistent with MEMS design tolerance rules and confirms the optimization results of the Taguchi Method. Normalized dimensionless calibration curves ($(V_f - V_0)/V_0$) were used instead of absolute voltage responses to eliminate baseline offsets among different sensor batches, providing a clearer comparison of linearity and sensitivity. Each sensor was calibrated three times, with an average measurement error of approximately 3%, ensuring that repeatability and sensitivity could be effectively demonstrated without excessive redundancy in the plots.

3. Results and Discussion

3.1. Back-End Integration of Integrated Three-in-One Micro-Sensor with FPC

In previous implementations, micro-sensors mounted on ceramic circuit boards were prone to mechanical stress due to the rigidity and weight of both the ceramic substrate and the attached conductive wires. When embedded into confined spaces, this often led to bending of the assembly, which caused delamination of the adhesive layer, resulting in damage to the conductive gel and silver paste interconnects—ultimately compromising the sensor's performance.

To address this issue, a flexible printed circuit (FPC) was developed as an alternative solution. The FPC uses polyimide (PI) as its substrate, offering excellent flexibility, chemical resistance, and thermal stability. The integrated three-in-one micro-sensor was carefully trimmed to a minimal footprint and then bonded onto the FPC to ensure a compact and robust connection. This integration approach significantly improves mechanical compliance and durability during installation and long-term operation. The completed back-end integration is illustrated in Figure 13.

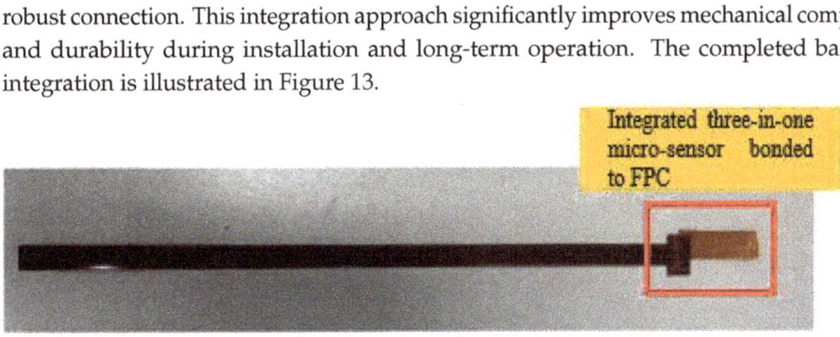

Figure 13. Back-end integration of integrated three-in-one micro-sensor with FPC.

3.2. Integrated Three-in-One Micro-Sensor Packaging

To ensure secure installation and long-term operation within the HVAC cold air duct, the sensor packaging was designed to interface with the manufacturer's existing mounting mechanism. Additionally, the packaging needed to withstand the oil-contaminated environment typically found inside such ducts. To meet these requirements, a customized package was fabricated using 3D printing technology.

The selected packaging material is poly(ethylene terephthalate-co-1,4-cyclohexylenedimethylene terephthalate), known for its excellent chemical resistance and mechanical strength [23]. The design allows the integrated three-in-one micro-sensor to be firmly fixed within the duct mechanism while protecting it from mechanical and chemical degradation. After assembly, the microsensor is inserted into the 3D-printed housing and securely locked in place, as illustrated in Figure 14.

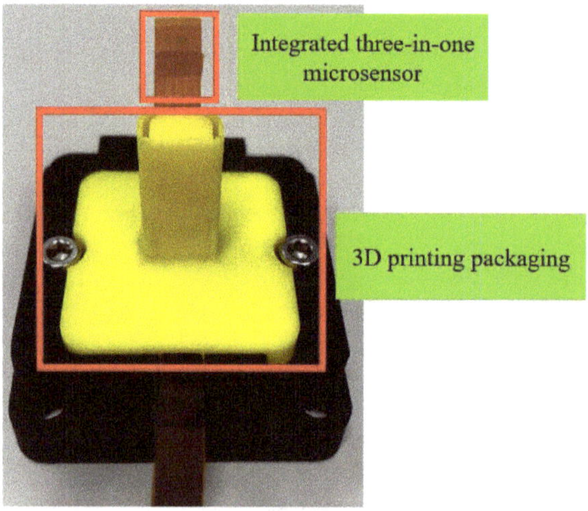

Figure 14. Packaging of integrated three-in-one micro-sensor.

3.3. *Performance Comparison of the Four Sets of Self-Made Integrated Three-in-One Micro-Sensors*

Prior to installation, the self-fabricated integrated three-in-one micro-sensor and the two commercially available velocity sensors underwent preliminary performance and oil-resistance tests to ensure suitability for the HVAC environment. To enhance protection against oil contamination, a layer of epoxy resin was applied to both the front and back sides of each sensor's solder joints. Once fully cured in ambient conditions, the epoxy formed a protective barrier against oil penetration.

After epoxy coating, the terminal blocks of all sensors were wired to a shared power supply, enabling simultaneous operation. The full circuit was then interfaced with an Arduino microcontroller for data acquisition. Finally, the sensors were installed into the cold air duct of the HVAC system, as illustrated in Figure 15.

Figure 15. Installing sensors in the HVAC cold air duct.

The monitoring data can be collected after the overall installation is completed. The wireless monitoring data communication protocol uses Message Queuing Telemetry Transport (MQTT) to connect the server installed in the factory and the internal network of the factory. Node-RED is used to connect the intranet to read data. The receiving time is set as collecting once every 15 s, as required by the manufacturer.

To evaluate the reliability of the four sets of custom-built integrated three-in-one micro-sensors, a comparison of the monitoring data from the four sets of micro-velocity sensors was conducted. Table 3 shows a comparison of specifications between the integrated three-in-one micro-sensor and the commercial sensor. As illustrated in Figure 16, the variation for the custom micro-velocity sensor (sensor 1) was ±300 mm/s, while for sensor 2, it was ±400 mm/s; for sensor 3, it was ±450 mm/s; for sensor 4, it was ±500 mm/s; for the F660 commercial velocity sensor, it was ±800 mm/s; and for the FS7.0.1L.195 commercial velocity sensor, it was ±1000 mm/s. Notably, the FS7.0.1L.195 commercial velocity sensor can only measure velocity, while the F660 commercial velocity sensor can only measure velocity and temperature. The self-made integrated three-in-one micro-sensor accounts for ambient temperature and humidity, utilizing different calibration curves to determine wind speed, which enhances its accuracy compared to commercial wind speed sensors. These results indicate that the precision of the custom micro-velocity sensor (sensor 1) exceeds that of the other three sets, demonstrating stable performance that enables more accurate real-time monitoring of data within HVAC cold air ducts. This improvement leads to better environmental management, increasing the quality and reliability of processed materials, thus confirming that the Taguchi Method effectively enhances the reliability of the custom-integrated three-in-one micro-sensor.

Table 3. A comparison of specifications between two commercially available velocity sensors and the integrated three-in-one micro-sensor.

	F660	**FS7.0.1L.195**	**Integrated Three-in-One Micro-Sensor**
Sensing range	0.15~20 m/s	0~100 m/s	0.15~20 m/s
Accuracy	±5%	<3%	±3%
Response time	400 ms	200 ms	1 ms
Measuring physical quantities	Velocity and temperature	Velocity	Velocity, temperature, and humidity
price	NT4000	NT560	NT535

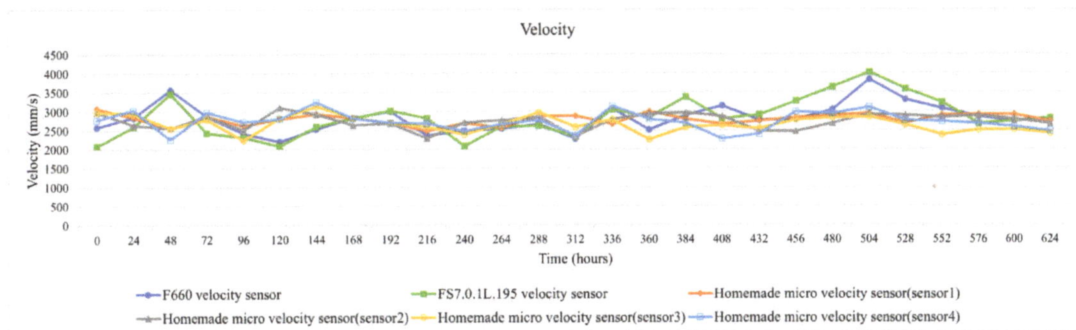

Figure 16. The 624 h monitoring data of commercial velocity sensors FS7.0.1L.195 and F660 and four sets of self-made micro-velocity sensors.

Although long-term durability testing, such as thermal cycling or environmental aging, was not conducted in this study, the application of the Taguchi Method has contributed to the enhancement of the sensor's durability through process robustness. Specifically, by identifying optimal fabrication parameters—such as electrode area, electrode thickness, and protective layer thickness—the method reduces the influence of structural defects and material inconsistencies that often lead to failure over time. This process-level optimization minimizes internal stress, delamination, and degradation caused by environmental variations such as temperature or humidity changes. As a result, although indirect, the improved stability and consistency of sensor output under different test conditions suggest that the Taguchi Method strengthens the underlying durability of the fabricated micro-sensors.

4. Conclusions

Currently, research on utilizing the Taguchi Method as a tool for optimizing sensor manufacturing processes remains limited. Therefore, this study applies the Taguchi Method to fine-tune the process parameters of sensor fabrication. Compared to previous research results, the method has demonstrated significant improvements in reliability and durability. Additionally, the reduction in sensor size not only lowers manufacturing costs but also facilitates sensor deployment in various locations. This proves that the Taguchi Method is highly effective in enhancing sensor manufacturing processes.

To enhance the quality and reliability of process materials, it is essential to monitor key environmental factors in real time within the cold air duct, such as wind speed, temperature, and humidity. Analyzing these three critical physical parameters enables the determination of optimal internal environment control settings.

This research successfully achieved real-time wireless monitoring of wind speed, temperature, and humidity within HVAC cold air ducts while also establishing a comprehensive database. A systematic approach was employed to identify the necessary sensors for monitoring the cold air duct's internal environment, including their optimal locations and quantities. Additionally, an evaluation and procurement process for commercially available wireless sensors with varying specifications was carried out. A wireless integrated three-in-one micro-sensor (measuring wind speed, temperature, and humidity) was developed, and both software and hardware were integrated to optimize the deployment and data collection of both commercially available wireless sensors and the self-developed three-in-one micro-sensors. Finally, joint field verification and data comparison were conducted to refine the internal environment control of HVAC cold air ducts, ultimately improving the quality and reliability of process materials.

Author Contributions: Methodology, J.-S.S.; data curation, G.-Q.H., C.-K.L., N.C. and C.-H.C.; writing—original draft, C.-Y.L. All authors have read and agreed to the published version of the manuscript.

Funding: The authors would like to thank the National Science and Technology Council of R.O.C. for its financial support through the grants MOST 110-2221-E-155-061 and 111-2221-E-155-048. The authors also would like to thank the Far Eastern Fibertech Co., Ltd. and YZU Fuel Cell Center and NENS Common Lab for providing access to their research facilities.

Data Availability Statement: The original contributions presented in this study are included in the article. Further inquiries can be directed to the corresponding author.

Conflicts of Interest: The authors declare no conflicts of interest. The funders had no role in the design of the study; in the collection, analyses, or interpretation of data; in the writing of the manuscript; or in the decision to publish the results.

References

1. JMP. Introduction to Design of Experiments (DOE): Classic Screening Design and Full Factorial Design. 2021. Available online: https://community.jmp.com/t5/JMP-Blog/%E5%AF%A6%E9%A9%97%E8%A8%AD%E8%A8%88-DOE-%E5%85%A5%E9%96%80%E7%B6%93%E5%85%B8%E7%AF%A9%E9%81%B8%E8%A8%AD%E8%A8%88%E8%88%87%E5%85%A8%E5%9B%A0%E5%AD%90%E8%A8%AD%E8%A8%88/ba-p/423195 (accessed on 8 October 2021).
2. Jankovic, A.; Chaudhary, G.; Goia, F. Designing the design of experiments (DOE)—An investigation on the influence of different factorial designs on the characterization of complex systems. *Energy Build.* **2021**, *250*, 111298. [CrossRef]
3. Bell, S. Experimental design. *Int. Encycl. Hum. Geogr.* **2009**, *10*, 672–675. [CrossRef]
4. Wang, Z.; Yi, Z.; Qin, M.; Huang, Q.-A.; Zhou, Z.-F. Linearity and sensitivity analysis of MEMS thermal wind sensor via analytical model. *IEEE Trans. Instrum. Meas.* **2023**, *72*, 1109–1118. [CrossRef]
5. Agrawal, V.K.; Patel, R.; Boolchandani, D.; Varma, T.; Rangra, K. Sensitivity and reliability enhancement of a MEMS based wind speed sensor. *Microelectron. Reliab.* **2020**, *104*, 2714–2720. [CrossRef]
6. Xu, Q.; Lu, Y.; Zhao, S.; Hu, N.; Jiang, Y.; Li, H.; Wang, Y.; Gao, H.; Li, Y.; Yuan, M.; et al. A wind vector detecting system based on triboelectric and photoelectric sensors for simultaneously monitoring wind speed and direction. *Nano Energy* **2021**, *89*, 106382. [CrossRef]
7. Yi, Z.; Wang, Y.; Qin, M.; Huang, Q. Research on dust effect for mems thermal wind sensors. *Sensors* **2023**, *23*, 5533. [CrossRef] [PubMed]
8. Yi, Z.; Wan, Y.; Qin, M.; Huang, Q.-A. Quadruple sensitivity improvement for wind speed sensor using dual-layer bended inductors. *Sens. Actuators A Phys.* **2020**, *303*, 111786. [CrossRef]
9. Mu, Y.; Liu, M.; Ma, Z.; Zhang, J. Resistance characteristic analysis based study on a novel damper torque airflow sensor for VAV terminals. *Build. Environ.* **2020**, *175*, 106813. [CrossRef]
10. Ligęza, P. Static and dynamic parameters of hot-wire sensors in a wide range of filament diameters as a criterion for optimal sensor selection in measurement process. *Measurement* **2020**, *151*, 107177. [CrossRef]
11. Hao, L.; Xiao, X.; Wu, Y.; Zhang, K.; Li, R.; Tian, H.; Ma, Y.; Ma, L. Fast response temperature sensor based on reduced graphene oxide through electron beam direct writing. *Sens. Actuators A Phys.* **2024**, *376*, 115669. [CrossRef]
12. Miao, F.; Zhang, X.; Tao, B.; Zhang, P. Wireless chipless RFID temperature and humidity sensor based on Fe_2O_3-Co_3O_4/SnO_2/rGO composites. *Mater. Sci. Eng. B* **2024**, *307*, 117549. [CrossRef]
13. Lin, S.; Wang, F.; Qu, Y.; Han, X.; Zhang, Y. Discharge splicing-free ultra-highly sensitive fiber-optic temperature sensor based on PDMS and the Vernier effect. *Sens. Actuators A Phys.* **2024**, *376*, 115653. [CrossRef]
14. Xu, Y.; Zhao, X.; Li, Y.; Qin, Z.; Pang, Y.; Liu, Z. Simultaneous measurement of relative humidity and temperature based on forward Brillouin scattering in polyimide-overlaid fiber. *Sens. Actuators B Chem.* **2021**, *348*, 130702. [CrossRef]
15. Pan, J.; Liu, S.; Zhang, H.; Lu, J. A flexible temperature sensor array with polyaniline/graphene–polyvinyl butyral thin film. *Sensors* **2019**, *19*, 4105. [CrossRef] [PubMed]
16. Tursunniyaz, M.; Meredith, A.; Andrews, J. Aerosol jet printed resistive temperature sensors with high sensitivity. *Sens. Actuators A Phys.* **2023**, *364*, 114777. [CrossRef]
17. Tursunniyaz, M.; Agarwal, V.; Meredith, A.; Andrews, J. Hybrid nanomaterial inks for printed resistive temperature sensors with tunable properties to maximize sensitivity. *Nanoscale* **2023**, *1*, 166118–166129. [CrossRef] [PubMed]
18. Liu, J.; Zhao, Z.; Fang, Z.; Liu, Z.; Zhu, Y.; Du, L. High-performance FBAR humidity sensor based on the PI film as the multifunctional layer. *Sens. Actuators B Chem.* **2020**, *308*, 127694. [CrossRef]
19. Mu, Y.; Jin, P.; Zheng, L.; Wang, C.; Hou, Y.; Liu, W.; Si, L.; Liu, Z. An enhanced MEMS-based polyimide capacitive-type relative-humidity sensor with halloysite nanotube as a modifier. *Microchem. J.* **2023**, *191*, 108934. [CrossRef]
20. Li, Z.; Dong, B.; Chen, E.; Li, Y.; Zhao, W.; Wang, Y.; Gao, C. High sensitivity FBG humidity sensor coated with graphene and polyimide films. *Opt. Fiber Technol.* **2021**, *66*, 102635. [CrossRef]
21. Zhu, J.; Cao, Y.; Chen, H.; Fan, B.; Zou, X.; Cheng, J.; Zhang, C. Rapid-response humidity sensors based on ultra-thin films stacked with single-layer graphene oxide. *Results Chem.* **2024**, *7*, 101444. [CrossRef]
22. Che, X.Y.; Mei, S.X.; Zhao, W.; Zhang, Y.C.; Zhang, X.M.; Cui, Z.; Fu, P.; Pang, X.C.; Liu, M.Y.; Ye, Y. Thermoplastic polyamide elastomer based flexible humidity sensor for breath monitoring. *Mater. Des.* **2023**, *235*, 112438. [CrossRef]
23. Zhang, H.-W.; Xu, X.; Huang, M.-L.; Wang, Y.-S.; Xu, Z.-Q.; Feng, Z.-S.; Zhang, Y. Interlayer cross-linked MXene enables ultra-stable printed paper-based flexible sensor for real-time humidity monitoring. *Chem. Eng. J.* **2024**, *495*, 153343. [CrossRef]

24. Lee, C.-Y.; Shieh, J.-S.; Chen, J.; Wang, X.-W.; Liu, C.-K.; Wei, C.-H. The application of a self-made integrated three-in-one microsensor and commercially available wind speed sensor to the cold air pipe of the heating, ventilation, and air conditioning in a factory for real-time wireless measurement. *Sensors* **2023**, *23*, 4471. [CrossRef] [PubMed]
25. Keshavarzi, M.; Hasani, J.Y. Design and optimization of fully differential capacitive MEMS accelerometer based on surface micromachining. *Microsyst. Technol.* **2019**, *25*, 1369–1377. [CrossRef]

processes

MDPI

Article

Optimal Scheduling of a Multi-Energy Hub with Integrated Demand Response Programs

Rana H. A. Zubo [1], Patrick S. Onen [2,3], Iqbal M Mujtaba [2], Geev Mokryani [4] and Raed Abd-Alhameed [2,5,*]

1 Medical Instrumentation Techniques Engineering, Technical Collage Kirkuk, Northern Technical University, Kirkuk 36001, Iraq; r.h.a.zubo@ntu.edu.iq
2 Faculty of Engineering and Digital Technologies, Bradford University, Bradford BD7 1DP, UK; patrick.onen@anchor.org.uk (P.S.O.); i.m.mujtaba@bradford.ac.uk (I.M.M.)
3 Anchor Company, Bradford BD1 2ST, UK
4 Future Power Systems Team at Jacobs, Manchester M2 5AD, UK; geev.mokryani@jacobs.com
5 Department of Information and Communication Engineering, Al-Farqadein University College, Basrah 61004, Iraq
* Correspondence: r.a.a.abd@bradford.ac.uk

Abstract: This paper presents an optimal scheduling framework for a multi-energy hub (EH) that integrates electricity, natural gas, wind energy, energy storage systems, and demand response (DR) programs. The EH incorporates key system components including transformers, converters, boilers, combined heat and power (CHP) units, and both thermal and electrical energy storage. A novel aspect of this work is the joint coordination of multi-carrier energy flows with DR flexibility, enabling consumers to actively shift or reduce loads in response to pricing signals while leveraging storage and renewable resources. The optimisation problem is formulated as a mixed-integer linear programming (MILP) model and solved using the CPLEX solver in GAMS. To evaluate system performance, five case studies are investigated under varying natural gas price conditions and hub configurations, including scenarios with and without DR and CHP. Results demonstrate that DR participation significantly reduces total operating costs (up to 6%), enhances renewable utilisation, and decreases peak demand (by around 6%), leading to a flatter demand curve and improved system reliability. The findings highlight the potential of integrated EHs with DR as a cost-effective and flexible solution for future low-carbon energy systems. Furthermore, the study provides insights into practical deployment challenges, including storage efficiency, communication infrastructure, and real-time scheduling requirements, paving the way for hardware-in-the-loop and pilot-scale validations.

Keywords: energy hub; electricity; heat; natural gas; demand response program

1. Introduction

The growing penetration of renewable energy resources and the rapid increase in multi-energy demand have created significant challenges for the secure and economic operation of energy systems [1]. Conventional electricity, natural gas, and thermal networks are often planned and operated independently, which limits opportunities for efficiency improvement, system flexibility, and resilience. The energy hub (EH) concept has therefore emerged as a promising paradigm for coordinating multiple energy carriers and conversion technologies within an integrated optimisation framework [2]. In particular, coupling electricity, natural gas, and thermal energy flows within a single hub allows operators to exploit cross-vector flexibility. For example, a combined heat and power (CHP) unit can simultaneously balance heat and electricity demand, while thermal and electrical energy storage systems (ESS) provide temporal flexibility. Such integration enables better utilisation of distributed generation resources, particularly in rural or remote areas, and reduces reliance on costly network reinforcements [3]. With the increasing variability of renewable energy such as wind, additional flexibility measures are required. Demand response (DR) programs provide an effective solution by encouraging consumers to shift or reduce loads in response to time-varying prices or incentives. Integrating DR into multi-energy hubs enhances system flexibility by enabling consumers to alter not only the timing but also the form of energy consumed, thereby improving reliability, reducing operating costs, and facilitating renewable integration [4]. Despite significant progress, several open questions remain. Most existing works have investigated either single-carrier systems (e.g., electricity-only with DR) or limited multi-carrier hubs without full integration of renewables, storage, and demand response. Furthermore, operational security aspects—both physical (energy balance, storage feasibility) and informatic (coordination and control underprice signals) are often simplified, limiting real-world applicability [5,6]. This study addresses these gaps by proposing a comprehensive optimal scheduling model for a multi-energy hub that integrates electricity, natural gas, wind energy, energy storage systems, and demand response. The model is formulated as a mixed-integer linear programming (MILP) problem to ensure tractability and scalability. Unlike prior works, the proposed framework evaluates hub performance across five operational case studies under varying fuel price scenarios and component configurations, providing new insights into the combined roles of DR, storage, and CHP in reducing costs and peak demand.

Related Literature Reviews

An energy hub (EH) is a relatively recent idea in energy system integration, and it offers a viable way to carry out future energy systems, especially in rural areas. Numerous techniques, each with its own set of objectives and limitations, have been used in recent research on the best way to operate an EH while taking the demand response program into account [7]. The demand response program problem for a residential energy hub with vehicle-to-grid (V2G) and renewable energy sources (RES) is formulated in ref. [8], demonstrating how the type of demand response program used severely impacts the customer's annual cost. To help the unit take part in demand response (DR) programs and bid in the day-ahead electricity market, the authors in [9] created an operational optimisation scheduling model, like MILP or stochastic programming. The findings demonstrated that optimal DR program participation resulted in load shifting to hours with low energy prices, which lowers unit costs and boosts network resilience. A dynamic pricing-based model of the active DR program for industrial users is presented [10]. An agent-based method has been presented for simulating customer behaviour to engage in DR programs and select the optimal behavioural pattern considering different industrial processes for the three cement plants. The findings show that active DR deployment reduces peak demand in industrial units and smooths the load curve throughout the system in comparison to passive DR.

More recent works have extended EH formulations to incorporate uncertainty. Maghsoodi et al. [6] developed a probabilistic scheduling model considering correlations between uncertain variables in EHs with renewable sources. Miao et al. [8] proposed a multi-energy inertia-based support strategy with gas network constraints, demonstrating new opportunities for resilience. Other studies, such as Germscheid et al. [9], examined industrial DR under uncertain short-term prices, while Song et al. [10] considered bi-level dispatch in integrated gas–electric systems.

While these studies provide valuable insights, three key limitations remain:

- Many models are confined to single-carrier systems or limited EH structures, without fully capturing the synergies between electricity, gas, heat, renewables, storage, and DR.
- The majority of existing DR studies emphasise cost minimisation, but neglect cross-carrier substitution effects that arise in integrated hubs.
- Few works critically examine practical constraints such as storage charging/discharging logic, efficiency losses across converters, and operational feasibility under varying market conditions.

This paper builds on and extends these studies by presenting a comprehensive EH scheduling framework that incorporates multi-carrier energy flows, wind generation, dual storage systems, and demand response, validated across five operational scenarios. Compared with recent works such as [11], which emphasise single-carrier DR optimisation, the proposed study advances the state of the art by explicitly modelling cross-vector flexibility and assessing its implications for system cost, reliability, and demand flattening.

The following is a summary of the paper's main contributions:

1. Development of an MILP-based optimisation framework for a decentralised multi-energy hub that simultaneously integrates multiple carriers (electricity, heat, natural gas), wind generation, storage, and DR.
2. Critical analysis of the interactions between storage systems, renewable resources, and DR under different market and technical conditions, highlighting their combined impact on operational cost and demand profile.
3. Introduction of case studies that demonstrate how DR participation under both high and low gas prices can reshape hub operations, reduce peak demand, and improve renewable utilisation.
4. Identification of research and practical implications for the real-world deployment of multi-energy hubs, including operational security and informatics challenges.
5. Investigation of the effects of low-cost gas for both DRS and non-DRS scenarios.
6. Assessment of DRS in the absence of CHP.

The remaining sections of the paper are structured as follows. Section 2 introduces the suggested mode's system description. Section 3 outlines the problem formulation of the proposed model, including the objective function. Section 4 case studies and simulation results analysis. Section 5 discusses the conclusion and future research direction.

2. Proposed Smart Energy Hub Model Description

This paper employs a comprehensive energy hub model. The model incorporates inputs, converters, and energy storage to meet different demands. It is applied to assess the effects of demand response programs in a decentralised smart energy hub. The structure of the proposed smart energy hub is shown in Figure 1. It is powered by wind, natural gas, and electricity networks. Energy carriers are converted through gas boilers, combining heat and power (CHP) units, transformers, and converters. The hub also incorporates energy storage systems for both heat and electricity. This paper considers the demand for natural gas, electricity, and heat. In addition, the proposed model includes demand response capabilities for electricity demand.

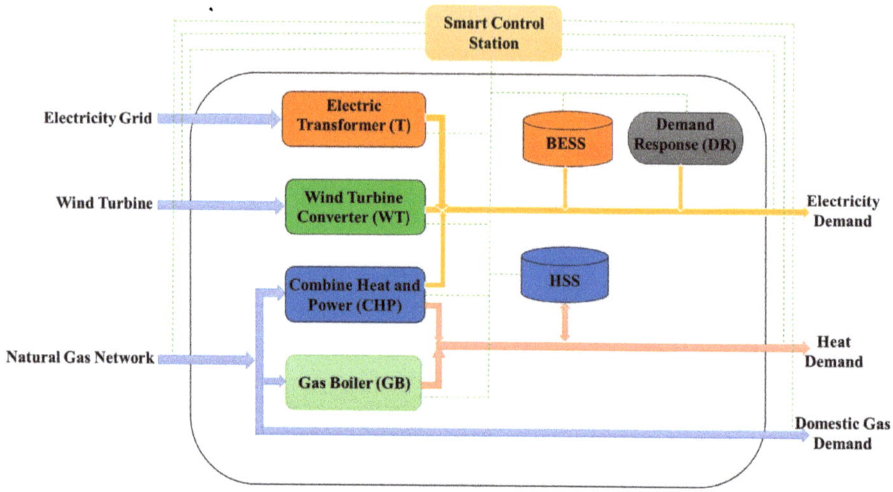

Figure 1. Structure of the proposed smart energy hub.

3. Problem Formulation

3.1. Objective Function

The objective function of the proposed model in this paper focuses on the price of energy purchased from different networks, the money that is earned from selling power to the network, the cost of charging and discharging heat storage devices and batteries, and lastly, the cost of the demand response (DR). Thus, the goal of this study is to minimise operating costs within a 24 h period while considering various constraints. Equation (1) represents the objective function of the best management of the suggested decentralised smart energy hub:

$$\begin{aligned} Min\ TOC = & \sum_{t=1}^{24} \left[\rho_t^{elec_{price}} P_t^{elec_{input}}\right] + \left[\rho_t^{WT_{price}} P_t^{WT_{input}}\right] \left[\rho_t^{gas_{price}} P_t^{gas_{input}}\right] \\ & + \left[\rho_t^{BESS_{price}^{unit}} \left(P_t^{BESS_{ch}} + P_t^{BESS_{dch}}\right)\right] \\ & + \left[\rho_t^{HSS_{price}^{unit}} \left(P_t^{HSS_{ch}} + P_t^{HSS_{dch}}\right)\right] \\ & + \left[\rho_t^{DR_{price}^{elec}} \left(P_t^{DR_{shup}} + P_t^{DR_{shdo}}\right)\right] - \left[\rho_t^{elec_{net}^{price}} P_t^{elec_{output}}\right] \end{aligned} \tag{1}$$

where $\rho_t^{elec_{price}}$ is the price of electricity purchased from the network at time t, $P_t^{elec_{input}}$ is purchased electric power from the network at time t, $P_t^{WT_{input}}$ is generated wind power at hour t, $\rho_t^{WT_{price}}$ is the price of wind power at time t, $\rho_t^{gas_{price}}$ is the price of natural gas purchased from the network at time t, $P_t^{gas_{input}}$ is natural gas purchased from the network at time t, $\rho_t^{BESS_{price}^{unit}}$ is the unit cost of battery storage, $P_t^{BESS_{ch}}/P_t^{BESS_{dch}}$ is the charged/discharged power of the battery storage system, $\rho_t^{HSS_{price}^{unit}}$ is the unit cost of heat storage systems, $P_t^{HSS_{ch}}/P_t^{HSS_{dch}}$ is the charged/discharged power of the heat storage system, $\rho_t^{DR_{price}^{elec}}$ is the cost of shifting electricity demand at hour t, $P_t^{DR_{shup}}/P_t^{DR_{shdo}}$ is the shift up/down of electricity demand by DR, and $\rho_t^{elec_{net}^{price}}$ is the price of selling electricity to the network at time t.

3.2. Constraints of Wind Power

The WT-rated power and wind speed determine the amount of electricity power generated by WT $\left(P_t^{WT_{input}}\right)$. When WT obtains a minimum wind speed (w_{ci}), it begins to generate power and keeps going until it reaches the rated wind speed (w_r). The WT generates power at the rated power of the WT upon receiving the rated wind speed. The WT will shut off if the wind speed is more than the maximum amount (w_{co}), or less than the minimum amount (w_{ci}). Z, x, and y are connected to the WT attributes.

$$P_t^{WT_{input}} = \begin{cases} 0 & w < w_{ci} \\ P_{Cap}^{WT}\left(z - yw_t + xw_t^2\right) & w_{ci} \leq w < w_r \\ 0 & w_r \leq w < w_{co} \end{cases} \tag{2}$$

3.3. Operation Constraints

The smart energy hub's optimal performance necessitates considering a few restrictions, satisfying which will result in the system operating at its best. These limitations are listed one at a time below.

3.3.1. Demand Constraints

The following formulation can be used to express the power equilibrium restrictions for meeting needs for heat, gas, and power.

3.3.2. The Smart Energy Hub Electricity Demand Constraint

The electricity requirement for the smart energy hub, as indicated by Equation (3), can be generated by the electrical network, CHP, WT, BESS, and DR. The electricity demand is shown by $P_t^{Elec_{dem}}$. The variables $P_t^{elec_{input}}$; $P_t^{chp_{input}^{elec}}$, and $P_t^{WT_{input}}$ denote the purchased electricity, gas for CHP, and wind power, respectively, from the network. The BESS charged and discharged power are represented by $P_t^{BESS_{dch}}$ and $P_t^{BESS_{ch}}$. $P_t^{DR_{shup}}$ and $P_t^{DR_{shdo}}$ indicate shifted downward and upward electricity demands, respectively. The electricity efficiency of transformers, the gas-to-electricity efficiency of CHP, and the electricity efficiency of converters are denoted by the values η_{elec}^{Trnsf}, $\eta_{gas_{elec}}^{chp}$, and η_{elec}^{WT}, respectively.

$$\begin{aligned} P_t^{Elec_{dem}} = \; & \eta_{elec}^{Trnsf}\left(1 - \alpha^{line}\right)P_t^{elec_{input}} + \eta_{gas_{elec}}^{chp} P_t^{chp_{input}^{elec}} + \eta_{elec}^{WT} P_t^{WT_{input}} \\ & + \eta_{dch}^{BESS} P_t^{BESS_{dch}} - \frac{1}{\eta_{ch}^{BESS}} P_t^{BESS_{ch}} + P_t^{DR_{shdo}} - P_t^{DR_{shup}} \\ & - P_t^{elec_{output}} \end{aligned} \tag{3}$$

where: $\alpha^{line} \in [0, 1)$: per-unit line loss factor between PCC and EH bus.

η_{elec}^{Trnsf}: transformer efficiency.

$\eta_{gas_{elec}}^{chp}$: CHP gas-to-electricity efficiency (net AC).

η_{elec}^{WT}: generator (converter/inverter) efficiencies in the wind conversion chain.

$P_t^{WT_{input}}$: Available wind power from the turbine power curve at time ttt.

η_{ch}^{BESS}, η_{dch}^{BESS}: battery charge/discharge efficiencies on the AC side.

3.3.3. The Smart Energy Hub Heat Demand Constraint

Equation (4) illustrates the hub heat requirement, which can be met by HSS, boiler, and CHP. The hub heat demand is shown by $P_t^{Heat_{Dem}}$. For CHP and boiler, the purchased gas power from the network is indicated by $P_t^{chp_{input}^{heat}}$ and $P_t^{boiler_{input}^{heat}}$. HSS is represented by the discharged and charged energies $P_t^{HSS_{dch}}$ and $P_t^{HSS_{ch}}$. The gas to heat efficiencies of CHP and boiler are denoted by $\eta_{gas_{heat}}^{chp}$ and $\eta_{gas_{heat}}^{boiler}$, respectively.

$$P_t^{Heat_{Dem}} = \left[\eta_{gas_{heat}}^{chp} P_t^{chp_{input}^{heat}} + \eta_{gas_{heat}}^{boiler} P_t^{boiler_{input}^{heat}} + P_t^{HSS_{dch}} - P_t^{HSS_{ch}}\right] \quad (4)$$

3.3.4. The Smart Energy Hub Gas Demand Constraint

Less than the gas needed for the boiler and CHP, purchased network gas can meet the gas requirement for the smart energy hub, as indicated by Equation (5). The gas demand for the smart energy hub is $P_t^{gas_{Dem}}$. $P_t^{gas_{input}}$ represents the gas power that was purchased from the network.

$$P_t^{gas_{Dem}} = \left[P_t^{gas_{input}} - P_t^{chp_{input}^{heat}} + P_t^{boiler_{input}^{heat}}\right] \quad (5)$$

3.4. *Network Constraints*

The network's capacity in the equations limits the amounts of bought power and gas from the networks and sold heat power to the networks, according to Equations (6) and (7). The maximum capacities of the electricity, gas, and heat networks are denoted by $P_t^{elec_{input}^{Max}}$ and $P_t^{gas_{input}^{Max}}$, respectively.

$$0 \leq P_t^{elec_{input}} \leq P_t^{elec_{input}^{Max}} \quad (6)$$

$$0 \leq P_t^{gas_{input}} \leq P_t^{gas_{input}^{Max}} \quad (7)$$

3.4.1. Converters Constraints

According to Equation (8) through (10), the installed optimised capabilities of the Transformer (P_{Cap}^{Trnsf}) , CHP (P_{Cap}^{chp}), and boiler (P_{Cap}^{boiler}) should be the limit on the purchased network power and network gas for CHP and boiler.

$$\eta_{elec}^{Trnsf} P_t^{elec_{input}} \leq P_{Cap}^{Trnsf} \quad (8)$$

$$\eta_{gas_{elec}}^{chp} P_t^{chp_{input}^{heat}} \leq P_{Cap}^{chp} \quad (9)$$

$$\eta_{gas_{heat}}^{boiler} P_t^{boiler_{input}^{heat}} \leq P_{Cap}^{boiler} \quad (10)$$

3.4.2. The Smart Energy Hub Components Constraints

P_{Cap}^{Trnsf}; P_{Cap}^{chp}; P_{Cap}^{boiler}; P_{Cap}^{WT}; $P_t^{BESS_{Cap}}$, and $P_t^{HSS_{Cap}}$ for transformer, CHP, boiler, WT, BESS, and HSS, respectively, represent the necessary and optimal capacities of the hub components. The maximum quantities that can be installed in the smart energy hub should set a limit on the optimised capacities of the hub's component parts. In Equations (11)–(16), the allowed maximum amounts for each hub component are as follows: $P_{Cap}^{Trnsf_{Max}}$; $P_{Cap}^{chp_{Max}}$; $P_{Cap}^{boiler_{Max}}$; $P_{Cap}^{WT_{Max}}$; $P_t^{BESS_{Cap}^{Max}}$, and $P_t^{HSS_{Cap}^{Max}}$.

$$P_{Cap}^{Trnsf} \leq P_{Cap}^{Trnsf_{Max}} \quad (11)$$

$$P_{Cap}^{chp} \leq P_{Cap}^{chp_{Max}} \quad (12)$$

$$P_{Cap}^{boiler} \leq P_{Cap}^{boiler_{Max}} \quad (13)$$

$$P_{Cap}^{WT} \leq P_{Cap}^{WT_{Max}} \tag{14}$$

$$P_t^{BESS_{Cap}} \leq P_t^{BESS_{Cap}^{Max}} \tag{15}$$

$$P_t^{HSS_{Cap}} \leq P_t^{HSS_{Cap}^{Max}} \tag{16}$$

3.5. Energy Storages Constraints

The model optimises the overall performance of the energy network by combining the restrictions of electricity and thermal energy storage to guarantee that the energy storage system runs safely, effectively, and within its specified capabilities.

3.5.1. Electrical Battery Storage Constraints

To store excess electricity created and release it when needed, BESS is used [12]. Based on the available power over the previous hour, the current amount of power that has been charged and discharged, and the amount of power loss of BESS in Equation (17), the available power in BESS (P_t^{BESS}) is estimated. In Equation (18), power loss ($P_t^{BESS_{loss}}$) is expressed. The BESS's available power should be restricted to the lowest and highest accessible power levels as stated in Equation (19). The variables $\alpha_{loss}^{BESS_{Min}}$ and $\alpha_{loss}^{BESS_{Max}}$ represent the minimum and maximum power amounts of the BESS. Equations (20) and (21) should additionally limit the power quantity of the BESS that is charged ($P_t^{BESS_{ch}}$) and discharged ($P_t^{BESS_{dch}}$). In Equation (22), binary variables of the BESS's charge ($I_{elec}^{BESS_{ch}}$) and discharge ($I_{elec}^{BESS_{dch}}$) prevent BESS from being charged and discharged simultaneously. The charge and discharge efficiencies of the BESS are represented by $\eta_{elec}^{BESS_{ch}}$ and $\eta_{elec}^{BESS_{dch}}$ respectively.

$$P_t^{BESS} = P_{t-1}^{BESS} + P_t^{BESS_{ch}} - P_t^{BESS_{dch}} - P_t^{BESS_{loss}} \tag{17}$$

$$P_t^{BESS_{loss}} = \alpha_{loss}^{BESS} P_t^{BESS} \tag{18}$$

$$\alpha_{loss}^{BESS_{Min}} P_t^{BESS} \leq P_t^{BESS} \leq \alpha_{loss}^{BESS_{Max}} P_t^{BESS} \tag{19}$$

$$\alpha_{loss}^{BESS_{Min}} \eta_{elec}^{BESS_{ch}} I_{elec}^{BESS_{ch}} \leq P_t^{BESS_{ch}} \leq \alpha_{loss}^{BESS_{Max}} \eta_{elec}^{BESS_{ch}} I_{elec}^{BESS_{ch}} \tag{20}$$

$$\alpha_{loss}^{BESS_{Min}} \eta_{elec}^{BESS_{dch}} I_{elec}^{BESS_{dch}} \leq P_t^{BESS_{dch}} \leq \alpha_{loss}^{BESS_{Max}} \eta_{elec}^{BESS_{dch}} I_{elec}^{BESS_{dch}} \tag{21}$$

$$0 \leq I_{elec}^{BESS_{ch}} + I_{elec}^{BESS_{dch}} \leq 1 \tag{22}$$

To account for minimum self discharge, the BESS includes a minimum drain constraint which prevent unrealistic idle states as show in Equation (23) below:

$$P_t^{BESS} \geq \alpha_{Min}^{BESS_{loss}} \tag{23}$$

3.5.2. Thermal Storage Constraints

The accessible energy of HSS (P_t^{HSS}) found in Equation (24) is calculated by considering the energy loss of HSS, the energy that was available an hour ago, as well as the energy that is now charged and discharged. Calculating heat energy loss ($P_t^{HSS_{loss}}$) is done using Equation (25). The least and maximum amounts of TS that can be accessed should be used to limit the available heat energy of HSS in Equation (26). The min/max factors of the min/max heat energy amount of HSS are represented by $\alpha_{loss}^{HSS_{Min}}$ and $\alpha_{loss}^{HSS_{Max}}$. Equations (27) and (28) limit the amounts of charged ($P_t^{HSS_{ch}}$) and discharged ($P_t^{HSS_{dch}}$) heat energy of HSS. ($I_{heat}^{HSS_{ch}}$) and ($I_{heat}^{HSS_{dch}}$) are binary variables of HSS that inhibit the charging and discharging performances in the interim in Equation (29). The charge and discharge efficiencies of HSS are shown by $\eta_{heat}^{HSS_{ch}}$ and $\eta_{heat}^{HSS_{dch}}$, respectively.

$$P_t^{HSS} = P_{t-1}^{HSS} + P_t^{HSS_{ch}} - P_t^{HSS_{dch}} - P_t^{HSS_{loss}} \tag{24}$$

$$P_t^{HSS_{loss}} = \alpha_{loss}^{HSS} P_t^{HSS} \tag{25}$$

$$\alpha_{loss}^{HSS_{Min}} P_t^{HSS} \leq P_t^{HSS} \leq \alpha_{loss}^{HSS_{Max}} P_t^{HSS} \tag{26}$$

$$\alpha_{loss}^{HSS_{Min}} \eta_{heat}^{HSS_{ch}} I_{heat}^{HSS_{ch}} \leq P_t^{HSS_{ch}} \leq \alpha_{loss}^{HSS_{Max}} \eta_{heat}^{HSS_{ch}} I_{heat}^{HSS_{ch}} \tag{27}$$

$$\alpha_{loss}^{HSS_{Min}} \eta_{heat}^{HSS_{dch}} I_{heat}^{HSS_{dch}} \leq P_t^{HSS_{dch}} \leq \alpha_{loss}^{HSS_{Max}} \eta_{heat}^{HSS_{dch}} I_{heat}^{HSS_{dch}} \tag{28}$$

$$0 \leq I_{heat}^{HSS_{ch}} + I_{heat}^{HSS_{dch}} \leq 1 \tag{29}$$

3.6. Demand Response Constraints

A portion of the electricity demand can be consumed during periods of low electricity demand and lowered during periods of high electricity demand. In Equation (30), the total decreased demand must match the entire increased demand. $P_t^{DR_{shdo}}$ and $P_t^{DR_{shup}}$ present shifted down and shifted up electricity demands respectively. The electricity load participation factors for shifting down and shifting up in that order are $LPF_{elec}^{DR_{Shdo}}$ and $LPF_{elec}^{DR_{Shup}}$. Equations (31) and (32) provide for the up and down shifting of the demand for electricity, respectively. The binary variables in Equation (33) that represent shifting up ($I_t^{DR_{shup}}$) and shifting down ($I_t^{DR_{shdo}}$) prevent simultaneous shifting of performances.

$$\sum_{t=1}^{24} P_t^{DR_{shup}} = \sum_{t=1}^{24} + P_t^{DR_{shdo}} \tag{30}$$

$$0 \leq P_t^{DR_{shup}} \leq LPF_{elec}^{DR_{Shup}} P_t^{DR_{dem}^{elec}} I_t^{DR_{shup}} \tag{31}$$

$$0 \leq P_t^{DR_{shdo}} \leq LPF_{elec}^{DR_{Shdo}} P_t^{DR_{dem}^{elec}} I_t^{DR_{shdo}} \tag{32}$$

$$0 \leq I_t^{DR_{shup}} + I_t^{DR_{shdo}} \leq 1 \tag{33}$$

4. Case Studies and Simulation Results Analysis

The hourly wind speed is shown in Figure 2. The demand response program has been taken into consideration for optimising the energy hub depicted in Figure 3 to minimise operating expenses. The energy hub operation was analysed using five case studies considering the availability of a demand response program. This study looks at the results of five different DR programs. The descriptions of a handful of case studies used in this investigation are used in this study. In the first and second scenarios, the impact of the energy hub's inclusion in the DR program is assessed assuming a moderately high price for natural gas. Both the third and fourth scenarios provide an adequate operation schedule for the energy hub with and without participation in the DR project, presuming low natural gas prices. The final scenario involves the removal of CHP from the energy hub and an evaluation of the effects of carrier interactions for the duration of the DR program. The centre of energy in Figure 3 shows different hourly load demands for heat, electricity, and natural gas. The hourly electricity cost are also shown in Figure 4. The MILP model of the problem of optimising the energy hub while taking the DR program into consideration has been solved using the CPLEX 22.1.0 solution method in the GAMS 37.2.0. software. The numerical results for the various occurrences are independently analysed in the following sections.

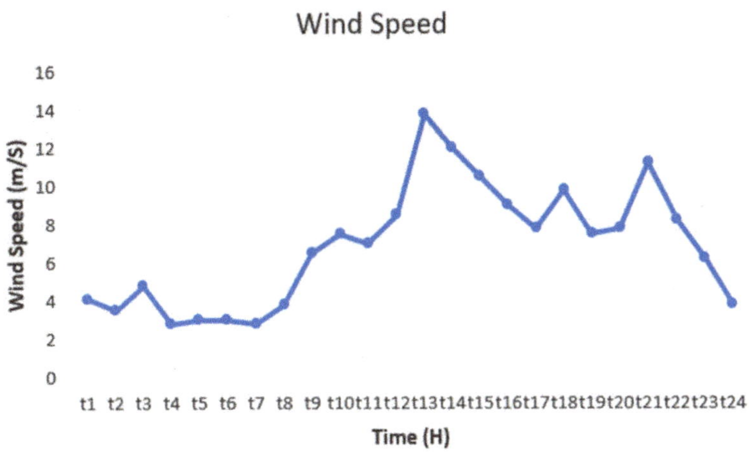

Figure 2. 24 Hour wind speed [13].

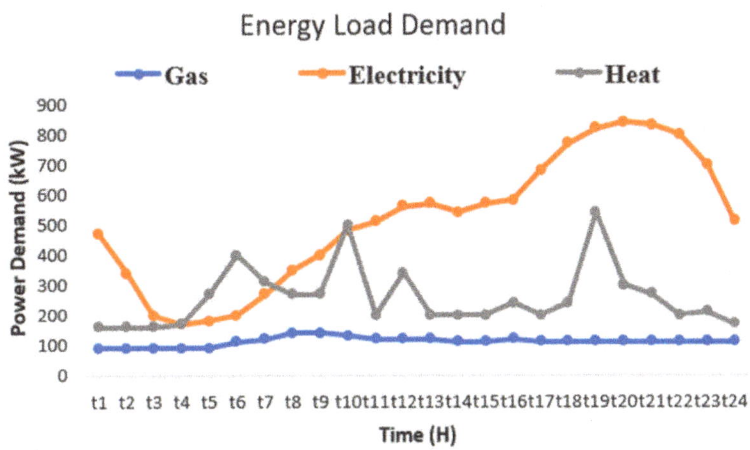

Figure 3. The smart energy hub's 24 h requirements for power, heat, and natural gas [14].

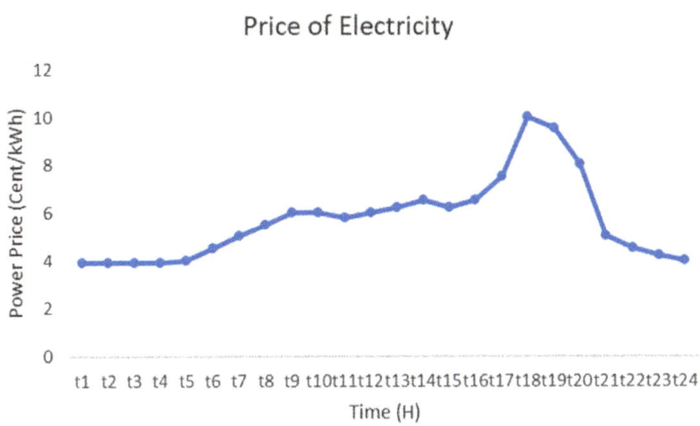

Figure 4. 24 h price of electricity [4].

4.1. An Energy Hub Case Study with Five Different Operational Scenarios

Five scenarios of a single energy hub were utilised to evaluate the suggested paradigm, as listed below.

- Scenario Number 1: Optimisation of the smart energy hub while taking the high cost of natural gas into consideration, but without addressing DR.
- Scenario Number 2: Optimising the smart energy hub while considering the concurrent DR program and high natural gas prices.
- Scenario Number 3: Optimising the smart energy hub while ignoring DR and taking into consideration the cheap price of natural gas.
- Scenario Number 4: Optimisation of the smart energy hub considering the simultaneous low natural gas price and DR program.
- Scenario Number 5: Optimisation of the smart energy hub without considering the implementation of CHP.

4.2. Simulation and Results Analysis

Scenario 1:

Given the perceived higher cost of natural gas in this scenario, the suggested energy hub should be billed at USD 0.055 for each kilowatt of gas it consumes from the network. Additionally, in this instance, the power and gas boiler network acts as a backup system if the CHP is not built, with the CHP functioning as the main source of heat and electricity at the proposed energy hub. The numerical results for this situation are shown in Figure 5. Figure 5 displays the optimal operating schedule for the energy hub's performance over the course of a day.

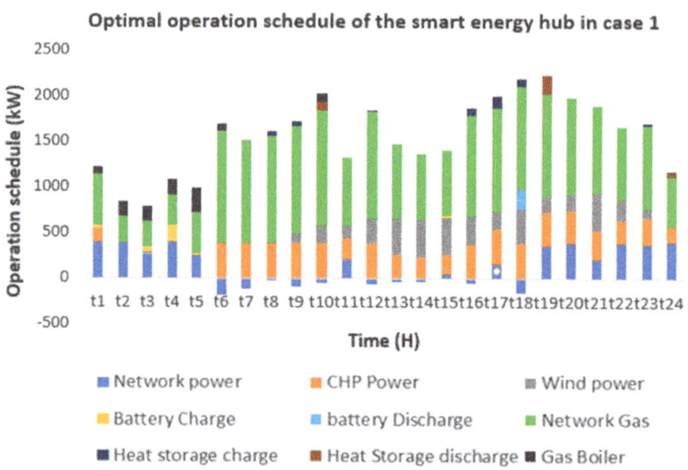

Figure 5. Scenario 1: Smart energy hub operation schedule.

In this scenario, the energy hub is made up of two different power-producing technologies: wind turbines and CHP. The CHP component of the system allows for the sale of any excess electricity to the network. The negative figures in the above table for the power exchanged with the network represent the sale of this power to the network. Due to the high cost of natural gas, the energy centre occasionally purchases electricity from the grid to make up for power shortages instead of using CHP.

The energy hub will make it possible to sell additional energy to the network, and the wind turbine in the SEH will allow production of local power from RES. Due to the CHP's limited capacity in the absence of a turbine, selling electricity to the grid is only practical early in the day when demand is low. Nevertheless, it is possible to sell energy all day, even in the afternoon when the market price of electricity is higher than it is in the morning, by adding a wind turbine. As a result, energy sales rise, increasing prices and SEH revenues.

A BESS has been designed in conjunction with the wind turbine to reduce the intermittent effects of this source on the hub energy's performance and to better match the power production with the consumption pattern. A sizable amount of the hourly electricity demand is satisfied by the BESS, which also charges during off-peak hours and discharges during the hour when market electricity costs are at their greatest (18:00 hrs). Peak hours are when wind turbines and CHP generate the most electricity, and this is also when BESS discharges are at their maximum. When the network is at its busiest, the energy hub can fulfil all its demand and even sell excess energy to it at the greatest price, thanks to the confluence of these factors.

The boiler fills in the gaps to provide heat when the CHP cannot meet demand during specific hours. Furthermore, excess energy is stored in the storage during periods of low heat demand, and during peak hours, the storage discharges some of its stored energy to offset part of the demand, hence reducing the SEH's operating expenses. The total running costs of the SEH in this case are USD 15,576.73 M/USD as seen in Table 1. The following case in point looks at the ramifications of providing the SEH with DR capabilities.

Table 1. Comparison of the total operation cost between Scenarios 1 and 2 and Scenarios 3 and 4, respectively.

Scenarios	Total Operation Cost (M/USD)
Scenario 1	15,576.73
Scenario 2	15,270.20
Scenario 3	65,094.20
Scenario 4	61,395.50

Scenario 2:

What sets this scenario apart from the previous one is the SEH's ability to participate in the DR program. The graphical findings for Scenario 2 are shown in Figure 6.

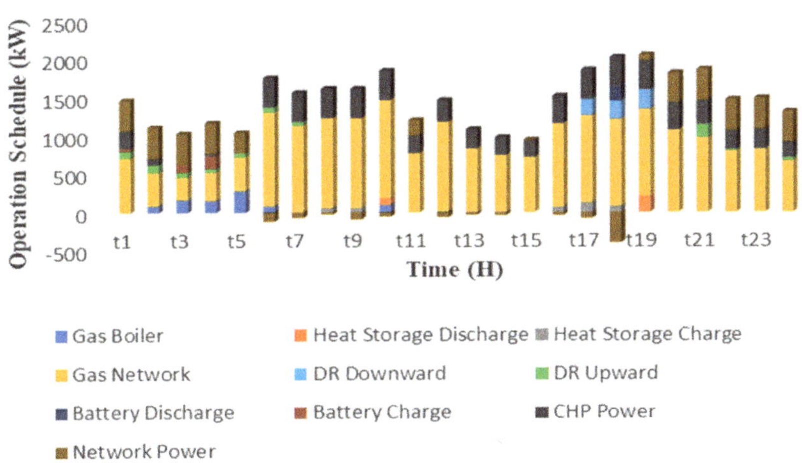

Figure 6. Scenario 2: Smart energy hub operation schedule.

Because it can move the load, the SEH in this scenario changes some of its demand from peak to off-peak hours. The demand for SEH thus decreases during peak hours and rises during off-peak hours. Consequently, the SEH uses its own energy-producing resources to produce more energy than it consumes during peak hours, and in certain situations, it is even able to sell the excess energy it generates back to the network. On the other hand, because of the load shift, CHP will be used more during off-peak hours, when it can help meet part of the additional demand. More people will purchase petrol from the network because of this. As shown in Figures 5 and 6, the quantity of gas obtained from the network has grown from 30,521.1 kW in instance 1 to 30,911.2 kW in Scenario 2. As for Figures 7 and 8. A graphical representation of the stated notions is presented in Figures 7 and 8. The amount of electricity generated by the CHP and the amount exchanged with the grid are displayed graphically in these results.

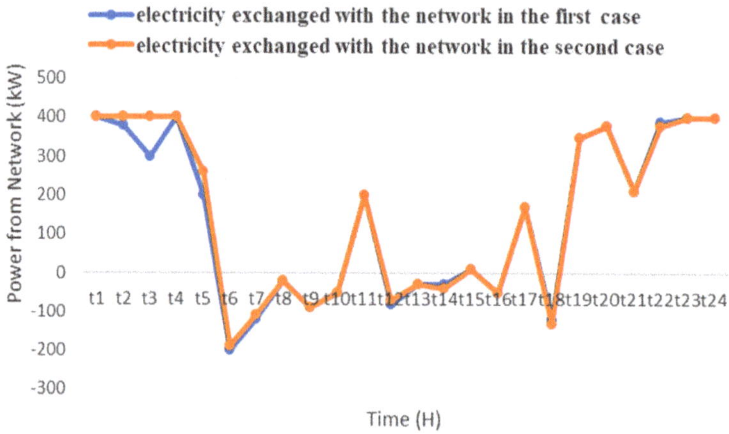

Figure 7. Scenarios 1 and 2, electricity exchange with network.

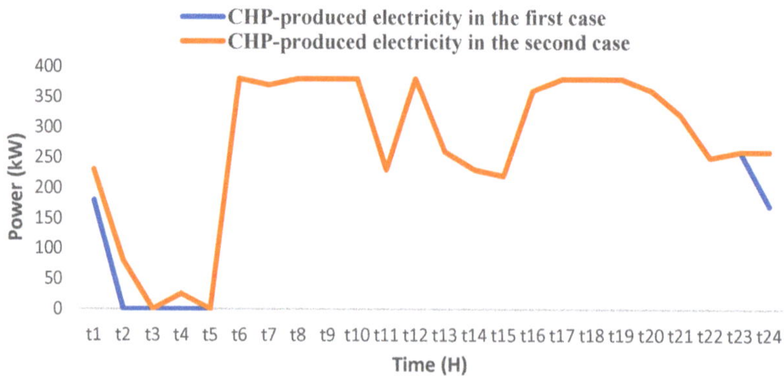

Figure 8. CHP-generated power in the first and second scenarios.

As can be seen, when demand shifted to off-peak hours, the amount of CHP produced increased during hours 1, 2, 4, and 24. This load shift has an effect on the power traded with the network curve in a way that raises purchases from the network at hours 2, 3, 21, and 22, and decreases sales to the network at hours 6 and 7. The selling of electricity to the network at hour 17, increased network sales at hour 18, and decreased network purchases at hour 19 were the outcomes of fluctuations and decreases in peak-hour demand. Running expenses for the SEH are 15,270.2 M/USD as seen in Table 1; because of these factors being combined, this is less than in the first instance.

Scenario 3:

In Scenario 3, natural gas is consistently more affordable than electricity. Furthermore, as the energy hub cannot participate in DR initiatives, all demand must be met in line with the intended pattern. The graphical findings for Scenario 3 are shown in Figure 9.

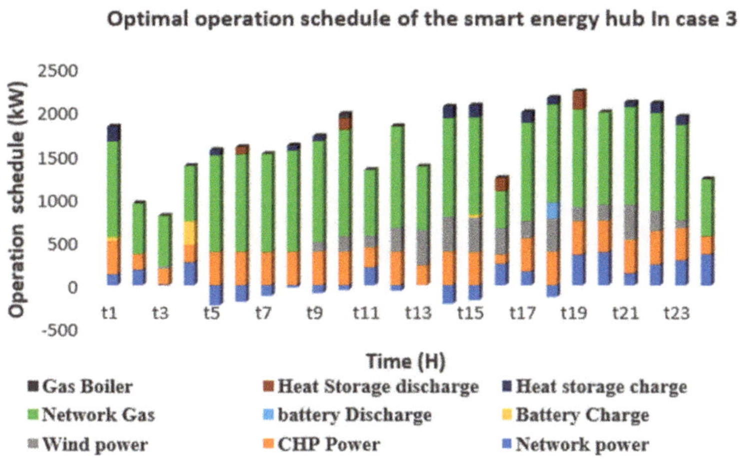

Figure 9. Scenario 3: Smart energy hub operation schedule.

The use of CHP has increased noticeably in this scenario compared to Scenario 1 since natural gas is less expensive than electricity. In this case, the energy hub that could use many energy carriers decides to use less expensive energy carriers to meet a specific need. According to the chart, the amount of power produced by CHP in this instance is 7823.7 kW, which is 1516.1 kW more than in Scenario 1. This increase in CHP use results in approximately 2799 kW more fuel purchases from the network than in the first scenario. In Figure 10, this increase is seen. The graph indicates that more petrol has been purchased from the network over most of the day.

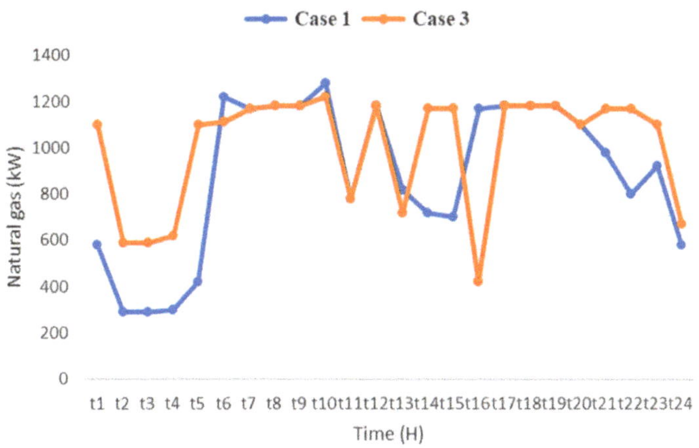

Figure 10. The amount of natural gas obtained in Scenarios 1 and 3 from the network.

Another problem is the decline in boiler manufacturing. The increased production of CHP reduces the need for auxiliary sources and the use of boilers. Additionally, as the amount of thermal energy generated by CHP rises, so does the requirement for thermal storage. This is because CHP production is linked with the thermal demand pattern. In this case, when the CHP production exceeds the heat requirement, the heat storage is replenished periodically during the day, and when it is released, it not only meets a significant amount of the thermal demand but also eliminates the need for boilers. Figures 5 and 9 illustrate that the heat storage has a charge and discharge power that is more than twice that of Scenario 1. Reliance on the electrical grid and the quantity of electricity network purchases will decrease with the adoption of natural gas as an energy source and its continuous use. For a more precise comparison, see Figure 11. Figure 11 shows how to exchange electrical power with the network in Scenarios 1 and 3.

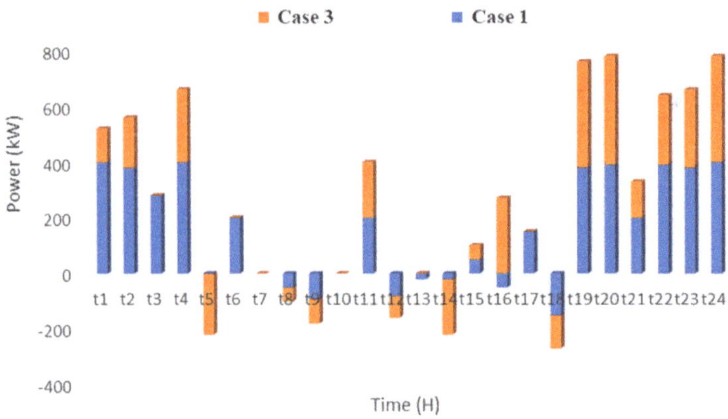

Figure 11. The energy transferred with the network in Scenarios 1 and 3.

The graph clearly shows a decrease in purchases from the electrical grid, especially in the morning and evening. This is seen in Scenario 1; purchasing electricity from the grid was more cost-effective at the time than producing power using natural gas. In contrast, CHP meets the need for these hours in Scenario 3, and because natural gas prices have decreased, some of this demand is fulfilled by purchases made from the network. Thanks to improved CHP power generation, it is now viable to sell electricity to the grid even during the middle of the day. The energy hub's operating costs have dropped significantly, to 65,094.2 M/USD as seen in Table 1, because of increased use of more cost-effective energy carriers, less reliance on the power grid, and enhanced ESS performance in this scenario. The following example, in which natural gas is a cheap energy source, looks at the effects of DR participation.

Scenario 4:

In this instance, the DR capacity is only utilised to produce electricity, even at a fair price for natural gas. The Scenario 3 results are shown graphically in Figure 12.

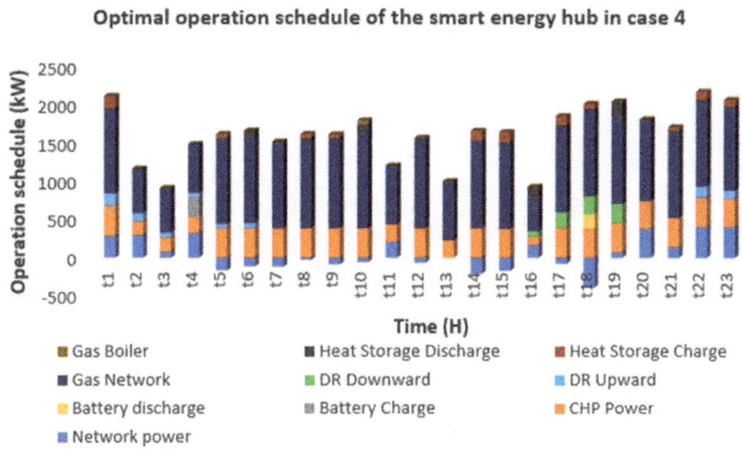

Figure 12. Scenario 4: Smart energy hub operation schedule.

In this instance, like in Scenario 2, some electricity consumption is shifted from network peak hours to off-peak hours. The effects of demand variations on the demand curve for electrical energy in Scenarios 2 and 4 are shown in Figure 13.

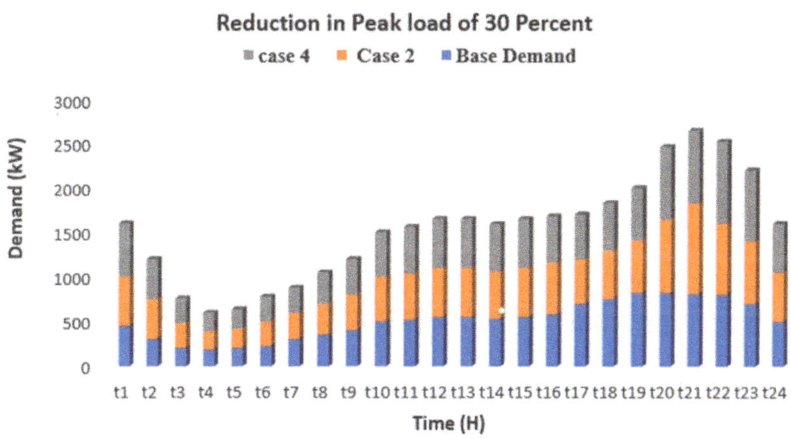

Figure 13. Effect of the DR program on the electricity consumption of the smart energy hub.

Demand variations lead to larger network sales during certain periods and lower network purchases during peak times. Figure 14 shows the electric power exchange curve for the network. This graph shows that the energy hub will be able to increase its capacity to sell excess electricity to the network and make money while simultaneously reducing its reliance on the network during peak hours. The energy hub will be able to depend less on the network during peak hours thanks to distributed power sources like wind turbines and combined heat pumps. By doing this, the network's stability will increase, and its peak demand will be reduced. The system's operational expenses are decreased because of the shifting demand, which is satisfied during off-peak hours. With a sum of 61,395.5 M/USD, as seen in Table 1, Scenario 4 has the smallest sum of all the examples studied thus far.

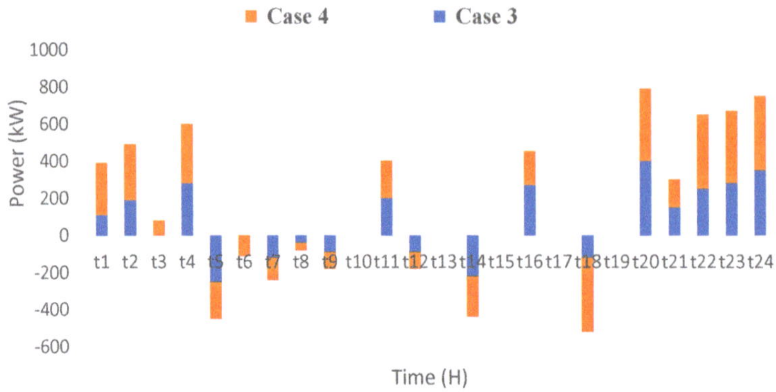

Figure 14. The energy exchanged with the network in Scenarios 3 and 4.

Scenario 5:

Here, we will assume that the energy hub under investigation does not have CHP and instead gets its power from the grid and wind turbines. Thermal energy will be produced by the boiler. In this scenario, the integration of several networks through multiple energy carriers is lost, making the direct acquisition of power-generating units from the relevant network necessary to satisfy fluctuating demands. We then contrast this scenario with two previous ones that had two high and two low natural gas prices. Figure 15 shows how to exchange electricity with the network in Scenario 5, as opposed to Scenarios 1 and 2. In Scenario 5, if the DR program is not carried out, the largest electricity purchase is scheduled to correspond with the energy's peak price (i.e., the network's peak demand), which elevates the network's peak demand and rises operational costs for customers.

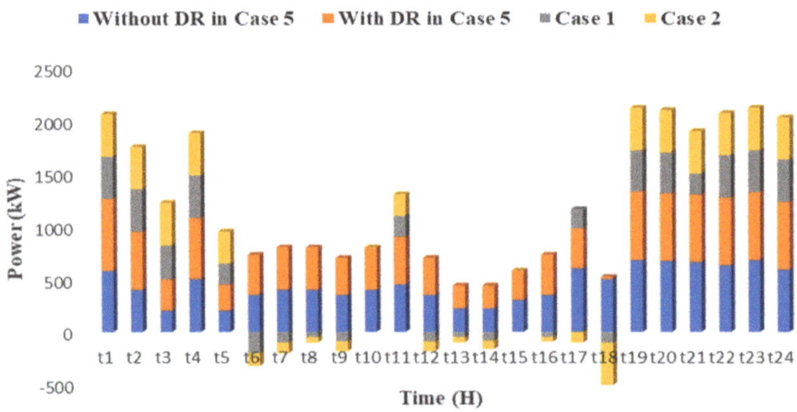

Figure 15. Scenarios 1, 2, and 5: The network traded electricity.

However, as DR capabilities increase and demand shifts to off-peak hours, the amount of electricity purchased from the network during peak hours decreases. During expensive periods, the system must pay a high price to meet demand because it still depends on the network's purchasing power. The system's capacity to transition between different energy carriers and the installation of CHP reduce energy purchases from the network significantly. The energy hub sells electricity at the full price and does not buy power from the network even during peak hours.

In this instance, the energy hub's participation in the DR program will result in higher sales of electricity to the grid at peak times and higher purchases made during off-peak hours, which will reduce operating expenses.

Utilising CHP reduces reliance on the electrical grid, as seen in Figures 5–16, which results in a higher gas buy volume from the gas network. The graph shows how the network's gas purchases grow significantly with the inclusion of CHP. When purchasing gas from the network first thing in the morning, when the cost of electricity and natural gas is almost equal, adding CHP has little effect. However, when the cost of electricity increases throughout the day, the energy hub is more eager to meet its needs from a less costly carrier, which causes it to produce more power through CHP and, as a result, buy more gas from the network.

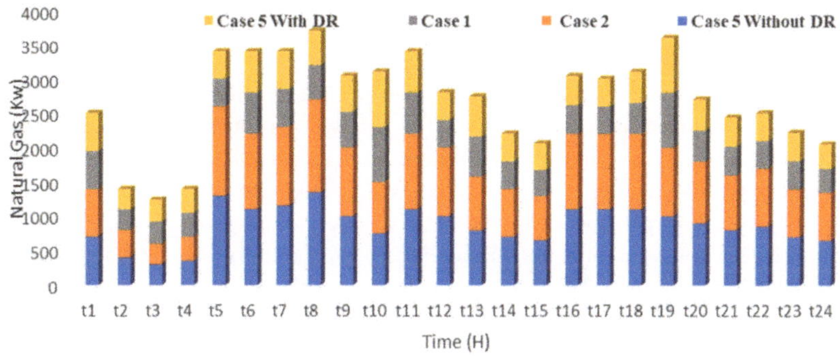

Figure 16. Purchase of natural gas from the network in Scenarios 1, 2, and 5.

If the electricity exchange curve for the low fuel price mode is taken into consideration, as shown in Figure 17, it can be demonstrated that the power purchase from the grid in the scenario of utilising CHP is cheaper than the scenario of eliminating CHP at all times of the day. It can be argued that the utilisation of natural gas for electricity production and the load shifting to low-cost times often cause the consumption curve in this scenario to flatten. Furthermore, even while petrol consumption rises during peak electricity hours and less electricity is bought from the grid during these periods, the cost of the consumers' energy bills decreases. From a financial perspective, the electrical grid's daily expenses decrease due to the flattening of the curve and reduced production and transmission costs, while the profit margin of the gas firm increases with increased sales of natural gas.

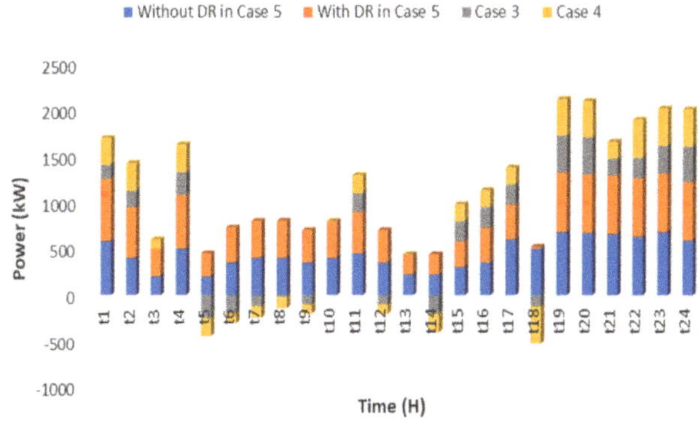

Figure 17. Scenarios 3, 4, and 5: Interchange of electricity with the network.

Thus, it can be said that the incorporation of DR programs into the architecture of smart energy hubs lowers utility profits, flattens the demand curve, and reduces peak demand while maintaining customer comfort.

5. Conclusions

This study developed and analysed an optimal scheduling model for a multi-energy hub (EH) integrating electricity, natural gas, wind energy, energy storage systems, and demand response (DR) programs. The problem was formulated as a mixed-integer linear programming (MILP) model and solved using CPLEX 22.1.0 in GAMS. 37.2.0. The model was tested under five operational scenarios with varying gas price conditions and component configurations, including cases with and without DR participation and CHP units. The results demonstrate that integrating DR into EH operations significantly improves operational efficiency. Specifically, DR enables demand shifting that reduces total operating costs by up to 6%, increases renewable energy utilisation, and lowers peak demand by approximately 6%, thereby flattening the demand curve. The coordinated operation of CHP, storage, and DR also enhances system flexibility by allowing substitution between electricity, gas, and heat carriers. This study's findings highlight several practical implications. First, while simulation results show substantial benefits, real-world deployment requires careful consideration of hardware constraints (e.g., response times of storage, minimum discharge levels), communication infrastructure, and cybersecurity of DR signals. Second, although MILP ensures tractable optimisation, implementation in real-time systems may demand heuristic or AI-based methods for faster scheduling under uncertainty. Finally, the demonstrated cost and reliability improvements suggest that EHs with DR could play an important role in future low-carbon energy systems, particularly in regions with high renewable penetration and volatile gas prices. In summary, the proposed model shows how multi-carrier coupling, storage integration, and DR flexibility can jointly reduce costs, enhance reliability, and improve renewable integration in smart energy hubs. Future research should focus on validating the framework in hardware-in-the-loop or pilot-scale demonstrations and exploring real-time control strategies using artificial intelligence or reinforcement learning to address dynamic market and network conditions.

Author Contributions: Conceptualization, R.H.A.Z., Software and Methodology, R.H.A.Z. and P.S.O., validation, R.H.A.Z., G.M. and R.A.-A., formal analysis, R.H.A.Z., P.S.O. and R.A.-A., writing—original draft preparation, R.H.A.Z., writing—review and editing, R.H.A.Z., P.S.O., R.A.-A., I.M.M. and G.M., supervision, G.M., R.A.-A. and I.M.M., project administration, R.A.-A. and I.M.M., funding acquisition, R.A.-A. All authors have read and agreed to the published version of the manuscript.

Funding: This work is partially supported by the UK Engineering and Physical Sciences Research Council (EPSRC) under grant EP/Y035135/1, and HORIZON-MSCA-2022-SE-01-01-ID: 101131501, Marie Skłodowska-Curie, Research and Innovation Staff Exchange (RISE), titled: 6G Terahertz Communications for Future Heterogeneous Wireless Network (6G-TERAFIT).

Data Availability Statement: The original contributions presented in this study are included in the article. Further inquiries can be directed to the first and corresponding author.

Conflicts of Interest: Author Geev Mokryani was employed by the company the Future Power Systems team. The remaining authors declare that the research was conducted in the absence of any commercial or financial relationships that could be construed as a potential conflict of interest.

References

1. Onen, P.S.; Mokryani, G.; Zubo, R.H. Planning of multi-vector energy systems with high penetration of renewable energy source: A comprehensive review. *Energies* **2022**, *15*, 5717. [CrossRef]
2. Wu, Q.; Zheng, J.; Jing, Z.; Zhou, X. Large-scale integrated energy systems. In *Energy Systems in Electrical Engineering*; Springer: Singapore, 2019. [CrossRef]
3. Amin, A.; Kem, O.; Gallegos, P.; Chervet, P.; Ksontini, F.; Mourshed, M. Demand response in buildings: Unlocking energy flexibility through district-level electro-thermal simulation. *Appl. Energy* **2022**, *305*, 117836. [CrossRef]
4. Onen, P.S.; Mokryani, G.; Zubo, R.H. Optimal Operation of Energy Hub with high penetration of renewable energy sources. In Proceedings of the 2nd International Multi-Disciplinary Conference Theme: Integrated Sciences and Technologies, IMDC-IST 2021, Sakarya, Turkey, 7–9 September 2021; European Alliance for Innovation: Vazovova, Slovakia, 2022; p. 26.
5. Dumas, J.; Dubois, A.; Thiran, P.; Jacques, P.; Contino, F.; Cornélusse, B.; Limpens, G. The energy return on investment of whole-energy systems: Application to Belgium. *Biophys. Econ. Sustain.* **2022**, *7*, 12. [CrossRef] [PubMed]
6. Maghsoodi, S.; Talavat, V.; Galvani, S. Probabilistic scheduling of a comprehensive energy hub integrated with renewable energy sources considering the correlation between uncertain variables. *Sustain. Energy Grids Netw.* **2023**, *36*, 101222. [CrossRef]
7. Liu, T.; Zhang, D.; Dai, H.; Wu, T. Intelligent modeling and optimization for smart energy hub. *IEEE Trans. Ind. Electron.* **2019**, *66*, 9898–9908. [CrossRef]
8. Miao, C.; Wang, Q.; Tang, Y. A multi-energy inertia-based power support strategy with gas network constraints. *Prot. Control Mod. Power Syst.* **2023**, *8*, 18. [CrossRef]
9. Germscheid, S.H.; Mitsos, A.; Dahmen, M. Demand response potential of industrial processes considering uncertain short-term electricity prices. *AIChE J.* **2022**, *68*, e17828. [CrossRef]
10. Song, X.; Wang, Y.; Zhang, Z.; Shen, C.; Peña-Mora, F. Economic-environmental equilibrium-based bi-level dispatch strategy towards integrated electricity and natural gas systems. *Appl. Energy* **2021**, *281*, 116142. [CrossRef]
11. Zhang, S.; Zhang, X.; Zhang, R.; Gu, W.; Cao, G. N-1 evaluation of integrated electricity and gas system considering cyber-physical interdependence. *IEEE Trans. Smart Grid* **2025**, *16*, 3728–3742. [CrossRef]
12. O'Malley, M.; Kroposki, B.; Hannegan, B.; Madsen, H.; Andersson, M.; D'haeseleer, W.; McGranaghan, M.F.; Dent, C.; Strbac, G.; Baskaran, S.; et al. *Energy Systems Integration: Defining and Describing the Value Proposition*; National Renewable Energy Lab. (NREL): Golden, CO, USA, 2016.
13. Yang, M.; Li, X.; Fan, F.; Wang, B.; Su, X.; Ma, C. Two-stage day-ahead multi-step prediction of wind power considering time-series information interaction. *Energy* **2024**, *312*, 133580. [CrossRef]
14. Onen, P.S.; Zubo, R.H.; Ali, N.T.; Mokryani, G.; Li, J.-P.; Abd-Alhameed, R. Stochastic Expansion planning Model for a coordinated Natural gas and Electricity Networks Coupled with Gas-fired Generators, Power-to-Gas Facilities and Renewable Power. *IEEE Access* **2024**, *12*, 105811–105830. [CrossRef]

processes

MDPI

Article

Exposing Vulnerabilities: Physical Adversarial Attacks on AI-Based Fault Diagnosis Models in Industrial Air-Cooling Systems

Stavros Bezyrgiannidis, Ioannis Polymeropoulos, Eleni Vrochidou and George A. Papakostas *

MLV Research Group, Department of Informatics, Democritus University of Thrace, 65404 Kavala, Greece; stbezyr@cs.duth.gr (S.B.); iwpolym@cs.duth.gr (I.P.); evrochid@cs.duth.gr (E.V.)
* Correspondence: gpapak@cs.duth.gr; Tel.: +30-2510462321

Abstract: Although neural network-based methods have significantly advanced the field of machine fault diagnosis, they remain vulnerable to physical adversarial attacks. This work investigates such attacks in the physical context of a real production line. Attacks simulate failures or irregularities arising from the maintenance or production department during the production process, a scenario commonly encountered in industrial environments. The experiments are conducted using data from vibration signals and operational parameters of a motor installed in an industrial air-cooling system used for staple fiber production. In this context, we propose the Mean Confusion Impact Index (MCII), a novel and simple robustness metric that measures the average misclassification confidence of models under adversarial physical attacks. By performing a series of hardware-level interventions, this work aims to demonstrate that even minor physical disturbances can lead to a significant reduction in the model's diagnostic accuracy. Additionally, a hybrid defense approach is proposed, which leverages deep feature representations extracted from the original classification model and integrates them with lightweight classifiers retrained on adversarial labeled data. Research findings underscore an important limitation in existing industrial artificial intelligence (AI)-based monitoring systems and introduce a practical, scalable framework for improving the physical resilience of machine fault diagnosis in real-world environments.

Keywords: physical adversarial attacks; fault diagnosis; deep learning; machine learning; industrial cooling systems; model robustness; predictive maintenance; artificial intelligence

1. Introduction

The deployment of deep learning (DL) models in industrial predictive maintenance has significantly improved the ability to detect early-stage mechanical faults, particularly in complex electromechanical systems such as centrifugal ventilators [1,2]. Despite the demonstrated accuracy of DL models under nominal conditions, their vulnerability to real-world, non-standard disturbances remains a critical concern [3]. In particular, physical adversarial attacks—defined as deliberate or accidental mechanical interventions that alter the condition of hardware components or sensor setups—pose a tangible yet underexplored threat in industrial settings. Unlike digital attacks, which perturb data algorithmically, physical attacks directly manipulate the environment or equipment. Indicative examples of attacks on a physical level include adversarial stickers on traffic signs that mislead vehicle perception systems [4,5], printed visual patterns that confuse robotic arms [6], and actuator-level perturbations that alter the dynamics of industrial processes without triggering conventional alarms [7,8]. These physical perturbations, whether malicious (e.g., sabotage) or accidental (e.g., human error), can confuse a trained model into confidently misclassifying normal behavior as a critical fault, potentially triggering false alarms or unwarranted shutdowns. To investigate these challenges under real operating conditions, the present work focuses on a single centrifugal ventilator within a fiber production facility, operated by Thrace Nonwovens & Geosynthetics S.A. (Thrace NG), located in Xanthi, Greece (Figure 1) [9].

Figure 1. Thrace Nonwovens & Geosynthetics S.A. [9].

The selected ventilator belongs to a fleet of ten parallel centrifugal fans comprising the plant's air-handling subsystem, where it contributes critically to airflow regulation during fiber extrusion and thermal bonding stages of production. The broader industrial setting under investigation is characterized by high production volumes, global export logistics, and frequent scheduled maintenance intervals—making it an ideal testbed for physically grounded robustness studies. Importantly, the existence of routine downtime allowed us to introduce controlled physical perturbations without affecting production. This enabled the capture of high-fidelity, attack-affected sensor data under operational dynamics, a scenario rarely available in academic research.

To this end, this work contributes to the growing field of robust industrial artificial intelligence (AI) by systematically evaluating the effect of five physical attack scenarios on DL fault classifiers, while proposing a hybrid defense mechanism. Specifically, we assess the susceptibility of ResNet50-1D and related models to misclassification when exposed to attacks such as misaligned sensors, impeller mass imbalance, or loose mechanical fittings. In this context, a novel custom metric is introduced, namely Mean Confusion Impact Index (MCII), to assess the adversarial robustness in machine learning models. Furthermore, we propose a hybrid defense strategy that combines deep feature extraction (via frozen ResNet50-1D backbone) with known ML classifiers retrained on attack data labels. Our findings demonstrate that such hybrid architectures can significantly improve model resilience and diagnostic accuracy under adversarial conditions.

The motivation behind this work stems from the critical gap between the demonstrated accuracy of deep learning-based fault diagnosis models under controlled or digital adversarial conditions and their untested reliability under real-world physical disturbances. In industrial environments, minor mechanical irregularities—such as sensor misalignments, loose screws, or eccentric shaft fittings—can easily occur during maintenance or operation, either unintentionally or through malicious tampering. These disturbances may not represent genuine machine faults, yet they can mislead diagnostic models into producing highly confident but incorrect classifications, ultimately resulting in false alarms, unscheduled production stoppages, or overlooked failures. Despite the increasing adoption of predictive maintenance solutions, little attention has been given to systematically studying such physical adversarial scenarios in actual factory settings. Addressing this gap is essential for building trustworthy industrial AI systems that can ensure operational continuity and safety.

The rest of the paper is structured as follows. Section 2 summarizes related work and comparatively highlights the contributions of this work. Materials and methods are presented in Section 3, including details on the proposed methodology and the experimental setup. Results are presented and discussed in Section 4, while defense Strategies are implied in Section 5. Finally, Section 6 discusses limitations and future work, and Section 7 concludes the paper.

2. Related Work and Contribution

Recent advances in deep learning have significantly improved the performance of industrial fault diagnosis systems. However, most models are highly sensitive to adversarial perturbations. While digital adversarial attacks—such as Fast Gradient Sign Method (FGSM) and Projected Gradient Descent (PGD)—have been extensively studied in the literature [10–13], the research on physical adversarial attacks in industrial systems remains comparatively sparse. In the following, related works addressing adversarial threats in vibration monitoring, sensor tampering, and mechanical systems are reviewed, aiming to highlight the contribution of this work.

2.1. Attacks on Machinery and Vibration Fault Diagnosis

Chen and Yan (2022) [14] investigated how fault diagnosis systems based on neural networks can be misled by imperceptible perturbations in vibration signals. Their experiments applied both time-domain and frequency-domain adversarial attacks—targeted and untargeted—and demonstrated a severe drop in classification accuracy across six common neural network architectures, including Convolutional Neural Networks (CNNs) and Long Short-Term Memories (LSTMs). Their results indicated that even minor changes to input signals could lead to highly confident misclassifications. Kim et al. (2024) [15] introduced the Spectrogram-Aware Ensemble Method (SAEM) to assess the real-world adversarial robustness of bearing fault diagnosis models. Unlike conventional methods such as FGSM or PGD, SAEM perturbs the spectral distribution of vibration signals, revealing hidden vulnerabilities in black-box attack scenarios and successfully deceiving multiple CNN-based classifiers. Zareapoor et al. (2020) [16] proposed an Oversampling Adversarial Network (MoGAN) designed to tackle the class imbalance problem in fault diagnosis. Their framework integrated a generator that produced synthetic fault samples and a discriminator that acted both as a classifier and a fault detector. This adversarial training approach improved the system's ability to detect minority-class faults and increased robustness under skewed data distributions.

Shaik et al. (2025) [17] evaluated adversarial attacks on anomaly detection models in industrial control systems (ICS), particularly using the Tennessee Eastman Process (FTEP) dataset. Their study demonstrated that LSTM and Gated Recurrent Unit (GRU) models can be manipulated by adversarial examples to misclassify one fault type as another. The authors implemented a structured adversarial attack framework using both white-box and black-box strategies, providing a detailed evaluation of vulnerability in fault classification. Villegas-Ch et al. (2023) [18] addressed the problem of physical tampering in industrial sensors by developing an anomaly detection-based method to spot sabotage or manipulation. Their approach, tested under real industrial conditions, showed high accuracy in identifying sensor-level anomalies caused by physical disturbances before they could affect broader system behavior. Finally, Nematirad et al. (2024) [19] explored an acoustic-based online monitoring method for detecting cooling fan malfunctions in air-forced transformers. Using CNNs and Random Forests on time-frequency spectrograms of audio signals, they were able to classify fan conditions as normal or faulty. The study showed that small physical changes—such as fan misalignment or mechanical deformation—can significantly alter acoustic signatures and deceive learning-based models.

2.2. Attacks on Industrial Soft Sensors and CPS

Kong and Ge (2021) [20] demonstrated that by directly maximizing or minimizing soft sensor outputs without requiring true labels, their adversarial attacks generated smooth, plausible perturbations that maintained output consistency, effectively deceiving operators and compromising industrial soft sensor reliability. Guo et al. (2024) [21] introduced a novel adversarial attack framework, namely KAGAN, which leveraged knowledge-aided generative adversarial networks to craft realistic and imperceptible perturbations against deep learning-based soft sensors. Unlike traditional gradient-based attacks, KAGAN bypassed the need for transfer gradient estimation, achieving stable, rational adversarial samples that maintain process consistency. Their results exposed significant vulnerabilities in the reliability and security of deep learning soft sensors deployed in complex industrial environments. Zizzo et al. (2019) [22] studied adversarial attacks on LSTM-based intrusion detectors in Industrial Control Systems (ICS) networks. By subtly altering multiple sensor inputs, they induced false fault alarms while bypassing detection. Sun et al. (2022) [23] demonstrated that multi-sensor fusion leveraging complementary signals (e.g., vibration and sound) within a CNN framework significantly improved fault diagnosis accuracy and robustness under noisy industrial conditions, highlighting the advantage of integrating diverse sensor data to enhance model generalization.

Farwell and Rohozinski (2018) [24] and Giraldo et al. (2018) [25] highlight significant vulnerabilities in cyber-physical systems (CPS), emphasizing that physical-layer attacks such as sensor tampering can bypass traditional cybersecurity measures. They stressed the necessity of integrating physical system behavior modeling with cybersecurity strategies to effectively detect and mitigate such attacks. Guo et al. (2017) [26] developed RIDS, a system that detected actuator and sensor spoofing in mobile robots using physical dynamics as validation. Kurakin et al. (2016) [27], Eykholt et al. (2018) [28] and Cao et al. (2020) [29] explored physical adversarial examples in computer vision systems, applying printed patterns and stickers to mislead classifiers. Though not mechanical, these studies underscore the real-world feasibility of adversarial deception. Nguyen et al. (2024) [30] extended this line of work by conducting an extensive analysis on vulnerabilities across multiple sensor modalities—including optical (e.g., facial recognition spoofing) and acoustic (e.g., ultrasound injection) channels. Their work highlighted challenges in real-world deployments, such as environmental variability, physical constraints, and sensor sensitivity, which mirror those found in vibration-based industrial monitoring systems.

2.3. Methodologies, Benchmarks, and Defense Strategies

Pozdnyakov et al. (2024) [31] performed a systematic adversarial benchmarking study using the Tennessee Eastman Process dataset. They evaluated the robustness of Multi-Layer Perceptron (MLP), GRU, and Temporal Convolutional Network (TCN) models under white-box and black-box attacks. GRUs were found to be the most robust, while MLPs were more prone to failure. The study also showed that adversarial training was beneficial in black-box scenarios but had a limited effect in white-box settings. Guo et al. (2024) [32] showed that while adversarial-based deep transfer learning (ADTL) techniques enhanced generalization to unseen machinery, they remained vulnerable to adversarial noise unless coupled with domain-aware defenses. Additionally, they emphasized the lack of standardized benchmarks and urged the incorporation of physical constraints and expert knowledge to ensure robustness in industrial deployments. Chen et al. (2024) [33] proposed the TFN model, a CNN with embedded time–frequency transform layers, which improved interpretability and showed enhanced robustness in noisy environments for fault diagnosis. Abadi (2021) [34] suggested that adversarial threats might compromise condition-based maintenance systems by falsifying inputs. Yin et al. (2023) [35] investigated transferable adversarial attacks in industrial intelligent systems and highlighted vulnerabilities across different models and environments, emphasizing the practical security risks in Industrial Internet of Things (IIoT) contexts.

In their recent work, Pozdnyakov et al. (2024) [36] systematically classified adversarial strategies and corresponding defenses—such as adversarial training, input denoising, and ensemble methods— establishing a modular framework that could support future hybrid or transfer-learning-based countermeasures in industrial AI systems. Wang et al. (2023) [37] conducted a comprehensive survey of physical adversarial examples, analyzing their characteristics, generation methods, real-world testbed implementations, and defense strategies. Their study also emphasized the challenges posed by physical-world constraints, such as environmental variability and sensor limitations, and highlighted the need for benchmarks that reflect real industrial setups. Ullah et al. (2024) [38] proposed an adaptive learning-driven multi-teacher knowledge distillation framework that improves model robustness against adversarial attacks by leveraging diversified knowledge from multiple pre-trained classifiers (teachers). Shi et al. (2022) [39] examined how multi-scale signal fusion increases resistance against perturbations in rotating equipment monitoring. Finally, Moosavi-Dezfooli et al. (2016) [40] introduced the DeepFool algorithm to compute minimal adversarial perturbations in image classification tasks. Although originally focused on vision systems, their work laid a theoretical foundation that inspired subsequent research on adversarial vulnerabilities in physical and industrial domains.

2.4. Comparison with Recent Approaches

To better highlight the novelty and practical contribution of this study, a comparative analysis with recent related works is provided. Table 1 summarizes the main characteristics and findings of state-of-the-art approaches and contrasts them with the present work.

Overall, the comparative analysis with the most recent related studies reveals several key distinctions between this study and recent approaches. First, whereas most prior works have evaluated adversarial robustness through digitally generated perturbations (e.g., FGSM/PGD, spectral-domain attacks, GAN-based samples) on laboratory or simulated datasets, the present work demonstrates that physical perturbations applied directly to industrial machinery induce equally systematic misclassifications. Second, while process-oriented benchmarks such as the Tennessee Eastman Process or soft-sensor simulations have provided valuable insights, they remain detached from the mechanical complexity of rotating equipment. In contrast, this study introduces a real-world benchmark in vibration-based fault diagnosis, capturing the interplay between mechanical faults, sensor dynamics, and diagnostic robustness. Third, existing defenses have primarily relied on adversarial training, robust neural architectures, or sensor fusion, which, although effective, are computationally heavy or limited to stochastic noise. The hybrid defense strategy proposed here—combining a frozen ResNet50-1D backbone with lightweight ML heads—offers a practical and computationally efficient alternative that maintains accuracy while reducing attack-induced confusion. Taken together, these differences highlight that the present study complements and extends recent works by bridging the gap between digital adversarial research and industrial practice, while also introducing a defense strategy tailored to operational requirements.

Table 1. Comparison of present work with recent related literature.

Ref.	Main Approach and Findings	Comparison with Present Work
[14]	Demonstrated that CNN and LSTM models suffer severe accuracy degradation when subjected to FGSM/PGD adversarial perturbations on bearing vibration data.	This work shows that similar vulnerabilities arise under physical perturbations (e.g., sensor misalignment, loose fittings) in an industrial ventilator, extending these findings from digital to real-world conditions.
[15]	Proposed a spectral-domain adversarial example method to craft more realistic black-box attacks on bearing datasets.	Unlike SAEM, which remains restricted to digital spectral manipulation, the present study demonstrates that comparable vulnerabilities emerge through real mechanical alterations under industrial operating conditions.
[16]	Introduced a GAN-based method for oversampling minority fault classes in bearing FD, mitigating class imbalance.	MoGAN addresses class distribution issues rather than adversarial robustness. The present work highlights how physical perturbations systematically bias diagnostic models, leading to consistent misclassifications.
[17]	Showed that LSTM and GRU anomaly detectors in ICS are highly vulnerable to adversarial inputs in simulations.	Similar vulnerabilities are identified here, but in industrial FD models (ResNet50-1D, BiLSTM) under physical adversarial conditions in a production plant.
[18]	Demonstrated that small-scale physical sabotage on ICS testbeds can bypass anomaly detection mechanisms.	While Villegas-Ch et al. addressed ICS security, this work applies the concept of physical adversarial interference specifically to vibration-based FD in real production equipment.
[20]	Proposed label-free black-box adversarial attacks on soft sensors, showing that regression outputs can be manipulated without ground-truth labels.	The present study extends this concept of stealthy perturbation to vibration-based FD, where physical interventions on sensors mislead classifiers without altering the true operational condition.
[21]	Developed a knowledge-aware GAN to generate smooth adversarial perturbations for soft sensor data in process simulations.	Whereas KAGAN focuses on simulated adversarial smoothness, the perturbations examined here arise naturally from mechanical interventions (eccentric shaft fit, loose bearing) in a real industrial system.
[23]	Proposed multi-sensor fusion (vibration and acoustic) to enhance robustness against random noise in bearings.	Similarly, this work integrates multi-feature inputs (vibration and operational features) but shows that such integration improves resilience against intentional physical adversarial perturbations, not only random noise.
[31]	Benchmarked defenses (adversarial training, denoising, ensembles) on the Tennessee Eastman Process; found adversarial training most effective in black-box scenarios.	That benchmark is process-simulation–oriented. This study introduces a new benchmark in rotating machinery, shifting robustness evaluation from simulated processes to real mechanical FD under physical attacks.
[32]	Reviewed adversarial transfer learning and emphasized its persistent vulnerabilities across domains.	This study empirically validates those observations: while the frozen ResNet50-1D backbone is vulnerable, robustness is enhanced by retraining lightweight ML heads (RF/LightGBM) with adversarial data, as quantified by the MCII metric.
[33]	Developed a Time–Frequency Network that improves noise robustness and interpretability in bearing FD.	TFN addresses stochastic disturbances, whereas the present study focuses on systematic adversarial perturbations in industrial machinery, offering complementary insights.

2.5. Contribution

Despite the growing interest in physical adversarial attacks, most works focus on simulations, synthetic perturbations, or purely digital threats. In contrast, our work implements five realistic physical attacks that emulate common operator mistakes and maintenance flaws—such as misaligned sensors, loose fasteners, and improper shaft installations. These scenarios reflect everyday plant conditions rather than hypothetical adversarial intents.

To our knowledge, this is the first work to systematically evaluate the robustness of deep learning models under real-world physical disturbances caused by human error in a ventilator motor system. Our framework provides a reproducible benchmark combining sensor-level perturbation with multi-feature input (vibration, current, temperature) and introduces physical realism into the adversarial robustness discourse. Moreover, in this work, the Mean Confusion Impact Index (MCII) is introduced, referring to a novel and simple robustness metric that aims to quantify the average misclassification confidence of models under adversarial physical attacks.

3. Materials and Methods

3.1. Proposed Methodology

This work aims to investigate the vulnerability of deep learning-based industrial fault diagnosis systems to real-world physical adversarial attacks and to propose a practical defense strategy. To this end, the study is guided by the following overarching research questions:

- **RQ1: How do real-world physical disturbances affect the reliability of AI-based fault diagnosis systems in industrial environments?** This question seeks to empirically evaluate how subtle physical interventions (e.g., sensor misalignment, mechanical imbalance) affect the predictive behavior and classification confidence of deep learning models under real production conditions.
- **RQ2: What defense strategy can enhance the resilience of such systems against adversarial conditions?** This question examines the effectiveness of modular, transfer-learning-based defense mechanisms in mitigating adversarial confusion and whether such architectures can maintain high diagnostic integrity under tampered scenarios.

The research roadmap adopted in this work is structured around a two-phase robustness framework that targets the evaluation and defense of deep learning-based fault diagnosis systems against real-world physical adversarial attacks, as illustrated in Figure 2.

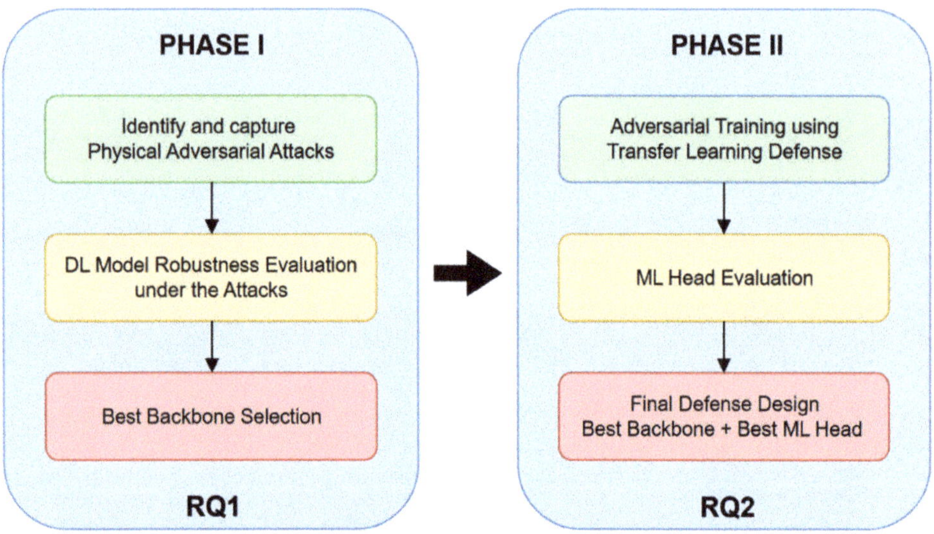

Figure 2. The proposed two-phase methodology framework block diagram.

PHASE I: In the first phase, we apply five physically grounded perturbations—designed to emulate common operator-induced mechanical anomalies—on an operational industrial ventilation system and systematically evaluate the response of state-of-the-art neural network models. In our previous work [41], we benchmarked various DL architectures on their ability to classify fault types across four operational air ventilators embedded in an active fiber production line. The study aimed to identify the most robust and accurate models for detecting faults such as bearing damage, impeller unbalancing, and support cracking, even under fluctuating operating conditions. The outcomes support the feasibility of deploying DL models for industrial fault detection beyond laboratory environments, thereby bridging the gap between academic research and production-level predictive maintenance. Therefore, the best four pretrained architectures previously validated for their diagnostic accuracy based on the results of [41], namely ResNet50-1D, CNN-1D, BiLSTM, and BiLSTM with Attention, are exposed to real attack scenarios such as sensor misalignment, loose mounting bolts, shaft eccentricity, and artificial rotor unbalancing. The aim is to assess which model exhibits the highest natural resilience under these perturbations by quantifying their vulnerability using the Mean Confusion Impact Index (MCII), a metric designed to measure how confidently each model misclassifies adversarial inputs as legitimate fault conditions. The model with the combination of the best fault diagnostic performance and lowest Mean Confusion Impact Index is selected as the most robust backbone candidate for subsequent defense enhancement.

PHASE II: In the second phase, the selected DL backbone is repurposed via transfer learning to construct an adversarial resilient hybrid classifier. Instead of retraining the entire network, we extract high-level latent features from the penultimate layer of the frozen backbone and retrain only the classification head using labeled data collected under physical attack conditions. A series of lightweight machine learning models—such as Random Forest, LightGBM, SVM, and XGBoost—are tested as alternative classifier heads. These models are evaluated based on their ability to suppress adversarial confusion and preserve classification integrity across all attack scenarios. The final defense mechanism is defined as the combination of the most resilient backbone model from phase I with the most robust classifier head from phase II, thereby creating a hybrid architecture specifically optimized to maintain fault diagnosis accuracy even under real-world adversarial interferences.

This two-stage methodology uniquely integrates empirical robustness evaluation with modular defense design and provides a systematic pipeline for enhancing model reliability in industrial predictive maintenance settings. By focusing on physically plausible perturbations rather than synthetic noise, and by combining deep feature extraction with classical decision-making mechanisms, the proposed framework addresses both practical deployment concerns and theoretical robustness gaps in current industrial AI systems.

3.2. Experimental Setup

The experimental setup consists of a 7.5 kW VFD-driven motor coupled with a VVB001 type vibration sensor, installed on the housing of the ventilator system. The motor's operational parameters, specifically rotational speed (rpm) and current (amps), are monitored in real time via a WinCC SCADA system, which communicates with the PLC controlling the motor drive unit using the Modbus-TCP/IP protocol. Simultaneously, the VVB001 sensor records vibration acceleration (m/s^2) and temperature (°C), with data acquisition handled by the moneo software (version 1.16). To enable integrated monitoring and centralized storage, both WinCC and moneo exchange data through the OPC-UA protocol, while Microsoft Message Queuing Services (MSMQ) facilitates the reliable transmission of collected information from WinCC to the Process Historian Acquisition server. This architecture ensures synchronized logging of both electrical and mechanical signals, enabling comprehensive analysis. Historical data are exported in .xlsx format to support subsequent processing and machine learning model development. Figure 3 illustrates the total equipment and network setup on the ventilator motor, along with the data flow diagram.

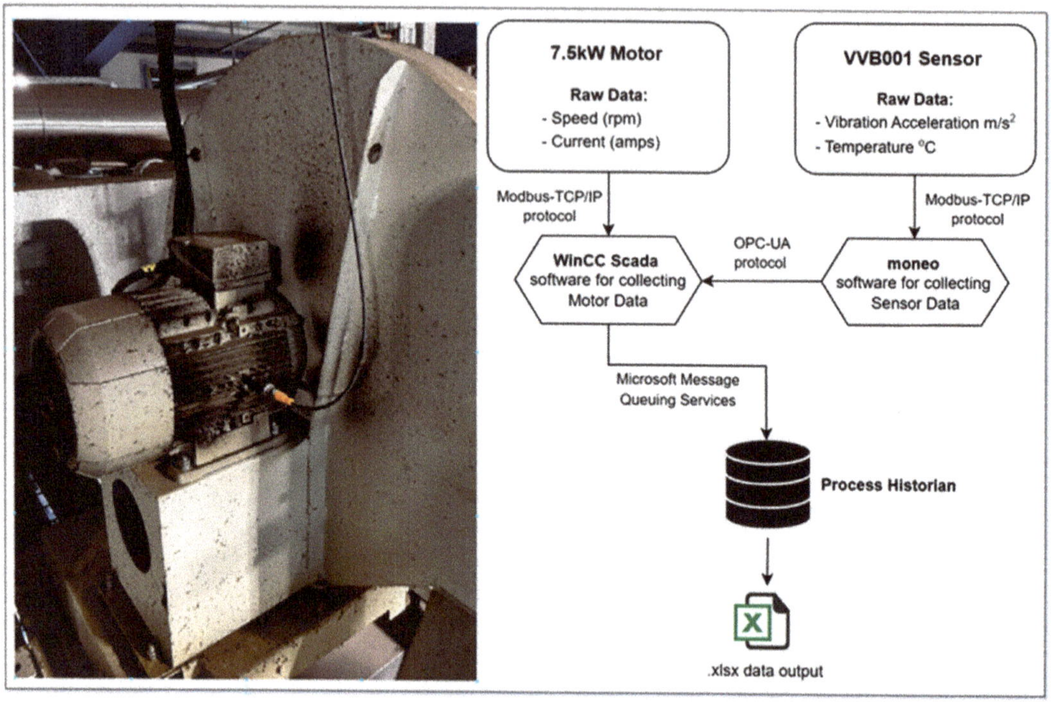

Figure 3. Equipment Setup and data flow diagram.

3.3. *Physical Attacks' Data Capturing*

A total of five different physical attacks were conducted on the ventilator motor, shown in Figure 4. The planning and execution of these interventions were carried out in collaboration with the plant's production and maintenance departments.

To enable a systematic investigation of adversarial robustness under real industrial conditions, a dedicated dataset was constructed during the execution of the five planned physical attack scenarios on the ventilator motor. Vibration signals were recorded in 4 s windows at 1 min intervals (25 kHz collection frequency), while additional operational parameters (motor speed, current, and temperature) were simultaneously logged through the integrated SCADA and sensor infrastructure (Figure 3). From the raw vibration signals, four key features (aRMS, aPeak, vRMS, Crest Factor) were extracted using the moneo software. Together with the logged operational parameters, these formed a set of seven representative features organized into a structured multivariate time series, aligned with the timeline of each attack and the corresponding normal operating states. The resulting dataset spans multiple days of scheduled production downtime interventions and constitutes a unique resource that combines both controlled adversarial physical perturbations and authentic industrial operating dynamics. The datasets of the attacks are openly available to the research community through our public repository, enabling transparent reproducibility and further advancements in industrial AI robustness research. A brief presentation of each physical attack scenario and the corresponding data-collection information is provided in the following (Figure 4):

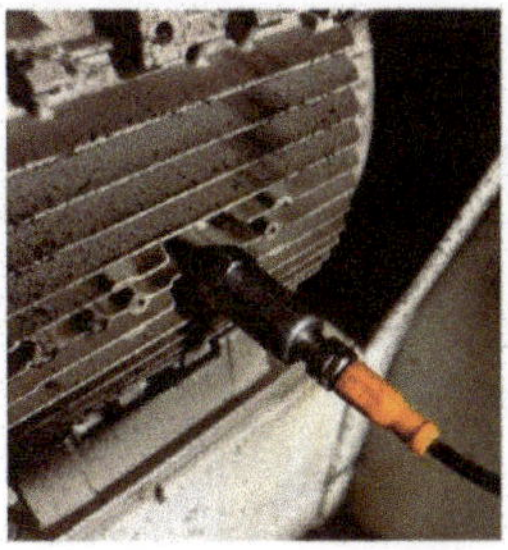

Period	2024/12/3 - 06:00 am 2024/12/18 - 05:41 am
Data	21.582 data points
Features	7 features
Interval	4-second samples aggregated to 1-min
Format	.xlsx files
Description	Deliberate displacement and angular offset of the vibration sensor

(**a**)

Period	2025/1/5 - 06:00 am 2025/2/1 - 05:59 am
Data	38,880 data points
Features	7 features
Interval	4-second samples aggregated to 1-min
Format	.xlsx files
Description	Untight screw of motor base foot

(**b**)

Period	2025/03/02 - 06:00 am 2025/03/30 - 05:04 am
Data	40.265 data points
Features	7 features
Interval	4-second samples aggregated to 1-min
Format	.xlsx files
Description	One of the four bolts securing the rear cover of the motor — where the drive-end bearing is housed — was intentionally removed.

(**c**)

Period	2025/02/02 - 06:00 am 2025/02/18 - 07:11 am
Data	23.112 data points
Features	7 features
Interval	4-second samples aggregated to 1-min
Format	.xlsx files
Description	A worn motor shaft was intentionally reassembled with a new bearing. A feeler gauge was used to assess the resulting clearance between the shaft and the bearing's inner ring, revealing a gap of 0.05 mm emulating an eccentric shaft fit attack.

(**d**)

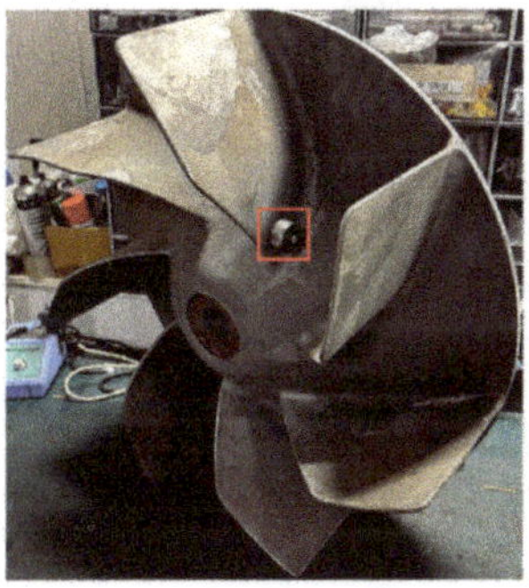

Period	2025/04/02 - 06:00 am 2025/04/19 - 12:33 pm
Data	24.874 data points
Features	7 features
Interval	4-second samples aggregated to 1-min
Format	.xlsx files
Description	A 92gr neodymium magnet was attached at the base of one blade of the motor impeller, to simulate an asymmetric mass distribution during rotation, thereby generating abnormal vibration patterns.

(**e**)

Figure 4. Application and data capturing of five physical attacks on the ventilator motor: (**a**) Misaligned Sensor; (**b**) Loose Motor Foot; (**c**) Loose Bearing Cap; (**d**) Eccentric Shaft fit; (**e**) Impeller Add Weight.

- **Misaligned Sensor:** The vibration sensor was deliberately repositioned away from its optimal mounting location. Specifically, the sensor was detached and reattached with a lateral offset of approximately 5 mm and a rotational misalignment of ~15° relative to its original axis, in order to introduce distortions in the captured signals (Figure 4a).
- **Loose Motor Foot:** One of the mounting bolts of the motor base was deliberately loosened. The bolt was unscrewed with a controlled torque to induce partial destabilization without complete detachment (Figure 4b).
- **Eccentric Shaft:** The shaft had sustained wear from a previously installed bearing inner ring. The old bearing had left a circumferential groove (shoulder/pit) on the shaft surface. A new bearing was subsequently mounted without restoring the original shaft geometry, resulting in an improper and eccentric fit. This led to non-uniform support

and slight misalignment between the shaft and bearing axis, producing asymmetric vibration patterns during operation (Figure 4c).

- **Loose Bearing Cap:** The bolts securing the bearing housing were partially unscrewed from their fully tightened position. This action reduced the preload on the bearing cover, causing minor mechanical instability without complete disassembly (Figure 4d).
- **Impeller Add Weight:** A small neodymium magnet was securely attached to the impeller at a location approximately 3 cm from the base of the blade. This eccentric mass placement introduced a controlled imbalance in the rotating system, altering its inertia and causing increased centrifugal loading. The modification was performed carefully to avoid mechanical damage, aiming to produce a detectable shift in the vibration profile. (Figure 4e).

Each physical attack was implemented during a corresponding scheduled 8 h production stop, which was allocated for spinnerets changeover operations. This ensured that the interventions could be carried out safely and without disrupting regular production (Figure 5).

Figure 5. Operation of spinneret changeover.

The design of the attack scenarios was guided by the most frequent faults recorded in the CMMS logs of the production line. In our previous study on ventilator fault diagnosis [41], three degradation modes—bearing failures, impeller unbalancing, and support cracking—were documented multiple times in the same equipment. Building on these findings, the adversarial interventions introduced in the present work were deliberately chosen to reproduce the vibration signatures of these recurrent degradation mechanisms, while remaining fully non-destructive and reversible. This reflects a necessary compromise: the attacks were applied during scheduled production stops in a way that avoided any risk of permanent machine damage yet still generated signals closely resembling the early manifestations of real industrial faults. In this way, the attack dataset complements the previously established fault dataset by bridging authentic long-term degradation and controlled adversarial manipulation, providing a meaningful benchmark for evaluating diagnostic model robustness.

From the raw 4 s vibration signal of the moneo sensor, four vibration-related features were extracted at 1 min intervals. These are part of a total of seven features, with the vibration features including:

- **Acceleration Root Mean Square (aRMS)**: Quantifies the overall vibration energy by averaging the squared acceleration values over time. It reflects the intensity of oscillatory motion in terms of acceleration. High aRMS indicates persistent mechanical stress, typically due to imbalance, misalignment, or progressive wear. It is sensitive to sustained abnormalities, rather than transient events.
- **Acceleration Peak (aPeak)**: Measures the maximum instantaneous acceleration within a given time window. It captures the strongest single vibration spike. High aPeak suggests the presence of sudden mechanical shocks, impacts, or momentary looseness, and it is highly sensitive to intermittent events that do not last long enough to significantly raise aRMS.
- **Velocity Root Mean Square (vRMS)**: Calculates the RMS of vibration velocity, a quantity that reflects how fast oscillatory motion occurs over time. It is especially relevant in the mid-frequency range of rotating machinery. It is a strong indicator of mechanical resonance, imbalance, and structural fatigue. It is often used to assess long-term degradation or imbalance conditions.
- **Crest Factor**: Crest Factor is the ratio of the peak acceleration to the RMS acceleration, representing the impulsiveness or spikiness of the vibration signal. High CF indicates sharp, transient disturbances, while low CF indicates smooth, consistent vibration, possibly due to steady-state imbalance.

The remaining three features are motor-related:

- **Motor Speed (rpm)** is included as a core input feature, since vibration response is inherently speed-dependent, providing any ML or DL model with RPM information enables it to learn fault-specific patterns across varying operating conditions, improving both accuracy and robustness in a real production environment. If the model does not "know" the motor speed, it cannot relate observed vibration frequencies to expected fault patterns under varying loads.
- **Motor Temperature (°C)** was included as a contextual feature to enrich the diagnostic capability of the model used. Although not a direct fault indicator in most cases, temperature provides valuable insight into thermal stress, load conditions, and the progression of mechanical wear. Its inclusion helps the model differentiate between vibration anomalies caused by faults and those driven by thermal operating conditions.
- **Motor Current (Amps)** was included as a final complementary feature reflecting the electromechanical interaction within the system. Mechanical faults such as imbalance, shaft misalignment, or bearing degradation often lead to increased or unstable current draw. Including current as a feature enables early detection of faults that impact torque or load characteristics and improves fault separability in conjunction with vibration-based features.

Figure 6 indicatively illustrates the vibration and motor load behavior of the seven features during the roughly 16 days "Eccentric Shaft Fit" attack for February.

This structured execution of five physical attack scenarios allowed us to induce real-world perturbations in the ventilator motor system and capture rich, feature-based datasets under controlled yet realistic conditions. These datasets serve as a critical foundation for evaluating how vulnerable our baseline classification models are to such physical interferences. In the following section, this experimental framework is used to rigorously

To assess their behavior under physical interference, first, each model is tested against real data collected during controlled physical attacks, including motor base loosening, sensor misalignment, and artificial unbalancing of the impeller. The core objective is to identify how often these physical attacks cause false classifications into actual fault categories, and to determine which architecture maintains the highest decision reliability. It should be emphasized that the confusion heatmaps presented in Figure 7 do not correspond to standard confusion matrices with a diagonal structure. This is because, at this stage, no retraining was performed on adversarial data; instead, the physical attack scenarios were directly tested on a pre-trained fault diagnosis model. Since attacks do not constitute genuine fault categories with explicit ground truths, no one-to-one mapping between rows and columns exists. As a result, the heatmaps illustrate how each perturbation is misinterpreted as an existing fault class rather than reporting correct-versus-incorrect classifications. This distinction underlines that the matrices capture attack-induced biases in prediction rather than conventional diagnostic accuracy. This approach provides a more meaningful evaluation of robustness, as it quantifies the degree to which attacks systematically bias predictions toward specific fault categories rather than measuring standard classification accuracy.

Next, Figure 8 illustrates the attack-to-fault misclassification heatmaps obtained when the four models are subjected to the five physical attack scenarios. It should be clarified that these are not typical confusion matrices of classification outcomes, but rather heatmaps specifically designed to illustrate the misclassifications induced by adversarial attack scenarios.

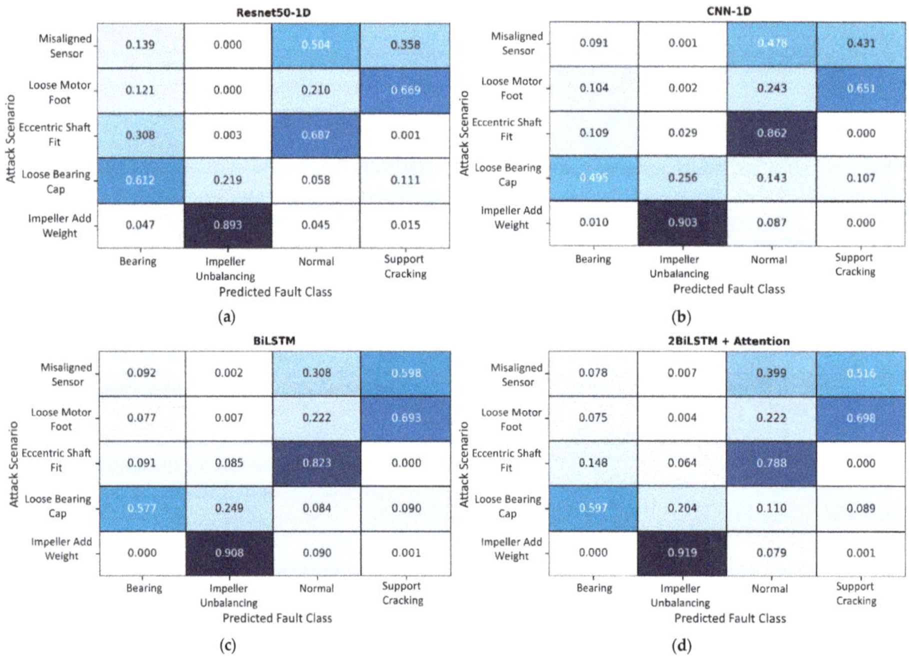

Figure 8. Attack-to-fault misclassification heatmaps of DL Models under five Physical Attack Scenarios (Misaligned Sensor → Normal, Loose Motor Foot → Support Cracking, Eccentric Shaft Fit→ Normal, Loose Bearing Cap → Bearing, Impeller Add Weight → Impeller Unbalancing): (**a**) ResNet50-1D; (**b**) CNN-1D; (**c**) BiLSTM; (**d**) BiLSTM with Attention. Color scale from white to dark blue represents the intensity of values from 0 to 1.

The normalized values in Figure 8 represent the proportion of predictions that each of the four DL models assigned to each fault class when subjected to the corresponding attack scenario. For example, on the Resnet50-1D fault classification model, a misaligned sensor is predominantly misclassified as Normal (50.4%) or Support Cracking (35.8%), indicating that sensor displacement closely mimics stable or structural loosening conditions. A loose motor foot is most frequently interpreted as Support Cracking (66.9%), reflecting the mechanical instability induced by base loosening. The eccentric shaft fit is overwhelmingly confused with Normal operation (68.7%), which suggests that this perturbation produces vibration signatures that resemble baseline conditions and can therefore escape detection. In contrast, a loose bearing cap is strongly predicted as a Bearing fault (61.2%), aligning with the mechanical stress imposed on the shaft-bearing interface. Finally, the added impeller weight is almost exclusively classified as Impeller Unbalancing (89.3%), since the induced imbalance produces vibration characteristics analogous to genuine impeller faults. These results collectively demonstrate that adversarial perturbations do not lead to random misclassifications but instead induce systematic biases toward specific fault classes, thereby exposing consistent vulnerabilities in the diagnostic decision process.

Overall, each attack-to-fault misclassification heatmap reveals the extent to which adversarial physical attacks mislead the four DL models into misclassifying attack scenarios as legitimate fault classes. Notably, CNN-1D and BiLSTM exhibit consistently high-confidence misclassifications—often assigning a dominant probability to a single fault class under attack. For example, the 'Impeller Add Weight' attack is almost entirely confused with Impeller Unbalancing across all models, indicating a critical vulnerability. The ResNet50-1D model, while slightly more balanced in its predictions across some attack types, still shows strong misclassification patterns, especially for 'Loose Bearing Cap'. The 2BiLSTM + Attention model appears marginally more diffused in its classification responses, particularly under 'Misaligned Sensor' and 'Eccentric Shaft Fit' attacks, where their predictions are more distributed across multiple classes, potentially reflecting higher robustness or lower certainty.

At this point, it is crucial to highlight that while all models suffer from varying degrees of adversarial confusion, their responses differ in confidence concentration, which motivates the need for a unified quantitative indicator to assess misclassification severity across architectures.

Previous works have introduced metrics such as n-MeRCI to assess the correlation between predicted uncertainty and error rates [42], and statistical models for confusion matrix uncertainty [43]. Inspired by these, we propose here the Mean Confusion Impact Index (*MCII*), a novel and simple robustness metric that measures the average misclassification confidence per model under adversarial physical attacks. While traditional confusion matrices offer qualitative insights into class-specific confusions, the *MCII* condenses the most confident misclassification per attack into a single interpretable percentage score. For a given normalized confusion matrix $\boldsymbol{C} \in \mathbb{R}^{\mathrm{nxm}}$ where each row corresponds to an attack type and each column to a predicted fault class, *MCII* is defined as:

$$MCII = \frac{1}{n} \times \sum_{i=1}^{n} max_{\mathrm{j}}\, \mathrm{C_{ij}} \tag{1}$$

where

- n = total number of attack scenarios;
- i = index of the attack scenario (row of the confusion heatmap);
- j = index of the predicted fault class (column of the confusion heatmap);
- $\mathrm{C_{ij}}$ = normalized proportion of predictions where attack scenario i is classified as fault class j;

- max_j C_{ij} = the maximum misclassification confidence for attack *i*, i.e., the fault class in which the model most strongly confuses that attack.

This captures the average strength with which a model confuses an attack scenario with a real fault, i.e., the degree to which the model is confidently deceived. A higher *MCII* indicates that the model is highly vulnerable, confidently mapping attacks to specific fault classes. Conversely, a lower *MCII* reveals a more robust model that exhibits higher uncertainty or distributes its predictions across multiple fault categories when facing attack inputs.

Although a variety of robustness indicators have been proposed in the literature, including entropy-based uncertainty scores [44], margin-based confidence gaps [45], calibration measures such as expected calibration error (ECE) [46], and attack success rate (ASR) [47], each has limitations in the context of industrial diagnostics. Entropy and margin metrics quantify how confident or uncertain a model is, but they do not reveal where the decision error is directed. Calibration errors provide insight into overall probability reliability but lack class-specific resolution. Similarly, ASR reduces adversarial outcomes to a binary success/failure metric, overlooking systematic misclassification patterns.

MCII addresses this gap by capturing the average confidence with which adversarial perturbations are mapped into specific fault classes. For example, in our experiments, the Impeller Add Weight attack was consistently misclassified as Impeller Unbalancing with an *MCII* of 0.89, indicating not just that the model failed, but that it did so in a systematic and confident manner towards one fault category. In contrast, the Misaligned Sensor attack produced lower *MCII* values (≈0.50), reflecting weaker but still directional bias toward the Normal class. This fault-oriented perspective is particularly relevant in predictive maintenance: while uncertainty scores indicate that a model is "less certain," *MCII* explicitly shows whether attacks mislead the system into confident but incorrect diagnoses, potentially triggering costly or unnecessary interventions. Thus, *MCII* complements entropy- and margin-based robustness indicators by providing a more interpretable, application-driven metric that links adversarial misclassifications directly to fault categories of industrial concern.

Table 2 summarizes both the baseline diagnostic performance and the behavior under physical attack conditions, for all models. The original performance metrics—including accuracy, precision, recall, and F1-score—are derived from the earlier evaluation conducted in the study by Polymeropoulos et al. [41], where ResNet50-1D emerged as the top-performing model in terms of classification accuracy on fault data.

Table 2. Summary of original performance and *MCII* for DL models under attack.

Model	Acc%	Loss%	Prec%	Rec%	F1%	MCII%
ResNet50	97.8	3.7	97.5	97.8	97.6	67.3
CNN-1D	97.3	8.4	98.0	94.8	96.4	67.8
BiLSTM	97.8	6.0	98.8	95.5	97.1	70.0
2BiLSTM+Attention	97.6	5.6	98.9	95.5	97.1	71.7

Performance metrics of original faults diagnosis (green values); Mean Confusion Impact Index of 5 Attacks (blue values).

In this work, we extend that evaluation by introducing *MCII* metric, to capture the average misleading impact of each physical attack on the model's predictions. The results indicate that although ResNet50-1D continues to demonstrate excellent diagnostic performance, it also maintains the lowest *MCII*, suggesting better resilience under adversarial physical conditions compared to the other models.

To complement the metrics reported in Table 2, Figure 9 presents a consolidated line graph plot that visualizes the mean confusion ratio per batch observed across all attack scenarios for each evaluated model. This metric reflects, for each incoming batch of 512 adversarial data points, the extent to which predictions are skewed toward a single dominant fault class, a phenomenon indicative of model confusion caused by physical interference.

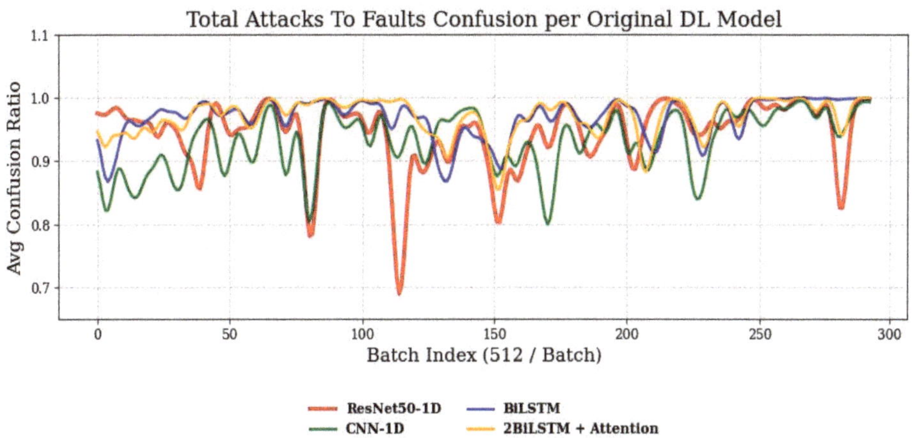

Figure 9. Batch-Wise Total Attack Confusion Trends Across the Four DL Models.

Notably, the red curve (ResNet50-1D) and the blue curve (CNN-1D) maintain the lowest average confusion levels per batch, indicating a higher degree of decision consistency even when exposed to adversarial perturbations. Conversely, BiLSTM and BiLSTM+Attention models exhibit steeper or more volatile confusion ratios, implying that their predictions are more easily destabilized by physical attacks.

The visual results support our findings of Table 2 and confirm that ResNet50-1D not only achieved top classification accuracy under normal conditions, in line with [41], but also exhibits greater robustness under adversarial interference, as evidenced by both its low *MCII* score and its batch-level confusion stability. Therefore, even though ResNet50-1D exhibits a significantly high confusion, it is identified as the most promising backbone architecture for developing a further defense mechanism against physical attacks on our ventilator motor.

5. Defense Strategy

5.1. Adversarial Training by Transfer Learning

Adversarial training is a widely recognized defense mechanism in the field of robust machine learning, originally developed to improve model performance against adversarial examples in image classification tasks [48]. The core idea involves incorporating adversarial perturbed inputs during training so that the model learns to resist deceptive alterations at inference time. While this strategy has proven effective in digital adversarial settings, its application in real-world physical attack scenarios—especially in industrial domains—remains relatively underexplored.

In this work, we adopt an adversarial training strategy tailored to physical attacks by leveraging transfer learning on a previously trained fault classification model. Specifically, we utilize ResNet50-1D architecture, which was validated as the most effective model for fault diagnosis in our system. Rather than retraining the entire network, we freeze the convolutional backbone and extract high-level feature representations from vibration signal windows. On top of this frozen backbone, we retrain only the classification head using labeled data collected under physical attack conditions.

This transfer learning approach offers multiple advantages. First, it allows for efficient adaptation to new data distributions (e.g., those induced by physical interference), without requiring large-scale retraining [49]. Second, by decoupling feature extraction from classification, we can test multiple lightweight machine learning classifiers (such as Random Forest, SVM, LightGBM, and XGBoost) as the head model [50]. This enhances flexibility and interpretability, while also improving robustness through ensemble-based or margin-based learning mechanisms.

Such hybrid architectures, combining deep feature extraction with classical machine learning heads, have recently gained traction as effective defense frameworks in adversarial learning contexts [51], particularly when labeled adversarial data are available for supervised retraining. Our proposed method thus represents a domain-specific adaptation of adversarial training via transfer learning, focused on improving resilience in real-world industrial fault classification systems affected by physical tampering.

Regarding the motivation for classifiers' selection, first, Logistic Regression was included as a baseline model. Despite its simplicity, it remains widely used for benchmarking in classification tasks because of its interpretability, computational efficiency, and ability to provide a transparent reference point against which the performance of more advanced models can be evaluated [52]. Next, the choice of Random Forest, LightGBM, XGBoost, and SVM was motivated by their complementary strengths in robustness and interpretability when applied to deep feature representations. Random Forest provides ensemble-based stability and resistance to overfitting [53,54], which is valuable under noisy or perturbed inputs. LightGBM and XGBoost, as gradient-boosting frameworks, are well known for their ability to capture complex, non-linear decision boundaries with high accuracy and computational efficiency [55,56]. SVM, in contrast, offers a margin-based learning mechanism that emphasizes generalization and robustness, particularly under limited or imbalanced training data [57]. By evaluating these diverse lightweight classifiers on top of the ResNet50-1D feature extractor, our study ensured that both ensemble-based and margin-based paradigms were represented, thereby enabling a balanced and fair assessment of which head model best suppresses adversarial confusion while maintaining diagnostic accuracy.

5.2. *Confusion Matrices, Heatmaps, and Evaluation Metrics*

To evaluate the effectiveness of the proposed defense strategy based on transfer learning, we first visualize the confusion matrices resulting from the application of five different machine learning classifiers (Logistic Regression, SVM, Random Forest, LightGBM, and XGBoost) as classification heads on top of the ResNet50-1D feature extractor. These confusion matrices represent how each model interprets vibration signals under physical attacks and highlight the degree to which attacks are still misclassified as real fault conditions.

After adversarial training, the evaluation is performed on an augmented dataset that contains both genuine fault conditions [41] and adversarial examples, each with explicit ground-truth labels. Unlike the attack-to-fault heatmaps—where no diagonal could be formed because attacks are not themselves fault classes—the presence of labeled instances for both faults and attacks enables the construction of standard confusion matrices with a meaningful diagonal. In this representation, the diagonal elements reflect correct classifications, while the off-diagonal entries indicate misclassifications between fault and attack categories. This shift marks a crucial step: moving from illustrating attack-induced misinterpretations on a pre-trained model to rigorously evaluating diagnostic accuracy and robustness in a retrained adversarial defense setting. Figure 10 presents a total of five confusion matrices (one per ML classifier).

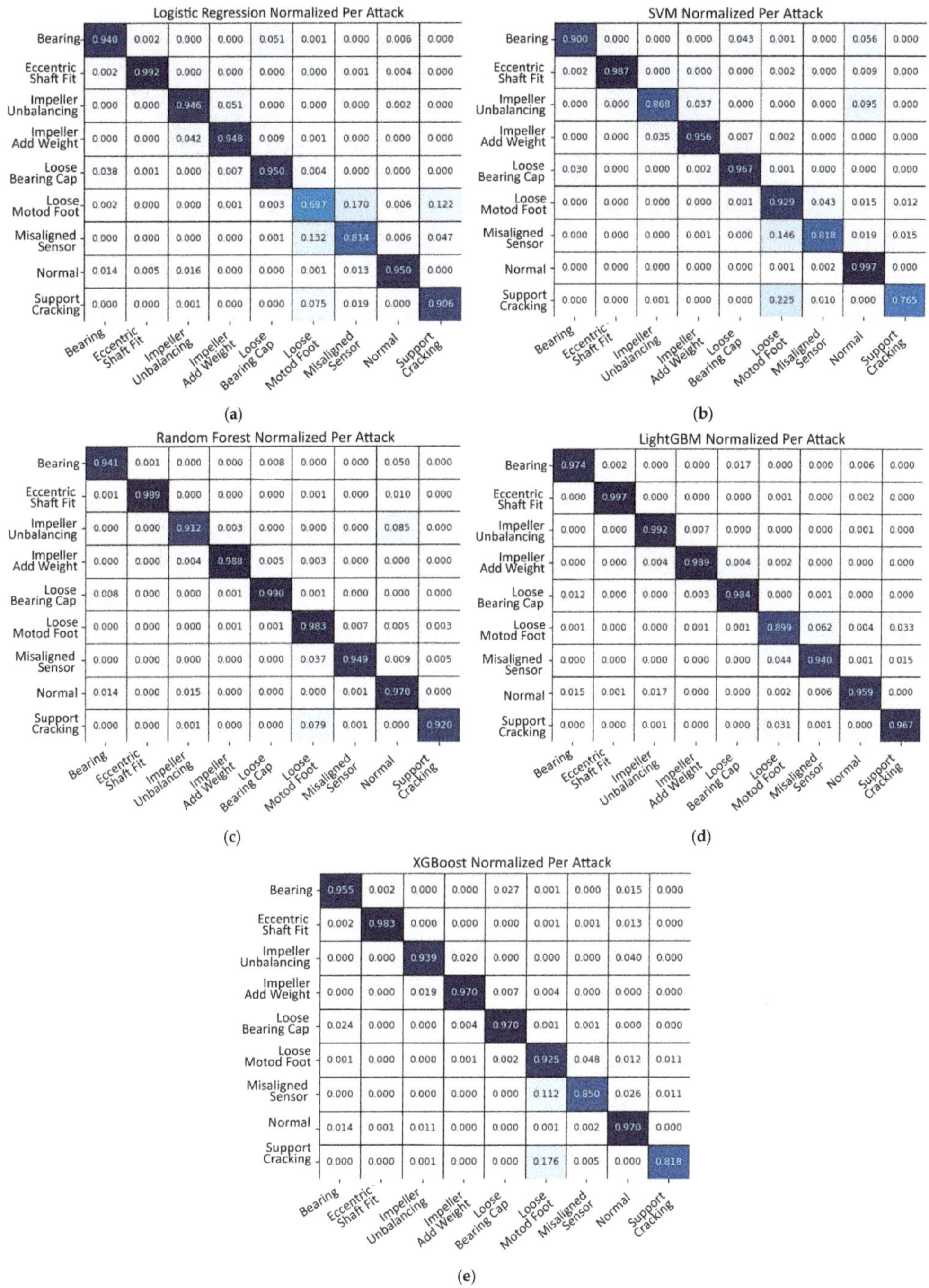

Figure 10. Normalized confusion matrices for five ML classifier architectures under physical attacks: (**a**) Logistic Regression; (**b**) SVM; (**c**) Random Forest; (**d**) LightGBM; (**e**) XGBoost. Color scale from white to dark blue represents the intensity of values from 0 to 1.

We observe that the ResNet50-1D + Random Forest and ResNet50-1D + LightGBM combinations produce the clearest and most diagonal-dominant matrices, indicating a strong ability to distinguish between real faults and adversarial distortions. In contrast, models such as ResNet50-1D + Logistic Regression and ResNet50-1D + SVM show higher confusion, particularly between impeller-related attacks and actual unbalancing faults, a trend previously observed in the standalone ResNet50-1D evaluation. This difference can be explained by the intrinsic properties of ensemble-based methods. Logistic Regression and SVM rely on a single global decision boundary, making them more sensitive to distribution shifts caused by adversarial perturbations [52,57]. In contrast, Random Forest reduces variance by averaging across multiple decision trees, while LightGBM adaptively reweights misclassified samples and captures complex non-linear interactions [53,56]. These mechanisms dilute the effect of localized perturbations, which explains their greater robustness in adversarial scenarios.

These qualitative insights are quantitatively supported by the summary metrics shown in Table 3. The models are evaluated based on classification accuracy, precision, recall, F1-score, and our newly introduced uncertainty metric *MCII*, which captures the average probability mass misallocated to incorrect fault classes under attack conditions.

Table 3. Performance + Attack Uncertainty metrics for all ML models.

Model	Accuracy%	Precision%	Recall%	F1-Score%	MCII%
ResNet50-1D + Log. Reg.	91.78	86.08	90.48	87.72	4.88
ResNet50-1D + SVM	94.81	93.94	90.97	92.31	2.30
ResNet50-1D + RF	96.53	96.78	96.00	96.37	0.14
ResNet50-1D + LGBM	96.30	93.94	96.68	95.21	0.90
ResNet50-1D + XGB	95.25	93.88	93.13	93.46	1.37

Notably, ResNet50-1D + Random Forest achieves the highest accuracy (96.53%) with a near-zero MCII score (0.14), followed closely by ResNet50-1D + LightGBM (96.30% accuracy and 0.90 MCII). This suggests that both models are highly effective at absorbing the impact of adversarial perturbations, preserving decision reliability even under physically tampered conditions.

Based on both confusion matrices and aggregate metrics, we focus our subsequent analysis on these two top-performing classifiers—Random Forest and LightGBM—to determine the most robust and practically viable defense mechanism. To further compare the resilience of the two top-performing classifiers, we analyzed in parallel the confusion matrices and batch-level confusion linear progression for each physical attack scenario (Figure 11).

More specifically, Figure 11 illustrates five representative attack-to-fault confusion pairs, one per attack type, before and after the application of each defense model. Each row includes a visualization of how the respective classifier misclassifies attack samples over time, allowing us to observe momentary peaks of confusion or stability across hundreds of batches.

From the graphs, a consistent pattern emerges: Random Forest as head (green curves on the left) exhibits superior robustness, maintaining almost zero confusion across all attack scenarios, regardless of the number of batches. In contrast, LightGBM as head (green curves on the right), although competitive, displays notable spikes in confusion ratios in multiple cases—particularly under attacks such as Loose Motor Foot (b)2 and Loose Bearing Cap (d)2. These fluctuations highlight occasional vulnerabilities in LightGBM's decision surface when faced with specific physical disturbances.

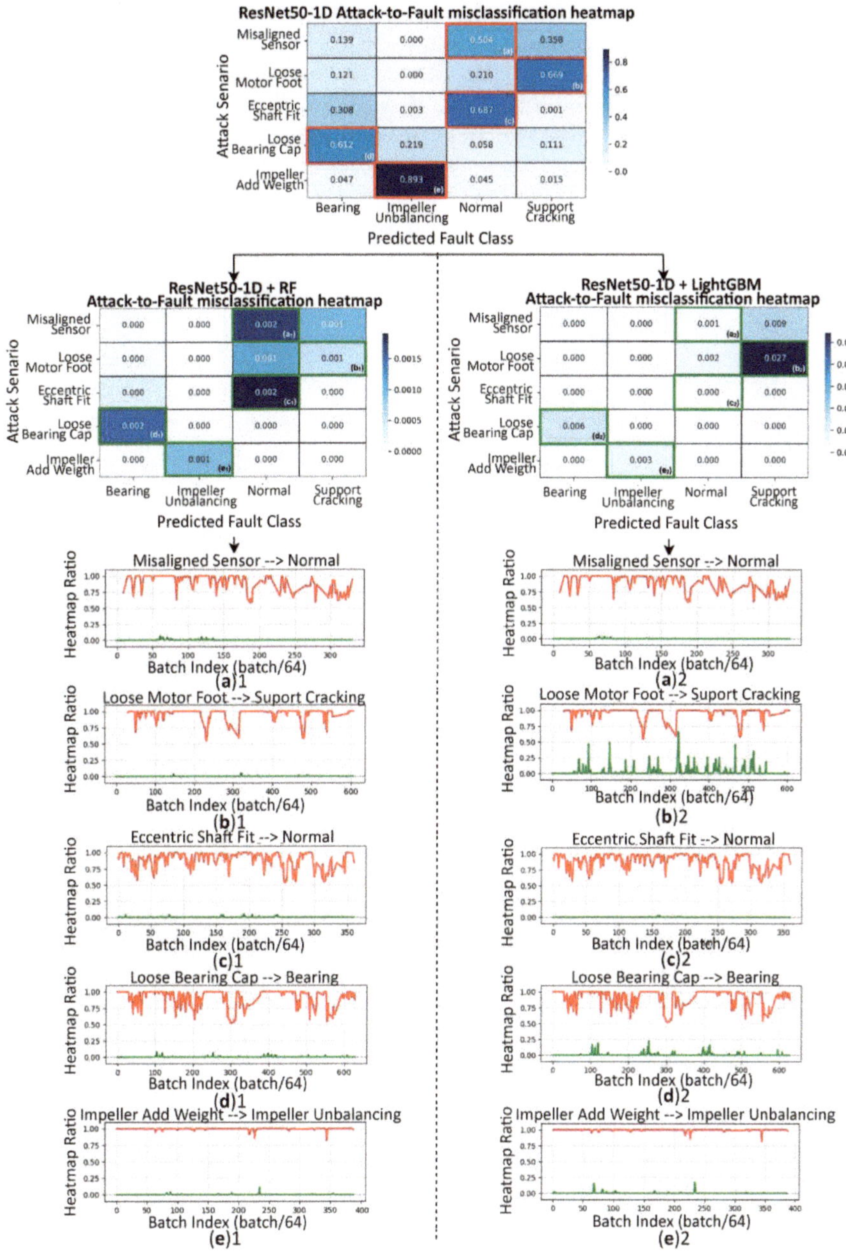

Figure 11. End-to-End Evaluation of Defence Effectiveness RF vs. LightGBM compared to original ResNet50-1D misclassification heatmap highlighting five major attack-induced misclassifications: For RF: (**a**)1 Misaligned Sensor → Normal; (**b**)1 Loose Motor Foot → Support Cracking; (**c**)1 Eccentric Shaft Fit → Normal; (**d**)1 Loose Bearing Cap → Bearing; (**e**)1 Impeller Add Weight → Impeller Unbalancing. For LightGBM: (**a**)2 Misaligned Sensor → Normal; (**b**)2 Loose Motor Foot → Support Cracking; (**c**)2 Eccentric Shaft Fit → Normal; (**d**)2 Loose Bearing Cap → Bearing; (**e**)2 Impeller Add Weight → Impeller Unbalancing. Red lines refer to attack heatmap ratio of original model while green lines refer to attack heatmap ratio after defence.

Finally, to complement the individual case-by-case confusion analyses, we further consolidate the results into a single comparative plot that illustrates the average heatmap ratio per batch across all five physical attacks (Figure 12). This final aggregated visualization provides a temporal overview of how each classifier (RF and LGBM) performs under continuous adversarial input over time. By collapsing all attack-fault heatmap events into a unified metric, we aim to capture the overall stability and resilience of each defense strategy throughout the entire evaluation horizon.

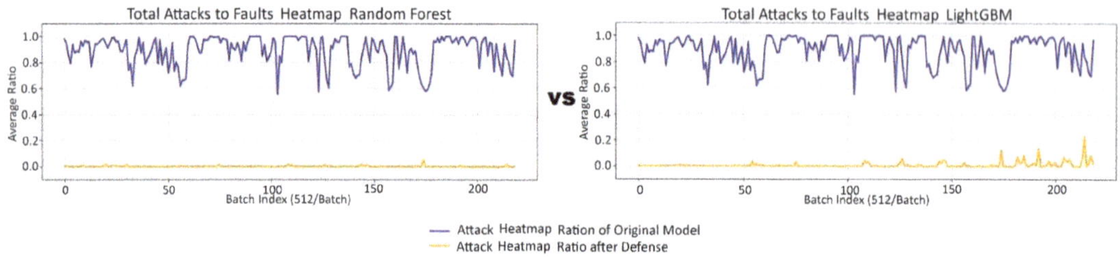

Figure 12. Average attack-to-fault heatmap ratio for the original pretrained ResNet50-1D model compared to the defended versions with Random Forest and LightGBM.

Concluding, multiple classifiers were trained and assessed on the Resnet50-1D backbone to determine their ability to resist physical attacks that cause confusion between real faults and adversarial perturbations. Through comparative analysis of heatmaps, average misclassification scores, and batch-level robustness plots, RF and LGBM emerged as the two most promising candidates. However, analyzing the individual and aggregated attacks to fault confusion graphs, RF consistently achieved the lowest confusion impact predictions, making it the most suitable and practical model for deployment in our physical defense framework. This lays the foundation for building an integrated and robust predictive maintenance system that remains resilient under adversarial physical interference.

The experimental results highlight the vulnerabilities of deep learning–based fault diagnosis models when subjected to physical adversarial attacks in real industrial environments. The attack-to-fault heatmaps demonstrated that perturbations do not cause random errors but instead induce systematic biases toward specific fault categories. For instance, mechanical disturbances such as loose bearing caps or added impeller weights consistently shifted predictions toward bearing-related and impeller-related faults, respectively. These findings reveal that physical attacks can exploit domain-specific failure signatures, making them particularly deceptive to diagnostic models.

The evaluation of adversarial training strategies further confirmed that robustness can be significantly improved by retraining lightweight classifier heads on adversarial enriched datasets. Among the tested methods, ensemble-based models such as Random Forest and LightGBM exhibited superior resilience, reducing attack-induced confusion while maintaining high diagnostic accuracy on genuine faults.

The confusion matrices, heatmaps, and performance metrics validated that these models are capable of simultaneously preserving reliability in fault detection and mitigating vulnerabilities to adversarial perturbations. Overall, the discussion underscores the dual challenge of ensuring both diagnostic accuracy and robustness. Our experimental findings provide evidence that while deep feature extractors such as ResNet50-1D retain strong representational capacity, their resilience critically depends on the choice of classifier head and the inclusion of adversarial data during training.

In addition to evaluating classification accuracy and robustness under adversarial perturbations, it is crucial to assess the computational cost of different classifier heads. In practical industrial environments, diagnostic models are expected not only to deliver reliable fault detection but also to operate within strict resource constraints, such as limited memory, training time, and real-time inference requirements. Models with high accuracy but excessive computational demands may become impractical for deployment on embedded systems or production lines where fast decision-making and efficient resource utilization are essential.

To capture these aspects, four complementary indicators were selected:

- **Training time** reflects the computational expense of preparing each classifier for deployment, highlighting how feasible it is to retrain models when new data becomes available [58].
- **Inference latency time** measures the time required to obtain predictions on the test set, directly linking to the responsiveness of the system in real-world operation [59].
- Complementing this, **inference throughput** expresses how many samples can be processed per second, which is critical in environments with high-frequency sensor data and large-scale monitoring tasks [60].
- Finally, **model size** quantifies the storage footprint of the trained classifier, determining whether it can be efficiently deployed on devices with limited memory, such as PLCs or embedded controllers [61].

Therefore, in Table 4, we systematically compare the training time, inference latency, inference throughput, and model footprint of all classifiers integrated with the ResNet50-1D backbone. This analysis provides a comprehensive view of computational efficiency, complementing the robustness evaluation, and highlights the trade-offs between diagnostic performance and the constraints imposed by potentially limited computational resources in real deployment scenarios.

Table 4. Comparison of four computational cost indicators across classifier heads.

Classifier Head	Training Time (s)	Inference Latency (s/Test)	Throughput (Samples/s)	Model Size
Random Forest	112.65	0.2858	280,220	276.15 MB
XGBoost	26.10	0.3239	247,291	3.10 MB
LightGBM	21.13	1.9187	41,741	3.00 MB
Logistic Regression	143.18	0.0252	3,174,606	8.17 KB
SVM	4435.05	518.81	154.4	39.18 MB

The comparison of computational cost across all classifier heads reveals clear trade-offs between robustness and efficiency. Ensemble-based methods (Random Forest, XGBoost, and LightGBM) demonstrated the most favorable balance, combining high diagnostic performance with acceptable training and inference costs. LightGBM, in particular, emerged as the most resource efficient candidate due to its fast training cycle and compact model size, while Random Forest, although the most robust, was significantly more resource-intensive. Logistic Regression achieved extreme efficiency in inference speed and footprint, but at the expense of robustness, whereas SVM proved computationally prohibitive, with very high training and inference times, which shows significant limitations in its applicability.

6. Discussion, Limitations, and Future Work

The findings of this study highlight a critical but often overlooked vulnerability in industrial AI deployments: the susceptibility of condition monitoring systems to hardware-level adversarial interference. By demonstrating both the risk and a viable mitigation strategy under real production constraints, this work provides a practical foundation for enhancing the trustworthiness of predictive maintenance systems in smart manufacturing environments.

While the proposed framework provides a novel and practical step toward enhancing the robustness of vibration-based fault diagnosis under physical adversarial attacks, several limitations should be acknowledged. The experimental evaluation was carried out on a single ventilator motor in one production line, which constrains the generalizability of the findings to other machines, fault types, or industrial domains.

In addition, the defense strategy relies on a transfer learning paradigm with a frozen ResNet50-1D backbone and retrained lightweight classifiers. Although this modular design offers efficiency and ease of deployment, its adaptability may be limited if feature distributions change significantly under different operating conditions.

Another consideration concerns the *MCII* metric, which, despite providing an interpretable measure of attack-induced confusion, should be regarded as a preliminary indicator. *MCII* does not account for temporal dynamics, uncertainty calibration, or gradual degradation effects, and no comparison with applying alternative robustness metrics was included, limiting the comprehensiveness of the evaluation.

This study also focused on supervised adversarial training, which assumes the availability of labeled attack data. In practice, such labels are costly and time-consuming to obtain, potentially constraining scalability in industrial settings.

A further limitation of this study is that the experiments were conducted with fixed pretrained backbones and a single training run per classifier head. Consequently, variance or standard deviation across multiple independent runs was not reported. This restriction was not only methodological but also practical: all adversarial interventions had to be carried out during scheduled production downtimes, and repeating them extensively would have caused unacceptable disruption to the manufacturing process. While stochastic effects were partially controlled by fixing random seeds and maintaining consistent training protocols, it is acknowledged that repeated experiments could provide additional insight into variability induced by random initialization or sampling. Future work should therefore consider larger-scale experimental campaigns with repeated runs, enabling confidence intervals to be reported alongside mean performance metrics and thus offering a more comprehensive statistical characterization of robustness.

Taken together, these limitations point to opportunities for future research, including validation across multiple ventilators and industrial contexts, benchmarking *MCII* against complementary robustness metrics, exploring unsupervised and domain-adaptive strategies to improve robustness under realistic production constraints, incorporating uncertainty-aware decision layers, and integrating anomaly detection mechanisms for unsupervised attack detection.

Future work would expand on this foundation by exploring additional physical attack scenarios under varying operational conditions, incorporating domain adaptation techniques to generalize robustness across different machines and environments, and investigating hybrid defense strategies that combine adversarial training with real-time anomaly detection. Furthermore, releasing the constructed dataset openly to the research community will enable reproducibility and foster comparative studies, supporting long-term progress in securing industrial AI systems.

7. Conclusions

This work has explored the resilience of deep learning-based fault diagnosis systems under real-world physical adversarial attacks in an industrial setting. Building upon our previous research—where ResNet50-1D was identified as the most accurate model for classifying common mechanical faults in centrifugal ventilators—we extended the evaluation to scenarios involving deliberate physical perturbations that mimic fault conditions without corresponding to true machine failures. These perturbations were designed to compromise the reliability of vibration-based fault classifiers by altering sensor input characteristics. Our experiments demonstrated that even high-performing DL models exhibit significant vulnerabilities under such conditions, often misclassifying attacks as legitimate faults with high confidence. To quantify this effect, we introduced the Mean Confusion Impact Index (MCII), a novel and simple robustness metric, which revealed the extent of decision compromise for each model.

Among the four evaluated architectures, ResNet50-1D consistently achieved the lowest *MCII*, followed by a clear visual distinction on a batch-wise level between the attack-fault pairs, confirming its superior robustness and positioning it as a suitable backbone for further defense development. Moreover, we proposed a hybrid defense strategy that couples a frozen ResNet50-1D feature extractor with retrained classification heads using adversarial labeled data. Through systematic evaluation across multiple classifiers, Random Forest and LightGBM emerged as the most effective defenses—reducing attack-to-fault confusion while maintaining high classification accuracy. Notably, Random Forest exhibited minimal confusion volatility in batch-wise temporal plots, demonstrating its strong resilience to physical tampering.

Author Contributions: Conceptualization, G.A.P.; methodology, G.A.P., S.B. and. I.P.; investigation, S.B. and I.P.; software, S.B. and I.P.; resources, S.B. and I.P.; writing—original draft preparation, S.B., I.P. and E.V.; writing—review and editing, S.B., I.P., E.V. and G.A.P.; visualization, G.A.P.; supervision, G.A.P. All authors have read and agreed to the published version of the manuscript.

Funding: This research received no external funding.

Data Availability Statement: The original data presented in the study are openly available in GitHub repository at https://github.com/MachineLearningVisionRG (accessed on 10 September 2025).

Acknowledgments: This work was supported by the MPhil program "Advanced Technologies in Informatics and Computers", hosted by the Department of Informatics, Democritus University of Thrace, Kavala, Greece. The authors would also like to express sincere gratitudes to their colleagues at Thrace Nonwovens & Geosynthetics for their valuable contributions and support throughout this work, particularly Sotiria Arampatzoglou (Process Engineer), Konstantinos Tsaousidis (Head of Manufacturing) and Diogenis Vasileiadis (Project Manager).

Conflicts of Interest: The authors declare no conflicts of interest.

Abbreviations

The following abbreviations are used in this manuscript:

MCII	Mean Confusion Impact Index
AI	Artificial intelligence
DL	Deep learning
FGSM	Fast Gradient Sign Method
PGD	Projected Gradient Descent
CNNs	Convolutional Neural Networks
LSTMs	Long Short-Term Memories
SAEM	Spectrogram-Aware Ensemble Method
ICS	Industrial control systems
FTEP	Tennessee Eastman Process
GRU	Gated Recurrent Unit
MLP	Multi-Layer Perceptron
MSMQ	Microsoft Message Queuing Services
aRMS	Acceleration Root Mean Square
aPeak	Acceleration Peak

vRMS Velocity Root Mean Square
CF Crest Factor

References

1. Salem, K.; AbdelGwad, E.; Kouta, H. Predicting Forced Blower Failures Using Machine Learning Algorithms and Vibration Data for Effective Maintenance Strategies. *J. Fail. Anal. Prev.* **2023**, *23*, 2191–2203. [CrossRef]
2. Maher, Y.; Danouj, B. Survey on Deep Learning Applied to Predictive Maintenance. *Int. J. Electr. Comput. Eng.* **2020**, *10*, 5592. [CrossRef]
3. Chakraborty, S.; Krishna, R.; Ding, Y.; Ray, B. Deep Learning Based Vulnerability Detection: Are We There Yet? *IEEE Trans. Softw. Eng.* **2022**, *48*, 3280–3296. [CrossRef]
4. Apostolidis, K.D.; Gkouvrikos, E.V.; Vrochidou, E.; Papakostas, G.A. Traffic Sign Recognition Robustness in Autonomous Vehicles Under Physical Adversarial Attacks. In *Cutting Edge Applications of Computational Intelligence Tools and Techniques. Studies in Computational Intelligence*; Daimi, K., Alsadoon, A., Coelho, L., Eds.; Springer: Cham, Switzerland, 2023; pp. 287–304.
5. Jia, W.; Lu, Z.; Zhang, H.; Liu, Z.; Wang, J.; Qu, G. Fooling the Eyes of Autonomous Vehicles: Robust Physical Adversarial Examples Against Traffic Sign Recognition Systems. In Proceedings of the 2022 Network and Distributed System Security Symposium, San Diego, CA, USA, 24–28 April 2022; Internet Society: Reston, VA, USA, 2022.
6. Jia, Y.; Poskitt, C.M.; Sun, J.; Chattopadhyay, S. Physical Adversarial Attack on a Robotic Arm. *IEEE Robot. Autom. Lett.* **2022**, *7*, 9334–9341. [CrossRef]
7. Nour, A.A.; Mehbodniya, A.; Webber, J.L.; Bostani, A.; Shah, B.; Ergashevich, B.Z. Optimizing Intrusion Detection in Industrial Cyber-Physical Systems through Transfer Learning Approaches. *Comput. Electr. Eng.* **2023**, *111*, 108929. [CrossRef]
8. Dionisopoulos, N.; Vrochidou, E.; Papakostas, G.A. Machine Learning Robustness in Predictive Maintenance Under Adversarial Attacks. In Proceedings of the Congress on Control, Robotics, and Mechatronics, Rajasthan, India, 25–26 March 2023; pp. 245–254.
9. Thrace Group Thrace Nonwovens; Geosynthetics, S.A. Available online: https://www.thracegroup.com/cz/en/companies/thrace-ng/ (accessed on 11 May 2025).
10. Naqvi, S.M.A.; Shabaz, M.; Khan, M.A.; Hassan, S.I. Adversarial Attacks on Visual Objects Using the Fast Gradient Sign Method. *J. Grid Comput.* **2023**, *21*, 52. [CrossRef]
11. Naseem, M.L. Trans-IFFT-FGSM: A Novel Fast Gradient Sign Method for Adversarial Attacks. *Multimed. Tools Appl.* **2024**, *83*, 72279–72299. [CrossRef]
12. Ayas, M.S.; Ayas, S.; Djouadi, S.M. Projected Gradient Descent Adversarial Attack and Its Defense on a Fault Diagnosis System. In Proceedings of the 2022 45th International Conference on Telecommunications and Signal Processing (TSP), Prague, Czech Republic, 13–15 July 2022; IEEE: New York, NY, USA, 2022; pp. 36–39.
13. Waghela, H.; Sen, J.; Rakshit, S. Robust Image Classification: Defensive Strategies against FGSM and PGD Adversarial Attacks. In Proceedings of the 2024 Asian Conference on Intelligent Technologies (ACOIT), Kolar, India, 6–7 September 2024; IEEE: New York, NY, USA, 2024; pp. 1–7.
14. Chen, J.; Yan, D. Adversarial Attacks on Machinery Fault Diagnosis. *arXiv* **2022**, arXiv:2110.02498. [CrossRef]
15. Kim, H.; Lee, S.; Lee, J.; Lee, W.; Son, Y. Evaluating Practical Adversarial Robustness of Fault Diagnosis Systems via Spectrogram-Aware Ensemble Method. *Eng. Appl. Artif. Intell.* **2024**, *130*, 107980. [CrossRef]
16. Zareapoor, M.; Shamsolmoali, P.; Yang, J. Oversampling Adversarial Network for Class-Imbalanced Fault Diagnosis. *Mech. Syst. Signal Process.* **2021**, *149*, 107175. [CrossRef]
17. Shaik, A.K.; Das, A.; Palleti, V.R. Study of Adversarial Attacks on Anomaly Detectors In Industrial Control Systems. *arXiv* **2025**, arXiv:2505.03120.
18. Villegas-Ch, W.; Govea, J.; Jaramillo-Alcazar, A. Tamper Detection in Industrial Sensors: An Approach Based on Anomaly Detection. *Sensors* **2023**, *23*, 8908. [CrossRef]
19. Nematirad, R.; Behrang, M.; Pahwa, A. Acoustic-Based Online Monitoring of Cooling Fan Malfunction in Air-Forced Transformers Using Learning Techniques. *IEEE Access* **2024**, *12*, 26384–26400. [CrossRef]
20. Kong, X.; Ge, Z. Adversarial Attacks on Neural-Network-Based Soft Sensors: Directly Attack Output. *IEEE Trans. Ind. Inform.* **2022**, *18*, 2443–2451. [CrossRef]
21. Guo, R.; Chen, Q.; Tong, S.; Liu, H. Knowledge-Aided Generative Adversarial Network: A Transfer Gradient-Less Adversarial Attack for Deep Learning-Based Soft Sensors. In Proceedings of the 2024 14th Asian Control Conference (ASCC), Dalian, China, 5–8 July 2024; 2024; pp. 1254–1259.
22. Zizzo, G.; Hankin, C.; Maffeis, S.; Jones, K. Adversarial Attacks on Time-Series Intrusion Detection for Industrial Control Systems. In Proceedings of the 2020 IEEE 19th International Conference on Trust, Security and Privacy in Computing and Communications (TrustCom), Guangzhou, China, 29 December 2020–1 January 2021; IEEE: New York, NY, USA, 2020; pp. 899–910.
23. Sun, J.; Gu, X.; He, J.; Yang, S.; Tu, Y.; Wu, C. A Robust Approach of Multi-Sensor Fusion for Fault Diagnosis Using Convolution Neural Network. *J. Dyn. Monit. Diagn.* **2022**, *1*, 103–110. [CrossRef]

24. Farwell, P.; Rohozinski, R. Threats to Industrial Cyber-Physical Systems. In Proceedings of the 10th International Conference on Cyber Conflict CyCon X, NATO Cooperative Cyber Defence Centre of Excellence (CCDCOE), Tallinn, Estonia, 29 May–1 June 2018; pp. 1–6.
25. Giraldo, J.; Urbina, D.; Cardenas, A.; Valente, J.; Faisal, M.; Ruths, J.; Tippenhauer, N.O.; Sandberg, H.; Candell, R. A Survey of Physics-Based Attack Detection in Cyber-Physical Systems. *ACM Comput. Surv.* **2019**, *51*, 1–36. [CrossRef]
26. Guo, P.; Kim, H.; Virani, N.; Xu, J.; Zhu, M.; Liu, P. Exploiting Physical Dynamics to Detect Actuator and Sensor Attacks in Mobile Robots. *arXiv* **2017**, arXiv:1708.01834. [CrossRef]
27. Kurakin, A.; Goodfellow, I.J.; Bengio, S. Adversarial Examples in the Physical World. In *Artificial Intelligence Safety and Security*; Chapman and Hall/CRC: First ed.; CRC Press/Taylor & Francis Group: Boca Raton, FL, USA, 2018; pp. 99–112.
28. Eykholt, K.; Evtimov, I.; Fernandes, E.; Li, B.; Rahmati, A.; Xiao, C.; Prakash, A.; Kohno, T.; Song, D. Robust Physical-World Attacks on Deep Learning Visual Classification. In Proceedings of the 2018 IEEE/CVF Conference on Computer Vision and Pattern Recognition, Salt Lake City, UT, USA, 18–23 June 2018; IEEE: New York, NY, USA; pp. 1625–1634.
29. Cao, Y.; Xiao, C.; Cyr, B.; Zhou, Y.; Park, W.; Rampazzi, S.; Chen, Q.A.; Fu, K.; Mao, Z.M. Adversarial Sensor Attack on LiDAR-Based Perception in Autonomous Driving. In Proceedings of the 2019 ACM SIGSAC Conference on Computer and Communications Security, London, UK, 11–15 November 2019; ACM: New York, NY, USA, 2019; pp. 2267–2281.
30. Nguyen, K.; Fernando, T.; Fookes, C.; Sridharan, S. Physical Adversarial Attacks for Surveillance: A Survey. *IEEE Trans. Neural Netw. Learn. Syst.* **2024**, *35*, 17036–17056. [CrossRef]
31. Pozdnyakov, V.; Kovalenko, A.; Makarov, I.; Drobyshevskiy, M.; Lukyanov, K. Adversarial Attacks and Defenses in Fault Detection and Diagnosis: A Comprehensive Benchmark on the Tennessee Eastman Process. *IEEE Open J. Ind. Electron. Soc.* **2024**, *5*, 428–440. [CrossRef]
32. Guo, Y.; Cheng, Z.; Zhang, J.; Sun, B.; Wang, Y. A Review on Adversarial–Based Deep Transfer Learning Mechanical Fault Diagnosis. *J. Big Data* **2024**, *11*, 151. [CrossRef]
33. Chen, Q.; Dong, X.; Tu, G.; Wang, D.; Cheng, C.; Zhao, B.; Peng, Z. TFN: An Interpretable Neural Network with Time-Frequency Transform Embedded for Intelligent Fault Diagnosis. *Mech. Syst. Signal Process.* **2024**, *207*, 110952. [CrossRef]
34. Abadi, H.H.N. Adversarial Machine Learning Attacks on Condition-Based Maintenance Capabilities. *arXiv* **2021**, arXiv:2101.12097. [CrossRef]
35. Yin, Z.; Zhuo, Y.; Ge, Z. Transfer Adversarial Attacks across Industrial Intelligent Systems. *Reliab. Eng. Syst. Saf.* **2023**, *237*, 109299. [CrossRef]
36. Pozdnyakov, V.; Kovalenko, A.; Makarov, I.; Drobyshevskiy, M.; Lukyanov, K. AADMIP: Adversarial Attacks and Defenses Modeling in Industrial Processes. In Proceedings of the Thirty-Third International Joint Conference on Artificial Intelligence, Jeju, Republic of Korea, 3–9 August 2024; pp. 8776–8779.
37. Wang, J.; Liu, X.; Hu, J.; Wang, D.; Wu, S.; Jiang, T.; Guo, Y.; Liu, A.; Zhou, J. Adversarial Examples in the Physical World: A Survey. *arXiv* **2024**, arXiv:2311.01473.
38. Ullah, H.; Zaidi, S.M.T.; Munir, A. Improving Adversarial Robustness Through Adaptive Learning-Driven Multi-Teacher Knowledge Distillation. *arXiv* **2025**, arXiv:2507.20996.
39. Shi, Y.; Deng, A.; Deng, M.; Xu, M.; Liu, Y.; Ding, X. A Novel Multiscale Feature Adversarial Fusion Network for Unsupervised Cross-Domain Fault Diagnosis. *Measurement* **2022**, *200*, 111616. [CrossRef]
40. Moosavi-Dezfooli, S.-M.; Fawzi, A.; Frossard, P. DeepFool: A Simple and Accurate Method to Fool Deep Neural Networks. In Proceedings of the 2016 IEEE Conference on Computer Vision and Pattern Recognition (CVPR), Las Vegas, NM, USA, 26 June–1 July 2016; IEEE: New York, NY, USA, 2016; pp. 2574–2582.
41. Polymeropoulos, I.; Bezyrgiannidis, S.; Vrochidou, E.; Papakostas, G.A. Bridging AI and Maintenance: Fault Diagnosis in Industrial Air-Cooling Systems Using Deep Learning and Sensor Data. *Under Rev.* **2025**, in press.
42. Moukari, M.; Simon, L.; Picard, S.; Jurie, F. N-MeRCI: A New Metric to Evaluate the Correlation Between Predictive Uncertainty and True Error. *arXiv* **2019**, arXiv:1908.07253.
43. Lovell, D.; Miller, D.; Capra, J.; Bradley, A.P. Never Mind the Metrics-What about the Uncertainty? Visualising Binary Confusion Matrix Metric Distributions to Put Performance in Perspective. *Proc. Mach. Learn. Res.* **2023**, *202*, 22702–22757.
44. Smith, L.; Gal, Y. Understanding Measures of Uncertainty for Adversarial Example Detection. *arXiv* **2018**, arXiv:1803.08533. [CrossRef]
45. Ren, P.; Xiao, Y.; Chang, X.; Huang, P.-Y.; Li, Z.; Gupta, B.B.; Chen, X.; Wang, X. A Survey of Deep Active Learning. *ACM Comput. Surv.* **2022**, *54*, 1–40. [CrossRef]
46. Guo, C.; Pleiss, G.; Sun, Y.; Weinberger, K.Q. On Calibration of Modern Neural Networks. In Proceedings of the 34th International Conference on Machine Learning, ICML 2017, Sydney, NSW, Australia, 6–11 August 2017; pp. 1321–1330.
47. Abomakhelb, A.; Jalil, K.A.; Buja, A.G.; Alhammadi, A.; Alenezi, A.M. A Comprehensive Review of Adversarial Attacks and Defense Strategies in Deep Neural Networks. *Technologies* **2025**, *13*, 202. [CrossRef]

48. Goodfellow, I.J.; Shlens, J.; Szegedy, C. Explaining and Harnessing Adversarial Examples. In Proceedings of the 3rd International Conference on Learning Representations, ICLR 2015, San Diego, CA, USA, 7–9 May 2015.
49. Pan, S.J.; Yang, Q. A Survey on Transfer Learning. *IEEE Trans. Knowl. Data Eng.* **2010**, *22*, 1345–1359. [CrossRef]
50. Madry, A.; Makelov, A.; Schmidt, L.; Tsipras, D.; Vladu, A. Towards Deep Learning Models Resistant to Adversarial Attacks. In Proceedings of the 6th International Conference on Learning Representations, ICLR 2018, Vancouver, BC, Canada, 30 April–3 May 2018.
51. Luo, P.; Yin, Z.; Zhang, Y.; Bai, C.; Zhang, P.; Liu, J. Increasing Interpretability and Feasibility of Data Driven-Based Unknown Fault Diagnosis in Permanent Magnet Synchronous Motors. *IEEE Trans. Energy Convers.* **2025**, *40*, 1422–1433. [CrossRef]
52. Hosmer, D.W.; Lemeshow, S. *Applied Logistic Regression*, 2nd ed.; John Wiley & Sons, Inc.: New York, NY, USA, 2000; ISBN 0471722146.
53. Breiman, L. Random Forests. *Mach. Learn.* **2001**, *45*, 5–32. [CrossRef]
54. Chen, F.; Zhang, L.; Liu, W.; Zhang, T.; Zhao, Z.; Wang, W.; Chen, D.; Wang, B. A Fault Diagnosis Method of Rotating Machinery Based on Improved Multiscale Attention Entropy and Random Forests. *Nonlinear Dyn.* **2023**, *112*, 1191–1220. [CrossRef]
55. Chen, T.; Guestrin, C. XGBoost: A Scalable Tree Boosting System. In Proceedings of the 22nd ACM SIGKDD International Conference on Knowledge Discovery and Data Mining, San Francisco, CA, USA, 13–17 August 2016; ACM: New York, NY, USA, 2016; pp. 785–794.
56. Ke, G.; Meng, Q.; Finley, T.; Wang, T.; Chen, W.; Ma, W.; Ye, Q.; Liu, T.Y. LightGBM: A Highly Efficient Gradient Boosting Decision Tree. In *Advances in Neural Information Processing Systems*; Curran Associates, Inc.: Nice, France, 2017; pp. 3149–3157.
57. Widodo, A.; Yang, B.-S. Support Vector Machine in Machine Condition Monitoring and Fault Diagnosis. *Mech. Syst. Signal Process.* **2007**, *21*, 2560–2574. [CrossRef]
58. Justus, D.; Brennan, J.; Bonner, S.; McGough, A.S. Predicting the Computational Cost of Deep Learning Models. In Proceedings of the 2018 IEEE International Conference on Big Data (Big Data), Seattle, WA, USA, 10–13 December 2018.
59. Han, S.; Mao, H.; Dally, W.J. Deep Compression: Compressing Deep Neural Networks with Pruning, Trained Quantization and Huffman Coding. In Proceedings of the 4th International Conference on Learning Representations, ICLR 2016-Conference Track Proceedings, San Juan, PR, USA, 2–4 May 2016.
60. Narayanan, D.; Shoeybi, M.; Casper, J.; LeGresley, P.; Patwary, M.; Korthikanti, V.; Vainbrand, D.; Kashinkunti, P.; Bernauer, J.; Catanzaro, B.; et al. Efficient Large-Scale Language Model Training on GPU Clusters Using Megatron-LM. In Proceedings of the International Conference for High Performance Computing, Networking, Storage and Analysis, Saint Louis, MS, USA, 14–19 November 2021; ACM: New York, NY, USA, 2021; pp. 1–15.
61. Sze, V.; Chen, Y.-H.; Yang, T.-J.; Emer, J.S. Efficient Processing of Deep Neural Networks: A Tutorial and Survey. *Proc. IEEE* **2017**, *105*, 2295–2329. [CrossRef]

processes MDPI

Article

Research on Bo-BiLSTM-Based Synchronous Load Transfer Control Technology for Distribution Networks

Cheng Long [1,2,*], Hua Zhang [1,2], Xueneng Su [1], Yiwen Gao [1] and Wei Luo [3]

[1] State Grid Sichuan Electric Power Research Institute, Chengdu 100144, China
[2] Power System Security and Operation Key Laboratory of Sichuan, Chengdu 610041, China
[3] School of Electrical Engineering, Sichuan University, Wangjiang Campus, Chengdu 610065, China
* Correspondence: 17828186876@163.com

Abstract: The operational modes and fault characteristics of distribution networks incorporating distributed generation are becoming increasingly complex. This complexity increases the difficulty of predicting switch control action times and leads to scattered samples with data scarcity. Consequently, it imposes higher demands on rapid fault isolation and load transfer control following system failures. To address this issue, this paper proposes a switch action time prediction and synchronous load transfer control method based on Bayesian optimization of bidirectional long short-term memory (Bo-BiLSTM) networks. A distribution network simulation model incorporating distributed generation was constructed using MATLAB/Simulink (R2023a). Three-phase voltage and current at the Point of Common Coupling (PCC) were extracted as feature parameters to establish a switch operation timing database. Bayesian optimization was employed to tune the BiLSTM hyperparameters, constructing the Bo-BiLSTM prediction model to achieve high-precision forecasting of switch operation times under fault conditions. Subsequently, a load-synchronized transfer control strategy was proposed based on the prediction results. A dynamic delay mechanism was designed to achieve "open first and then close" sequential coordinated control. Physical experiments verified that the time difference between opening and closing was controlled within 2–12 milliseconds (ms), meeting the engineering requirement of less than 20 ms. The results demonstrate that the proposed control method enhances switch operation time prediction accuracy while effectively supporting rapid fault isolation and seamless load transfer in distribution networks, thereby improving system reliability and control precision.

Keywords: distributed power sources; Bayesian optimization; switching action time; time prediction; load synchronous transfer control

1. Introduction

With the large-scale integration of distributed generation (DG) into distribution networks, system operation modes and fault characteristics have become increasingly complex, placing higher demands on rapid fault isolation and load transfer control post-fault [1]. As the core component enabling distribution network automation and fault self-healing, the precise prediction of switching equipment response times is crucial for ensuring reliable operation of protection and control systems and minimizing outage duration [2]. However, DG integration significantly alters fault current magnitude and direction, resulting in highly dispersed and uncertain switch operation time samples. Concurrently, the scarcity of training data for predictive models is exacerbated by the challenges in collecting real-world fault data. These factors cause a significant decline in the accuracy of prediction methods based on traditional mechanism models or shallow learning algorithms, making it difficult to meet the stringent requirements for real-time control and reliability in smart distribution grids.

In recent years, deep learning techniques have achieved remarkable success in the field of time-series data forecasting. Long short-term memory (LSTM) networks and their variants have been widely applied in areas such as remaining equipment life prediction [3,4] and power load forecasting due to their exceptional ability to capture long-term sequential dependencies. However, standard LSTM models suffer from blindness in hyperparameter selection and predominantly employ unidirectional encoding, making it difficult to fully exploit the bidirectional contextual information inherent in time-series data. Although bidirectional long short-term memory (BiLSTM) networks enhance feature extraction through bidirectional encoding, their performance remains heavily dependent on optimized hyperparameter configurations [5]. Bayesian optimization (BO), as an efficient global hyperparameter optimization framework, can adaptively search for optimal hyperparameter combinations of complex models with minimal evaluation iterations, significantly improving model convergence speed and prediction accuracy [6].

Recent research has significantly advanced the application of complex artificial intelligence models in power systems. For instance, Reference [7] successfully integrated convolutional neural networks (CNNs) with bidirectional long short-term memory (BiLSTM) networks for equipment remaining life prediction, while Reference [8] employed Bayesian optimization of long short-term memory (LSTM) networks for distribution system state estimation. Concurrently, studies such as [9] focus on enhancing fault classification accuracy through hybrid deep learning models. Collectively, these achievements confirm a clear trend: integrating advanced neural networks with intelligent optimization has become a key research paradigm for enhancing the perception and predictive capabilities of power systems.

While previous studies have explored LSTM and BiLSTM networks for fault diagnosis and switch time prediction and Bayesian methods for model optimization [10], a significant gap remains. Existing work primarily focuses on improving prediction accuracy alone, often overlooking the critical next step: the seamless integration of these high-fidelity predictions into a real-time, reliable control strategy for load transfer. The challenge of closing the loop from "accurate prediction" to "precise control" in a single framework, especially under the constraints of small sample sizes and high data dispersion, has not been fully addressed.

Based on this, this paper addresses the challenge of high-precision prediction of switch operation times in distribution networks under DG integration and further applies the prediction results to the precise control of synchronous load transfer. The main contributions of this study are as follows:

1. A Bo-BiLSTM-based prediction model for switching operation timing in distribution networks is proposed. By adaptively obtaining optimal hyperparameters for the BiLSTM network through Bayesian optimization algorithms, the model's predictive accuracy and generalization capability are effectively enhanced under conditions of small sample sizes and high data dispersion, laying the foundation for subsequent implementation of control strategies.
2. A comprehensive database of switch operation time characteristics was established. A detailed simulation model of the distribution network incorporating DG was constructed using MATLAB/Simulink. This model simulated 620 operational scenarios encompassing various short-circuit fault types, different fault locations, and transition resistances, providing a robust data foundation for model training and control logic validation.

3. We innovatively integrate switch timing prediction results with load transfer control strategies. Based on the predicted time parameters, a synchronous transfer control strategy was designed following the principle of "open first and then close" with dynamic delay, Δt, as its core. Physical experiments validated that this strategy can strictly control the opening and closing time difference within a 20 ms safety margin, with experimental results ranging from 2 to 12 ms. This fundamentally eliminates the risk of circulating currents caused by asynchronous closing, achieving seamless load transfer control after a fault.

This study not only provides a novel data-driven solution for predicting switching control action time in distribution networks but also establishes a comprehensive technical framework spanning from "precise prediction" to "precise control." It holds significant theoretical value and engineering significance for enhancing the power supply reliability and self-healing control capabilities of distribution networks under high-penetration DG access.

Figure 1 illustrates the overall workflow of this study, which forms a closed loop from data-driven prediction and model optimization to control strategy and experimental validation.

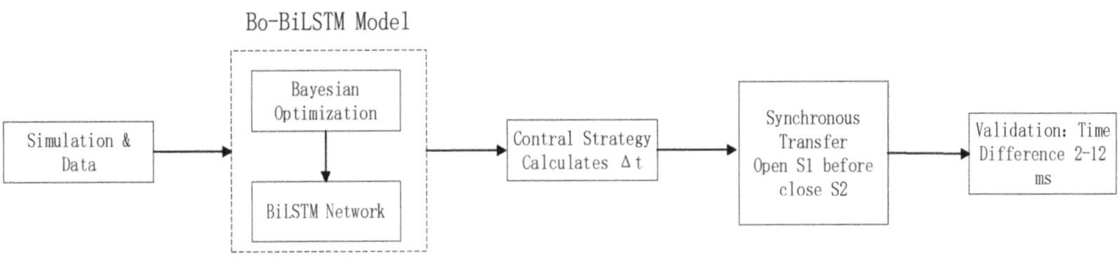

Figure 1. Schematic diagram of the research framework.

2. System Modeling and Feature Database Construction

2.1. System Modeling

This paper investigates the switching sequence of distribution network switches during different feeder faults in a distribution system with a distributed photovoltaic (PV) connection to a double-feeder network. A typical distribution network model incorporating a distributed PV system is constructed, featuring a 35 kV system voltage, one distributed PV generator, and *L*1 and *L*2, representing loads. The structure is illustrated in Figure 2.

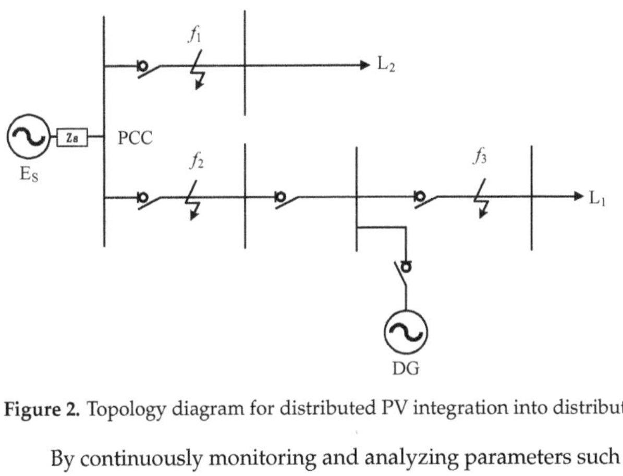

Figure 2. Topology diagram for distributed PV integration into distribution networks.

By continuously monitoring and analyzing parameters such as current and voltage at the Point of Common Coupling (PCC), various faults occurring in the distribution system can be promptly detected and addressed, ensuring reliable power supply. The operational conditions of the grid and the specific fault type determine the switching action time [11].

This paper selects fault types, locations, and transition resistance values based on power system engineering principles and distribution network fault scenarios. Fault types include single-phase grounding, two-phase short circuit, two-phase grounding short circuit, and three-phase short circuit. Fault locations are chosen at the feeder head, midpoint, and feeder tail. The study evaluates the impact of fault point distance from the power source and distributed power sources on protection system operation and switching sequence timing, covering scenarios with varying contributions from main grid and distributed power source fault currents. The grounding transition resistance values range from 0.001 Ω to 130 Ω, testing the model's sensitivity and accuracy under various fault conditions. Key system equipment and model parameters are listed in Table 1.

Table 1. Table of system parameters.

Parameter	Value
Distributed power supply access location	Distribution network feeder non-terminal bus
Line positive sequence impedance/(Ω·km^{-1})	0.27 + j0.335
Active power/kW	$P_{L1} = P_{L2} = 300$
Reactive power/kvar	$Q_{L1} = Q_{L2} = 130$
Fault type	Single-phase short circuit, two-phase short circuit, two-phase ground short circuit, three-phase short circuit, three-phase short circuit grounding
Fault location	f_1, f_2, f_3
Grounding transition resistance/Ω	0.001~130
Duration of the fault/s	0.5
Photovoltaic capacity/kW	213.15
L1, L2, length/km	300, 100
Power supply rated voltage/kV	35
Power supply rated capacity/kVA	630
Zs/Ω	0.8929 + j5.209

This paper established a distribution network system simulation model using MATLAB/Simulink. It obtained voltage and current conditions at the Point of Common Coupling (PCC) during different types of faults occurring at frequencies $f1$, $f2$, and $f3$. Subsequently, it recorded and analyzed the switching operation times of distribution network switches from both voltage and current perspectives after integrating distributed power sources into the distribution network. This provided historical data support for subsequent predictive control.

2.2. Construction of a Database for Time Characteristics of Distribution Network Switch Operations

Distribution networks can be broadly categorized into 12 operational conditions: single-phase short circuits (AG, BG, CG), two-phase short circuits (AB, AC, BC), two-phase grounded short circuits (ABG, ACG, BCG), three-phase short circuits (ABC), three-phase grounded short circuits (ABCG), and non-fault states. Simulations were conducted for different distribution network conditions, collecting three-phase voltage and current data at the PCC for each scenario to construct a database of switch operation time characteristics for the distribution network. In power system transient analysis, voltage and current peaks serve as the most direct and stable key physical quantities characterizing abrupt system state changes. Therefore, these peak values were employed as inputs for the credibility network classifier, as shown in Table 2.

Table 2. Time-characteristic parameters of grid switch operation.

Variables	Parameter
V_A	Phase A voltage peak
V_B	Phase B voltage peak
V_C	Phase C voltage peak
I_A	Phase A current peak
I_B	Phase B current peak
I_C	Phase C current peak
T	Switching time

The peak voltage and peak current values in Table 2 were calculated based on simulation sampling data. This study obtained 620 operating conditions through simulation, denoted as O_1–O_{620}. The six characteristic data points and switching action times under different operating conditions are shown in Appendix A, Table A1.

To eliminate the influence of feature dimensions and enhance model training stability, this study employs the Z-Score normalization method to process the feature matrix. This approach transforms each feature column into a distribution with a mean of 0 and a standard deviation of 1, enabling comparability among different features while preserving the distribution characteristics of the original data. It is suitable for subsequent gradient-optimized deep learning models. Taking O_1 as an example, its normalization process follows the method below:

$$V_i' = \frac{V_i - \mu}{\sigma} \tag{1}$$

In the formula, *Vi* and *Vi′* denote the normalized and pre-normalized values of the *i*-th operational state symptom attribute, respectively; μ denotes the mean of each feature column across all samples; and σ denotes the standard deviation, calculated for S_2 through S_{620}. The normalized database is shown in Appendix A, Table A2.

The normalization of the feature matrix involves six operational status indicators for each operating condition in the database, collectively forming a feature vector. A training dataset comprising 620 × 70% = 434 feature vectors is randomly selected from the database, while the remaining 620 × 30% = 186 vectors serve as the test dataset.

3. Bayesian Optimization-Based BiLSTM Forecasting Model

3.1. BiLSTM Principles and Applicability

The LSTM algorithm is used to process and predict highly time-dependent, strongly coupled events, but LSTM network models exhibit lower adaptability [12,13]. Unlike LSTM networks, BiLSTM consists of two independent forward and backward LSTM networks, one for forward propagation and the other for backward propagation. In each time step, the forward LSTM processes the input sequence in forward order, while the backward LSTM processes it in reverse order. This allows each LSTM unit to simultaneously access both preceding and subsequent contextual information. Furthermore, compared to other methods—such as random forests and gradient boosters, which are powerful for tabular data but primarily model static correlations—BiLSTM inherently incorporates temporal state mechanisms [14]. These mechanisms are crucial for simulating causal, time-dependent processes from fault states to protective actions. Similarly, purely unsupervised learning approaches are unsuitable for regression-based prediction tasks. Standard recurrent neural networks encounter gradient vanishing issues when capturing long-range dependencies. The LSTM architecture and its gating mechanism were specifically developed to overcome this limitation, making it a superior choice for learning the potentially long and complex causal pathways inherent in power system fault responses. Therefore, BiLSTM was selected to most effectively capture the rich temporal dependencies present in the data [15].

This study employed BiLSTM to capture the temporal characteristics of electrical quantities before and after failures, performing bidirectional training on sample failure data. The schematic diagram of the BiLSTM structure is shown in Figure 3 [16].

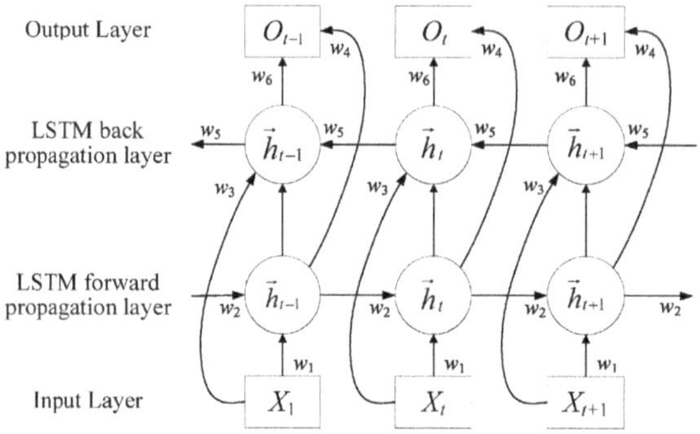

Figure 3. BiLSTM architecture diagram.

3.2. Bayesian Optimization Principle

Bayesian optimization employs Gaussian processes as probabilistic surrogate models, enhancing algorithmic robustness through continuous iteration on existing sample data [17]. When determining new prediction points, Bayesian algorithms require maximizing the utility function. Thus, while not applicable to every optimization problem, Bayesian algorithms constitute an effective global optimization approach. Bayesian optimization comprises Gaussian process regression and acquisition function computation. Gaussian process regression yields the mean and variance of the function value, while the acquisition function constructs a sampling strategy based on these parameters to determine the sampling points for the current iteration.

(1) Gaussian Process Regression

Gaussian process regression (GPR) is a nonparametric Bayesian method for solving regression problems [18]. GPR is based on a Gaussian process prior distribution, which is adjusted by observational data to yield a posterior distribution [19]. Its model assumptions comprise both noise and Gaussian process prior components, with solutions derived through Bayesian inference.

(2) Collection Function

The sampling function generates the next observation point to be predicted. Based on the mean and covariance dataset D calculated by Gaussian process regression, it determines the next sample point after iteration. Its primary role is to accelerate convergence by leveraging previous observations while simultaneously exploring regions of high uncertainty within the decision space, thereby avoiding getting stuck in local optima.

The Bayesian optimization process is meticulously configured to ensure efficient and reproducible hyperparameter search. The search targets four key BiLSTM hyperparameters within the following ranges: the number of hidden units is optimized as an integer between 32 and 128; the initial learning rate is searched on a logarithmic scale between 1×10^{-4} and 1×10^{-2}; the L2 regularization factor is explored on a logarithmic scale between 1×10^{-5} and 1×10^{-2}; and the gradient threshold is adjusted within [0.5, 2] to prevent gradient explosion. A Gaussian process with a Matern 5/2 kernel was selected as the surrogate model for its flexibility in capturing diverse function behaviors. Optimization is guided by an expectation-maximization (EM) sampling function to effectively balance exploration and exploitation. The process is capped at 50 iterations, representing a trade-off between computational cost and performance. Convergence is assessed by monitoring the relative improvement in the optimal observation target value; the search is deemed converged if this improvement falls below a 0.2% tolerance for ten consecutive iterations.

3.3. Bo-BiLSTM Network Parameter Optimization

The hyperparameters of BiLSTM are difficult to set manually, so a Bayesian optimization (BO) approach is employed to automatically identify the optimal hyperparameter combination, aiming to achieve higher prediction accuracy. Given the large number of parameters in the BiLSTM network, the BO algorithm is used to optimize several key parameters within the network. This ensures that the optimized parameters achieve the best possible fit with the network, enabling high-precision data prediction.

The hyperparameter selection formula for Bayesian optimization is as follows:

$$x^* = \text{argmin} f(x), x \in X \tag{2}$$

In the formula, f(x) denotes the optimization function; x represents the set of optimization parameters; X is the set of optimization parameter sets; and x* is a set of optimization functions.

Define the Error function as the prediction error rate function for the BiLSTM model, with the formula as follows:

$$\text{Error} = 1 - \frac{R}{N} \tag{3}$$

In the formula, N represents the total number of sample groups; R denotes the number of sample groups in N where the predicted values match the actual loss values.

The BiLSTM weight update formula is

$$W_{t+1} = W_t - \alpha \frac{1}{n} \frac{\partial L}{\partial W_{t+1}} \tag{4}$$

In the formula, α denotes the learning rate; n represents the mini-batch size. It can be seen from the formula that α and n influence the weight updates of the network, with α affecting the step size per iteration and n determining the network's loss function.

3.4. Switch Operation Time Prediction Method for Distribution Networks Based on Bo-BiLSTM Network

The modeling and testing process of the Bo-BiLSTM network is illustrated in Figure 4. First, switch test data from distribution network simulations is acquired, smoothed, and divided into training and test sets. Next, the Bo algorithm is applied to tune hyperparameters on the training set, with the optimal settings fed back to reconstruct the BiLSTM network. Finally, predictions are made on the test set data.

The optimization process of the model can be divided into the following steps.

(1) Model initialization: Define the function f(x) and the definition domain of x to be optimized in the Bayesian optimization algorithm, set the objective function and the maximum number of iterations, etc.; divide the processed failure characteristics dataset into test set kernel and training set; and train the BiLSTM model.
(2) Select the BiLSTM model failure prediction error rate to obtain the initial observation of the Bayesian algorithm.
(3) The Gaussian process is applied to estimate the initial observation value to determine the next observation point.
(4) Select the optimal parameters according to the acquisition function until the end of the iteration.

(5) The optimal parameters obtained by the Bayesian algorithm are brought into the BiLSTM network for parameter reconstruction.
(6) The optimal parameters are brought into the BiLSTM model to predict the test set data, and the predicted results are obtained and analyzed.

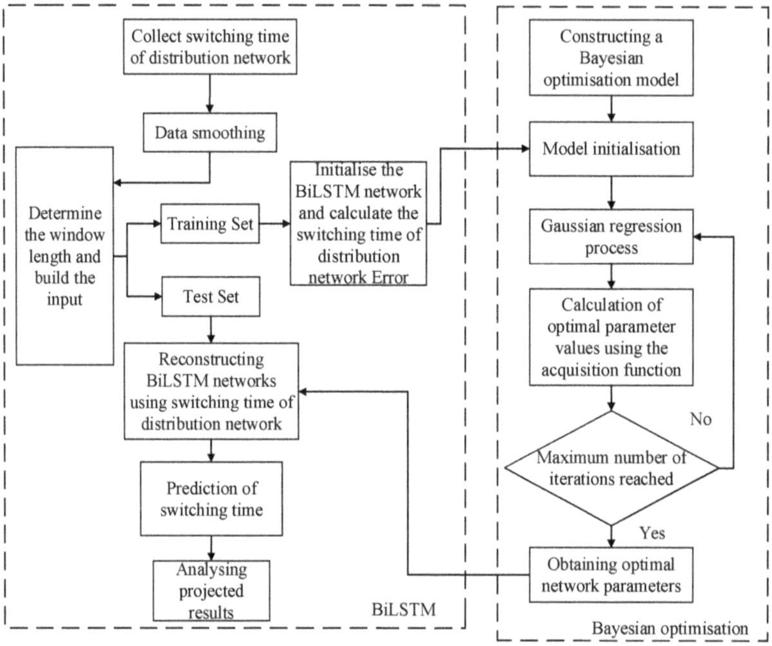

Figure 4. Bo-BiLSTM prediction model workflow.

4. Analysis of BiLSTM Prediction Model Results

4.1. Model Prediction Standard

The goodness-of-fit coefficient (R^2), root mean square error (RMSE), and mean absolute error (MAE) can be used to evaluate the performance of time-series forecasting algorithms. R^2 indicates the degree of fit between the data and the regression model; the RMSE represents the average value of errors, reflecting the model's stability; and the MAE accurately reflects the magnitude of actual prediction errors. Their expressions are as follows:

$$R^2 = 1 - \frac{\sum_{i=1}^{N} \left(x_{\text{tri}} - x_{\text{pri}}\right)^2}{\sum_{i=1}^{N} \left(x_{\text{tri}} - \hat{x}_{\text{tri}}\right)^2} \tag{5}$$

$$RMSE = \sqrt{\left(\sum_{i=1}^{N} \left|x_{\text{tri}} - x_{\text{pri}}\right|^2\right)/N} \tag{6}$$

$$MAE = \left(\sum_{i=1}^{N} \left|x_{\text{tri}} - x_{\text{pri}}\right|\right)/N \tag{7}$$

In the formula, x_{tri} is the true value, $\hat{x}_{tri}$ is the average of the true value, x_{pri} is the predicted value, and N is the number of predictions.

4.2. Bayesian Optimization Search Results

The Bayesian optimization algorithm repeatedly refines the target value through computational operations of the collection function. By continuously computing the objective function, it seeks the global optimum to optimize the BiLSTM network, ensuring more precise prediction results. The Bayesian optimization results are shown in Figure 5.

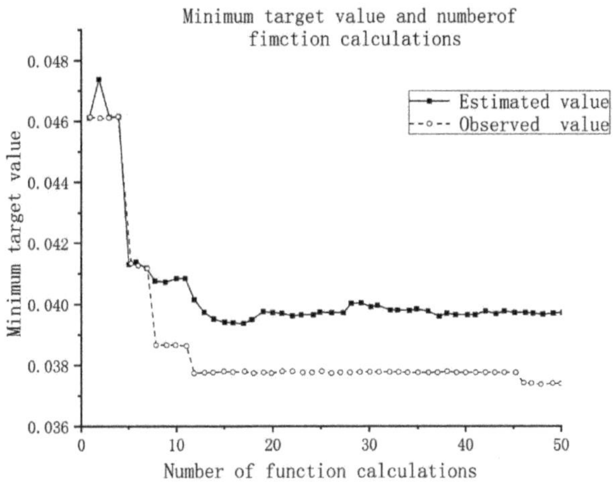

Figure 5. Bayesian optimization.

Figure 5 illustrates the convergence characteristics of the Bayesian optimization algorithm during the hyperparameter auto-tuning process, reflecting the iterative trajectory of hyperparameter tuning within the constructed Bo-BiLSTM prediction model. The horizontal axis represents the number of function evaluations, while the vertical axis shows the minimum objective value. The figure also compares the algorithm's estimated values of the objective function with the actual observed values throughout the iteration process. As iterations increase to the preset 50, the two curves gradually converge and eventually stabilize. After 50 function evaluations, the minimum objective value decreases from an initial value of approximately 0.048 to 0.0378. This value falls within a tolerance of less than 0.2% compared to the estimated minimum objective value of 0.0395, indicating that the search has converged.

To clearly present the final model structure determined through Bayesian optimization, Table 3 lists the core hyperparameters of the BiLSTM network and their optimal values. This set of parameters was automatically identified by the Bayesian optimization algorithm through 50 iterative evaluations within a predefined search space, directly corresponding to the configuration where the objective function converged to its minimum value, as shown in Figure 5.

Specifically, the optimized number of hidden units was set to 96, achieving a balance between model capacity and computational efficiency; the initial learning rate was 3.16×10^{-3}, facilitating a stable descent of the "adam" optimizer during early training. The L2 regularization coefficient was set to 2.45×10^{-4} to effectively suppress overfitting without excessively compromising fitting capability; the gradient threshold was set to 1.2 to prevent gradient explosion during training. This set of parameters collectively formed the final architecture of the Bo-BiLSTM network used for switch timing prediction, serving as the performance foundation for achieving high-accuracy predictions.

Table 3. The optimal hyperparameter set for the BiLSTM network obtained through Bayesian optimization.

Hyperparameters	Search Scope	Optimum Value	Note
Number of hidden cells	[32, 128]	96	Control network capacity and expressiveness
Initial learning rate	$[1 \times 10^{-4}, 1 \times 10^{-2}]$	3.16×10^{-3}	Initial step size of the "adam" optimizer
L2 regularization coefficient	$[1 \times 10^{-5}, 1 \times 10^{-2}]$	2.45×10^{-4}	Prevent overfitting and enhance generalization capabilities
Gradient threshold	[0.5, 2]	1.2	Trim gradients to prevent training instability
Maximum number of iterations	-	50	Bayesian optimization iteration upper bound
Convergence tolerance	-	0.2%	Stop when the relative improvement of the target value falls below this threshold for 10 consecutive iterations.

4.3. Distribution Network Switching Time Prediction Results

To evaluate the performance of the proposed Bo-BiLSTM model, we employed a widely recognized robust benchmark—Extreme Gradient Boosting (XGBoost). As an efficient and practical implementation of gradient boosting algorithms, XGBoost has consistently achieved top results in numerous machine learning competitions and structured data regression tasks, establishing itself as a powerful non-deep learning benchmark. The hyperparameters of the XGBoost model—including the learning rate, maximum tree depth, and number of estimators—were optimized using the same Bayesian optimization framework as that in BiLSTM to ensure a fair and rigorous comparison. The models were trained and tested on the same datasets described in Section 2.2.

The first 70% of the distribution network switch time series was used as the training set, while the remaining 30% served as the test set. The delay step size was set to 15, the solver was configured as "adam", and the initial learning rate was 0.005. Training was conducted for 250 epochs with a learning rate decay factor of 0.2 and a decay cycle of 125. To prevent gradient explosion, the gradient threshold was set to 1.

The prediction results of the three network models for distribution network switching times are shown in Figure 6.

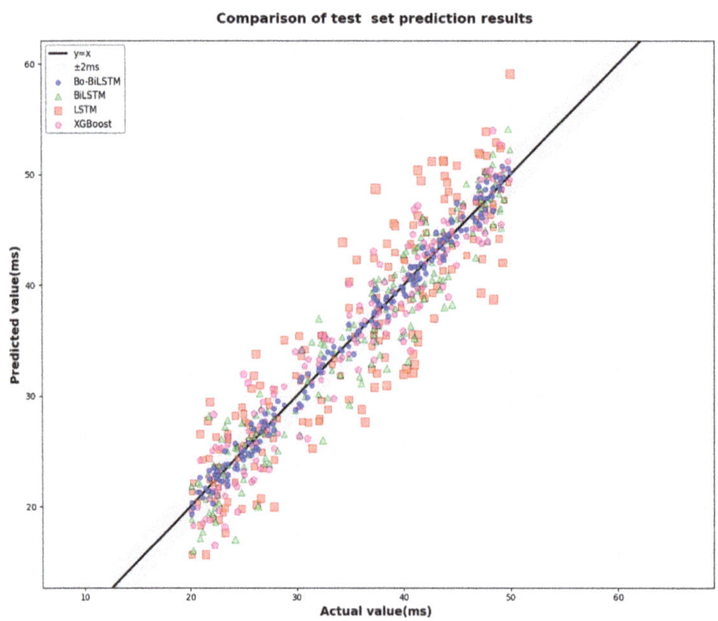

Figure 6. Comparison of prediction results from different network models on the test set.

This figure compares the prediction results of four different network models against actual simulation data on the same test set. Visual verification via the coordinate axes clearly shows that the Bo-BiLSTM model's prediction curve most closely aligns with actual values, demonstrating the highest consistency and stability. The BiLSTM and LSTM models follow in performance, while XGBoost exhibits noticeable deviations under certain operating conditions. This indicates that the fusion architecture integrating the attention mechanism (Bo) with the bidirectional long short-term memory network (BiLSTM) effectively enhances the ability to capture complex time-series fault features and improves prediction accuracy.

The performance of the prediction models is quantitatively evaluated using three standard metrics: R^2, the RMSE, and the MAE. A superior model is characterized by an R^2 value closer to 1, indicating a better fit to data, while simultaneously achieving lower RMSE and MAE values, which reflect higher prediction accuracy and smaller error magnitudes. A comparison of these metrics for the Bo-BiLSTM, BiLSTM, LSTM, and XGBoost models is presented in Table 4.

Table 4. Comparison of prediction metrics for distribution network switch operation time prediction models.

Model	R^2	RMSE	MAE
Bo-BiLSTM	0.91464	0.59996	0.49855
BiLSTM	0.61988	1.1900	1.0219
LSTM	0.54715	1.3022	1.1131
XGBoost	0.80122	0.85011	0.70144

As shown in Table 4, for the prediction metrics of the distribution network switch operation time prediction model, the Bo-BiLSTM network achieved a model fit quality of 0.91464, representing improvements of 0.29476 and 0.36749 compared to the BiLSTM and LSTM networks, respectively. The root mean square error was 0.59996, representing reductions of 0.59004 and 0.702224 compared to the BiLSTM and LSTM networks, respectively. The mean absolute error was 0.49855, showing decreases of 0.52335 and 0.61455 relative to the BiLSTM and LSTM networks.

Bo-BiLSTM achieved approximately 11.34% higher R^2 scores than XGBoost, while reducing the root mean square error (RMSE) and mean absolute error (MAE) by approximately 29.43% and 28.92%, respectively, demonstrating a significant competitive advantage. This performance gap highlights the key strengths of our proposed architecture.

To more intuitively demonstrate the superiority of the Bo-BiLSTM model in time prediction, the prediction error was calculated between the predicted and actual switch operation times for the 100th sample point in the distribution network. The prediction errors for switch operation times under different network models are shown in Table 5.

Table 5. Bo-BiLSTM network model prediction error table.

Model	True Value (ms)	Predicted Value (ms)	Absolute Value Error (ms)
Bo-BiLSTM	36.88883	36.12697	0.76186
BiLSTM	36.88883	35.54005	1.34878
LSTM	36.88883	34.72493	2.1639
XGBoost	36.88883	35.21045	1.67838

The absolute error values for the four network models, as shown in the table, are 0.76186, 1.34878, 2.1639, and 1.67838, respectively. The Bo-BiLSTM network exhibits the smallest absolute error in its predictions, indicating superior predictive performance.

It should be noted that the validation of this study is primarily based on the constructed typical distribution network model and its derived dataset. Although the model's effectiveness was demonstrated through training–test set partitioning and physical experiments, future work will include testing in distribution network cases with different topologies and higher penetration rates of distributed generation (DG) to further evaluate the model's broad applicability.

4.4. Discussion on Prediction Error Accumulation

An observation from the prediction results is that the discrepancy between the actual and predicted values tends to increase with the sample point index along the testing timeline. This phenomenon, often referred to as error accumulation, is a known challenge in sequence-to-sequence prediction tasks and can be attributed to the following factors:

1. Temporal Dependency and Autoregressive Nature: The BiLSTM model makes predictions based on a sequence of previous time steps. In an autoregressive setting, the current prediction relies on past observed or predicted values. As the prediction horizon extends into the future, the model increasingly depends on its own previous predictions rather than ground-truth data. Any small error made at an early step can therefore be propagated and amplified in subsequent predictions, leading to a gradual divergence from the actual series.
2. Attenuation of Long-range Contextual Information: Although BiLSTM is designed to capture long-term dependencies, its ability to retain and utilize information from the distant past is still finite. For very long sequences or sequences with complex, non-stationary patterns, the influence of the initial, most informative states of the network may diminish over time. This makes it increasingly difficult for the model to correct its course based on the foundational context.
3. Limitations of the Training Dataset: Our model was trained on a finite set of 620 operational scenarios. While this provides a solid foundation, the testing sequence might encounter a subtle, cumulative combination of features that represents an edge case or a gradual shift in system dynamics not fully captured in the training distribution. The model, having not been explicitly trained on such a prolonged, specific trajectory, struggles to maintain perfect accuracy.

It is important to note that despite this observed accumulation, the Bo-BiLSTM model demonstrated superior robustness compared to the baseline BiLSTM and LSTM models. As quantified in Tables 4 and 5, the absolute magnitude of its errors remained significantly lower throughout the sequence. This indicates that the Bayesian optimization successfully found a parameter set that minimized the initial error and slowed the rate of accumulation, thereby enhancing the model's practical utility for a sufficient prediction horizon relevant to the subsequent control action.

5. Time-Based Model and Experimental Validation for Full-Process Control of Load Transfer Using Bo-BiLSTM

Accurate switch timing prediction provides the prerequisite for achieving precise load transfer control.

This chapter will design a synchronous control strategy based on the switching times predicted by the Bo-BiLSTM model to ensure the safety and rapidity of the transfer process. The core control objective of the model is to accurately estimate the total time control interval from the issuance of the master station command to the final switching action, while ensuring compliance with the "open first and then close" sequence and time difference constraints even under extreme scenarios. By integrating mathematical modeling with engineering experience, we can systematically evaluate the boundary conditions of time parameters and their impact mechanisms on the overall control process, providing theoretical support for strategy design. This strategy requires, based on precise time synchronization, that the time interval from complete open ping to complete closing be within 20 ms, disregarding communication time from the master station to the terminal.

5.1. Time Constraints and Problem Modeling

Control Objective: The total time interval from the completion of the original power switch (S1) opening to the completion of the target power switch (S2) closing must satisfy the safety constraint:

$$0 < T_{total} < 20 \text{ ms}$$

Known Conditions:

Based on the Bo-BiLSTM model predictions, the closing operation time of the target switch S2 (T_close) is shorter than the opening operation time of the original switch S1 (T_open).

Master Station Command Sequence:

To achieve the "open first and then close" logic and compensate for the time difference, the master station issues commands with a dynamic delay (Δt):

t_0: Issue the closing command to the target switch (S2).

$t_0 + \Delta$: Issue the opening command to the original switch (S1).

Mathematical Model:

Completion time of closing operation (S2):

$$t\_close\_end = t_0 + T\_close \tag{8}$$

Completion time of opening operation (S1):

$$T_{total} = t\_close\_end - t\_open\_end \tag{9}$$

The total time interval is defined as the period from the end of the opening to the end of the closing:

$$T_{total} = t\_close\_end - t\_open\_end \tag{10}$$

In the formula, the physical meaning of T_{total} is the time interval from the complete opening of the main contacts of the original power switch to the complete closing of the main contacts of the target power switch. This interval represents the duration during which the load side is actually deprived of power supply during the load transfer process.

The control objective can be formulated as the following constraint:

$$0 < T_{total} < 20 \text{ ms}$$

5.2. Control Strategy Design: Determination of Dynamic Delay Time Δt

The core of the control strategy is to calculate the dynamic delay Δt such that Ttotal always falls within the safe range, even with prediction errors. Δt is a time offset precalculated by the master station and embedded within the control command. Its implementation does not depend on the interval between sequential commands issued by the master station but rather relies on high-precision distributed time synchronization.

Derivation of the Constraint for Δt:

From Equation (10) and the control objective $0 < T_{total} < 20$ ms, we can derive the feasible range for Δt:

1. For Ttotal > 0: T_close − T_open − Δt > 0 => Δt < T_close − T_open

This ensures that the opening operation completes before the closing operation, preventing paralleling.

2. For Ttotal < 20 ms: T_close − T_open − Δt < 20 => Δt > T_close − T_open − 20

   This ensures the load transfer is rapid enough to minimize the outage time.
   Combining both inequalities, the safe operating range for the dynamic delay is

$$T\_close - T\_open - 20 < \Delta t < T\_close - T\_open$$

To ensure robustness against minor prediction inaccuracies and provide a safety margin, the dynamic delay is set to the midpoint of the feasible range:

$$\Delta t \ = (T\_close \ - \ T\_open) - 10 \text{ ms} \tag{11}$$

This setting provides a 10 ms control margin on both sides, ensuring that even if the actual switching times deviate slightly from the predictions, T_{total} is very likely to remain within the 0–20 ms safety window.

5.3. *Physical Experiment Verification and Result Analysis*

Based on the 'dynamic delay Δt' control strategy established in Section 5.2, this section details how we constructed a physical experimental platform to validate the strategy's temporal accuracy and reliability. Through satellite-synchronized timing and multi-device collaborative data acquisition, we quantified whether the time difference between closing and opening operations met engineering requirements. The specific experimental design and analysis of results are as follows.

5.3.1. Experimental Platform Setup

The test equipment included a satellite signal simulator (GSS7000, Divibe Technology Co., Ltd., Shenzhen, China), time synchronization tester (YZ-9910, Chengdu Yinzhong Digital Equipment Co., Ltd., Chengdu, China), oscilloscope (Fluke-190, Fluke Test Instruments (Shanghai) Co., Ltd., Shanghai, China), distribution terminal, and vacuum circuit breaker. Equipment parameters are listed in Table 6 below.

Table 6. Equipment specifications.

Equipment Name	Model	Core Parameters	Effect
Satellite Signal Simulator	GSS7000	Time synchronization accuracy ≤ 10 ns, supporting BeiDou/GPS dual-mode	Provides a high-precision, coherent clock reference
Time Synchronization Tester	YZ-9910	Time resolution: 1 ns; supports contact action moment capture	Records the precise moment of switch closure and opening
Oscilloscope	Fluke-190	Sampling rate ≥ 100 MS/s; bandwidth ≥ 200 MHz	Auxiliary verification switch contact action waveform
Distribution Terminal Unit	Custom Edition	Timing accuracy ≤ 50 ns; supports absolute timestamp instruction execution	Receives commands from the master station and control switch operation
Vacuum Circuit Breaker	VS1-12, Shanghai Jikai Electric Power Technology Co., Ltd., Shanghai, China	Opening time ≤ 50 ms; closing time ≤ 60 ms (nominal value)	Simulated switch gear for actual distribution networks

The experimental test platform is shown in Figure 7.

The experiment simulated load transfer scenarios following distribution network faults. Both the original power source switch S1 and the target power source switch S2 were controlled by distribution terminals, with both terminals acquiring synchronized clocks via the GSS7000. When a main grid fault occurred, S1 had to be opened first, followed by closing S2, to ensure no circulating currents were present.

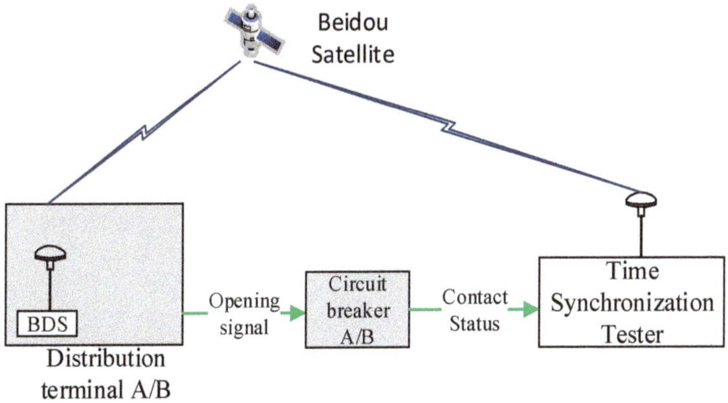

Figure 7. Distribution network load synchronous transfer control time accuracy test.

5.3.2. Experimental Protocol Design

To more comprehensively characterize the experimental conditions, this physical test was conducted under loaded conditions. The applied load matched the simulation model settings (active power of 300 kW, reactive power of 130 kvar), corresponding to a steady-state current of approximately 850 A. The experiment employed a VS1-12 vacuum circuit breaker, whose rated parameters (12 kV/1250 A/31.5 kA) fully encompassed the experimental current range. To prioritize verification of the timing coordination mechanism within the "prediction-control" closed-loop system, the most typical single-phase ground fault scenario in distribution networks was selected as the representative test case. Monitoring was performed using a high-sampling-rate oscilloscope.

The experimental procedure is as follows:

(1) Before starting the test, first ensure that both power distribution terminals are synchronized and operating stably for over 10 min under the synchronized time reference output by the satellite signal simulator to establish a reliable synchronized state.

(2) During testing, the master station system issues control commands with absolute timestamps to distribution terminal A (to open the original power source switch S1) and distribution terminal B (to close the target power source switch S2) based on the switching times predicted by the Bo-BiLSTM model and the calculated dynamic delay, following the "open first and then close" timing logic.

(3) Critical timing points during the switching operation—specifically, the precise moment when breaker A's contacts open and the precise moment when breaker B's contacts close—are captured and recorded in parallel by a high-precision time synchronization tester. This tester also receives the time reference from the satellite signal simulator, ensuring that its recorded data aligns perfectly with the distribution terminal's timeline to eliminate system timing errors. To quantitatively assess control precision, the time difference between Circuit Breaker A's opening moment and Circuit Breaker B's closing moment is calculated. This value represents the actual opening-closing time interval.

(4) The aforementioned test procedure is repeated 100 times to obtain statistically significant results and verify the consistency and reliability of the control strategy across different operations.

5.3.3. Experimental Data Analysis

In all 100 physical experiments, the closing-to-opening time difference consistently fell within the range of 2 ms to 12 ms, with each result strictly meeting the engineering safety threshold of less than 20 ms. The experimental findings conclusively demonstrate that the proposed control strategy reliably maintained the time difference within a narrow range far below the safety upper limit, fundamentally eliminating the risk of circulating currents caused by asynchronous closing. For specific data, see Appendix A, Table A3.

5.3.4. Experimental Conclusions

This paper validated the Bo-BiLSTM-based synchronous load transfer control method for distribution networks through 100 repeated experiments. The experimental results demonstrated the following:

1. High Precision and Safety: In all experimental rounds, the time difference between opening and closing operations was strictly controlled within 2–12 ms. This result not only fully meets the engineering safety requirement of less than 20 ms but also demonstrates that the system possesses sufficient performance margin, fundamentally avoiding the risk of circulating currents caused by asynchronous closing.
2. Exceptional Reliability: No noticeable contact bounce was observed in all 100 repeated operations. All 100 experiments successfully achieved safe transfer, demonstrating that this control strategy exhibited high reliability and consistency rather than merely coincidental effectiveness.
3. Effectiveness of the Predictive-Control Closed Loop: The experiment successfully validated the effectiveness of the dynamic delay control strategy designed based on Bo-BiLSTM high-precision time prediction, achieving a closed loop from "precise sensing" to "precise control." This provides a reliable technical solution for seamless and rapid load transfer in distribution networks following faults.

6. Conclusions

This study proposes a switch timing prediction and synchronous control method for distribution networks based on Bayesian optimization of bidirectional long short-term memory (Bo-BiLSTM) networks. The main contributions of this work are summarized as follows:

1. A Novel Prediction Model for Switching Times

We propose a Bo-BiLSTM-based prediction model that automatically adapts hyperparameters of bidirectional long short-term memory networks by integrating Bayesian optimization techniques. This approach effectively addresses challenges posed by high data dispersion and scarce training samples, significantly improving prediction accuracy and generalization capabilities compared to standard BiLSTM and LSTM models as well as XGBoost models.

2. A Comprehensive and Publicly Accessible Feature Database

We constructed a robust simulation model of a distribution network with distributed generation using MATLAB/Simulink. A comprehensive database containing 620 operational scenarios was established, encompassing various short-circuit fault types, locations, and transition resistances. This database provides a solid foundation for model training and future research in the field.

3. An Innovative Integrated Framework from Prediction to Control

We innovatively bridged the gap between accurate prediction and practical control by integrating the switch timing predictions into a load transfer control strategy. A "open first and then close" synchronous transfer control strategy with a dynamic delay mechanism was designed, ensuring seamless and safe load transfer.

4. Experimental Validation with High Practicality

The effectiveness and reliability of the proposed control strategy were rigorously validated through physical experiments. Experimental results demonstrated that the time difference between opening and closing was consistently controlled within a range of 2–12 ms, fully meeting the stringent engineering requirement of less than 20 ms and effectively mitigating the risk of circulating currents.

In summary, this method provides a reliable fault response and rapid self-healing control solution for distribution networks with high penetration of distributed energy. Future research can be further deepened in the following areas: First, advancing the model's application to real-world data is central to our next steps. We will focus on collecting high-resolution operational data from multiple actual distribution network feeders to validate and optimize the model's predictive performance in complex real-world environments. Second, we will explore more advanced data augmentation and few-shot learning techniques to address the inherent challenge of scarce fault data in practical engineering scenarios, ensuring that the model maintains high accuracy and reliability even with limited data.

Author Contributions: Conceptualization, H.Z. and C.L.; methodology, C.L.; validation, H.Z., C.L., and W.L.; formal analysis, X.S.; investigation, Y.G.; resources, Y.G.; data curation, C.L.; writing—original draft, W.L. and C.L.; writing—review and editing, X.S.; visualization, Y.G.; supervision, H.Z.; project administration, H.Z.; funding acquisition, H.Z. All authors have read and agreed to the published version of the manuscript.

Funding: This research was funded by the project of State Grid Sichuan Electric Power Company Limited, grant number 52199723001R.

Data Availability Statement: The original contributions presented in this study are included in the article. Further inquiries can be directed to the corresponding author.

Acknowledgments: We are grateful to the reviewers for their comprehensive review of this manuscript, as well as for their insightful comments and valuable suggestions that have significantly contributed to enhancing the quality of our work.

Conflicts of Interest: Author Cheng Long was employed by the company State Grid Sichuan Electric Power Research Institute. The remaining authors declare that the research was conducted in the absence of any commercial or financial relationships that could be construed as potential conflicts of interest.

Appendix A

Table A1. Fault characteristic data under different operating conditions.

Operating Order Number	V_A/(V)	V_B/(V)	V_C/(V)	I_A/(A)	I_B/(A)	I_C/(A)	T/(ms)
O_1	28,140	28,760	28,780	106.4	40.78	35.07	26.04
O_2	28,750	27,500	28,790	33.38	276.7	42.67	37.60
O_3	28,730	28,740	28,730	35.99	36.01	35.99	31.70
O_4	28,730	28,750	26,540	36.92	36.35	388.9	16.85
O_5	28,740	28,790	27,970	40.8	37.22	119.2	27.73
⋮	⋮	⋮	⋮	⋮	⋮	⋮	⋮
O_{620}	28,730	27,250	25,710	35.96	574.3	542.7	36.889

Table A2. Standardized database of feature matrices.

	V_A	V_B	V_C	I_A	I_B	I_C
O_1	0.4144	0.9822	0.9911	−0.8120	−0.9799	−0.9945
O_2	0.6390	0.4209	0.9956	−0.9988	−0.3763	−0.9751
O_3	0.6317	0.9733	0.9689	−0.9921	−0.9921	−0.9921
O_4	0.6317	0.9777	−0.0044	−0.9898	−0.9912	−0.0893
O_5	0.6354	0.9955	0.6311	−0.9798	−0.9890	−0.7793
⋮	⋮	⋮	⋮	⋮	⋮	⋮
O_{620}	0.6317	0.3096	−0.3733	−0.9922	0.3850	0.3042

Table A3. Accuracy test data for timed action of distribution terminals.

No.	The Moment When Circuit Breaker A Opens	The Closing Time of Circuit Breaker B	Time Difference (ms)	Is the Difference Between the Switch Operation Time and the Model Prediction Time Less than 0.1 ms?
1	8:50:27.244	8:50:27.251	7	Yes
2	8:52:51.831	8:52:51.841	10	Yes
3	9:04:10.433	9:04:10.443	10	Yes
4	9:09:11.311	9:09:11.316	5	Yes
5	9:15:10.211	9:15:10.223	12	Yes
6	9:20:20.430	9:20:20.433	3	Yes
7	9:25:40.556	9:25:40.559	3	Yes
8	9:29:30.466	9:29:30.477	11	Yes
9	9:35:10.876	9:35:10.879	3	Yes
10	9:40:20.431	9:40:20.438	7	Yes
11	10:50:27.244	10:50:27.251	7	Yes
12	10:20:20.430	10:20:20.438	8	Yes
13	10:25:47.556	10:25:47.559	3	Yes
14	10:29:30.466	10:29:30.477	11	Yes
15	10:15:10.211	10:15:10.223	12	Yes
16	10:20:20.430	10:20:20.439	9	Yes
17	10:25:40.556	10:25:40.558	2	Yes
18	10:29:30.466	10:29:30.477	11	Yes
19	10:35:10.876	10:35:10.879	3	Yes
20	10:40:20.431	10:40:20.438	7	Yes
⋮	⋮	⋮	⋮	⋮
100	21:24:40.556	21:24:40.558	2	Yes

References

1. Jiang, J.; Xin, P.; Xiao, N.; Li, Q. A Multi-Domain Protection for Reliable Slicing in Network Coding Based 5G/B5G RAN Enabled Power Distribution Network. In Proceedings of the 2024 22nd International Conference on Optical Communications and Networks (ICOCN), Harbin, China, 26–29 July 2024; pp. 1–3. [CrossRef]
2. Li, X.; Yao, Z.; Zhou, L.; Zhou, S. Dispersed operating time control of a mechanical switch actuated by an ultrasonic motor. *J. Vibroeng* **2018**, *20*, 321–331. [CrossRef]
3. Zhang, J.; Wang, P.; Yan, R.; Gao, R.X. Long short-term memory for machine remaining life prediction. *J. Manuf. Syst.* **2018**, *48*, 78–86. [CrossRef]
4. Gugulothu, N.; Tv, V.; Malhotra, P.; Vig, L.; Agarwal, P.; Shroff, G. Predicting remaining useful life using time series embed-dings based on recurrent neural networks. *Int. J. Progn. Health Manag.* **2018**, *9*. [CrossRef]
5. Lu, Q.; Polyzos, K.D.; Li, B.; Giannakis, G.B. Surrogate modeling for Bayesian optimization beyond a single Gaussian process. *IEEE Trans. Pattern Anal. Mach. Intell.* **2023**, *45*, 2891–2904. [CrossRef] [PubMed]
6. Burke, J.; King, S. Edge tracing using Gaussian process regression. *IEEE Trans. Image Process.* **2021**, *31*, 138–148. [CrossRef] [PubMed]
7. Hochreiter, S.; Schmidhuber, J. Long short-term memory. *Neural Comput.* **1997**, *9*, 1735–1780. [CrossRef] [PubMed]
8. Babu, G.S.; Zhao, P.; Li, X.-L. Deep Convolutional Neural Network Based Regression Approach for Estimation of Remaining Useful Life. In *Database Systems for Advanced Applications, Proceedings of the 21st International Conference, DASFAA 2016, Dallas, TX, USA, 16–19 April 2016*; Navathe, S., Wu, W., Shekhar, S., Du, X., Wang, X., Xiong, H., Eds.; Lecture Notes in Computer Science; Springer: Cham, Switzerland, 2016; Volume 9642. [CrossRef]
9. Guo, L.; Lei, Y.; Li, N.; Xing, S. Deep convolution feature learning for health indicator construction of bearings. In Proceedings of the 2017 Prognostics and System Health Management Conference (PHM-Harbin), Harbin, China, 9–12 July 2017; pp. 318–323. [CrossRef]

10. Zhang, S.; Wang, L.; Zhu, M.; Chen, S.; Zhang, H.; Zeng, Z. A Bi-directional LSTM Ship Trajectory Prediction Method based on Attention Mechanism. In Proceedings of the 2021 IEEE 5th Advanced Information Technology, Electronic and Automation Control Conference (IAEAC), Chongqing, China, 12–14 March 2021; pp. 1987–1993. [CrossRef]
11. Guan, L.; Qiao, F.; Zhai, X.; Wang, D. Model Evolution Mechanism for Incremental Fault Diagnosis. *IEEE Trans. Instrum. Meas.* **2022**, *71*, 3522111. [CrossRef]
12. Liu, X.; Yan, F.; Song, C.; Yin, X.; Wang, Y. A research on Bayesian optimisation of bilstm neural network algorithms-application to photovoltaic power prediction. In Proceedings of the 2024 4th International Conference on Neural Networks, Information and Communication Engineering (NNICE), Guangzhou, China, 19–21 January 2024; pp. 1544–1547. [CrossRef]
13. Dong, E.; Guo, Y. The Bayesian CNN-LSTM Mixed Hybrid Algorithm Model of the Photovoltaic Short-term Output Forecasting. In Proceedings of the 2023 3rd International Conference on New Energy and Power Engineering (ICNEPE), Huzhou, China, 24–26 November 2023; pp. 242–246. [CrossRef]
14. Ma, Y.; He, Y.; Wang, L.; Zhang, J. Probabilistic reconstruction for spatiotemporal sensor data integrated with Gaussian process regression. *Probabilistic Eng. Mech.* **2022**, *69*, 103264. [CrossRef]
15. Hodson, T.O. Root mean square error (RMSE) or mean absolute error (MAE): When to use them or not. *Geosci. Model Dev. Discuss.* **2022**, *15*, 5481–5487. [CrossRef]
16. Alturki, Y.A.; Alhussainy, A.A.; Alghamdi, S.M.; Rawa, M. A novel point of common coupling direct power control method for grid integration of renewable energy sources. *Energies* **2024**, *17*, 5111. [CrossRef]
17. Raj, R.D.A.; Bhattacharjee, S. Short-circuit fault analysis of three-phase grid integrated photovoltaic power system. In Proceedings of the 2020 International Conference on Power Electronics & IoT Applications in Renewable Energy and its Control (PARC), Mathura, India, 28–29 February 2020; pp. 231–236. [CrossRef]
18. Jiang, W.-J.; Kim, C.-W.; Goi, Y.; Zhang, F.-L. Data normalization and anomaly detection in a steel plate-girder bridge using LSTM. *ASCE-ASME J. Risk Uncertain. Eng. Syst. Part A Civ. Eng.* **2022**, *8*, 2022. [CrossRef]
19. Ran, J.; Cui, Y.; Xiang, K.; Song, Y. Improved runoff forecasting based on time-varying model averaging method and deep learning. *PLoS ONE* **2022**, *17*, e0274004. [CrossRef]

processes

MDPI

Article

Intelligent Disassembly System for PCB Components Integrating Multimodal Large Language Model and Multi-Agent Framework

Li Wang [1,*], Liu Ouyang [1], Huiying Weng [1], Xiang Chen [1], Anna Wang [1] and Kexin Zhang [2,*]

[1] State Grid Zhejiang Electric Power Co., Ltd., Hangzhou 310015, China
[2] Institute of Cyber-Systems and Control, Zhejiang University, Hangzhou 310027, China
* Correspondence: wang_li@zj.sgcc.com.cn (L.W.); zhangkexin@zju.edu.cn (K.Z.)

Abstract

The escalating volume of waste electrical and electronic equipment (WEEE) poses a significant global environmental challenge. The disassembly of printed circuit boards (PCBs), a critical step for resource recovery, remains inefficient due to limitations in the adaptability and dexterity of existing automated systems. This paper proposes an intelligent disassembly system for PCB components that integrates a multimodal large language model (MLLM) with a multi-agent framework. The MLLM serves as the system's cognitive core, enabling high-level visual-language understanding and task planning by converting images into semantic descriptions and generating disassembly strategies. A state-of-the-art object detection algorithm (YOLOv13) is incorporated to provide fine-grained component localization. This high-level intelligence is seamlessly connected to low-level execution through a multi-agent framework that orchestrates collaborative dual robotic arms. One arm controls a heater for precise solder melting, while the other performs fine "probing-grasping" actions guided by real-time force feedback. Experiments were conducted on 30 decommissioned smart electricity meter PCBs, evaluating the system on recognition rate, capture rate, melting rate, and time consumption for seven component types. Results demonstrate that the system achieved a 100% melting rate across all components and high recognition rates (90–100%), validating its strengths in perception and thermal control. However, the capture rate varied significantly, highlighting the grasping of small, low-profile components as the primary bottleneck. This research presents a significant step towards autonomous, non-destructive e-waste recycling by effectively combining high-level cognitive intelligence with low-level robotic control, while also clearly identifying key areas for future improvement.

Keywords: waste printed circuit boards; waste of electrical and electronic equipment; automatic disassembly; multimodal large language model; multi-agent system

1. Introduction

The issue of Waste Electrical and Electronic Equipment (WEEE or e-waste) indeed constitutes a major global challenge due to its rapidly increasing volume worldwide [1]. As highlighted in the UN Digital Economy Report 2024 [2], the negative environmental footprint of digital transformation, including the surge in e-waste, necessitates urgent circular economy interventions. This waste stream encompasses a wide variety of devices, including mobile phones, computers, televisions, refrigerators, household appliances, lamps, medical equipment, and photovoltaic panels [3–5]. E-waste contains hazardous substances that can cause significant environmental and health problems if not managed properly. Furthermore, modern electronic products contain rare and valuable resources that can be recycled and reused, contributing to a circular economy and enhancing the security of supply for critical raw materials [6–8].

Dismantling or disassembly is widely regarded by numerous researchers as the initial and pivotal stage in the product recovery lifecycle. This is because the effectiveness of the disassembly operation significantly influences the outcomes of subsequent stages in the recycling process [9]. An effective disassembly process can also help reduce the excessive use of reagents in chemical processes used to separate various valuable metals. Additionally, it presents a non-destructive alternative to techniques like shredding, which, while faster, limits the usability of recovered materials due to its detrimental impact on the structural integrity of components [10].

With the high-quality development of smart grids and the arrival of large-scale replacement cycles for smart electricity meters, the environmentally sound treatment and resource recovery of massive quantities of decommissioned meters have become a critical bottleneck for the green development of the power industry. This paper, therefore, focuses on exploring an efficient disassembly and recycling strategy for waste electricity meter circuit boards, aiming to recover valuable resources and reduce environmental pollution.

The core value and the primary recycling challenge of electricity meters lie in their internal precise circuit boards. If components such as chips, sensors, and connectors integrated on these boards can be disassembled non-destructively and recovered at high value, it would yield significant economic and environmental benefits.

In the field of PCB component disassembly and recycling, while current automation research has made preliminary progress, it still faces several core technical challenges. For instance, Santos et al. [11] developed a robotic system for automated disassembly of PCB components. This system successfully achieved desoldering and extraction of larger components with a near 100% success rate, utilizing a custom push-pull tool and precise force-motion control. The study clearly delineated the disassembly process into six stages: approach, contact, melting, grasping, transport, and release, and recorded key force data. However, this approach also reveals limitations: its tool requires customization based on component size, is prone to collisions with densely packed smaller components, and exhibits a relatively low success rate in the grasping phase. This highlights the shortcomings of existing technologies in terms of adaptability, dexterity, and intelligent decision-making capabilities. Current disposal methods predominantly rely on manual labor or semi-automated equipment, which suffer from low efficiency, high safety risks, and a high probability of damaging valuable components. Although researchers have proposed automation solutions based on machine vision and robotic arms, effectively achieving tasks like automatic screw removal and board handling, they remain inadequate for component-level fine disassembly tasks. These include unplugging various interfaces, handling different soldering methods, and manipulating densely arranged fragile components. The flexibility, adaptability, and intelligent decision-making capabilities of existing technologies fall short of the high requirements for non-destructive disassembly. Traditional rule-based or pre-programmed systems often rely on rigid sequences, which struggle to adapt to the diverse and non-standardized layouts of decommissioned PCBs. By utilizing a Multimodal Large Language Model (MLLM) as the intelligent core, our system achieves a transition from rigid automation to cognitive-driven reasoning, allowing for a deeper semantic understanding of complex board layouts. To bridge this gap, we utilize a Multi-modal Large Language Model (MLLM) as the "intelligent core," enabling the system to transition from rigid automation to cognitive-driven reasoning.

https://doi.org/10.3390/pr14020227

To address the challenges mentioned above, this paper proposes an intelligent disassembly system and method for PCB components based on a multimodal large model and a multi-agent framework. The core innovation of this research lies in the deep integration of high-level cognitive intelligence with low-level dexterous control. Specifically, the system utilizes a multimodal large model as its "intelligent core," endowing the robot with powerful visual-language understanding and task planning capabilities. This enables it to parse complex natural language instructions and decompose them into specific, executable action sequences. Building upon this, we designed a multi-agent framework to control the dual dexterous hands performing the disassembly tasks collaboratively. One arm is responsible for applying heat, while the other utilizes force feedback to perform fine "probing-grasping" actions. A dedicated classification and registration agent could be employed for OCR recognition, categorization, and data logging of the disassembled chips.

In summary, our innovations are as follows:

- We have proposed a novel intelligent PCB disassembly system, exploring the systemic feasibility of using robotic arms to achieve the entire process of waste PCB recycling, including identification and disassembly to recycling and storage.
- Addressing the challenges of lacking flexibility and intelligent decision-making in existing rule-based robotic disassembly systems for waste PCBs, we proposed a solution that utilizes multimodal large models for intelligent perception of circuit boards and employs multi-agent systems to achieve flexible decision-making for the multi-stage disassembly process.
- Experimental results demonstrate that our proposed system can achieve efficient identification of components such as chips and diodes, with a recovery rate of up to 80% for various types of components.

2. Related Works

2.1. Traditional Disassembly Methods

Waste printed circuit boards (WPCBs), as important secondary resource carriers, contain precious metal components such as gold, silver, platinum, and palladium [12,13]. The distribution characteristics of these metals are closely related to the types of components. Taking electronic pins as an example, their precious metal content is relatively low. Currently, the resource recovery treatment of WPCBs mainly revolves around three technical pathways: thermal disassembly [14,15], mechanical separation [13,16], and chemical treatment [17,18], which provide effective approaches for the sustainable recycling of precious metals.

2.1.1. Thermal Disassembly Technology

Thermal treatment of WPCBs using customized or commercial heating devices is a common method. Heat sources can include electricity, gas, or open flames, but in practical applications, issues such as insufficient temperature control precision and lack of exhaust gas treatment systems often exist, posing potential threats to operator safety and the ecological environment. Modern industrial equipment has begun to employ engineering control systems to effectively treat toxic gases through processes such as adsorption, separation, and decomposition.

Thermo-fluid heating technology involves immersing WPCBs in a silicone oil or paraffin oil bath at 215–250 °C [19–21], combined with vibrators and ultrasonic generators to improve the separation efficiency of electrochemical cells (ECs). The wastewater generated by this process requires specialized treatment. Although ionic liquids (ILs) offer advantages in stability and safety, their cost control, recyclability, and process stability remain unresolved issues [22,23].

https://doi.org/10.3390/pr14020227

Infrared radiation technology uses radiators arranged along a conveyor belt to liquefy solder, enabling the removal of electronic components through gravity separation or robotic grasping. Residual components are then cleared by brushing or shearing [24]. Park's team achieved a 94% disassembly rate using a rotating steel brush structure at 250 °C and a speed of 0.33 cm/s [25].

High-temperature centrifugal method achieves efficient recovery of metals such as copper, lead, and tin under conditions of 1400 rpm and 240 °C. Experiments show that under 1300 °C conditions, the recovery rate of multiple metals can exceed 90%, but the temperature range must be strictly controlled to prevent the release of toxic substances [26,27]. Hot air disassembly technology uses a 260 °C portable hot air gun. Although its efficiency is lower than that of infrared methods, it effectively avoids the risk of component overheating. The industrial waste heat and steam coupling system developed by Chen et al. significantly improved temperature control precision [24,28].

The solder bath technique involves briefly immersing WPCBs in molten solder at 230 °C but carries the risk of overheating. In China, manual hammering disassembly is commonly used. Even with ventilation equipment, health risks persist, and molten metal splashing can form hazardous waste [29].

2.1.2. Mechanical Separation Technology

Informal recycling sectors often use tools such as chisels and hammers for manual disassembly, focusing on recovering high-value components like aluminum heat sinks, batteries, and capacitors. However, this operational model significantly reduces the resource value of the remaining materials, affecting the economic benefits of formal recycling enterprises [30]. Surface cutting blade technology, as a core equipment of automated recycling systems, can efficiently liberate the PCB structure while maintaining the integrity of components, providing key technical support for improving recovery rates [10,31].

2.1.3. Chemical Treatment Technology

Hydrometallurgical technology achieves efficient metal leaching under optimized conditions using strong acid media such as hydrochloric acid and nitric acid [32]. Chloride leaching systems utilize copper(II) ions to promote metal dissolution, creating conditions for subsequent electrochemical recovery. Bioleaching technology relies on microbial action to extract metals such as gold and copper, significantly reducing chemical usage. Electrodeposition technology improves the selectivity and efficiency of metal recovery through ionic liquids or molten salt media [33]. Compared with thermal and mechanical methods, chemical methods offer four main advantages: (1) Significant reduction in toxic gas emissions. (2) Effective control of energy consumption costs. (3) Minimized thermal damage to reusable components. (4) Reduced impact of residual solder on subsequent processes. However, chemical treatment methods are associated with higher costs, which hinders the alignment of economic and environmental benefits in WPCB source processing.

2.2. Robotic Disassembly Fundamentals

The process of removing components from a PCB requires the robotic tool to contact both the board and its individual components [11]. Therefore, it is imperative to control the robot's forces and positions to ensure operational safety and effectiveness [34]. The tools used for desoldering, grasping, and transporting PCB components face a series of obstacles due to the relatively small size of the components. Automating this process requires the integration of diverse components into architectures that ensure functionality, namely real-time force and motion control mechanisms for the robot.

https://doi.org/10.3390/pr14020227

2.2.1. Control Methods in Robotics

In robotics literature, force control is often categorized into two core strategies: passive and active. The essence of passive strategy lies in constraining the applied forces within a predefined range to ensure the smooth execution of tasks. The active strategy adjusts contact forces in real time based on the efficiency level required for the task, which necessitates accurate perception of force magnitude. Thus, real-time awareness of contact forces is crucial for effective force control [35]. Active control can be further divided into direct and indirect types according to their implementation principles. Direct control separately handles the force and position control loops. Indirect control aims to enable the robot's end-effector to exhibit the desired compliant dynamic behavior during interaction with the environment [36], with impedance control and admittance control being two typical forms of indirect control for this purpose [37]. At the implementation level, explicit approaches measure environmental contact forces directly using force sensors, while implicit approaches estimate the forces indirectly through other physical quantities. For robotic tasks that involve contact with the external environment, such as polishing or printed circuit board (PCB) disassembly, the compliance of the robot allows it to flexibly adapt to the workspace [38,39]. The grasping and manipulation of PCB components can be accomplished using various grippers or customized tools, which may employ mechanically actuated methods with multiple contact points or rely on principles such as vacuum suction or magnetic actuation [40,41].

2.2.2. Recent Advances and Persistent Challenges

The automatic disassembly of e-waste, specifically the removal of PCBs from mobile phones, was addressed in the European project ADIR [42]. The proposed approach covers multiple stages, such as image processing, robotic handling, and automated sorting of parts into different material streams. More recently, a system was developed for the automated recovery of CPUs from cell phones [43]. This system utilizes eye-to-hand visual servoing, employs hot air for desoldering, and uses a gripper for component transfer. Separately, in 2019, Apple unveiled a machine that performs automatic iPhone disassembly, where components are separated according to their material composition [44]. Recently, Santos et al. [11] developed a robotic system for automating the disassembly of PCB components. While it achieved high success rates for large components using custom tools, it lacks adaptability for densely packed small components and requires manual tool adjustment. Similarly, Park's team utilized a mechanical brushing method reaching a 94% disassembly rate [25], but posed higher risks of component damage compared to our precision thermal-force control. However, a significant challenge remains: the lack of a fine, intelligent, and universally adaptable method for separating components from PCBs without compromising their integrity, especially for small, densely packed components. Our work aims to bridge this gap through the proposed intelligent system.

3. Materials and Methods

3.1. The Role of Multi-Agent and Multi-Modal in Waste Printed Circuit Board Disassembly

The integration of Multi-agent systems and Multi-modal learning represents a transformative paradigm for intelligent robotic disassembly of Waste Printed Circuit Boards (WPCBs). This integrated architectural approach is of particular benefit to tasks requiring complex coordination and high-level cognitive understanding, which are fundamental challenges in the disassembly of WPCBs. The term "Multi-agent system" in this context refers to a coordinated framework where multiple software agents, each specializing in a specific function (e.g., visual comprehension, temperature control, robotic manipulation), collaborate to achieve the common objective of non-destructive disassembly. It is clear that

https://doi.org/10.3390/pr14020227

traditional methods, including early automation attempts, rely on rigid, pre-programmed sequences or inefficient manual procedures. This inevitably results in low adaptability, high component damage rates, and limited economic viability. Unlike conventional hard-coded algorithms, MLLM-based reasoning allows the system to parse natural language instructions and perform autonomous task planning. By integrating RAG technology, the framework can dynamically retrieve disassembly strategies from a knowledge base, mimicking human-like expertise in handling varying heating profiles and grasping forces, which is essential for universal applicability across different products.

Concurrently, Multi-modal learning serves as the intelligent core that empowers these agents. It enables the system to process and fuse diverse data types—primarily visual information from high-definition cameras and thermal data from infrared sensors—to achieve a human-like understanding of the disassembly scene. This deep semantic understanding allows the system to parse complex board layouts, identify component types, and crucially, generate context-aware disassembly strategies. The real-time processing of multi-modal data markedly enhances the system's perception and decision-making capabilities, facilitating more dexterous and precise manipulation of components.

The synergy between these two technologies is the key to enhancing PCB disassembly. The multi-agent framework provides the structural basis for distributed and concurrent task execution, while multi-modal learning supplies the necessary intelligence for situational awareness and planning. This cohesive integration is revolutionizing the WPCB recycling process, boosting its efficiency, sustainability, and economic viability by enabling a level of flexibility and intelligence previously unattainable with conventional automation.

This chapter elaborates on an intelligent PCB component disassembly system based on a multi-modal large model and a multi-agent framework. Its core innovation lies in constructing a hierarchical intelligent architecture and applying it to the field of circuit board disassembly. This architecture achieves human-like fine disassembly by integrating high-level semantic understanding with low-level imitation of human skills. The overall architecture of the system is shown in Figure 1, which includes the following core parts: the Perception and Strategy Module, the Task Allocation Module, the Collaborative Disassembly Module, and the Identification and Registration Module.

3.2. Perception and Strategy Module

The Perception and Strategy Module consists of a hardware-side multi-modal perception unit and a visual comprehension and planning agent. Since all components are arranged on a single side of the PCB, they can all be captured in a single image by a camera. Therefore, our multi-modal perception unit includes a high-definition camera. Furthermore, to achieve precise component disassembly, we plan to use a hot air gun to melt the solder and a gripping tool for component retrieval. Consequently, we have also placed a thermal imager next to the camera to obtain real-time temperature readings of the components.

The visual comprehension and planning agent first invokes the sensors of the perception unit to receive image input $X_{img} \in R^{W \times H \times C}$. The agent utilizes the multi-modal large model to achieve deep semantic understanding of the image, converting it into a corresponding textual description $D_W = \{D_i\}_{i=1}^{N}$. This description includes the position and category of electronic components, achieving a coarse-grained understanding of the image. While multimodal large language models (MLLMs) offer holistic image understanding, their localization precision is often insufficient. To compensate for this, we integrated the YOLOv13 [45] object detection algorithm, which features hypergraph-enhanced adaptive visual perception, to obtain high-precision location descriptions (D_L).

$$D_W = MLLM(Q, X_{img})$$

https://doi.org/10.3390/pr14020227

$$D_L = YOLO(X_{img})$$

where Q is the prompt used to guide the MLLM in obtaining a standardized description. Subsequently, the holistic semantic description D_W and the fine-grained location data D_L are synergistically integrated to form a comprehensive and precise representation of the components on the circuit board image, named D.

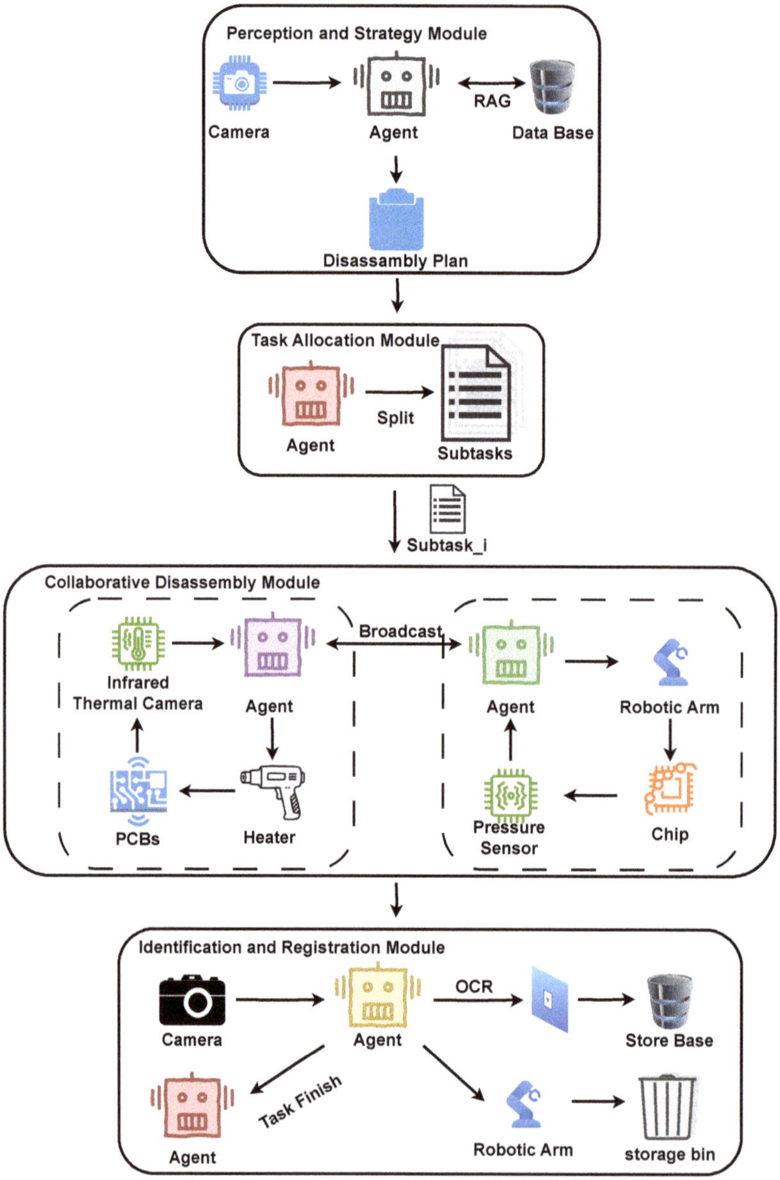

Figure 1. Architectural diagram of the disassembly strategy for electricity meter circuit boards proposed in this paper, composed of the Perception and Strategy Module, Task Allocation Module, Collaborative Disassembly Module, and Identification and Registration Module.

https://doi.org/10.3390/pr14020227

Subsequently, based on D, the agent retrieves disassembly methods and information S for the corresponding component from a chip knowledge base using RAG methods. To ensure the system's universality in the absence of specific board design documents, we utilized Adaptive RAG [46] to dynamically retrieve disassembly methods from a general chip knowledge base, allowing the system to adapt to unknown PCB layouts. Both D and S are used as the prompt for an LLM to generate the overall component disassembly strategy P, which includes the target coordinates, package type, required temperature profile, and grasping strategy.

$$P = LLM(D, S)$$

3.3. Task Allocation Module

The Task Allocation Module consists of an orchestrator agent. This agent decomposes the overall disassembly strategy P into multiple subtasks and allocates them to the corresponding executive agents in the subsequent modules. This enables the orderly allocation of the task plan from the overall level down to the individual modules.

3.4. Collaborative Disassembly Module

The Collaborative Disassembly Module consists of a temperature control agent, a robotic arm control agent, a heater, and a robotic arm. During the complete disassembly process, after the temperature control agent receives a temperature control task from the Task Allocation Module, it initiates a closed-loop temperature control process. The temperature map of the circuit board, captured by the infrared thermal imager, is input into the multi-modal large model by the agent to obtain a semantic modeling of the temperature distribution. The agent then compares this with the required temperature profile from the assigned task, deriving the control signal for the heater in the next heating phase. This signal is sent to the heater for execution, achieving closed-loop temperature control. The temperature control agent ensures the target solder pad area reaches the melting point while keeping other areas within safe temperature thresholds. Once the target pad temperature is achieved, the temperature control agent broadcasts a signal to the robotic arm control agent and the orchestrator agent to initiate the robotic arm disassembly sequence and obtain the instructions and parameters for the next step.

At this point, the agents will control the heater and the robotic arm to move to the target component position based on the instructions. The heater is controlled by the temperature control agent, and the robotic arm is controlled by the robotic arm control agent. During the heating process, the robotic arm applies a periodic pulling force of less than 0.05 N to the component to check the melting state of the solder. When the pressure sensor on the robotic arm detects a sudden change in pressure, it indicates the solder is fully melted. The robotic arm control agent then sends an instruction to the temperature control agent to coordinate gripping the target component and stop heating. This completes one full component removal cycle.

3.4.1. Identification and Registration Module

The Identification and Registration Module is responsible for identifying and logging the disassembled components. The robotic arm from the Collaborative Disassembly Module moves the gripped component to a dedicated camera. The registration agent performs OCR on the component image I_2 captured by the registration camera to obtain the component's model and parameter information. This information is cross-verified with nearby components on the PCB and the knowledge base. Subsequently, the registration agent records data such as the component model, source (PCB ID), disassembly time, and thermal profile during heating into a database. Then, the registration agent controls the robotic arm to place the component into the corresponding storage bin and sends a completion mes-

https://doi.org/10.3390/pr14020227

sage to the orchestrator agent. At this point, the system has completed the identification, disassembly, recording, and storage steps for one component. The orchestrator agent will then proceed to send the disassembly task for the next component until all components are disassembled and stored.

3.4.2. Experimental Setup

The experimental setup comprises two main parts: Knowledge Base Construction, Model Pretraining and Experimental Design. In this experiment, Gemini 2.5 [47] was employed as the model for the agents and multi-modal model via API calls. All experiments were conducted on a setup equipped with four RTX 4090 GPUs. The core framework was developed using Python 3.10. During the pre-training of the YOLOv13 model, hyperparameters such as learning rate and epochs were optimized following the protocol in Reference [48] to ensure convergence on the specific PCB component detection task. The multi-agent core and semantic understanding were implemented via Gemini 2.5 API calls, while Chroma served as the vector database for Knowledge Base management. The disassembled circuit board used in the experiment is shown in Figure 2.

Knowledge Base Construction. To build a specialized vector knowledge base for electricity meter circuit boards, we aggregated a comprehensive set of documents. This included board design documents, component datasheets, maintenance guides, historical manual disassembly records, and repair reports. These materials contain invaluable information such as component parameters, soldering techniques, locations of potential sensitive components, and historical failure points. A crucial part of this collection is the Bill of Materials (BOM), which details every component on the board with information like reference designators, model numbers, and package types. We also incorporated data parsed by experts from schematic diagrams and PCB layout files, including netlists, component coordinates, and pad dimensions.

This diverse dataset underwent a rigorous process of text extraction (from images where necessary), data cleaning, and segmentation to form a complete and structured data collection. Subsequently, we utilized the Gemini embedding API to convert the textual information into vector representations. These vectors were then stored in a vector database using Chroma. Each vector is associated with its corresponding component data, facilitating manual verification and retrieval.

(a) (b) (c)

Figure 2. Schematic diagrams of the different types of electricity meter circuit boards used in the experiment. The circuit boards of different models vary significantly in board size, electronic component layout, and device types. (**a**) Single-phase electricity meter (**b**) Single-phase electricity meter (**c**) Three-phase electricity meter.

https://doi.org/10.3390/pr14020227

Model Pretraining. Building upon the established methodology, we pre-trained the YOLOv13 model for the specific task of component detection on circuit boards, employing the same training protocol as detailed in Reference [48].

Experimental Design. To objectively and comprehensively evaluate the performance of the intelligent PCB disassembly system based on the multimodal large model and multi-agent framework proposed in this paper, we have designed a comprehensive experimental validation scheme.

The experimental sample consists of 30 different decommissioned electricity meter PCB boards of the same model. To ensure a substantial and comparable dataset across all test conditions, the experimental setup involved selecting 20 target components from each PCB board for disassembly. Detailed information regarding these target components is provided in Table 1.

Table 1. Description of the target electronic components selected for disassembly in the experiment.These components exhibit significant variations in size, shape, function, and disassembly difficulty, thereby demonstrating the comprehensiveness and generalizability of the experimental data selection.

Component	Type	Reason
Integrated circuit (IC)	Composite	An integrated circuit (IC) is fabricated on a semiconductor substrate, typically silicon, which is doped with impurities to form p-n junctions. These components are interconnected via metallic pathways, usually made from alloys. The packaging material may consist of plastic or ceramic.
Aluminum Capacitor (AC)	Composite	This component comprises an aluminum foil and an aluminum oxide coating, along with a liquid or gel electrolyte. These distinct materials are physically combined to enable its function.
Tantalum Capacitor (TC)	Composite	It is composed of tantalum powder and an electrolyte. The tantalum powder forms a porous structure with a large surface area, coated with an oxide layer that acts as a dielectric; this structure is physically integrated with the electrolyte.
Diode	Alloy/Composite	It is constructed from semiconductor materials, such as silicon or germanium, which are doped with impurities and feature metal contacts that may be alloys. The use of doped semiconductors and metallic connections is essential to its function.
Transistor	Alloy/Composite	Like diodes, transistors are manufactured from semiconductor materials doped with impurities and equipped with metal (alloy) contacts. These materials physically influence the flow of electrons.
Resistor	Composite	This component can be produced using a carbon film, metal film, or metal oxide film coated onto an insulating substrate. The body and leads employ various materials, which are physically connected to form the resistive device.
Inductor	Composite	An inductor consists of a copper coil wound around a magnetic core, which may be air, ferrite, or an iron alloy. The coil and core are physically separate but interact magnetically.

Throughout the disassembly process, system logs, high-speed cameras, and precision measuring instruments were synchronized to record detailed data for each task. This meticulously collected dataset serves as the foundational raw material for all subsequent calculations and analysis, ensuring the validity and reliability of the evaluation.

https://doi.org/10.3390/pr14020227

4. Results

We conducted comprehensive testing and evaluation of the proposed disassembly system on the specific task of disassembling and recycling electricity meter circuit boards. The primary objective was to assess the system's practical performance in a controlled environment, with a particular focus on its accuracy in handling individual electronic components. The evaluation protocol was designed to quantify the system's precision at each critical stage: from the initial identification and localization of components on the board to their final successful recovery. Each component processed by the system was meticulously tracked and its outcome was recorded. The performance was rigorously measured based on the recognition and recovery accuracy for various types of electronic components. Due to limitations in experimental equipment and technology, we conducted experiments using only the proposed method.

The quantitative results of this assessment, detailing the accuracy statistics for each component type, are systematically compiled and presented in Table 2 below. Overall, the system demonstrates exceptional reliability in the initial recognition and melting phases but exhibits variable performance in the subsequent pick-up stage.

Table 2. Experimental Results for the Disassembly of Various Components Achieved by the Proposed Framework.

Component	Recognition Rate (%)	Capture Rate (%)	Melting Rate (%)	Average Time Consumption (s)
IC	100	80	100	168.5
AC	100	70	100	112.3
TC	93.3	36.6	100	109.6
Diode	96.7	46.7	100	98.1
Transistor	100	66.7	100	100.4
Resistor	90.0	56.7	100	106.7
Inductor	93.3	70	100	124.0

5. Discussion

As shown in Table 2, the system achieved a 100% recognition rate for Integrated Circuits (ICs), Aluminum Capacitors (ACs), and Transistors, with rates for other components all exceeding 90%. This validates the effectiveness of the strategy combining the Multimodal Large Language Model (MLLM) with the YOLOv13 algorithm in complex PCB scenarios, enabling accurate parsing of component layout and type. The slightly lower recognition rates for Resistors (90.0%) and Tantalum Capacitors (TCs, 93.3%) can be attributed to their small size, faint markings, or low contrast against the background, posing greater challenges to the vision system's resolution.

What's more, The system achieved a 100% melting rate for all components, which is fundamental to successful disassembly. This accomplishment stems from the effectiveness of the closed-loop temperature control strategy. As illustrated in the system architecture (Figure 1), the temperature control agent, guided by real-time feedback from the thermal imager, precisely regulates the heater output. This ensures the target solder pad reaches the melting point while preventing overheating in surrounding areas.

The significant variance in capture success rate (Capture Rate) is the most critical finding of this experiment, revealing the system's main current bottleneck. Capture rate is highly correlated with the physical characteristics of the components (size, shape, weight, solder joint strength). Success rates are higher for larger, more robust components like Integrated Circuits (ICs, 80%) and Inductors (70%). In contrast, capture rates plummet for

https://doi.org/10.3390/pr14020227

small, low-profile components like Tantalum Capacitors (TCs, 36.6%) and Diodes (46.7%). This bottleneck originates from the limitations of the current grasping strategy.

The significant variance in the capture success rate, as presented in Table 2, reveals a strong correlation between a component's physical characteristics and the system's execution efficacy. Larger components with robust geometries, such as Integrated Circuits (80%) and Inductors (70%), provide sufficient surface area for stable grasping and exhibit higher tolerance to positional offsets. Conversely, the capture rate drops significantly for components with smaller dimensions and high solder-joint-to-volume ratios, suggesting that physical scale and surface geometry are the primary determinants of grasping stability in autonomous disassembly.

We have also added a more detailed case of the robotic arm grasping chips during the experiment. Figure 3 illustrates the moment when the robotic arm uses YOLO solely for chip recognition, while Figure 4 shows the pressure feedback curve from the pressure sensor during the grasping process.

As suggested by the force curve diagram (Figure 4), the "probing-grasping" phase is where most failures occur for miniature components. Specifically, for Tantalum Capacitors (36.6% capture rate), the failure is attributed to two primary physical factors: (1) the current end-effector lacks the ultra-fine force resolution required to counteract the surface tension and residual adhesion of molten solder without displacing the component; (2) the lack of adaptive tactile sensing makes it difficult to maintain a stable equilibrium between the pulling force and the solder's phase-change resistance. This mechanical bottleneck indicates that super-fine force control and specialized micro-grippers are essential for handling low-profile, fragile components.

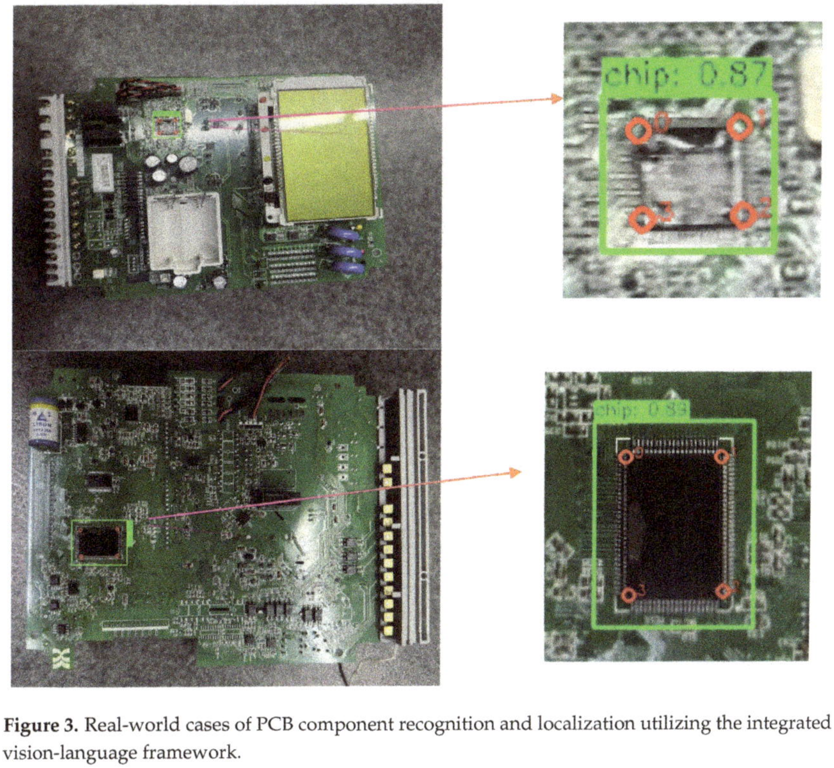

Figure 3. Real-world cases of PCB component recognition and localization utilizing the integrated vision-language framework.

https://doi.org/10.3390/pr14020227

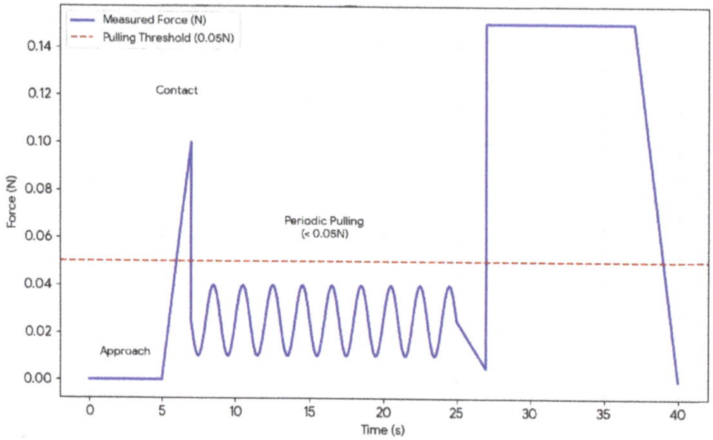

Figure 4. Schematic diagram of the pressure sensor feedback curve during the robotic arm's chip grasping process.

The analysis of disassembly time (Table 2) reflects a clear logical relationship between task complexity and system efficiency, governed primarily by the component's thermal mass and pin count. Integrated Circuits (ICs) require the longest duration (168.5 s) because their high thermal mass necessitates extended heating periods to ensure simultaneous melting of all pins, combined with more intricate path planning for the dual arms. In contrast, Diodes (98.1 s) and Resistors (106.7 s), characterized by low thermal mass and simple lead structures, allow for rapid heat transfer and simplified grasping trajectories, thus optimizing the overall disassembly throughput.

The experimental results strongly validate the advantages of the proposed system in high-level semantic understanding and precise thermal control. The collaborative operation of the multimodal large model and the multi-agent framework provides a viable technological pathway for intelligent PCB disassembly. However, the lack of adaptability in the grasping phase remains the key constraint for overall system performance.

6. Conclusions

This study successfully designed and implemented an intelligent PCB component disassembly system based on a multimodal large model and a multi-agent framework. By deeply integrating high-level semantic understanding with low-level dexterous control, the system provides an innovative solution to the challenges of fine-grained recycling of electronic waste. The core contribution of this research lies in the construction of a hierarchical intelligent architecture. This architecture utilizes a multimodal large model as the "intelligent core," endowing the system with powerful visual-language understanding and task planning capabilities, enabling it to adapt to complex scenarios involving different circuit boards and components. Furthermore, the multi-agent framework facilitates the collaborative operation of dual dexterous hands. The integration of force-feedback control and closed-loop temperature control ensures the controllability and safety of the disassembly process. Experimental results demonstrate the system's high reliability and stability in component recognition and solder joint melting, achieving a 100% melting rate and generally high recognition rates, thereby validating the effectiveness of the proposed architecture at the perception and decision-making levels. However, the experimental

https://doi.org/10.3390/pr14020227

results also reveal the system's current main limitation: a relatively low success rate in picking up small, and low-profile components such as tantalum capacitors. This bottleneck clearly identifies the direction for future optimization. It is essential to focus on researching end-effector designs tailored to the geometric features of specific components and to develop more refined grasping and manipulation strategies.

In summary, this research provides a viable technological pathway for achieving efficient and non-destructive recycling of PCB components. By continuously optimizing the grasping and manipulation module and extending this intelligent system to a broader range of e-waste processing scenarios, it holds significant potential for substantially improving resource recovery efficiency and advancing the circular economy. In conclusion, our system achieved a 100% melting rate across all components and high recognition rates ranging from 90% to 100%, validating the proposed hierarchical architecture. Qualitatively, the integration of high-level cognitive intelligence with low-level dexterous control offers a viable pathway for autonomous recycling, although future work must focus on optimizing end-effector designs for miniature component retrieval. Future work will concentrate on overcoming the identified challenges to further enhance the system's robustness and universality.

Author Contributions: Conceptualization, L.W.; methodology, L.O.; software, H.W.; validation, X.C.; formal analysis, A.W.; investigation, K.Z.; resources, A.W.; data curation, X.C.; writing original draft preparation, L.W.; writing review and editing, L.O. and H.W.; visualization, X.C.; supervision, K.Z. and A.W.: project administration, L.W.; funding acquisition, K.Z. All authors have read and agreed to the published version of the manuscript.

Funding: This paper is funded by the State Grid Zhejiang Electric Power Science and Technology Project (5211WF250006) Research on Key Component Reverse Disassembly Technology of Scrap Electric Energy Meter Circuit Boards Based on Embodied Intelligence.

Data Availability Statement: The raw data supporting the conclusions of this article will be made available by the authors on request.

Conflicts of Interest: Authors Li Wang, Liu Ouyang, Huiying Weng, Xiang Chen, and Anna Wang were employed by State Grid Zhejiang Electric Power Co., Ltd. The remaining authors declare that the research was conducted in the absence of any commercial or financial relationships that could be construed as a potential conflicts of interest.

References

1. Adenuga, A.A.; Amos, O.D.; Olajide, O.D.; Eludoyin, A.O.; Idowu, O.O. Environmental impact and health risk assessment of potentially toxic metals emanating from different anthropogenic activities related to E-wastes. *Heliyon* **2022**, *8*, e10296. [CrossRef] [PubMed]
2. United Nations Conference on Trade and Development. *Digital Economy Report 2024: Shaping an Environmentally Sustainable and Inclusive Digital Future*; Technical Report UNCTAD/DER/2024; United Nations: Geneva, Switzerland, 2024.
3. de Oliveira Neto, J.F.; Monteiro, M.; Silva, M.M.; Miranda, R.; Santos, S.M. Household practices regarding e-waste management: A case study from Brazil. *Environ. Technol. Innov.* **2022**, *28*, 102723. [CrossRef]
4. Abdelbasir, S.M.; Hassan, S.S.; Kamel, A.H.; El-Nasr, R.S. Status of electronic waste recycling techniques: A review. *Environ. Sci. Pollut. Res.* **2018**, *25*, 16533–16547. [CrossRef] [PubMed]
5. Tunali, M.; Tunali, M.M.; Yenigun, O. Characterization of different types of electronic waste: Heavy metal, precious metal and rare earth element content by comparing different digestion methods. *J. Mater. Cycles Waste Manag.* **2020**, *23*, 149–157. [CrossRef]
6. Mudali, U.K.; Patil, M.; Saravanabhavan, R.; Saraswat, V. Review on e-waste recycling: Part I—A prospective urban mining opportunity and challenges. *Trans. Indian Natl. Acad. Eng.* **2021**, *6*, 547–568. [CrossRef]
7. Thakur, P.; Kumar, S. Evaluation of e-waste status, management strategies, and legislations. *Int. J. Environ. Sci. Technol.* **2022**, *19*, 6957–6966. [CrossRef]
8. Sahle-Demessie, E.; Mezgebe, B.; Dietrich, J.; Shan, Y.; Harmon, S.; Lee, C.C. Material recovery from electronic waste using pyrolysis: Emissions measurements and risk assessment. *J. Environ. Chem. Eng.* **2021**, *9*, 104943. [CrossRef]

https://doi.org/10.3390/pr14020227

9. Kopacek, B. Intelligent Disassembly of components from printed circuit boards to enable re-use and more efficient recovery of critical metals. *IFAC-PapersOnLine* **2016**, *49*, 190–195. [CrossRef]
10. Maurice, A.A.; Dinh, K.N.; Charpentier, N.M.; Brambilla, A.; Gabriel, J.C.P. Dismantling of Printed Circuit Boards Enabling Electronic Components Sorting and Their Subsequent Treatment Open Improved Elemental Sustainability Opportunities. *Sustainability* **2021**, *13*, 10357. [CrossRef]
11. Santos, S.; Marques, L.; Neto, P. A robotic system to automate the disassembly of pcb components. In *Proceedings of the Iberian Robotics Conference*; Springer: Berlin/Heidelberg, Germany, 2023; pp. 456–465.
12. Bizzo, W.A.; Figueiredo, R.A.; De Andrade, V.F. Characterization of Printed Circuit Boards for Metal and Energy Recovery after Milling and Mechanical Separation. *Materials* **2014**, *7*, 4555–4566. [CrossRef]
13. Vakilchap, F.; Mousavi, S.M. Structural study and metal speciation assessments of waste PCBs and environmental implications: Outlooks for choosing efficient recycling routes. *Waste Manag.* **2022**, *151*, 181–194. [CrossRef]
14. An, T.; Huang, Y.; Li, G.; He, Z.; Chen, J.; Zhang, C. Pollution profiles and health risk assessment of VOCs emitted during e-waste dismantling processes associated with different dismantling methods. *Environ. Int.* **2014**, *73*, 186–194. [CrossRef] [PubMed]
15. Ren, Z.; Bi, X.; Huang, B.; Liu, M.; Sheng, G.; Fu, J. Hydroxylated PBDEs and brominated phenolic compounds in particulate matters emitted during recycling of waste printed circuit boards in a typical e-waste workshop of South China. *Environ. Pollut.* **2013**, *177*, 71–77. [CrossRef] [PubMed]
16. de Oliveira Neto, J.F.; Silva, M.M.; Machado Santos, S. A mini-review of E-waste management in Brazil: Perspectives and challenges. *CLEAN-Air Water* **2019**, *47*, 1900152. [CrossRef]
17. Vlasopoulos, D.; Mendrinou, P.; Oustadakis, P.; Kousi, P.; Stergiou, A.; Karamoutsos, S.D.; Hatzikioseyian, A.; Tsakiridis, P.E.; Remoundaki, E.; Agatzini-Leonardou, S. Hydrometallurgical recovery of silver and gold from waste printed circuit boards and treatment of the wastewater in a biofilm reactor: An integrated pilot application. *J. Environ. Manag.* **2023**, *344*, 118334. [CrossRef] [PubMed]
18. Han, J.; Duan, C.; Lu, Q.; Jiang, H.; Fan, X.; Wen, P.; Ju, Y. Improvement of the crushing effect of waste printed circuit boards by co-heating swelling with organic solvent. *J. Clean. Prod.* **2019**, *214*, 70–78. [CrossRef]
19. Zeng, X.; Li, J.; Xie, H.; Liu, L. A novel dismantling process of waste printed circuit boards using water-soluble ionic liquid. *Chemosphere* **2013**, *93*, 1288–1294. [CrossRef]
20. Ghosh, B.; Ghosh, M.; Parhi, P.; Mukherjee, P.; Mishra, B. Waste printed circuit boards recycling: An extensive assessment of current status. *J. Clean. Prod.* **2015**, *94*, 5–19. [CrossRef]
21. Huang, H.; Pan, J.; Liu, Z.; Song, S.; Liu, G. Study on disassembling approaches of electronic components mounted on PCBs. In *Advances in Life Cycle Engineering for Sustainable Manufacturing Businesses, Proceedings of the 14th CIRP Conference on Life Cycle Engineering, Waseda University, Tokyo, Japan, 11–13 June 2007*; Springer: Berlin/Heidelberg, Germany, 2007; pp. 263–266.
22. Cao, Y.; Mu, T. Comprehensive investigation on the thermal stability of 66 ionic liquids by thermogravimetric analysis. *Ind. Eng. Chem. Res.* **2014**, *53*, 8651–8664. [CrossRef]
23. Zhu, P.; Chen, Y.; Wang, L.Y.; Zhou, M. A new technology for recycling solder from waste printed circuit boards using ionic liquid. *Waste Manag. Res.* **2012**, *30*, 1222–1226. [CrossRef]
24. Duan, H.; Hou, K.; Li, J.; Zhu, X. Examining the technology acceptance for dismantling of waste printed circuit boards in light of recycling and environmental concerns. *J. Environ. Manag.* **2011**, *92*, 392–399. [CrossRef]
25. Park, S.; Kim, S.; Han, Y.; Park, J. Apparatus for electronic component disassembly from printed circuit board assembly in e-wastes. *Int. J. Miner. Process.* **2015**, *144*, 11–15. [CrossRef]
26. Zhou, Y.; Qiu, K. A new technology for recycling materials from waste printed circuit boards. *J. Hazard. Mater.* **2010**, *175*, 823–828. [CrossRef]
27. Meng, L.; Zhong, Y.; Guo, L.; Wang, Z.; Chen, K.; Guo, Z. High-temperature centrifugal separation of Cu from waste printed circuit boards. *J. Clean. Prod.* **2018**, *199*, 831–839. [CrossRef]
28. Duan, H.; Li, J.; Liu, Y.; Yamazaki, N.; Jiang, W. Characterization and inventory of PCDD/Fs and PBDD/Fs emissions from the incineration of waste printed circuit board. *Environ. Sci. Technol.* **2011**, *45*, 6322–6328. [CrossRef] [PubMed]
29. Wang, J.; Guo, J.; Xu, Z. An environmentally friendly technology of disassembling electronic components from waste printed circuit boards. *Waste Manag.* **2016**, *53*, 218–224. [CrossRef]
30. Yang, C.; Tan, Q.; Liu, L.; Dong, Q.; Li, J. Recycling tin from electronic waste: A problem that needs more attention. *ACS Sustain. Chem. Eng.* **2017**, *5*, 9586–9598. [CrossRef]
31. Vermeșan, H.; Tiuc, A.E.; Purcar, M. Advanced recovery techniques for waste materials from IT and telecommunication equipment printed circuit boards. *Sustainability* **2019**, *12*, 74. [CrossRef]
32. Carneiro, M.F.C.; Leão, V.A. The role of sodium chloride on surface properties of chalcopyrite leached with ferric sulphate. *Hydrometallurgy* **2007**, *87*, 73–82. [CrossRef]
33. Altinkaya, P.; Liipo, J.; Kolehmainen, E.; Haapalainen, M.; Leikola, M.; Lundström, M. Leaching of trace amounts of metals from flotation tailings in cupric chloride solutions. *Mining, Metall. Explor.* **2019**, *36*, 335–342. [CrossRef]

https://doi.org/10.3390/pr14020227

34. Zhu, M.; Gong, D.; Zhao, Y.; Chen, J.; Qi, J.; Song, S. Compliant Force Control for Robots: A Survey. *Mathematics* **2025**, *13*, 2204. [CrossRef]
35. Mendes, N.; Neto, P.; Pires, J.N.; Loureiro, A. An optimal fuzzy-PI force/motion controller to increase industrial robot autonomy. *Int. J. Adv. Manuf. Technol.* **2013**, *68*, 435–441. [CrossRef]
36. Calanca, A.; Muradore, R.; Fiorini, P. A Review of Algorithms for Compliant Control of Stiff and Fixed-Compliance Robots. *IEEE/ASME Trans. Mechatron.* **2016**, *21*, 613–624. [CrossRef]
37. Sadun, A.; Jalani, J.; Sukor, J. An overview of active compliance control for a robotic hand. *FME Trans.* **2016**, *11*, 11872–11876.
38. Huang, J.; Pham, D.; Wang, Y.W.; Ji, C.; Xu, W.; Liu, Q.; Zhou, Z. A strategy for human-robot collaboration in taking products apart for remanufacture. *FME Trans.* **2019**, *47*, 731–738. [CrossRef]
39. Su, S.; Pham, D.T.; Ji, C.; Wang, Y.; Huang, J.; Zhou, W.; Wang, H. Design of a compliant device for peg-hole separation in robotic disassembly. *Int. J. Adv. Manuf. Technol.* **2022**, *124*, 3011–3019. [CrossRef]
40. Fantoni, G.; Santochi, M.; Dini, G.; Tracht, K.; Scholz-Reiter, B.; Fleischer, J.; Kristoffer Lien, T.; Seliger, G.; Reinhart, G.; Franke, J.; et al. Grasping devices and methods in automated production processes. *CIRP Ann.* **2014**, *63*, 679–701. [CrossRef]
41. Ruggeri, S.; Fontana, G.; Basile, V.; Valori, M.; Fassi, I. Micro-robotic Handling Solutions for PCB (re-)Manufacturing. *Procedia Manuf.* **2017**, *11*, 441–448. [CrossRef]
42. Noll, R.; Ambrosch, R.; Bergmann, K.; Britten, S.; Brumm, H.; Chmielarz, A.; Connemann, S.; Eschen, M.; Frank, A.; Fricke-Begemann, C.; et al. Next generation urban mining-Automated disassembly, separation and recovery of valuable materials from electronic equipment: Overview of R&D approaches and first results of the European project ADIR. In Proceedings of the EMC, ADIR, Zrenjnin, Serbia, 16–17 June 2017; p. 1.
43. He, C.; Jin, Z.; Gu, R.; Qu, H. Automatic disassembly and recovery device for mobile phone circuit board CPU based on machine vision. *J. Phys. Conf. Ser.* **2020**, *1684*, 012137. [CrossRef]
44. Li, J.; Zhang, X.; Feng, P. Detection Method of End-of-Life Mobile Phone Components Based on Image Processing. *Sustainability* **2022**, *14*, 12915. [CrossRef]
45. Lei, M.; Li, S.; Wu, Y.; Hu, H.; Zhou, Y.; Zheng, X.; Ding, G.; Du, S.; Wu, Z.; Gao, Y. YOLOv13: Real-Time Object Detection with Hypergraph-Enhanced Adaptive Visual Perception. *arXiv* **2025**, arXiv:2506.17733.
46. Jeong, S.; Baek, J.; Cho, S.; Hwang, S.J.; Park, J. Adaptive-RAG: Learning to Adapt Retrieval-Augmented Large Language Models through Question Complexity. *arXiv* **2024**, arXiv:2403.14403.
47. DeepMind, G. Gemini API Documentation. Technical Report, Google. 2024. Available online: https://ai.google.dev/gemini-api/docs (accessed on 10 November 2025).
48. Mohsin, M.; Rovetta, S.; Masulli, F.; Cabri, A. Automated Disassembly of Waste Printed Circuit Boards: The Role of Edge Computing and IoT. *Computers* **2025**, *14*, 62. [CrossRef]

https://doi.org/10.3390/pr14020227

processes

MDPI

Article

Neuro-Symbolic Verification for Preventing LLM Hallucinations in Process Control

Boris Galitsky [1,*] and Alexander Rybalov [2]

[1] Knowledge Trail Inc., San Jose, CA 95127, USA
[2] LAMBDA Laboratory, Tel Aviv University, Ramat Aviv, Tel Aviv 6997801, Israel; alxndr_r@yahoo.com
* Correspondence: bgalitsky@hotmail.com

Abstract

Large Language Models (LLMs) are increasingly used in industrial monitoring and decision support, yet they remain prone to process-control hallucinations—diagnoses and explanations that sound plausible but conflict with physical constraints, sensor data, or plant dynamics. This paper investigates hallucination as a failure of abductive reasoning, where missing premises, weak mechanistic support, or counter-evidence lead an LLM to propose incorrect causal narratives for faults such as pump restriction, valve stiction, fouling, or reactor runaway. We develop a neuro-symbolic framework in which Abductive Logic Programming (ALP) evaluates the coherence of model-generated explanations, counter-abduction generates rival hypotheses that test whether the explanation can be defeated, and Discourse-weighted ALP (D-ALP) incorporates nucleus–satellite structure from operator notes and alarm logs to weight competing explanations. Using our 500-scenario Process-Control Hallucination Dataset, we assess LLM reasoning across mechanistic, evidential, and contrastive dimensions. Results show that abductive and counter-abductive operators substantially reduce explanation-level hallucinations and improve alignment with physical process behavior, particularly in "easy-but-wrong" cases where a superficially attractive explanation contradicts historian trends or counter-evidence. These findings demonstrate that abductive reasoning provides a practical and verifiable foundation for improving LLM reliability in safety-critical process-control environments.

Keywords: abductive reasoning; process-control hallucinations; counter-abduction; discourse-weighted ALP; fault diagnosis; neuro-symbolic LLM verification

1. Introduction

Large Language Models (LLMs) have recently been explored not only as conversational systems, but increasingly as autonomous decision-making components in process automation and industrial control. Their ability to parse technical language, generate control strategies, and interface with operations via natural language instructions makes them attractive as supervisory agents for tasks such as batch sequencing, alarm response, optimization, and procedure generation. However, this promise is constrained by a persistent and under-studied failure mode: hallucinated control commands—outputs that appear syntactically valid, operationally confident, and often persuasive, yet are physically infeasible, unsafe, or inconsistent with the underlying plant state.

Process systems engineering (PSE) is a core discipline in chemical engineering and is dedicated to the systematic design, modeling, optimization, and control of industrial processes. Traditionally, PSE has relied on mechanistic modeling approaches, in which complex physical and chemical processes are mathematically formalized to predict system behavior under varying operating conditions [1,2]. Over time, the field has progressively incorporated data-driven methodologies, including optimization, estimation, and advanced control strategies, to improve decision-making under uncertainty [3]. PSE underpins critical industrial sectors such as chemical manufacturing, energy systems, and wastewater treatment, where efficiency, safety, cost reduction, and sustainability are paramount [4]. With the advent of Industry 4.0, digital technologies—particularly AI, large-scale data analytics, and automation—have accelerated a shift from purely physics-based models toward hybrid and learning-based approaches that support predictive analytics and real-time decision assistance [5].

Recently, LLMs have emerged as a promising interface layer for PSE, enabling natural-language interaction with process data, automated explanation generation, and decision support across monitoring, diagnosis, and control tasks. The integration of LLMs into PSE revolutionizes process modeling, control, and optimization by enhancing automation, decision-making, and knowledge extraction. By enhancing process efficiency, operational reliability, and resource optimization, LLM-enabled PSE contributes directly to industrial sustainability goals and broader societal benefits. The role of LLMs in key research domains of PSE, including data availability, process modeling, process control, and process optimization, is shown in Figure 1.

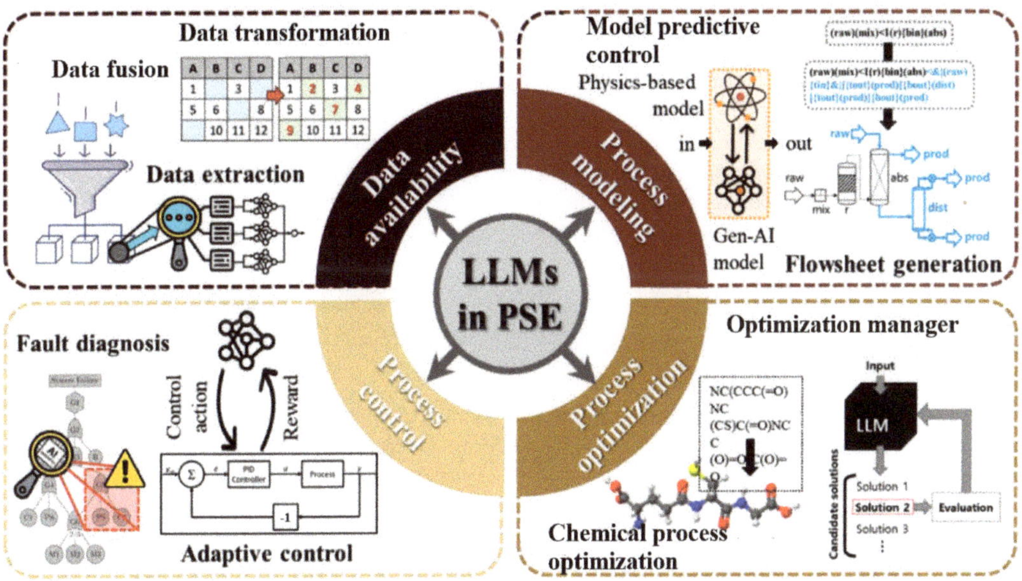

Figure 1. The role of LLMs in key research domains of process systems engineering.

The expanding role of LLMs across the PSE lifecycle are summarized—from data extraction and transformation, through hybrid physics–AI process modeling and flowsheet generation, to fault diagnosis, adaptive control, and process optimization. While this integration enables natural-language interfaces, automated modeling, and decision support under Industry 4.0, it also highlights where hallucinations can arise: during data interpretation, explanation of process behavior, fault attribution, and optimization recommendations. The chart motivates the need for verification mechanisms in LLM-enabled PSE, as unsupported or fluent-but-incorrect outputs at any of these stages can

https://doi.org/10.3390/pr14020322

propagate into unsafe control actions or suboptimal operational decisions in safety-critical process environments.

However, unlike traditional model-based or statistical tools, LLMs are susceptible to hallucinations: fluent yet unsupported or incorrect explanations, diagnoses, or control recommendations. In safety-critical process control environments, such hallucinations pose significant risks to operational reliability, safety, and regulatory compliance. Addressing hallucinations in LLM-enabled PSE is therefore essential not only for improving technical performance, but also for ensuring trustworthy human–AI collaboration, safe automation of cognitive tasks, and sustainable industrial operation [6].

Most research on hallucinations has focused on factual or semantic errors in natural language generation for knowledge-centric tasks. In contrast, hallucinations in automation differ fundamentally: they manifest as incorrect control logic, fabricated causal assumptions, or implausible action sequences issued to real equipment, where consequences extend beyond textual error into material risk. Existing hallucination-detection methods developed for summarization, QA, or fact-verification are not directly applicable to this domain. White-box approaches, which rely on internal model activations [7,8], lack portability to deployed industrial models. Gray-box indicators such as token probability or entropy [9] poorly correlate with physical safety—high-confidence text may still violate pressure limits, fail to respect interlock logic, or misinterpret sensor drift. Classical black-box techniques [10,11] detect factual deviation but do not reason about thermodynamics, control constraints, or equipment sequencing. Even heuristic ensemble techniques such as self-consistency or sampling consensus [12] struggle when hallucinated instructions remain logically fluent while being operationally wrong.

To address this gap, we propose a neuro-symbolic abductive framework for detecting hallucination-driven control errors. Rather than validating commands through factual lookup or statistical confidence, the framework treats each LLM-generated control action as a hypothesis about how the plant should evolve. Counter-abduction is then used to generate competing hypotheses—alternative control sequences, explanations of observed sensor values, or safety-compliant causal interpretations. A command is considered hallucinated when it fails to survive this challenge: when a rival hypothesis aligns more closely with physical constraints, interlock logic, energy and mass balances, or risk-minimizing operation.

By defining hallucination as defeat under counter-abductive reasoning, this method shifts the detection problem from lexical correctness to causal soundness and physical validity. It operates entirely at the black-box level—neither model internals nor curated fact databases are required—yet it uncovers subtle control-logic hallucinations that remain undetected by probability-based or factual-verification approaches [13]. Importantly, this positions abduction not merely as a language-interpretation mechanism, but as a defensive layer protecting automation systems from confidently incorrect AI-generated actions.

1.1. Example: Hallucination ≈ Abductive Failure + Counter-Abductive Defeat

To illustrate how hallucinations manifest in automation systems, we analyze a concrete failure case in which an LLM generates a fluent but unsafe control recommendation. The example highlights how abductive reasoning and counter-abductive challenge serve as a basis for hallucination diagnosis within industrial environments.

Consider a continuous stirred-tank reactor (CSTR) operating at 92 °C, above its recommended safety threshold of 85 °C. The system queries an LLM-based supervisory agent: "Recommend control actions to reduce reactor temperature to safe operating limits."

https://doi.org/10.3390/pr14020322

The model responds with the following sequence (H_0—Hallucinated control plan):

1. Increase feed pump rate to dilute the reaction mixture.
2. Reduce agitation speed to minimize heat generation.
3. Maintain coolant flow constant for system stability.

The recommendation is linguistically coherent and uses appropriate process-control terminology. However, several actions contradict basic thermodynamic reasoning. Increasing feed rate can introduce more reactive material and exacerbate heat release; reducing agitation impairs heat transfer to the cooling jacket; maintaining coolant flow unchanged ignores the most direct lever for temperature reduction. Thus, although H_0 is plausible in form, it is physically unsound in function.

Abductive formulation acts as follow. Abduction asks whether a hypothesis exists under which the model's output would represent a coherent explanation of how cooling might be achieved. A repair hypothesis can be formulated as:

H_0 '. *Increased feed acts as thermal dilution, and reduced agitation slows reaction rate sufficiently to offset heat accumulation. We then test H_0' against established process knowledge (Table 1).*

Table 1. Evaluation criteria.

Evaluation Criterion	Result
Dilution decreases temperature faster than volume expansion increases heat load	Typically false for exothermic kinetics
Reduced agitation improves safety	Opposes heat-transfer principles; likely to worsen response
Constant coolant flow is sufficient	Violates standard safe-operation strategy

H_0' therefore fails to explain the evidence without contradictions.

Counter-abductive defeat and hallucination recognition works as follows. Counter-abduction generates an alternative control hypothesis H^* grounded in mass and energy balance:

H^*—Physically valid response:

1. Increase coolant flow or enhance heat-transfer coefficient.
2. Reduce or suspend reactant feed.
3. Maintain or increase agitation to improve heat removal.

H^* aligns with core chemical-engineering principles and operational safety logic. Since H^* explains system behavior more coherently and without contradiction—while H_0 cannot be reconciled with known thermodynamics—H_0 is defeated under counter-abductive challenge.

A process-automation hallucination is therefore defined as *a candidate control hypothesis that fails to survive counter-abductive evaluation against physically grounded alternatives.*

This mechanism establishes hallucination not as a linguistic artifact, but as a breakdown in causal reasoning and constraint-consistent action synthesis, enabling detection at the level of process logic rather than surface text.

1.2. Constrained Control

Constrained control is a cornerstone of industrial process control, reflecting the necessity to operate systems within explicit physical, safety, and operational limits such as actuator saturation, pressure and temperature bounds, and regulatory constraints. Among constrained control strategies, Model Predictive Control (MPC) has become the dominant industrial paradigm because it formulates control actions as a constrained optimization problem over a prediction horizon, explicitly enforcing input and state constraints while optimizing performance objectives [14]. Complementary tactics—including anti-windup

https://doi.org/10.3390/pr14020322

mechanisms for PID controllers, reference governors for constraint-safe set-point adjustment, and soft-constraint formulations with penalty terms—are widely used to ensure robust and safe operation under disturbances, model mismatch, and uncertainty [15,16]. Together, these methods form a safety envelope that limits the consequences of imperfect models and noisy measurements.

As LLMs are increasingly introduced into process control workflows—for supervision, fault explanation, operator assistance, or high-level decision support—the role of constrained control becomes even more critical. LLMs can generate fluent control rationales or recommendations that are not grounded in process physics or constraint logic, a phenomenon commonly referred to as hallucination [17]. In safety-critical control settings, such hallucinations may suggest infeasible set-points, unsafe operating regions, or incorrect interpretations of constraint violations. Consequently, LLM-driven control assistance must remain subordinate to constraint-aware control layers such as MPC and reference governors, and should be coupled with verification mechanisms that check consistency against process models and enforced constraints before any operational influence [18]. Without such safeguards, hallucinations risk undermining both operational safety and operator trust in AI-assisted control systems.

1.3. Contribution

To the best of our knowledge, this is the first work to systematically demonstrate how abductive reasoning can be used to detect and correct hallucinated control actions generated by LLMs in industrial process automation. We show that abduction provides a principled mechanism for constructing and testing operational hypotheses, enabling the system to recognize when an LLM's recommended actions rely on unsupported causal assumptions, violate physical constraints, or contradict established process logic. By extending this with counter-abduction—and grounding evaluation in thermodynamic, kinetic, and interlock-aware reasoning—we expose failure modes that are invisible to probability-based confidence scores, retrieval augmentation, or ensemble sampling methods. This positions abduction as a generalizable safety layer for LLM-driven automation, improving factual grounding, operational reliability, and the interpretability of control decisions in real industrial environments.

2. Abduction vs. Deduction and Induction

Charles S. Peirce [19,20] identified three fundamental modes of inference—deduction, induction, and abduction—that together constitute the foundation of human reasoning and scientific discovery. These reasoning forms differ in their epistemic orientation, logical direction, and ampliative strength (adds new information not contained in the premises). Peirce emphasized that deduction is non-ampliative and certain, while both induction and abduction are ampliative and uncertain, expanding knowledge beyond what is explicitly contained in the premises.

Deduction operates by applying general rules to specific cases to derive necessary conclusions. It preserves truth but does not introduce new information: if the premises are true, the conclusion must also be true. Deductive reasoning thus provides logical certainty but no epistemic novelty.

Induction, by contrast, generalizes from specific instances to formulate general laws or statistical regularities. It infers generality from repetition: from observing many cases of a phenomenon, one concludes that it holds universally or probabilistically. Inductive inference is inherently uncertain and probabilistic—its strength depends on the representativeness and size of the observed sample.

https://doi.org/10.3390/pr14020322

Abduction—sometimes called inference to the best explanation—reverses this direction of reasoning [21]. It begins with an observed or surprising fact and seeks the most plausible hypothesis that could account for it. Abduction is therefore explanatory and hypothetical in nature: it introduces a new conceptual assumption that, if true, would make the observed fact intelligible.

Peirce maintained that induction and abduction are irreducible to one another, despite both being ampliative. Their irreducibility lies in their distinct epistemic roles within inquiry. Abduction functions as the logic of discovery, generating explanatory hypotheses to make sense of observations. Induction functions as the logic of verification, evaluating the empirical adequacy of those hypotheses through repeated testing. Deduction mediates between the two, deriving predictions from the abductively generated hypotheses for inductive testing.

This cyclical relationship—*abduction → deduction → induction*—constitutes what Peirce regarded as the *complete logic of scientific inquiry*. In modern contexts such as abductive logic programming, probabilistic reasoning, and neuro-symbolic AI, this triad continues to serve as a conceptual framework for balancing hypothesis generation, logical inference, and empirical validation (Table 2).

Table 2. Forms of reasoning.

Form of Reasoning	Direction of Inference	Goal/Function	Type of Conclusion	Certainty	Example
Deduction	General → Particular	Apply rule to case	Necessarily true if premises are true	Certain	All metals expand when heated → Iron expands when heated
Induction	Particular → General	Formulate general law from observations	Probably true, subject to testing	Probabilistic	Iron, copper, and aluminum expand → All metals expand
Abduction	Particular → Hypothesis	Propose best explanation for observed fact	Plausible, explanatory conjecture	Uncertain	The surface of a reactor is wet → there was a leak and the humidity is high

3. Abduction as a Structural Corrective Layer for Reasoning About Processes

Chain-of-Thought (CoT) prompting has rapidly emerged as a default approach for eliciting multi-step operational reasoning from large language models. In process automation, CoT is often used to make the model articulate the intermediate control logic behind a recommended sequence of actions—e.g., why it suggests adjusting feed rate, coolant flow, or agitation in a particular order. However, empirical evidence shows that CoT frequently produces *narrative reasoning* rather than *causal reasoning*. Because CoT unfolds autoregressively, each step is conditioned primarily on previous text rather than on physical laws, kinetic models, interlock logic, or plant constraints. This leads to characteristic failure modes in automation: invented process states, circular justification of control actions, step sequences that violate safe operating procedures, and high variability when the same problem is phrased differently. The result is a reasoning chain that may appear methodical and domain-aware, yet lacks grounding in thermodynamics, mass transfer, or safety envelopes.

https://doi.org/10.3390/pr14020322

Abductive reasoning offers a structural countermeasure to these weaknesses. Unlike deduction, which derives consequences from known rules, or induction, which extrapolates from prior experience, abduction seeks the *most plausible causal hypothesis* for a given set of process observations—sensor readings, alarm states, rate deviations, or temperature excursions. Integrated with LLMs, abduction provides a corrective layer that forces generative control rationale to align with physical constraints and interlock logic. Rather than merely checking CoT after generation, abductive integration reshapes the reasoning trajectory itself, yielding multi-step control explanations that are coherent, defeasible, and grounded in process reality.

In a neuro-symbolic pipeline, the role of abduction is to constrain the model's reasoning space, reveal implicit assumptions, and ensure that the chain as a whole satisfies the explanatory minimality principles characteristic of abductive logic programming and related frameworks (e.g., probabilistic logic programming, argumentation-based abduction, and paraconsistent abduction). The resulting system treats CoT not as a static artifact but as a dynamic structure subject to revision, hypothesis insertion, and consistency checking. This greatly mitigates classical CoT hallucinations, particularly those involving unjustified intermediate premises.

LLMs demonstrate several recurrent weaknesses when generating multi-step control logic for industrial processes:

1. Local coherence without global constraint compliance. Autoregressive reasoning produces steps that sound plausible in isolation, yet the overall control plan may violate material balances, safety limits, or standard operating procedures, leading to hidden contradictions across the sequence:
2. Operational drift. A control recommendation may begin sensibly—e.g., increasing coolant flow—but gradually diverge into irrelevant or unsafe actions (changing feed composition, reducing agitation, bypassing interlocks) as the reasoning chain extends.
3. Fabricated plant-state assumptions. To maintain fluency, models often insert implicit premises—unverified sensor readings, nonexistent equipment states, or speculative causal relationships—not grounded in actual process data.
4. Irreversibility of reasoning. Chain-of-thought control plans are monotonic: once a step is generated (e.g., reduce feed), the model rarely retracts or corrects it even when subsequent reasoning contradicts that action or new evidence invalidates it.
5. Non-minimal, overly verbose action paths. Generated control sequences frequently include redundant or unnecessary steps, increasing execution risk and inflating the opportunity space for hallucination and unsafe actuation.

These failure patterns stem from the lack of symbolic, constraint-aware structure in LLM-generated control logic—the core instance of the language-model fallacy, where linguistic plausibility is mistaken for physical validity. Abductive reasoning directly targets these failure modes, introducing a mechanism to test, revise, and compete control hypotheses against thermodynamic, interlock, and mass-balance realities.

Abductive CoT

Abductive models enforce minimality: explanations should contain no unnecessary assumptions. This principle acts as a structural regularizer on CoT, pruning verbose or extraneous content and discouraging speculative detours. Minimality also makes verification more tractable because the reasoning chain becomes closer to a canonical explanation.

Moreover, minimality reduces one of the main sources of hallucination in CoT systems: the inclusion of tangential premises or loosely associated facts. A minimal abductive

https://doi.org/10.3390/pr14020322

explanation is not only easier to inspect but also more robust to adversarial perturbations and paraphrased prompts.

A coherent architecture for Abductive CoT emerges from combining these elements (Figure 2):

1. Initial CoT generation by the LLM.
2. Logical extraction converting text into predicates or defeasible rules.
3. Abductive solver evaluates consistency, minimality, and coherence.
4. Hypothesis generation to fill explanatory gaps.
5. Feedback to LLM prompting revision or alternative reasoning paths.
6. Discourse-aware weighting using RST to distinguish central from peripheral content.
7. Final, verified CoT chain that satisfies explanatory constraints.

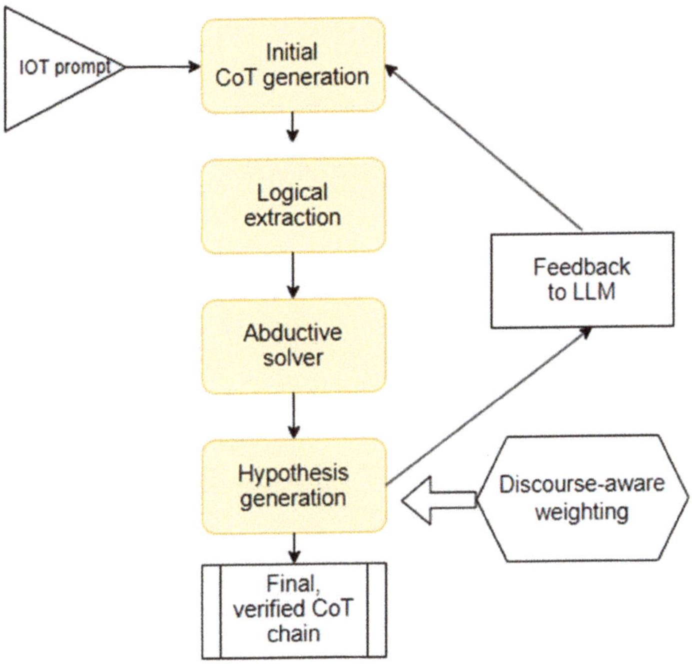

Figure 2. Abductive support for CoT.

This loop is compatible with multiple logical formalisms, including probabilistic abduction, argumentation-based abduction, and paraconsistent abductive reasoning—allowing different degrees of uncertainty, conflict tolerance, and rule expressiveness. The core advantage is that the LLM no longer bears the full burden of reasoning; instead, it operates within a scaffold of symbolic constraints.

4. Abductive Logic Programming

In Abductive logic programming (ALP), one allows some predicates (called *abducibles*) to be "hypothesized" so as to explain observations or to achieve goals, subject to integrity constraints. An abductive explanation is a set of ground abducible facts Δ such that:

1. $P \cup \Delta \models G$ (i.e., the goal/observation G is entailed);
2. $P \cup \Delta \models IC$ (the integrity constraints are satisfied);
3. $P \cup \Delta P$ is consistent.

https://doi.org/10.3390/pr14020322

Here <*P,A,IC*> is the abductive logic program: *P* is the normal logic program, *A* the set of abducible predicates, and *IC* the constraints.

There are many ALP systems available (Table 3).

Table 3. Abductive logic programming systems.

Name	Approach/Features	Notes/Strengths	Limitations/Caveats
ACLP	Integrates abduction with constraint solving (built over ECLiPSe CLP)	Good fit when you need both abduction and constraints (e.g., planning, scheduling).	Performance can degrade for large or complex abductive tasks.
CIFF/IFF-based systems	Use a variant of the IFF proof procedure extended with abductive reasoning and constraints	More expressive handling of integrity constraints, etc., widely referenced in ALP literature	As with many meta-interpreters, efficiency is a concern for large domains.
A-system	A Prolog-based abductive system	One of the classical ALP systems.	Might not scale to very large problems; also dependent on the Prolog engine.
SCIFF	An extension of ALP tailored for specifying and checking protocols (e.g., interaction, contracts)	Good for normative reasoning, protocol compliance monitoring.	Specialized; might require tailoring for more general domains.
ABDUAL	A system combining abduction and tabling techniques [22]	Helps in improving efficiency, avoiding redundant recomputation.	Implementation complexity; tradeoffs in memory vs. speed.
DLV (with abductive diagnosis front-end)	DLV is a disjunctive ASP/nonmonotonic reasoning system; it supports a front end for abductive diagnosis tasks.	Leverages efficient ASP back ends; good for problems reducible to abductive diagnosis.	May require rephrasing of your problem into the dialect ASP supports; constraints of DLV's language.
ToyElim	A more general system for operator elimination (e.g., quantifier elimination, projection, and forgetting) which can express abductive explanations.	Elegant, theoretically grounded in classical logic; may serve as a backend or bridge.	It is a prototype; may not be optimized for large logic programming tasks.

There are Prolog-based approaches/tools that support or partially support abductive reasoning/ALP. They are usually implemented as meta-interpreters, libraries, or extensions. We mention three families of approaches:

- Aleph (with "abduce" mode). Aleph is primarily an Inductive Logic Programming (ILP) system. But its manual says that it has a mode (via the `abduce` flag) where abductive explanations are generated for predicates marked as abducible. The abductive part in Aleph is limited: it assumes abducible explanations must be *ground*, and you may need to limit the number of abducibles (via `max_abducibles`) for efficiency [23].
- Meta-interpreter/CHR implementations in Prolog. Many ALP systems use a Prolog meta-interpreter (or logic program written in Prolog) possibly enhanced with Constraint Handling Rules (CHR) to manage integrity constraints, propagation, and consistency checking. Since SWI-Prolog supports CHR (via its CHR library/attributed variables), you can port or build an abductive system using CHR in SWI [24].

https://doi.org/10.3390/pr14020322

It is possible to build a meta-interpreter for ALP directly. The general approach:

1. declare which predicates are abducibles,
2. write a meta-interpreter that, when trying to prove a goal, allows adding abducible atoms hypotheses,
3. maintain integrity constraints and check them,
4. control search (pruning, minimality, consistency).

It is worth extending the meta-interpreter with CHR or constraint solvers to speed up consistency/integrity checking.

Some recent proposals aim to make ALP systems more efficient (e.g., by eliminating CHR overhead) or compile them, but they may not yet have full, robust SWI-Prolog ports. Also, SWI Prolog has features like attributed variables, constraint libraries, and delimited control (in newer versions) which facilitates more advanced meta-programming approaches useful in ALP. Several methodological and computational challenges are associated with the use of ALP:

1. Scalability remains a central issue. Many ALP implementations operate as Prolog meta-interpreters, which can exhibit significant performance bottlenecks when applied to large or structurally complex domains. Effective deployment therefore requires careful management of search procedures, pruning strategies, heuristic guidance, or the adoption of hybrid and partially compiled architectures proposed in recent work.
2. Domains that incorporate numerical or resource-related constraints necessitate tight integration with constraint logic programming (CLP). Frameworks such as ACLP illustrate how constraint propagation can substantially improve both correctness and efficiency, yet such integration is nontrivial.
3. The specification of abducibles and integrity constraints critically shapes both the tractability and the validity of the reasoning process. Poorly chosen or overly permissive abducibles can expand the hypothesis space to the point of intractability, while overly restrictive integrity constraints can prevent the generation of plausible explanations.
4. Although many abductive tasks can be reformulated as Answer Set Programming (ASP) problems and thus leverage highly optimized ASP solvers, doing so typically requires nontrivial representational transformations [25]. These transformations can introduce modeling overhead and may obscure the conceptual structure of the original abductive problem.

Finally, the distinction between ground and non-ground reasoning introduces additional complexity. Systems optimized for propositional, fully grounded settings often achieve superior performance, whereas support for variables, unification, and non-ground abductive hypotheses tends to complicate search and reduce scalability. Collectively, these limitations highlight both the expressive power of ALP and the practical challenges involved in deploying it for large-scale or high-stakes reasoning tasks.

The computational pipeline for Abductive Logic Programming (ALP), illustrated in Figure 3, proceeds through the following stages.

1. Discourse Parsing. The input explanation text is first segmented into Elementary Discourse Units (EDUs) and organized into a discourse tree capturing nucleus–satellite relations and rhetorical roles. For high-fidelity, low-volume analyses we employ GPT-5 for discourse parsing, which provides robust identification of causal, evidential, and contrastive relations. For large-scale evaluation, we use a lightweight wrapper around the discourse parser of [26], trading some granularity for scalability while preserving core rhetorical structures.

https://doi.org/10.3390/pr14020322

2. Fact Extraction. Each EDU is then mapped into one or more logical literals representing causal relations, evidential support, contextual qualifiers, or contrastive constraints. This step grounds natural-language explanations in a symbolic representation suitable for reasoning, ensuring that all inferred relations are explicitly traceable.
3. Weight Assignment. Discourse roles are converted into numerical weights, with nucleus units receiving higher scores than satellites. These weights reflect the relative explanatory importance of different discourse elements, prioritizing causal and evidential statements over background or elaborative content, while allowing contrastive units to introduce defeaters.
4. Abductive Search. The weighted literals are passed to an abductive reasoning engine implemented using SWI-Prolog and a probabilistic ALP framework (e.g., ProbLog). The solver generates candidate hypotheses that explain the observed facts while minimizing abductive cost and respecting discourse-weighted constraints.
5. Ranking and Selection. Finally, candidate explanations are ranked by their aggregated weighted scores, and the top-k hypotheses are returned. These ranked hypotheses provide both the most plausible explanations and an explicit ordering of alternatives, enabling counter-abductive analysis and hallucination detection.

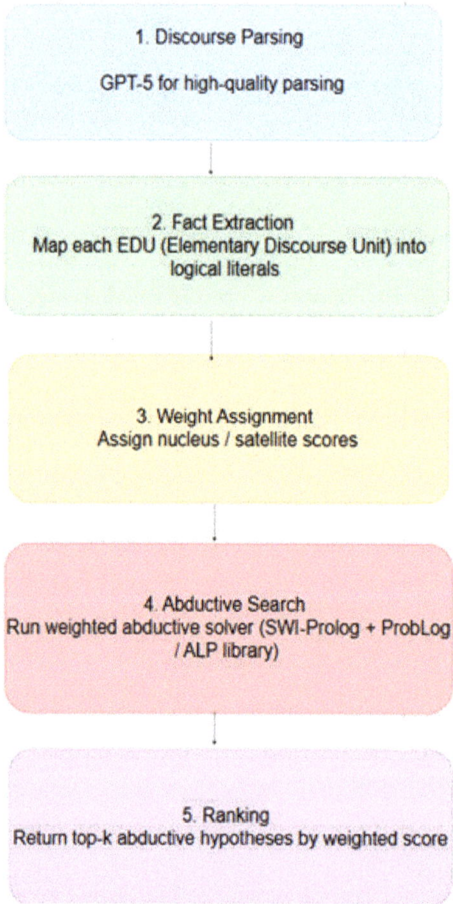

Figure 3. Computational pipeline.

https://doi.org/10.3390/pr14020322

5. Counter-Abduction

The role of counter-abduction in neuro-symbolic reasoning is best understood by tracing its origins to classical accounts of abductive inference and modern theories of confirmation. Abduction denotes the inferential move in which a reasoner proposes a hypothesis H that, if true, would render a surprising observation E intelligible. Peirce emphasized that abduction is neither deductively valid nor inductively warranted; its justification lies in explanatory plausibility rather than certainty.

However, in both human and machine reasoning, the first abductive hypothesis is often not the most reliable. This motivates the introduction of *counter-abduction*, a concept developed implicitly in sociological methodology and more formally in abductive logic programming [27]. Counter-abduction refers to the generation of alternative hypotheses that likewise explain the evidence, thereby challenging the primacy of the initial explanation. For example, while an explosion may abductively explain a loud bang and visible smoke, counter-abductive alternatives—such as a car backfire combined with smoke from a barbecue—demonstrate that multiple explanations can account for the same phenomena [28].

To evaluate these competing hypotheses, the framework draws on *confirmation theory*, which provides probabilistic and logical tools for assessing evidential support [29]. In Bayesian terms, evidence E confirms hypothesis H if it increases its probability, i.e., if $P(H \mid E) > P(H)$. Probability-increase measures such as $d(H,E) = P(H \mid E) - P(H)$ and ratio-based measures such as $r(H,E) = P(H \mid E)/P(H)$ quantify the extent of confirmation [30]. Likelihood-based measures, including the likelihood ratio $P(E \mid H)/P(E \mid \neg H)$, further assess how much more expected the evidence is under the hypothesis than under alternatives. These tools allow structured comparison of hypotheses $\{H_1, H_2, \ldots\}$ generated via abduction and counter-abduction.

Chemical Process Example

Consider a continuous stirred-tank reactor (CSTR) exhibiting an unexpected temperature rise accompanied by reduced conversion efficiency. An LLM-based diagnostic component might abductively propose the hypothesis H_1: "the temperature controller is mis-tuned, causing oscillatory overheating." While H_1 plausibly explains the observed temperature deviation, counter-abductive reasoning generates alternative hypotheses, such as H_2: "an unmeasured increase in reactant feed concentration has elevated the reaction rate," or H_3: "fouling in the heat exchanger has reduced heat removal capacity." All three hypotheses can account for the same evidence E (temperature increase and efficiency loss), yet they imply fundamentally different corrective actions.

Using counter-abduction, the framework evaluates which hypothesis is best supported by additional process evidence and integrity constraints. For example, stable controller outputs may weaken H_1, while unchanged feed analyzer readings may partially disconfirm H_2. Meanwhile, a gradual increase in coolant-side pressure drop provides confirmatory evidence for H_3. Likelihood ratios and probability updates allow the system to rank these explanations, exposing cases where the initial abductive hypothesis—although fluent and operationally plausible—lacks sufficient evidential support. In the context of hallucination detection, this mechanism is critical: an LLM-generated explanation may satisfy process constraints and appear coherent, yet be rejected through counter-abduction because a rival hypothesis better explains the observed data.

This example illustrates how counter-abduction functions as a safeguard against premature or overconfident explanations in process control settings. By explicitly generating and comparing competing causal accounts, the framework prevents single-hypothesis dominance and enables principled rejection of unsupported narratives. When integrated with LLM-

https://doi.org/10.3390/pr14020322

assisted diagnostics, counter-abduction thus plays a central role in detecting explanation-level hallucinations, ensuring that accepted interpretations remain grounded in process physics, measurements, and causal structure rather than linguistic plausibility alone.

6. Discourse in Abductive Logic Programming

ALP is designed to generate hypotheses (abducibles) that, when added to a knowledge base, explain observations. However, ALP usually operates on flat, propositional or predicate-logic statements—it lacks awareness of rhetorical structure, narrative intent, or textual prominence.

Discourse analysis, especially based on RST, gives us a hierarchy of rhetorical relations between text segments—e.g., *Cause–Effect, Condition, Evidence, Contrast, Elaboration*. Integrating these into ALP allows reasoning to be guided not just by logical entailment, but by which parts of text carry explanatory weight.

Conceptual integration is shown in Table 4.

Table 4. Extending the features of ALP with discourse information.

Aspect	In Classical ALP	With Discourse-Aware ALP
Observation	A set of atomic facts or predicates.	Clauses extracted from *nucleus* discourse segments (main claims).
Abducibles	Candidate explanatory literals.	Hypotheses aligned with *satellite* segments, weighted by rhetorical relation (e.g., Evidence ↑, Elaboration ↓).
Explanatory Preference	Minimality or cost-based.	Weighted abductive preference: prioritize hypotheses supported by nucleus–satellite strength and coherence relations.
Conflict Resolution	Based on logical consistency.	Also guided by discourse coherence: conflicting explanations that preserve discourse flow are preferred.

We now proceed to the example: "The unit triggers a high-temperature alarm at the outlet and a low-flow alarm on the coolant line. Since the deviation began right after a post-maintenance startup and the coolant flow is unstable, a pump failure is likely."

Discourse parsing output: Nucleus (primary symptoms/alarms):

"The outlet high-temperature alarm is active and the coolant low-flow alarm is active."

Satellite (Cause–Effect/context):

"Since the event began after post-maintenance startup and the coolant flow shows instability."

Claim (Evaluation):

"Pump failure is likely."

ALP representation is as follows:

https://doi.org/10.3390/pr14020322

```
% --- Background Knowledge (Process Control Causal Relationships) ---

cause(high_feed_temp, increased_outlet_temp).
cause(low_coolant_flow, increased_outlet_temp).
cause(fouling, reduced_heat_transfer).
cause(reduced_heat_transfer, increased_outlet_temp).

cause(pump_failure, low_coolant_flow).
cause(valve_stiction, fluctuating_flow_rate).

symptom(increased_outlet_temp, high_temp_alarm).
symptom(low_coolant_flow, low_flow_alarm).
symptom(fluctuating_flow_rate, oscillatory_control_signal).

% Useful intermediate rules
triggers(X, Y) :- cause(X, Y).
leads_to(X, Y) :- cause(X, Y).
indicates(Observation, Fault) :- symptom(Fault, Observation).

% --- Observations from Process Historian / SCADA ---
obs(high_temp_alarm).
obs(low_flow_alarm).
obs(post_maintenance_startup).     % relevant contextual fact
obs(flow_rate_instability).
% Discourse weighting
nucleus_weight(1.0).
satellite_weight(0.6).
% Abductive rule (discourse-aware)
explain(Obs, Hyp) :-
    nucleus(Obs, Nuc), satellite(Obs, Sat),
    abduct(Hyp),
    satisfies(Nuc, Hyp, W1),
    satisfies(Sat, Hyp, W2),
    Score is W1*1.0 + W2*0.6,
    Score > Threshold.
```

Here the *nucleus* (joint pain, swelling) gives hard constraints, while the *satellite* (seafood meal cause) provides softer evidence with lower weight [31]. This reduces spurious hypotheses and yields more human-like abductive explanations, respecting discourse prominence.

6.1. *Discourse-Weighted ALP (D-ALP)*

Let $P = (\Pi, \Delta, A)$ be a standard abductive logic program:

- Π—strict rules;
- Δ—defeasible rules;
- A—set of abducibles.

We extend it with a discourse weighting function w: $L \in [0, 1]$ over literals L derived from RST trees:

- $w(l) = 1.0$ if l originates from a nucleus clause;
- $0 < w(l) < 1$ if l originates from a satellite clause;
- $w(l) = 0$ if l appears in background or elaborative relations.

Then the abductive explanation $E \subseteq A$ minimizes:

https://doi.org/10.3390/pr14020322

$$\mathrm{Cos}(E) = \sum_{l \in E} (1 - w(l)) \cdot penalty(l)$$

subject to $\Pi \cup E \models O$.

Thus discourse prominence directly affects the search space and preference ordering among explanations.

The penalty function *penalty*(l) quantifies how "expensive" it is to *abduce* literal l—i.e., to assume l as true when it is not derivable from the strict rules Π. It represents epistemic risk: how far l departs from evidence, domain priors, or discourse plausibility.

$$\mathrm{penalty}(l) = \alpha \cdot \rho(l) + \beta \cdot \kappa(l) + \gamma \cdot \delta(l),$$

where

- $\rho(l)$ is a rule distance—number of rule applications needed to derive l (depth in derivation tree).
- $\kappa(l)$ is a conflict measure—degree to which l contradicts existing facts or competing hypotheses.
- $\delta(l)$ is a discourse mismatch measure—how incompatible l is with its rhetorical context (e.g., "Contrast", "Condition").
- Constants α, β, γ control the importance of logical vs. discourse penalties (often $\alpha = 0.5$, $\beta = 0.3$, $\gamma = 0.2$).

Thus, penalty(l) is higher for:

- hypotheses that are logically remote;
- contradict evidence;
- misalign with the discourse flow.

Assume a process event log produces three competing fault hypotheses: coolant-flow restriction, heat-exchanger fouling, and sensor drift. Each is associated with different abductive characteristics. The system identifies coolant-flow restriction as the primary (nucleus) hypothesis: it has low rule distance from observations ($\rho \approx 0.1$), no direct contradictory evidence ($\kappa \approx 0$), and minimal discourse mismatch with the operator narrative ($\delta \approx 0.1$), yielding a small normalized penalty of roughly 0.12.

Heat-exchanger fouling appears as a competing hypothesis: although similarly plausible in principle ($\rho \approx 0.1$), it carries observable conflict—recent maintenance contradicted the expectation of fouling—and a moderate discourse mismatch ($\kappa \approx 0.4$, $\delta \approx 0.2$). This raises the penalty to approximately 0.22.

The third hypothesis, sensor drift, functions as an irrelevant satellite explanation. It is weakly connected to the alarm pattern and has greater inconsistency with recorded temperature and flow signals ($\rho \approx 0.3$, $\kappa \approx 0.5$, $\delta \approx 0.7$). Its penalty rises to nearly 0.43, making it the least supported explanation of the three.

Computing discourse-weighted abductive cost (D-ALP) produces the following pattern of preference:

- If *w*(*coolant-flow-restriction*) = 1.0, the cost approaches 0;
- If *w*(*fouling*) = 0.7, the cost is approximately 0.066;
- If *w*(*sensor-drift*) = 0.4, the cost increases to ≈0.26.

Thus, D-ALP selects coolant-flow restriction as the most plausible root cause—lowest penalty, strongest discourse alignment, and highest diagnostic confidence.

6.2. Combining Discourse Analysis with Explainable Reasoning

We combine discourse analysis with explainable reasoning frameworks such as ALP, argumentation, or probabilistic logic. We propose how the structure of a discourse tree can

https://doi.org/10.3390/pr14020322

be used to extract, classify, and weight different forms of explanations from text or model outputs. When a discourse tree with EDUs and relations is mapped into logical structures, each subtree becomes a candidate explanatory unit—either *causal, evidential, conditional,* or *contrastive*. Thus, the discourse tree serves as a blueprint for structured explanation extraction. The mapping is shown in Table 5.

Table 5. From discourse to explanation types.

Discourse Relation	Typical Explanation Type	Logical Template
Cause/Result	Causal explanation	$E \leftarrow$ *cause(X,Y)*
Evidence/Justify	Evidential explanation	$E \leftarrow$ *supports(X,Y)*
Condition/Hypothetical	Conditional explanation	$E \leftarrow$ *if(X) then(Y)*
Contrast/Antithesis	Counterfactual or contrastive explanation	$E \leftarrow$ *prefer(X,Y)*
Elaboration/Background	Contextual explanation	$E \leftarrow$ *context(X,Y)*
Purpose/Motivation	Goal-oriented explanation	$E \leftarrow$ *intent(X,Y)*

In this framework, discourse relations are not treated as merely rhetorical structures but are assigned explicit weights reflecting their diagnostic importance in explanation verification. Relations linked to the nucleus of the discourse—such as *Cause/Result* and *Evidence/Justify*—receive the highest weights, as they directly support the core explanatory claim and correspond to physically grounded causal or evidential links in the process. Satellite relations, including *Elaboration/Background* and *Purpose/Motivation*, are weighted lower because they provide contextual or operational background rather than direct causal support. *Condition/Hypothetical* relations receive intermediate weights, reflecting their role in qualifying explanations without asserting actual causation. *Contrast/Antithesis* relations are treated asymmetrically: while they do not increase support for a hypothesis, they contribute negatively by enabling counter-abductive defeat when a competing explanation better accounts for the observed data. By weighting nucleus and satellite relations differently, the system prioritizes explanations that are causally and evidentially grounded while remaining sensitive to contextual qualifiers and explicit counter-evidence, thereby improving robustness against superficially plausible but incorrect explanations.

Given a discourse tree with nodes—EDUs, the following steps are performed:

1. Parse the text → obtain discourse tree.
2. Identify nucleus/satellite pairs → mark *nucleus* as potential hypothesis/conclusion, *satellite* as premise/explanation.
3. Label each edge with its rhetorical relation.
4. Translate edges to logical relations, as follows.

```
cause(X, Y). % from Cause relation
evidence(X, Y). % from Evidence relation
condition(X, Y).
contrast(X, Y).
```

Subtrees are aggregated according to the following rules:

- A *deep* subtree (multiple Evidence nodes) → composite explanation;
- A *flat* tree (multiple Causes) → multi-factor explanation;
- A *branch with Contrast* → comparative explanation.

Consider the following diagnostic narrative: "The reactor outlet temperature has risen sharply. Since this increase followed a drop in coolant flow, and heat-transfer effi-

https://doi.org/10.3390/pr14020322

ciency is low, pump restriction is suspected. However, vibration readings do not indicate impeller damage."

Discourse Tree (simplified):
```
          [Pump restriction suspected]
                    /          \
          (Evidence)              (Contrast)
          /        \                  \
[Low coolant]  [Low heat-transfer]    [No vibration anomaly]]
```

From this, we extract:

```
explain(pump_restriction,cause(low_coolant_flow, increased_temperature)).
explain(pump_restriction, evidence(low_heat_transfer_efficiency)).
explain_not(pump_restriction, contrast(no_vibration_detected)).
```

These can be ranked or combined into a structured abductive hypothesis:

E ={cause(A,B),evidence(C),¬contrast(D)}, which forms a multi-faceted explanation for the same claim.

Different *forms of explanation* (causal, evidential, contextual, counterfactual) can be systematically extracted by pattern-matching substructures of the discourse tree (Table 6).

Table 6. The shape and depth of the discourse tree are important for ALP construction.

Structural Feature	Explanation Property	Interpretation
Depth (levels below nucleus)	Explanation completeness	deeper = more elaborated reasoning chain
Branching factor	Explanation diversity	multiple satellites = multi-evidence explanation
Contrastive nodes count	Explanation contestability	High → balanced/counterfactual reasoning
Causal path length	Explanation causal depth	long paths correspond to mechanistic explanations

When an ALP is formed:

1. Each explanation type defines a different abductive schema:
   - *Causal explanations* suggest new rules (A → B).
   - *Evidential explanations* add support to hypotheses.
   - *Contrastive explanations* add defeaters or alternative hypotheses.
2. The discourse tree structure provides hierarchical weighting for hypothesis generation.
3. The abductive search space becomes tree-constrained rather than flat—drastically improving interpretability and efficiency.
   By leveraging discourse trees, we can automatically construct:
   - Level 1 (Direct): nucleus-satellite explanation (*"Acidic PH because of uric acid ↑"*)
   - Level 2 (Intermediate): composite explanation combining several satellites
   - Level 3 (Meta): contrastive or counterfactual explanation derived from opposing branches
   - Level 4 (Global): entire discourse tree summarized as an explanatory graph, suitable for argumentation or visualization.

7. Counter-Abduction Hallucination Detection Algorithm

While abduction has been widely studied as an explanatory mechanism, the complementary concept of *counter-abduction*—the generation and evaluation of rival hypotheses against an initial abductive explanation—has received less attention in the context of LLM reasoning.

In a neuro-symbolic setting, counter-abduction plays a crucial role in hallucination detection. It introduces an adversarial explanatory pressure on the LLM-generated CoT,

https://doi.org/10.3390/pr14020322

challenging its assumptions, testing its logical validity, and evaluating whether alternative explanations better satisfy coherence and minimality conditions. This section formalizes the relationship between counter-abduction and hallucination, illustrates how counter-abductive reasoning exposes invented premises and inconsistencies, and presents a unified algorithm for counter-abductive hallucination detection within the Abductive Chain-of-Thought (A-CoT) framework.

The algorithm takes as input an LLM-generated CoT, a symbolic knowledge base KB, a discourse-structure representation RST, and a set of abductive/counter-abductive generators. It outputs (1) whether the CoT is hallucinated, (2) the defeat score, and (3) a revised reasoning chain if needed.

The algorithm for counter-abductive hallucination detection is as follows:

Input:

- CoT explanation E_0;
- Knowledge base KB;
- Discourse structure RST with weights $w(\cdot)$;
- Abductive hypothesis generator A;
- Counter-abductive generator C;
- Logical verifiers $V = \{V_1, \ldots, V_n\}$.

Output:

- Hallucination label $H \in$ {True,False};
- Defeat score $D(E_0)$;
- Suggested corrected explanation E^*.

Procedure:

```
1. Parse E0 into atomic propositions P = extract(E0)

2. Identify surprising or inconsistent elements:
      S ← detect_surprises(P, KB, RST)

3. Generate abductive hypotheses for each surprise:
      H_A ← U{s ∈ S} A(s, KB)

4. Generate counter-abductive rival explanations:
      H_C ← U{s ∈ S} C(s, KB)

5. For each explanation h ∈ H_A U H_C:
      - Evaluate consistency score:
            Cons(h) = consistency_check(h, KB)
      - Evaluate discourse alignment score:
            Disc(h) = Σ_{p ∈ h} w(RST(p))
      - Combine:
            Score(h) = α·Cons(h) + β·Disc(h)

6. Compute defeat of the original explanation:
      D(E0) = aggregate_{Vi ∈ V} Vi.compare(E0, H_C)

      where each Vi produces:
            Vi.compare(E0, H_C) =
                  1 if ∃ h ∈ H_C such that Score(h) > Score(E0)
                  0 otherwise

7. If D(E0) > τ (default τ = 0.5):
        H ← True            // hallucinated
        E* ← argmax_{h ∈ H_C} Score(h)
   Else:
        H ← False
        E* ← E0

8. Return (H, D(E0), E*)
```

The steps of counter abduction computation procedure are shown in Figure 4.

https://doi.org/10.3390/pr14020322

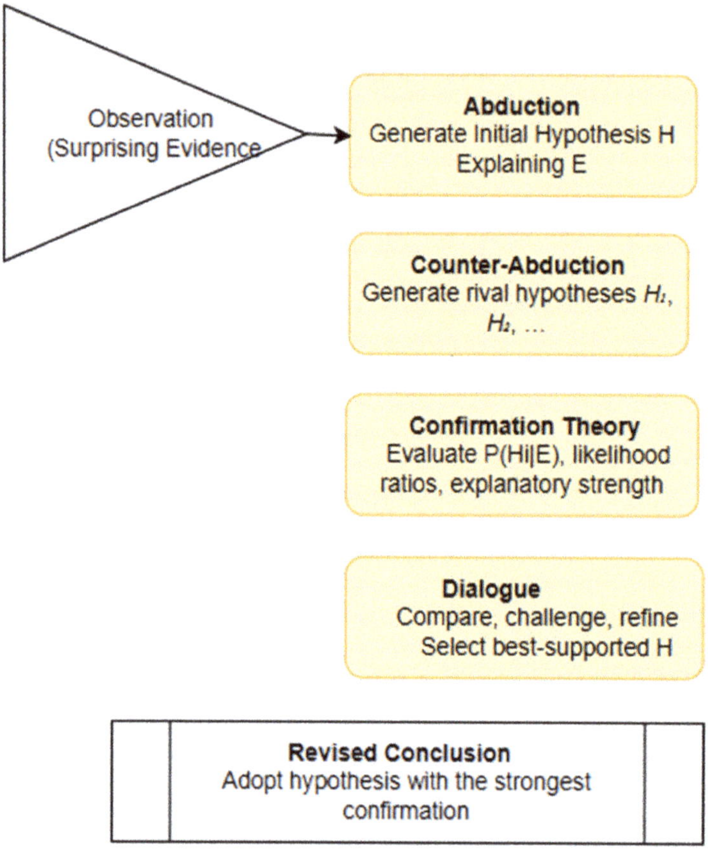

Figure 4. Steps of counter abduction.

8. Evaluation

We present a comprehensive evaluation of the proposed Discourse-Based Abductive Logic Programming (D-ALP) framework. The evaluation examines both computational performance and human-perception aspects. The computational evaluation measures the accuracy, efficiency, and consistency of the abductive reasoning process when discourse structure is integrated. The human evaluation investigates how users perceive clarity, coherence, and trust in discourse-aware explanations.

Our focus is hallucination detection for model answers using four logical assessment methods as validators. Each validator assesses whether an answer's central claim is defeated by the argument-validation system. All datasets were annotated with rhetorical roles such as nucleus and satellite, and rhetorical relations such as *Cause, Evidence, Contrast, Condition, and Elaboration*. These annotations guide D-ALP's discourse weighting.

We define a hallucination as a claim whose defeat probability exceeds 0.5. This cautious threshold is motivated by safety-critical domains, where we prefer to reject answers that are defeated with substantial probability.

8.1. Evaluation Dataset

To evaluate explanation-level hallucinations in industrial settings, we constructed a new Process-Control Hallucination Dataset consisting of 500 synthetic but realistic diagnostic scenarios derived from common faults in chemical and energy systems (e.g.,

https://doi.org/10.3390/pr14020322

pump restriction, valve stiction, heat exchanger fouling, compressor surge, sensor drift, and reactor runaway). Each scenario includes: (i) a structured set of process signals and contextual information (temperatures, flows, vibration, controller patterns, maintenance notes); (ii) the true underlying fault; (iii) a correct mechanistic explanation grounded in process physics; (iv) a plausible but incorrect hallucinated explanation that an LLM might generate; (v) an easy-but-wrong explanation, designed to appear superficially credible while contradicting operational evidence; and (vi) explicit counter-evidence supporting counter-abductive reasoning.

A sample record includes: Each item includes:

1. Context (signals + logs);
2. True fault;
3. Correct mechanistic explanation;
4. Hallucinated explanation (plausible but wrong);
5. Easy-but-wrong explanation (dangerous, highly seductive);
6. Counter-evidence (for counter-abduction).

The record example is below:


```
{
  "context": {
    "temp_out": 145,
    "flow_coolant": 40%,
    "vibration": normal,
    "pressure_drop": minimal,
    "maintenance_log": "cleaned heat exchanger two days ago"
  },
  "true_fault": "pump_restriction",
  "correct_explanation": "The outlet temperature rise follows coolant flow reduction, not fouling. No vibration anomaly supports pump impeller integrity.",
  "hallucinated_explanation": "Heat exchanger fouling is increasing thermal resistance."   // appears plausible but incorrect
}
```


The dataset (https://github.com/bgalitsky/halluc_in_health/blob/master/data/process_control_hallucination_500.json, accessed on 16 January 2026) mirrors real diagnostic practice in process control, where multiple fault hypotheses may appear consistent with partial observations, and where misleading yet intuitive narratives frequently arise. By encoding mechanistic, evidential, and contrastive structures, the dataset enables fine-grained evaluation of explanation fidelity—not just fault identification. This resource therefore provides a controlled testbed for comparing black-box, retrieval-based, deliberation-based, and neuro-symbolic systems. In particular, it allows us to probe whether a model can distinguish between valid causal chains and structurally incorrect but plausible explanations, a key challenge in high-stakes autonomous monitoring and decision support.

We observe the following hallucination types (Table 7).

The dataset is constructed by combining synthetic and, where available, real process-control data to create diagnostically realistic scenarios. Synthetic plant data are generated from simplified but physically grounded models such as two-stream heat exchangers and closed-loop cooling systems, using simulation environments including Modelica (OpenModelica: latest release 1.22.2), Aspen HYSYS (V. 15), or MATLAB Simulink (v. 24.2) to produce time-series signals, alarms, and contextual events. For each simulated fault or operating condition, we generate paired explanations: a correct mechanistic explanation consistent with process physics, a plausible but incorrect hallucinated explanation that fits surface patterns, and harder cases containing conflicting or partial evidence designed to trigger misleading inferences. Each scenario is then annotated using a discourse-aware abductive representation, explicitly encoding causal links, supporting evidence, and contrastive signals that weaken a hypothesis. Finally, scenarios are labeled by difficulty, ranging from single-cause faults to multi-causal and deceptive cases with strong hallucination triggers. In evaluation, an LLM is tasked with producing both a fault diagnosis and its underlying reasoning, which is then assessed for whether the explanation is grounded in the encoded causal structure or constitutes a hallucinated narrative.

Table 7. Hallucination types in our dataset.

Type	Example
Mechanistic hallucination	Inferring fouling from high ΔT without flow or pressure drop evidence
Temporal hallucination	Reversing cause-effect (ex. "temperature rise caused the flow reduction")
Symmetry hallucination	Assuming conditions upstream mirror downstream readings
Sensor substitution hallucination	Treating one sensor's reading as if it measured a different variable
Counterfactual blindness	Ignoring conflict evidence like stable vibration or normal pump power draw

8.2. Experimental Setup

Five systems were compared (Table 8). GPT 5.1 was used for evaluation.

Table 8. Approaches to hallucination detection evaluated in this study.

System	Description
Baseline ALP	Classical abductive logic programming with uniform rule weights
ProbALP	Probabilistic ALP without discourse cues
D-ALP-Depth	Depth-weighted abduction using nucleus/satellite hierarchy
D-ALP-Rel	Relation-aware abduction using rhetorical relation types
Full D-ALP	Combined hierarchical and relational weighting (proposed system)
Abductive RAG	Framework that integrates abductive inference into retrieval-augmented LLMs [32]

Evaluation metrics included logical accuracy (F1), average reasoning time, the proportion of defeated hypotheses (logical inconsistency), and human ratings of clarity, coherence, and trust. All numerical results were averaged across ten runs and three datasets.

8.3. Hallucination Detection

We evaluate on three claim-verification datasets that we derive from existing QA/NLI resources: TruthfulHalluc (from TruthfulQA, general factual QA [33]), MedHalluc (from MedQA, clinical narratives and PubMedQA [34], eSNLI_Halluc (from eSNLI, inference with injected inconsistencies [35], and HotPotQA [36]. For each source, we convert items into question–answer (QA) style pairs [37] and then inject controlled inconsistencies by appending randomly sampled, semantically incompatible attributes (facts, circumstances, symptoms). These perturbations create positive "hallucination" cases; unmodified items

https://doi.org/10.3390/pr14020322

serve as negatives. Our focus is hallucination detection for model answers using four logical assessment methods as validators.

Table 9 shows F1 of hallucination detection for our datasets and our development of ALP. One can observe that Full D-ALP outperforms Abductive RAG system on HotPotQA dataset.

Table 9. Hallucination detection performance.

Dataset	Baseline ALP	ProbALP	D-ALP-Depth	D-ALP-Rel	Full D-ALP	Abductive RAG
TruthfulHalluc	0.63	0.66	0.70	0.72	**0.79**	
MedHalluc	0.63	0.68	0.71	0.75	**0.83**	
eSNLI_Halluc	0.60	0.68	0.71	0.72	**0.77**	
HotPotQA	0.65	0.64	0.68	0.72	**0.80**	0.75
Process-Control Hallucination	0.57	0.55	0.60	0.62	**0.72**	
Average	0.63	0.66	0.70	0.73	0.80	

Full D-ALP achieved 12–18 percentage points improvement over the baseline across datasets, confirming that discourse guidance improves abductive inference precision.

We observe in Table 10 that discourse structure pruned peripheral hypotheses and shortened inference time by over 20% compared to the baseline. Best performance is shown in bold.

Table 10. Efficiency of Full D-ALP in comparison with reduces ALPs.

System	Avg. Reasoning Time (s)	Search Space Reduction (%)
Baseline ALP	1.00	–
ProbALP	1.35	–
D-ALP-Depth	0.88	−12
D-ALP-Rel	0.82	−18
Full D-ALP	**0.79**	**−21**

We proceed to evaluation of logical consistency in Table 11. Best performance is shown in bold.

Table 11. Logical consistency.

System	Defeated Hypotheses (%)
Baseline ALP	19
ProbALP	15
D-ALP-Depth	13
D-ALP-Rel	10
Full D-ALP	**7**

Discourse-aware reasoning reduced logically inconsistent explanations by roughly two-thirds relative to the baseline.

8.4. Ablation Study

Baseline is the plain abductive logic program without any discourse information. Δ *Accuracy* means how many percentage points higher the correct-explanation rate became after adding each feature. Δ *Consistency* means how many percentage points fewer logically self-contradictory or "defeated" explanations appeared. In Table 12:

https://doi.org/10.3390/pr14020322

- D-ALP-Depth (+7%, +6%)—when only the nucleus/satellite hierarchy is used, the system already improves moderately. Prioritizing nucleus clauses focuses reasoning on the central facts and avoids side-track hypotheses.
- D-ALP-Rel (+10%, +9%)—adding rhetorical-relation knowledge (Cause, Evidence, Contrast, etc.) helps more: the model learns what type of relation best supports an abductive link, giving stronger logical and discourse coherence.
- Full D-ALP (+15%, +12%)—combining both depth and relation cues yields the largest gain. The two features reinforce one another: hierarchy tells the engine *which* statements matter most, relation type tells it *how* they connect.

Table 12. Ablation study of discourse features.

Variant	Δ *Accuracy* vs. Baseline (%)	Δ *Consistency* vs. Baseline (%)
D-ALP-Depth	+7	+6
D-ALP-Rel	+10	+9
Full D-ALP	**+15**	**+12**

8.5. Human Evaluation

A user study with 36 participants was conducted to assess explanation interpretability (Table 13). Participants included 12 engineers, 12 AI researchers, and 12 general users. Each participant reviewed 20 explanations per dataset (2160 total ratings). Explanations were rated on clarity, coherence, and trust (1 = poor, 5 = excellent). Best performance is shown in bold.

Table 13. Assessment of explanation.

System	Clarity	Coherence	Trust
Baseline ALP	3.1	2.9	2.8
ProbALP	3.3	3.0	3.0
D-ALP-Depth	4.0	3.8	3.9
D-ALP-Rel	4.2	4.1	4.1
Full D-ALP	**4.6**	**4.5**	**4.4**

Full D-ALP explanations were perceived as clearer and more coherent. Participants described them as "human-like," "logical," and "educational."

Trust calibration was measured as the change in self-reported confidence before and after viewing explanations (Table 14). Best performance is shown in bold.

Table 14. Trust calibration.

System	Trust Before	Trust After	Δ (Calibrated Trust)
Baseline ALP	0.58	0.65	+0.07
ProbALP	0.57	0.67	+0.10
Full D-ALP	**0.55**	**0.78**	**+0.23**

Trust rose by 23% when discourse-structured reasoning was displayed, indicating higher user confidence grounded in transparency.

Alignment between system-generated and human discourse trees was also measured using F1 for nuclearity and rhetorical relations (Table 15).

https://doi.org/10.3390/pr14020322

Table 15. Alignment between machine and human discourse.

Dataset	Nucleus F1	Relation F1
TruthfulHalluc	0.73	0.69
MedHalluc	0.77	0.72
eSNLI_Halluc	0.70	0.65
Average	**0.73**	**0.69**

Higher alignment correlated strongly (r = 0.68) with human trust ratings. The results show that incorporating discourse structure enhances both reasoning performance and user understanding (Table 16). Discourse relations prioritize central claims and down-weight peripheral elaborations, leading to process control commands that are both logically precise and narratively coherent. Logical accuracy improved by up to 18%, while human clarity and trust rose by nearly 50%. The presence of rhetorical markers such as "because" and "however" increased users' sense that the model was reasoning transparently.

Table 16. Assessment for reasoning performance and user understanding.

Evaluation Aspect	Best System (Full D-ALP)	Improvement Over Baseline
Logical Accuracy	0.80 (average)	+16%
Runtime Efficiency	0.79 s/inference	−21%
Consistency	7% defeated	−63%
Human Clarity	4.6/5	+48%
Human Trust	4.4/5	+57%
Trust Calibration	+0.23	+0.16

Error analysis indicated that residual weaknesses stemmed from misclassification of rhetorical relations, particularly confusing Condition and Evidence, and from underestimation of multi-nuclear claims. Nonetheless, D-ALP consistently outperformed all baselines in accuracy, efficiency, and human trust.

Discourse-based abductive logic programming delivers process control commands that are both computationally efficient and intuitively meaningful. The framework advances explainable AI by integrating rhetorical structure into logic reasoning, yielding outcomes that are verifiable, interpretable, and trusted by humans. Future evaluations will extend to adaptive weighting learned from user feedback, cross-domain transfer to legal and educational corpora, and longitudinal studies of sustained trust.

8.6. Counter-Abduction and Hallucination Mitigation

An additional evaluation focused on how counter-abduction enhances D-ALP's capacity to detect and correct hallucinations in process control. Counter-abduction is defined here as the process of generating and testing *alternative hypotheses* that could back up the same control commands but contradict the current explanation. When integrated into D-ALP, counter-abduction acts as a built-in hallucination filter, comparing competing explanatory paths and identifying cases where the main abductive hypothesis lacks sufficient discourse or logical support.

To test this mechanism, the system was extended to compute for each hypothesis H a counter-hypothesis H' that minimizes overlap in supporting discourse nodes. A hypothesis was flagged as *hallucinatory* if its counter-hypothesis achieved equal or higher discourse-weighted plausibility (Table 17).

Adding counter-abductive reasoning increased hallucination detection accuracy by approximately 6 percentage points across datasets. The improvement was most pronounced

https://doi.org/10.3390/pr14020322

in narratives with ambiguous causal or evidential cues, where rival explanations helped expose unsupported assertions.

Table 17. Evaluation of counter-abduction.

Dataset	Baseline ALP Hallucination Accuracy	Full D-ALP (No Counter-Abduction)	D-ALP + Counter-Abduction
TruthfulHalluc	0.64	0.79	**0.86**
MedHalluc	0.66	0.83	**0.88**
eSNLI_Halluc	0.63	0.77	**0.84**
Average	**0.64**	**0.80**	**0.86**

Participants in a follow-up evaluation (N = 20) were shown pairs of system explanations—one original and one with counter-abductive contrast. Average trust in the system's factual grounding increased from 4.1 to 4.5 on the five-point scale. Annotators reported that explicitly seeing "why not-X" arguments enhanced their confidence that the reasoning engine could self-check for over-interpretation.

Qualitative review indicated that counter-abduction excels in identifying discourse-peripheral claims (often satellites or elaborations) that were incorrectly promoted to nucleus status by LLMs (Table 18). By generating rival hypotheses anchored in contrasting discourse relations—such as *Contrast*, *Condition*, or *Exception*—the system effectively invalidates spurious explanations before presentation to the user.

Table 18. Contribution of counter-abduction.

Evaluation Metric	Full D-ALP	D-ALP + Counter-Abduction
Hallucination Detection F1	0.81	**0.87**
False Positive Rate	0.14	**0.09**
Human Trust in Factuality	4.1	**4.5**

These findings demonstrate that counter-abduction provides a critical verification layer complementing the discourse-weighted abductive process. Together, they yield an interpretable mechanism for hallucination suppression that aligns with human reasoning: first hypothesize (abduction), then challenge (counter-abduction), and finally accept only the explanation that survives logical and rhetorical scrutiny.

9. Practical Deployment

The proposed abduction-based hallucination detection framework is not intended to operate inside the high-frequency inner control loop, but rather as a semantic verification layer integrated into supervisory and decision-support levels of a closed-loop process control architecture. In a standard hierarchy, the real-time control layer (e.g., PID or MPC) remains fully responsible for actuation and constraint enforcement, while the abductive verifier interfaces with LLM components that generate diagnoses, explanations, or recommendations. LLM outputs are treated as hypotheses, which are then evaluated against process knowledge, integrity constraints, and observed data using abductive and counter-abductive reasoning before being surfaced to operators or passed to supervisory logic (Figure 5).

In real-time settings, the framework operates event-driven rather than continuously. Verification is triggered when (i) an LLM proposes a fault explanation, root cause, or mitigation strategy; (ii) a recommendation would modify set-points, safety margins, or

https://doi.org/10.3390/pr14020322

operating modes; or (iii) anomalous conditions are detected that require human interpretation. Because abductive verification is applied selectively and at supervisory time scales (seconds to minutes), its computational overhead does not interfere with real-time control performance. The output of the verifier is a structured validation signal—accepted, rejected, or revised hypothesis—along with explicit justification or competing explanations.

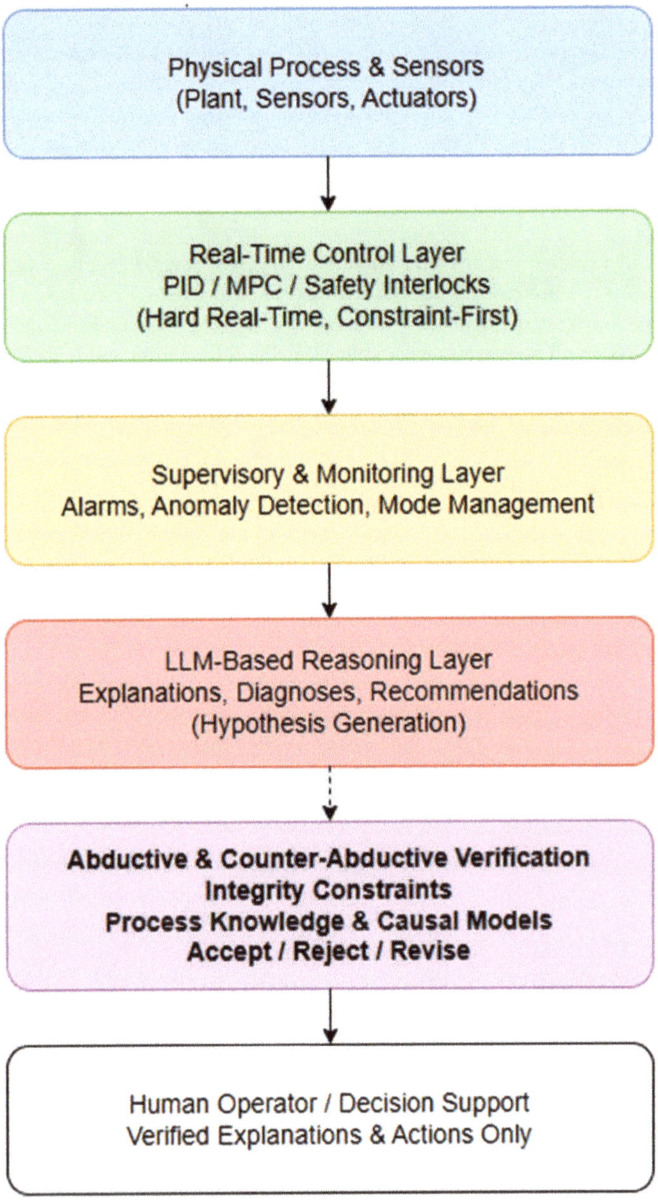

Figure 5. Integration of abduction-based hallucination detection into a closed-loop process control architecture. Real-time constrained controllers operate independently of LLM components, while abductive verification functions as a supervisory semantic safety layer that validates LLM-generated explanations and recommendations before they influence operators or supervisory decisions.

https://doi.org/10.3390/pr14020322

Usage scenarios are as follows:

1. Online monitoring and diagnosis. In monitoring applications, LLMs may summarize alarms, interpret multivariate trends, or propose fault diagnoses. The abductive verifier evaluates whether these interpretations are consistent with process constraints, causal dependencies, and available measurements. Hallucinated diagnoses—such as unsupported fault attributions or impossible causal chains—are flagged before being presented to operators, reducing cognitive overload and false confidence.
2. Decision support and supervisory control. In supervisory settings, LLMs may suggest operational adjustments, maintenance actions, or recovery strategies. Abductive verification checks whether such recommendations are supported by evidence and whether alternative explanations better explain the observed process behavior. Only verified recommendations are forwarded to supervisory controllers or operators, ensuring that downstream constrained controllers are not guided by semantically invalid inputs.
3. Human–AI collaboration and incident analysis. During abnormal events or post-incident reviews, the framework supports transparent human–AI interaction by exposing why a generated explanation is accepted or rejected. Counter-abductive reasoning makes implicit assumptions explicit, enabling operators and engineers to assess credibility rather than relying on fluent narratives. This strengthens trust without granting the LLM authority over control actions.

Software Implementation

The proposed framework is implemented using a modular neuro-symbolic pipeline that combines large language models with symbolic reasoning components. Logical reasoning and abductive inference are implemented in SWI-Prolog (v9.x), chosen for its mature support of meta-interpreters, constraint handling (CHR), and probabilistic extensions. Probabilistic abductive reasoning is realized using ProbLog, enabling uncertainty-aware hypothesis ranking and defeat scoring.

Discourse analysis is performed in two modes depending on scale. For small-scale, high-fidelity experiments, GPT-based discourse parsing is used to extract EDUs and rhetorical relations. For large-scale evaluation, a lightweight implementation based on the RST discourse parser [26] is employed to improve efficiency while preserving nucleus–satellite structure. Discourse trees are mapped to logical literals with associated weights and passed to the abductive solver.

The LLM component (GPT-5.1 in evaluation) is accessed via API and used exclusively for generation, not verification. All hallucination detection, counter-abduction, and ranking are performed externally in the symbolic layer, ensuring a black-box integration independent of model internals. Experiments were run on standard Linux servers; no specialized hardware (e.g., GPUs) is required for the reasoning layer. Configuration files specify abducibles, integrity constraints, discourse weights, and scoring parameters to ensure reproducibility across datasets and domains.

10. Related Work and Discussions

Although hallucination research has primarily focused on natural language tasks, the underlying phenomena—producing coherent but unsupported explanations—have clear analogs in industrial diagnostics and process control. Recent work has begun to explore these risks. FaultExplainer [38] demonstrates that LLMs used for fault diagnosis may generate plausible but incorrect mechanistic narratives, exposing the need for structured verification in safety-critical domains. Traditional fault detection and diagnosis (FDD) research provides a complementary foundation: model-based and data-driven

https://doi.org/10.3390/pr14020322

systems [39,40] emphasize consistency between observations, physical models, and causal pathways, thereby offering principles for identifying hallucination-like reasoning errors.

Explainable AI (XAI) for industrial systems further highlights the challenge of ensuring that diagnostic outputs remain grounded in sensor reality. Jang et al. [41] show that attribution-based explanations may fail to reflect true causal drivers, a form of mismatch analogous to hallucination in generative models. Work on rule induction for automated FDD [42] similarly notes that AI-generated rules may diverge from physical constraints unless tightly validated.

At the same time, research in NLP provides conceptual and methodological tools for detecting unsupported reasoning—hallucination taxonomies [17], verification-based prompting, and self-evaluation frameworks [43] highlight mechanisms through which LLMs can self-correct or expose faulty intermediate steps. While these studies do not address industrial processes directly, they motivate the need for domain-specific verification frameworks, such as abductive and counter-abductive reasoning, that integrate physical constraints and discourse evidence.

Taken together, these lines of work indicate a growing recognition that detection of ungrounded explanations—whether in language or process control—requires combining model-based reasoning, evidential validation, and structured hypothesis testing. Our work builds on this insight by proposing a unified abductive framework tailored to process-control hallucinations, where incorrect causal explanations can have significant operational and safety implications.

In abductive logic programming [22], the structure of the inference cycle:

1. Abductive generation: The system proposes candidate explanations—hypotheses that, when combined with background knowledge, entail the observed data.
2. Deductive testing: The implications of each hypothesis are deduced and checked against constraints.
3. Inductive evaluation: Empirical or probabilistic measures assess which hypotheses remain consistent with evidence.

Here, the availability constraint manifests through the space of abductive hypotheses the system can construct—defined by its symbolic vocabulary, its logical rules, and its computational resources. Thus, the inferential behavior of an ALP system corresponds not to an ideal IBE, which presupposes access to all possible explanations, but to an IBAE process bounded by representational and algorithmic feasibility.

Inductive inferences in the narrow sense are well-investigated, but their inferential power is limited. With their help it is possible to reason from regularities observed in the past to unobserved or future instances of these regularities, but it is not possible to infer conclusions containing new concepts expressing un observed properties that are not contained in the premises. These inferences are conceptually creative. They are needed in science whenever one reasons from the observed phenomena to theoretical concepts. Theoretical concepts in science describe unobservable properties (e.g., electric forces) or structures (e.g., electrons) that explain the observed phenomena in a unified way. It was an important insight of the post-positivistic philosophy of science that these theoretical concepts cannot be reduced to observable concepts via chains of definitions [29]. Thus the justification of explanations introducing theoretical concepts cannot be based on conceptual analysis, but must take the form of an ampliative inference.

Modern neuro-symbolic AI systems [44] face similar epistemic limitations. Even when neural networks are used to expand the hypothesis space by generating candidate patterns or latent variables, the symbolic reasoning layer can only evaluate those that fit within its logic schema. The model therefore performs an approximation to abduction, balancing between *expressive generativity* (neural) and *logical evaluability* (symbolic). This interplay

https://doi.org/10.3390/pr14020322

mirrors the philosophical IBAE trade-off: the system can only infer the best explanation among those currently expressible and computationally tractable.

Furthermore, the evaluation criterion—what counts as the "best" explanation—must also be contextually defined. In philosophy, explanatory virtues include simplicity, coherence, and unification [45]; in AI, analogous metrics include likelihood, posterior probability, minimality, or information gain. As in human inquiry, a system's "rational acceptability" depends on whether its best available explanation satisfies the methodological standards relevant to its domain—e.g., causal adequacy in science, logical consistency in expert systems, or interpretability in explainable AI.

HaluCheck is introduced as a visualization framework for assessing and prominently displaying the likelihood of hallucination in model outputs [10]. It allows users to select among multiple LLMs, providing flexibility for different tasks and preferences. The system integrates a diverse set of hallucination-evaluation metrics, enabling users to compute and compare likelihood scores using alternative methods. By allowing users to switch between these evaluators, HaluCheck supports experimentation with different assessment strategies and helps identify the most effective approach for a given use case.

Within the abductive framework developed in this paper, visualization plays an especially important role. Abductive and counter-abductive reasoning naturally generate sets of competing hypotheses, partial explanations, defeaters, and discourse-based evidence links. These structures are difficult to interpret when presented linearly, but they become immediately understandable when rendered visually. A visualization system such as HaluCheck can display the abductive reasoning graph—showing which evidence nodes support the main hypothesis, which discourse relations strengthen or weaken it, and where counter-hypotheses expose possible hallucinations. This makes the internal logic of hallucination detection transparent: users can see when an explanation relies on low-weight satellite information, when a rival hypothesis achieves a better discourse-weighted cost, or when inconsistency triggers a defeat. Visualizing these abductive and counter-abductive interactions thus strengthens interpretability, enables human verification of the hallucination score, and supports calibrated trust by revealing why a response is flagged as hallucinated rather than providing a numerical score alone.

Post-LLM Generation Verification

The concern regarding post-generation verification versus constraint-first synthesis is valid from a classical process control perspective, where embedding physical and operational constraints directly into control generation is essential for real-time safety guarantees. However, the proposed abductive verification framework is intentionally positioned at a different layer of the control stack. Constraint-first approaches, such as MPC and safety filters, ensure feasibility with respect to known constraints but implicitly assume that upstream reasoning—diagnoses, explanations, and recommendations—is semantically correct. LLM-generated outputs may satisfy all explicit constraints while remaining causally unsupported or incorrect. Abductive and counter-abductive reasoning specifically target these epistemic failures by evaluating whether a generated hypothesis is actually warranted by process evidence.

A key advantage of post-generation abductive verification is its model-agnostic and non-intrusive nature. Embedding detailed process constraints directly into LLM generation requires tight coupling to domain-specific formalisms and repeated re-engineering as models or constraints evolve, which is difficult in regulated industrial environments. In contrast, abductive verification operates as an external semantic checker that can be applied across different LLMs and deployments without modifying existing, certified control

https://doi.org/10.3390/pr14020322

logic. This makes the approach particularly suitable for incremental adoption in industrial settings where established constrained control architectures must remain intact.

Finally, concerns about computational overhead and latency are mitigated by the intended scope of application. The framework is not designed for high-frequency inner-loop control, but for supervisory, diagnostic, and decision-support tasks, where reaction times are significantly longer. In these contexts, the cost of post-generation verification is small relative to the risk of acting on hallucinated explanations or recommendations. Moreover, verification can be selectively triggered only when outputs affect control-relevant decisions. Viewed in this way, post-generation abductive verification complements constraint-first control by providing a semantic safety layer that enhances trust, transparency, and robustness in LLM-enabled process control systems.

11. Conclusions

The discourse-aware abductive framework introduced in this work provides a principled foundation for constructing and verifying complex explanations generated by large language models. By integrating abductive inference with rhetorical structure analysis, the approach enables systems to distinguish central, hypothesis-bearing content from peripheral or contextual material, thereby strengthening both explanatory precision and hallucination detection. The value of this integration is evident across multiple application domains.

Counter-abduction is thus a foundational component of hallucination-resistant neuro-symbolic reasoning. By positioning rival explanations as defeaters of LLM-generated CoTs, counter-abductive reasoning transforms narrative reasoning into a competitive, evidence-driven process grounded in logic and discourse structure. This provides a unified theoretical and computational basis for hallucination detection and correction across medical analysis, legal reasoning, scientific interpretation, and general-purpose CoT verification.

In medical narratives, weighting discourse nuclei over satellite descriptions allows the system to focus abductive diagnosis generation on patient-relevant complaints rather than tangential remarks, improving causal hypothesis extraction. In legal reasoning, the framework supports more transparent argument evaluation by giving precedence to claims occurring in the conclusion or main argument segments while attenuating the influence of background information. In scientific writing, it enhances the identification of robust causal explanations by prioritizing claims derived from results and discussion sections over speculative or forward-looking commentary. Finally, in LLM verification, discourse-aware abductive logic programming offers a structured mechanism for identifying hallucinations: statements originating in low-weight, peripheral text segments can be selectively discounted, while central claims undergo rigorous consistency checking.

Taken together, these applications demonstrate that combining abductive reasoning with discourse structure provides a versatile and effective method for improving reasoning fidelity, ensuring interpretability, and increasing trust in neuro-symbolic systems across diverse high-stakes domains.

Advantages are:

- Increases interpretability: abductive hypotheses are justified by discourse roles.
- Improves precision: ignores peripheral text when generating explanations.
- Enables alignment with human reasoning: since humans emphasize nuclei when forming explanations.
- Supports hallucination detection: contradictions in nucleus-derived claims outweigh peripheral inconsistencies.

"Counter-abduction strength of confirmation metrics dialogue" refers to a structured, interactive process where

https://doi.org/10.3390/pr14020322

1. Abductive reasoning proposes initial explanations for observed evidence.
2. Counter-abduction introduces competing explanations.
3. Confirmation metrics quantitatively assess how well the evidence supports each hypothesis.
4. Dialog facilitates the comparison and discussion of these assessments to arrive at the most plausible explanation.

This framework is powerful in any scenario requiring rigorous evaluation of competing hypotheses, ensuring that conclusions are well-supported by evidence. It combines logical reasoning, probabilistic assessment, and collaborative discussion to navigate complex, uncertain situations effectively.

Our evaluation demonstrates that incorporating discourse structure and counter-abductive reasoning significantly improves both the logical validity and operational credibility of AI-generated diagnostics in chemical process control. Discourse-weighted Abductive Logic Programming (D-ALP) not only produces fault explanations that are consistent with process physics and observed signals, but also actively tests these explanations against competing fault hypotheses, thereby exposing and reducing hallucinated causal narratives. The results indicate that explanation robustness—rather than surface plausibility alone—is critical for reliable decision support in safety-critical chemical systems. More broadly, abductive hallucination detection can be deployed as a verification layer atop existing white-box, gray-box, and black-box diagnostic approaches [46,47], allowing it to complement data-driven, model-based, and hybrid monitoring techniques [48]. Together, these findings position discourse-aware abductive reasoning as a practical foundation for verifiable and trustworthy neuro-symbolic AI in industrial process monitoring and control.

Safety-aware planning and constrained control methods, although not dependent on LLMs, address a closely related fundamental objective: guaranteeing the feasibility, safety, and regulatory compliance of decisions in process control environments. Constraint-first synthesis techniques—such as model predictive control, reference governors, and formal safety filters—achieve this goal by explicitly restricting the admissible control space and enforcing physical and operational limits by construction [14,16]. These approaches are highly effective at preventing unsafe control actions but implicitly assume that upstream inputs—process diagnoses, explanations, objectives, or operator guidance—are themselves valid and well grounded in process knowledge. As a result, they do not directly address epistemic failures in reasoning or explanation that may arise when LLM-based components are introduced into the control workflow.

In contrast, the abductive verification framework proposed in this work operates at the semantic reasoning layer, targeting hallucinations in explanations, diagnoses, and recommendations generated by LLM-assisted process control systems. Rather than synthesizing control actions under constraints, our method evaluates whether a proposed claim is logically supported by process knowledge, causal structure, and observed evidence, and exposes unsupported outputs through counter-abductive reasoning [17,21]. This distinction clarifies the novelty and scope of the contribution: abductive verification complements, rather than replaces, constraint-first control by guarding against fluent but incorrect reasoning that may pass syntactic or feasibility checks yet remain semantically invalid. By integrating abductive hallucination detection as a verification layer above established constrained control architectures, the proposed approach strengthens trust, interpretability, and safety in LLM-enabled process control without disrupting existing control enforcement mechanisms.

https://doi.org/10.3390/pr14020322

Limitations

While the proposed abductive and counter-abductive framework demonstrates clear advantages in detecting explanation-level hallucinations in process control, several limitations should be acknowledged. First, the current evaluation relies primarily on synthetic and semi-synthetic process scenarios, which, although physically grounded and carefully constructed, cannot fully capture the variability, noise, and undocumented contingencies present in real industrial plants. Second, the quality of abductive inference depends on the completeness and correctness of the underlying process ontology; missing causal links or oversimplified fault models may lead to incomplete hypothesis spaces or false defeasibility. Third, discourse parsing—particularly when using high-capacity LLMs—introduces computational overhead that may limit real-time deployment in high-frequency monitoring settings, although lighter parsers mitigate this at some cost to granularity. Finally, while the framework is designed to operate as a verification layer atop white-box, gray-box, and black-box diagnostic systems, its performance is sensitive to the quality of upstream signals and explanations. Addressing these limitations will require broader validation on real plant data, expansion of domain ontologies, and further optimization for online industrial use.

Author Contributions: Conceptualization, B.G.; Methodology, B.G.; Software, B.G. and A.R.; Validation, B.G.; Formal analysis, B.G. and A.R.; Investigation, B.G.; Data curation, B.G.; Writing—original draft, B.G.; Writing—review and editing, A.R. All authors have read and agreed to the published version of the manuscript.

Funding: This research received no external funding.

Data Availability Statement: Code is available at https://github.com/bgalitsky/halluc_in_health/tree/master/abduction (accessed on 12 January 2026). The original data presented in the study are openly available in: https://github.com/bgalitsky/halluc_in_health/tree/master/process_control (accessed on 12 January 2026).

Acknowledgments: During the preparation of this manuscript/study, the authors used GPT 5.1 and Qwen 3 for the purposes of English grammar improvement. The authors have reviewed and edited the output and take full responsibility for the content of this publication.

Conflicts of Interest: Author Boris Galitsky was employed by the Knowledge Trail Inc. The remaining authors declare that the research was conducted in the absence of any commercial or financial relationships that could be construed as a potential conflict of interest.

References

1. Qin, S.J.; Badgwell, T.A. A survey of industrial model predictive control technology. *Control Eng. Pract.* **2003**, *11*, 733–764. [CrossRef]
2. Galitsky, B. A neuro-symbolic system for distributed chemical engineering simulations. In *Applications of Neuro-Symbolic AI*; Springer: Cham, Switzerland, 2026; *in press*.
3. Venkatasubramanian, V. A review of process fault detection and diagnosis: Part I–III. *Comput. Chem. Eng.* **2003**, *27*, 293–346. [CrossRef]
4. Lee, J.; Bagheri, B.; Kao, H.A. A cyber-physical systems architecture for Industry 4.0-based manufacturing systems. *Manuf. Lett.* **2015**, *3*, 18–23. [CrossRef]
5. Peres, R.S.; Jia, X.; Lee, J.; Sun, K.; Colombo, A.W.; Barata, J. Industrial artificial intelligence in industry 4.0—Systematic review, challenges and outlook. *IEEE Access* **2020**, *8*, 220121–220139. [CrossRef]
6. Decardi-Nelson, B.; Alshehri, A.S.; Ajagekar, A.; You, F. Generative AI and process systems engineering: The next frontier. *Comput. Chem. Eng.* **2024**, *187*, 108723. [CrossRef]
7. Azaria, A.; Mitchell, T. The internal state of an LLM knows when it's lying. *arXiv* **2023**, arXiv:2304.13734.
8. Su, W.; Wang, C.; Ai, Q.; Hu, Y.; Wu, Z.; Zhou, Y.; Liu, Y. Unsupervised real-time hallucination detection based on the internal states of large language models. In *Findings of the Association for Computational Linguistics: ACL 2024*; Ku, L.W., Martins, A., Srikumar, V., Eds.; Association for Computational Linguistics: Bangkok, Thailand, 2024; pp. 14379–14391.

https://doi.org/10.3390/pr14020322

9. Varshney, N.; Yao, W.; Zhang, H.; Chen, J.; Yu, D. A stitch in time saves nine: Detecting and mitigating hallucinations of llms by validating low-confidence generation. *arXiv* **2023**, arXiv:2307.03987. [CrossRef]
10. Heo, S.; Son, S.; Park, H. Halucheck: Explainable and verifiable automation for detecting hallucinations in llm responses. *Expert Syst. Appl.* **2025**, *272*, 126712. [CrossRef]
11. Kossen, J.; Han, J.; Razzak, M.; Schut, L.; Malik, S.; Gal, Y. Semantic entropy probes: Robust and cheap hallucination detection in LLMs. *arXiv* **2024**, arXiv:2406.15927. [CrossRef]
12. Galitsky, B. Improving open domain content generation by text mining and alignment. In *AI for Health Applications and Management*; Galitsky, B., Goldberg, S., Eds.; Elsevier: Amsterdam, The Netherlands, 2021.
13. Shi, X.; Xue, S.; Wang, K.; Zhou, F.; Zhang, J.; Zhou, J.; Tan, C.; Mei, H. Language Models Can Improve Event Prediction by Few-Shot Abductive Reasoning. *Adv. Neural Inf. Process. Syst.* **2023**, *36*, 29532–29557.
14. Rawlings, J.B.; Mayne, D.Q. *Model Predictive Control: Theory and Design*; Nob Hill Publishing: San Francisco, CA, USA, 2009.
15. Åström, K.J.; Hägglund, T. *PID Controllers: Theory, Design, and Tuning*; ISA: Singapore, 1995.
16. Gilbert, E.G.; Kolmanovsky, I.; Tan, K.T. Discrete-time reference governors and the nonlinear control of systems with state and control constraints. *Int. J. Robust Nonlinear Control* **1995**, *5*, 487–504. [CrossRef]
17. Ji, Z.; Lee, N.; Frieske, R.; Yu, T.; Su, D.; Xu, Y.; Ishii, E.; Bang, Y.J.; Madotto, A.; Fung, P. Survey of Hallucination in Natural Language Generation. *ACM Comput. Surv.* **2023**, *55*, 12. [CrossRef]
18. Rawlings, J.B.; Meadows, E.S.; Muske, K.R. Nonlinear model predictive control: A tutorial and survey. *IFAC Proc. Vol.* **1994**, *27*, 185–197. [CrossRef]
19. Peirce, C.S. Illustrations of the Logic of Science. Popular Science Monthly. 1878. Volume 13. Available online: https://en.wikisource.org/wiki/Popular_Science_Monthly/Volume_13/August_1878/Illustrations_of_the_Logic_of_Science_VI (accessed on 12 January 2026).
20. Peirce, C.S. Lectures on Pragmatism. In *Collected Papers of Charles Sanders Peirce*; Hartshorne, C., Weiss, P., Eds.; Harvard University Press: Cambridge, MA, USA, 1903; Volume 5.
21. Harman, G. The inference to the best explanation. *Philos. Rev.* **1965**, *74*, 88–95. [CrossRef]
22. Kakas, A.C.; Mancarella, P. Generalized Abduction. *J. Log. Comput.* **1990**, *1*, 389–407.
23. Swi-Prolog. 2025. Available online: https://www.swi-prolog.org/ (accessed on 12 January 2026).
24. Christiansen, H. Executable specifications for hypothesis-based reasoning with Prolog and Constraint Handling Rules. *J. Appl. Log.* **2009**, *7*, 341–362. [CrossRef]
25. Galitsky, B.; Rybalov, A. Extending large language model capabilities beyond reasoning. In *Healthcare Applications of Neuro-Symbolic Artificial Intelligence*; Elsevier: Amsterdam, The Netherlands, 2025; Chapter 4; pp. 81–107.
26. Jansen, P.; Surdeanu, M.; Clark, P. Discourse complements lexical semantics for non-factoid answer reranking. In Proceedings of the 52th Annual Meeting ACL, Baltimore, MD, USA, 22–27 June 2014. [CrossRef]
27. Kakas, A.C.; Kowalski, R.A.; Toni, F. Abductive logic programming. *J. Log. Comput.* **1992**, *2*, 719–770. [CrossRef]
28. Haig, B.D. *Investigating the Psychological World: Scientific Method in the Behavioral Sciences*; MIT Press: Cambridge, MA, USA, 2014.
29. Carnap, R. *The Logical Foundations of Probability*, 2nd ed.; University of Chicago Press: Chicago, IL, USA, 1962.
30. Crupi, V.; Tentori, K.; González, M. On Bayesian measures of evidential support. *Philos. Sci.* **2007**, *74*, 229–252. [CrossRef]
31. Galitsky, B. Identifying large language model hallucinations in health communication. In *Healthcare Applications of Neuro-Symbolic Artificial Intelligence*; Elsevier: Amsterdam, The Netherlands, 2025; Chapter 8; pp. 283–329.
32. Lin, S. Abductive Inference in Retrieval-Augmented Language Models: Generating and Validating Missing Premises. *arXiv* **2025**, arXiv:2511.04020v1. [CrossRef]
33. Lin, S.; Hilton, J.; Evans, O. TruthfulQA: Measuring How Models Mimic Human Falsehoods. In Proceedings of the 60th Annual Meeting of the Association for Computational Linguistics, Dublin, Ireland, 22–27 May 2022; Volume 1, pp. 3214–3252. [CrossRef]
34. Jin, D.; Pan, E.; Oufattole, N.; Weng, W.-H.; Fang, H.; Szolovits, P. What disease does this patient have? A large-scale open domain question answering dataset from medical exams. *arXiv* **2020**, arXiv:2009.13081. [CrossRef]
35. Camburu, O.-M.; Rocktäschel, T.; Lukasiewicz, T.; Blunsom, P. e-SNLI: Natural language inference with natural language explanations. *arXiv* **2018**, arXiv:1812.01193. [CrossRef]
36. Yang, Z.; Qi, P.; Zhang, S.; Bengio, Y.; Cohen, W.W.; Salakhutdinov, R.; Manning, C.D. HotpotQA: A dataset for diverse, explainable multi-hop question answering. In Proceedings of the 2018 Conference on Empirical Methods in Natural Language Processing (EMNLP), Brussels, Belgium, 31 October–4 November 2018; ACL Anthology. Association for Computational Linguistics: Brussels, Belgium, 2018; pp. 2369–2380. [CrossRef]
37. Rasool, Z.; Kurniawan, S.; Balugo, S.; Barnett, S.; Vasa, R.; Chesser, C.; Hampstead, B.M.; Belleville, S.; Mouzakis, K.; Bahar-Fuchs, A. Evaluating LLMs on document-based QA: Exact answer selection and numerical extraction using CogTale dataset. *Nat. Lang. Process. J.* **2024**, *8*, 100083. [CrossRef]
38. Khan, A.; Nahar, R.; Chen, H.; Flores, G.E.C.; Li, C. FaultExplainer: Leveraging large language models for interpretable fault detection and diagnosis. *Comput. Chem. Eng.* **2025**, *199*, 109152. [CrossRef]

https://doi.org/10.3390/pr14020322

39. Venkatasubramanian, V. The promise of artificial intelligence in chemical engineering: Is it here, finally? *AIChE J.* **2019**, *65*, 466–478. [CrossRef]
40. Isermann, R. *Fault-Diagnosis Systems: An Introduction from Fault Detection to Fault Tolerance*; Springer: Berlin/Heidelberg, Germany, 2006.
41. Jang, K.; Pilario, K.E.; Lee, N.; Moon, I.; Na, J. Explainable Artificial Intelligence for Fault Diagnosis of Industrial Processes. *IEEE Trans. Ind. Inform.* **2023**, *21*, 4–11. [CrossRef]
42. Fischer, J. AI-Powered Rule Generation for Automated Fault Detection and Diagnostics. Doctoral Dissertation, Technische Universität Wien, Vienna, Austria, 2025.
43. Miao, N.; Teh, W.Y.; Rainforth, T. Selfcheck: Using LLMs to zero-shot check their own step-by-step reasoning. *arXiv* **2023**, arXiv:abs/2308.00436.
44. Bader, S.; Hitzler, P. Dimensions of Neural-Symbolic Integration—A Structured Survey. *arXiv* **2005**, arXiv:cs/0511042.
45. Lipton, P. *Inference to the Best Explanation*, 2nd ed.; Routledge: London, UK, 2004.
46. Xie, L.; Liu, H.; Zeng, J.; Tang, X. A Survey of Calibration Process for Black-Box LLMs. *arXiv* **2024**, arXiv:2412.12767. [CrossRef]
47. Wu, J.; Dong, N.; Liu, F.; Yang, S.; Hu, J. Feature hallucination via Maximum A Posteriori for few-shot learning. *Knowl.-Based Syst.* **2021**, *225*, 107129. [CrossRef]
48. Galitsky, B.; Tsirlin, A. Step Wise Approximation of CBOW Reduces Hallucinations in Tail Cases. 2025. Available online: https://www.preprints.org/manuscript/202507.0670 (accessed on 12 January 2026).

https://doi.org/10.3390/pr14020322

processes

MDPI

Article

Research and Application of Intelligent Control System for Uniform Pellet Distribution

Tingting Liao [1,2], Xiaoxin Zeng [1,2,*], Xudong Li [1,2], Zongping Li [1,2], Jianming Zhang [3], Chen Liu [1,2] and Weisong Wu [4]

1 ZhongYe ChangTian International Engineering Co., Ltd., Changsha 410006, China; liaotign@163.com (T.L.); lxd950202@gmail.com (X.L.); lizongping_cie@163.com (Z.L.)
2 National Research Center for Equipment System Engineering Technology of Sintering Pellet, Changsha 410006, China
3 School of Computer and Communication Engineering, Changsha University of Science and Technology, Yuntang Campus, Changsha 410114, China
4 ZhongYe ChangTian (Changsha) Intelligent Technology Co., Ltd., Changsha 410114, China; 13054195704@139.com
* Correspondence: zxx1852004@163.com; Tel.: +86-13549688483

Abstract

In pellet production, the uniformity of material distribution directly affects the subsequent roasting effect and the quality of finished products. Aiming at the problems of uneven distribution in traditional shuttle distribution systems, such as material stacking at both ends of the wide belt, insufficient parameter matching leading to uneven distribution, and reliance on manual adjustment which makes it difficult to adapt to dynamic working conditions, this paper proposes an intelligent control method based on Integral Simulation and Gradient Descent optimization (IS-GD). Firstly, this method combines the structure and operating parameters of the distribution equipment and accurately simulates the material distribution law on the wide belt during the reciprocating movement of the shuttle through integral technology. Based on the simulation results, longitudinal and lateral uniformity discriminant functions are constructed, and a phased gradient descent optimization strategy is adopted to dynamically adjust the shuttle belt speed, walking speed, and operating parameters of each stage with the goal of minimizing the uniformity index. Experimental results show that this method achieves a significant improvement in lateral distribution uniformity without affecting the stability of longitudinal distribution. This research provides reliable technical support for intelligent distribution control in pellet production and helps to improve the roasting quality and production efficiency of pellets.

Keywords: pellet; shuttle; integral simulation; gradient descent optimization

1. Introduction

Against the backdrop of the green and low-carbon transformation in the iron and steel industry, pellet ore, as a high-quality burden for blast furnace ironmaking, has been continuously increasing its proportion in the blast furnace burden structure due to its advantages such as low process energy consumption and excellent metallurgical properties. It has thus become one of the important approaches to achieving low-carbon smelting [1,2]. In mainstream pellet production processes including the belt roaster and grate-kiln system, the uniformity of green pellet distribution directly determines the bed permeability, the uniformity of roasting heating, as well as the strength and output of finished pellets [3,4].

At present, the mainstream distribution methods in pellet production include oscillating belts, shuttle car, and so on. Among them, shuttle car systems are widely used in new construction and renovation projects due to their advantages such as compact structure and adaptability to large-scale production. However, traditional shuttle car systems have significant drawbacks: on the one hand, during the reciprocating movement of the shuttle car, the long commutation braking time and multiple falling times of green pellets easily lead to material accumulation at both ends of the wide belt, forming a "basin"-shaped material profile with poor distribution uniformity [3]; on the other hand, the insufficient matching between its speed parameters (such as the traveling speed of the shuttle car and the belt speed) and the operating speed of the wide belt is prone to causing material strip overlap or gaps, increasing the system operating load [4]. To improve distribution uniformity, compared with the traditional method of manual adjustment based on production output, some control models have been developed in recent years relying on technologies such as neural networks and big data analysis. Zhou Jianping [5] proposed using multi-point thickness measurement data of the cross-section to judge the uniformity of distribution and finding the speed combination of distribution equipment with the most uniform distribution through big data analysis. Li Dan et al. [6] proposed establishing the relationship between the deviation value of material thickness and parameters such as the stopping margin of the wide belt and the reciprocating car by detecting the material layer distribution on the wide belt with sensors, thereby realizing distribution control. He Shuangping et al. [7]. proposed dividing the pendulum distributor into equal spacing arc lengths and controlling the oscillating belt to run for the same time within the same arc length distance, so as to achieve uniform distribution.

Based on the existing literature, there are few studies on the relationship between material quantity distribution and operating parameters during the shuttle car distribution process. Some researchers have attempted to investigate the feeding distribution under different equipment parameters via discrete element simulation, with the main adjusted parameters being those of the wide belt and roller screen [8–10]. The lateral feeding uniformity on the wide belt is mainly affected by the uniformity of the reciprocating lateral feeding of the shuttle feeder. Therefore, this paper studies the causes of uneven feeding based on the equipment structure of a selected plant, establishes the relationship between equipment operating parameters and the material receiving quantity at each position under this equipment structure, and optimizes the operating parameters of shuttle car feeding using the gradient descent optimization algorithm. The aim is to provide technical support for uniform feeding in pellet production.

2. Problem Analysis and Technology Introduction

2.1. Problem Analysis of Uneven Distribution by Shuttle Car

Shuttle feeding has two structural types: full-car reciprocating and head reciprocating. To facilitate the verification of the system's application, this technology selects a domestic steel plant for research and application. The structure in this project is shown in Figure 1: green pellets produced by the pelletizing disc are collected onto the green pellet belt, which transports them to the shuttle car. Through forward and backward reciprocating movement, the shuttle car distributes the green pellets onto the wide belt.

The shuttle car consists of a movable belt that operates unidirectionally. It continuously conveys the pellet ore transported by the green pellet belt. The entire shuttle car reciprocates perpendicular to the running direction of the wide belt to control the position where green pellets fall onto the wide belt. When the shuttle car moves forward, its relative speed with respect to the green pellet belt is the sum of the shuttle car's moving speed and the shuttle belt's speed: $v_b + v_m$; when moving backward, the relative speed is the difference between

https://doi.org/10.3390/pr14030490

the shuttle belt's speed and the shuttle car's moving speed: $v_b - v_m$. Therefore, even if the feeding quantity of the green pellet belt is stable, the material receiving quantity on the shuttle belt fluctuates during the shuttle car's reciprocating movement, leading to unstable material discharge onto the wide belt. As shown in Figure 2, within the same time interval, the moving distance of the material receiving position during the forward process is greater than that during the backward process, resulting in more material accumulation in the backward process. Thus, the material loaded per unit length on the shuttle car during the backward process is greater than that during the forward process. Even if the discharge speed is the same during both forward and backward movements of the shuttle car, the discharge quantity differs due to changes in the material loaded on the belt. Direction switching of the shuttle car requires start-up and braking time, and continuous material discharge during the commutation process causes material accumulation at both ends of the wide belt.

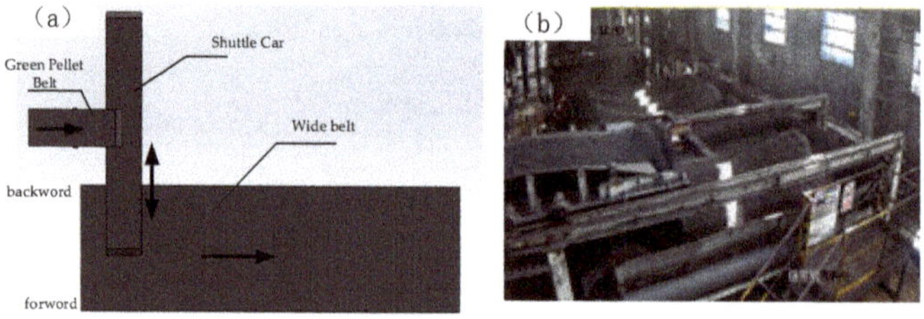

Figure 1. Structure of green ball belt and shuttle car distributing device. (**a**) schematic diagram; (**b**) physical diagram.

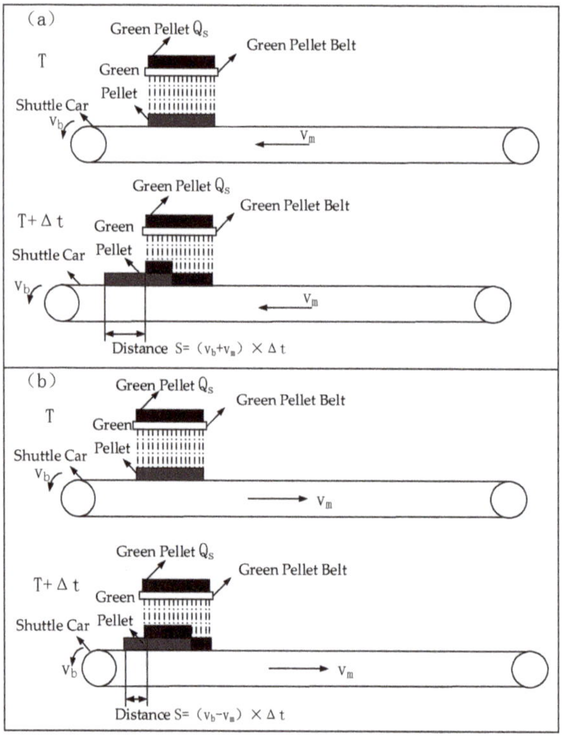

Figure 2. Schematic of material receiving during the forward and backward operation of the shuttle distributor. (**a**) shuttle car forward movement; (**b**) shuttle car backward movement.

https://doi.org/10.3390/pr14030490

Due to the equipment structure inherently leading to uneven distribution in shuttle feeding, this paper first uses integral technology to simulate the distribution process. Based on the simulation results, it analyzes the uniformity and inversely optimizes the distribution parameters, realizing the rapid search for appropriate distribution parameters under a given production volume to ensure stable production.

2.2. Integral-Based Wide Belt Distribution Simulation Model

Due to the variation in material quantity during shuttle feeding, uneven distribution on the trolley will ultimately occur. To understand the distribution status, multiple radar level gauges are currently mostly installed uniformly along the cross-sectional direction of the trolley at the position where green pellets enter the roasting port to monitor the uniformity of trolley distribution [5]. The number of radar level gauges determines the monitoring accuracy of the material layer thickness on the cross-section, but generally only four to six measuring points are adopted in practice due to cost constraints. Since the distribution uniformity on the trolley is closely related to that on the wide belt, this paper proposes an integral-based distribution simulation approach.

2.2.1. Calculation of Shuttle Car Traveling Frequency

According to the shuttle car's operation in the applied project, its operation is controlled by multiple proximity switches. As shown in Figure 3, six proximity switches are installed on the shuttle car's running track. Deceleration proximity switches are used to set the advance deceleration frequency for the shuttle car during forward and backward movements, preventing it from failing to stop safely and overshooting the stop position when reaching the stop proximity switches. The limit proximity switches mark the maximum allowable distance the shuttle car can reach at both ends. Exceeding this position may cause the shuttle car to derail, leading to emergency shutdown due to accidents.

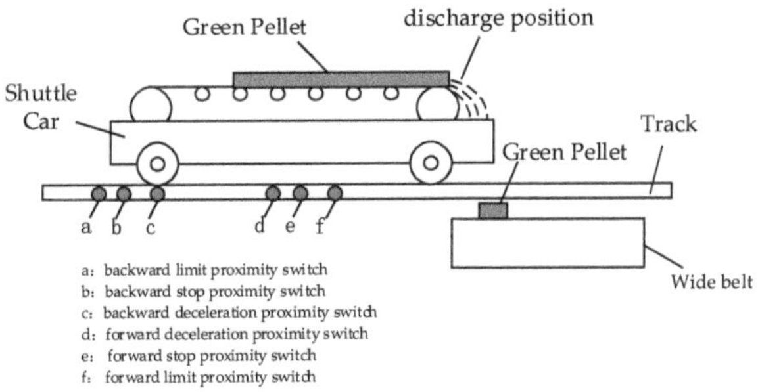

Figure 3. Schematic of the shuttle car structure in the applied engineering project.

Based on the layout of the proximity switches, the operation of the shuttle car can be divided into eight states: shuttle car forward acceleration, shuttle car forward constant speed, shuttle car forward deceleration, shuttle car forward stop, shuttle car backward acceleration, shuttle car backward constant speed, shuttle car backward deceleration, and shuttle car backward stop. Thus, the setting parameters related to the shuttle car's operation are as follows: shuttle car forward start-up time go_{time}, shuttle car forward constant speed frequency go_f, shuttle car forward deceleration frequency go_{dec_f}, shuttle car forward stop time go_{stop_t}, shuttle car backward start-up time $back_{time}$, shuttle car backward constant speed frequency $back_f$, shuttle car backward deceleration frequency $back_{dec\_f}$, and shuttle car backward stop time $back_{stop\_t}$, as shown in Figure 4.

https://doi.org/10.3390/pr14030490

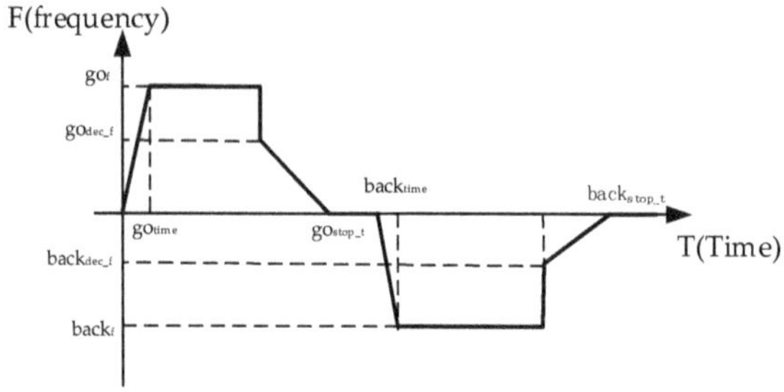

Figure 4. Schematic of the shuttle car's operating frequency variation.

Let the shuttle car's travel sampling frequency be Δt. Based on the above operating parameters, the shuttle car's frequency values at each time point can be calculated as follows:

(1) Frequency of the shuttle car's forward start-up phase

The acceleration and deceleration slope of the shuttle car's motor speed changes linearly; thus, the frequency value vector F_{go1} of the shuttle car's forward start-up phase is obtained:

$$F_{go1} = \left[\frac{go_f}{(go_{time}/\Delta t)}, \frac{go_f \times 2}{(go_{time}/\Delta t)}, \frac{go_f \times 3}{(go_{time}/\Delta t)}, \ldots, go_f\right] \tag{1}$$

(2) Frequency of the shuttle car's forward constant speed phase

First, calculate the operating distance of the start-up phase s_{go1}:

$$s_{go1} = \frac{v_{go}}{2} \times go_{time} \tag{2}$$

Among them, v_{go} represents the speed of the shuttle car's forward constant speed phase, which is calculated based on the conversion formula $v = (f \times \pi \times D)/(i \times p)$ between motor frequency and speed. In the formula, f is the motor frequency (hz); p denotes the number of motor pole pairs; v represents the shuttle car's travel speed (m/s); D is the diameter of the motor's drive drum (m); i denotes the proportional relationship between the motor's input speed and output speed.

The moving distance of the constant speed phase is $s_{go2} = s_{cd} + (s_{bc} - s_{go1})$. Among them, s_{bc} is the distance between the backward stop proximity switch and the backward deceleration proximity switch, and s_{cd} is the distance between the backward deceleration proximity switch and the forward deceleration proximity switch. The frequency value vector is F_{go2}:

$$F_{go2} = [go_f, go_f, \ldots, go_f] \tag{3}$$

The length of the vector F_{go2} is $\frac{s_{go2}}{v_{go} \times \Delta t}$.

(3) Frequency of the shuttle car's forward deceleration phase

The shuttle car triggers the deceleration operation when it reaches the forward deceleration proximity switch, decelerates linearly according to the set deceleration frequency, and its speed becomes 0 when it reaches the position of the forward stop proximity switch. The speed at the deceleration point is $v_{go\_dec} = \left(go_{dec\_f} \times \pi \times D\right)/(i \times p)$, the operating time of the forward deceleration phase is $go_{dec\_time} = \frac{s_{de} \times 2}{v_{go\_dec}}$, and the forward deceleration frequency vector is F_{go3}:

$$F_{go3} = \left[go_{\text{dec}\_f}, (go_{\text{dec}\_f} - \frac{go_{\text{dec}\_f}}{(go_{dec\_time}/\Delta t)}), (go_{\text{dec}\_f} - \frac{go_f \times 2}{(go_{time}/\Delta t)}), \ldots, 0\right] \tag{4}$$

(4) Frequency of the shuttle car's forward stop phase

https://doi.org/10.3390/pr14030490

To control the thickness of the material layer at both ends, a stop time is set for the shuttle car at both ends. The frequency vector $F_{go4} = [0, 0, \ldots, 0]$ has a length of $\frac{go_{stop\_t}}{\Delta t}$.

Then, the frequency distribution of the shuttle car at every Δt time during the forward movement is $F_{go} = [F_{go1}, F_{go2}, F_{go3}, F_{go4}]$. Following the same logic above, the frequency distribution vector of the shuttle car during the backward movement is calculated as F_{back}.

2.2.2. Simulation Calculation of Material Receiving Capacity for Shuttle Cars and Wide Belts

Based on the dimensions of the equipment, corresponding arrays are established to store the material quantity of green pellets distributed per unit length on each equipment. The array of green pellet belt width is W_{sq}, the array of shuttle car length is L_{sc}, and the array of wide belt width is W_k. The lengths of the arrays are $N1$, $N2$, and $N3$, respectively, with all initial values set to 0. The starting coordinate of material receiving for the shuttle car belt is SS_{start}, and the starting coordinate of material receiving for the wide belt is SK_{start}. The arrays obtain the comprehensive production throughput from the system and calculate the feeding quantity q_{sq} of the green pellet belt within time Δt. The green pellet quantity per unit length on the green pellet belt is q'_{sq}:

$$q'_{sq} = q_{sq} / W_{sq} \tag{5}$$

The material initially received by the shuttle car in its initial state:

$$L_{sc}^{start}[SS_{i\_start}, SS_{i\_start} + W_{sq}] = q'_{sq} \tag{6}$$

At the j-th moment during the forward movement, the running speed v_{sc_b} of the shuttle car belt is obtained based on its current frequency. The forward moving speed v_{sc_m} of the shuttle car is calculated using $F_{go}[j]$, thereby deriving the moving distance d_{sc_m} of the shuttle car and the moving distance d_{sc_b} of the shuttle car belt compared to the previous moment. All values in the shuttle car array L_{sc}^{j-1} at the (j − 1)-th moment are shifted forward by a length of d_{sc_b} as a whole. The material corresponding to the shifted distance falls onto the wide belt (denoted as $q_{sc} = sum(L_{sc}^{j-1}[0, d_{sc\_b} - 1])$), and the tail of the shuttle car array is assigned a value of $L_{sc}^{j-1}[N2 - d_{sc\_b} - 1, N2 - 1] = 0$ to simulate the change in material distribution caused by the movement of the shuttle car belt. At the j-th moment, the starting coordinate of material receiving for the shuttle car belt is $SS_j = SS_{j-1} + d_{sc\_m}$. The material quantity in the shuttle car array at the j-th moment is $L_{sc}^{j}[SS_j, SS_j + W_{sq}] = L_{sc}^{j-1}[SS_j, SS_j + W_{sq}] + q'_{sq}$. This is used to simulate the change in the shuttle car's material receiving position caused by its movement and the distribution change brought by the newly added material. The material receiving quantity at each position of the wide belt is $W_k[SK_{j-1}, SK_{j-1} + d_{sc\_m}] = q_{sc} / d_{sc\_m}$.

At the j-th moment during the shuttle car's backward movement, the running speed v_{sc_b} of the shuttle car belt is obtained based on its current frequency. The backward moving speed v_{sc_m} of the shuttle car is calculated using $F_{back}[j]$, thereby deriving the moving distance d_{sc_m} and d_{sc_b}. Similarly, all values in the shuttle car array L_{sc}^{j-1} at the (j − 1)-th moment are shifted forward by a length of d_{sc_b} as a whole. The material corresponding to the shifted distance falls onto the wide belt (denoted as $q_{sc} = sum(L_{sc}^{j-1}[0, d_{sc\_b} - 1])$), and the tail of the shuttle car array is assigned a value of $L_{sc}^{j-1}[N2 - d_{sc\_b} - 1, N2 - 1] = 0$ to simulate the change in material distribution caused by the movement of the shuttle car belt. At the j-th moment, the starting coordinate of material receiving for the shuttle car belt is $SS_j = SS_{j-1} - d_{sc\_m}$. The material quantity in the shuttle car array at the j-th moment is $L_{sc}^{j}[SS_j, SS_j + W_{sq}] = L_{sc}^{j-1}[SS_j, SS_j + W_{sq}] + q'_{sq}$. The material receiving quantity at each position of the wide belt is $W_k[SK_{j-1}, SK_{j-1} - d_{sc\_m}] = q_{sc} / d_{sc\_m}$.

https://doi.org/10.3390/pr14030490

Subsequently, the shuttle car operates in a forward–backward cycle. Through the above steps, the material distribution on the shuttle car and the wide belt of the distributing system under the current parameters can be simulated.

2.3. Gradient Descent-Based Distribution Optimization Control Model

In actual working conditions, the shuttle car distribution process has problems such as longitudinal stacking deviation and uneven transverse distribution. Traditional control methods rely on the setting of empirical parameters and have difficulty meeting the dynamic adjustment needs under complex working conditions. As an efficient continuous optimization method, the gradient descent algorithm has shown excellent parameter optimization capabilities in the field of industrial control [11,12]. In addition to equipment operating parameters, the feeding condition is also affected by factors such as the vibration energy of equipment [13,14]. For example, the roller screen exerts a certain dispersing effect during operation. To reduce the complexity of the model, vibration energy is not incorporated into the model in this paper. Meanwhile, during the production process of large-scale equipment, frequent adjustments to equipment operating parameters are prohibited to reduce production fluctuations and extend equipment service life when the production volume is preset. Therefore, after obtaining the preset production volume, this paper proposes a gradient descent optimization model based on operation phase division, which is tailored to the feeding characteristics of the dual processes of the shuttle car's forward and backward movements. By establishing the mapping relationship among production volume, operating parameters of the acceleration, constant-speed and deceleration phases, and feeding uniformity indicators, a multi-objective optimization function is constructed. The gradient descent algorithm is then employed to achieve the collaborative optimization of control parameters in each phase, ultimately realizing the simultaneous improvement of both longitudinal and transverse uniformity.

2.3.1. The Principle of the Gradient Descent Algorithm

Gradient descent is an iterative method based on first-order optimization. Its core idea is to continuously update parameters along the negative direction of the gradient of the objective function, thereby gradually approaching the minimum value of the function. As shown in Figure 5, for a continuously differentiable objective function $f(\theta)$, gradient descent is updated through the following iterative formula:

$$\theta_{t+1} = \theta_t - \alpha \nabla f(\theta_t) \tag{7}$$

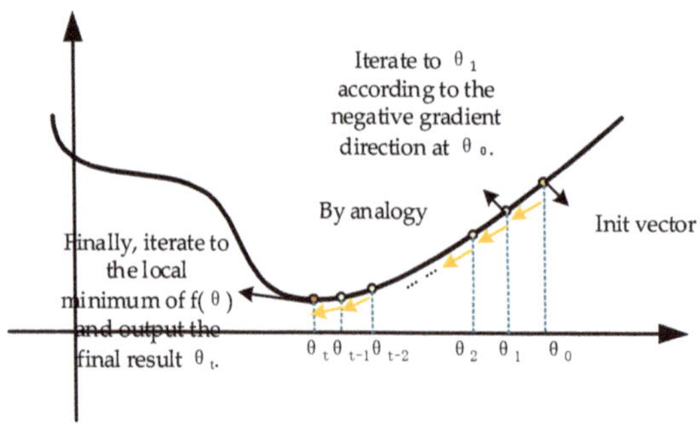

Figure 5. Schematic of gradient descent.

https://doi.org/10.3390/pr14030490

Among them, θ is the parameter vector, the gradient $\nabla f(\theta)$ represents the direction of the maximum rate of change of the function at that point, α is the step size (learning rate) during each iterative update, and t denotes the number of iterations.

Common gradient descent methods include Batch Gradient Descent (BGD), Stochastic Gradient Descent (SGD), and Mini-Batch Gradient Descent (MBGD). In Batch Gradient Descent, the gradient is calculated using the entire dataset in each iteration, ensuring stable convergence but incurring high computational costs. Stochastic Gradient Descent randomly selects one sample per iteration to compute the gradient, offering fast calculation speed yet exhibiting significant fluctuations during the convergence process. Mini-Batch Gradient Descent combines the advantages of both approaches; by calculating the gradient using a subset of samples, it balances stability and convergence speed. As indicated in Equation (7), the learning rate plays a crucial role in gradient descent. A too-small step size leads to slow convergence; excessively refined searching may not only result in low efficiency but also fail to find a favorable local optimal solution. A too-large step size, while shortening the convergence time, may cause drastic parameter fluctuations and even deviate from the optimal range.

2.3.2. Gradient Descent Optimization Model for Fabric Based on Operation Phase Division

Since the operating parameters for regulating the shuttle car's fabric distribution vary across different phases, for instance, when the material discharge amount is insufficient at both ends of the shuttle car, priority can be given to adjusting the front or rear stopping time of the shuttle car. In contrast, the main adjustable parameter for the material discharge amount during the uniform speed phase of the shuttle car is its traveling speed. Combining the characteristics of gradient descent and the requirements of fabric distribution optimization, this paper proposes a gradient descent optimization model for fabric distribution based on operation phase division. First, through simulation calculations, the fabric distribution vector on the wide belt during the shuttle car's forward movement W_k^{go} and backward movement W_k^{back}, under the current parameters can be obtained, respectively. According to the operating distances of the shuttle car in acceleration, uniform speed, and deceleration phases, W_k^{go} and W_k^{back} are each divided into three segments:$W_k^{go\_1}$, $W_k^{go\_2}$, $W_k^{go\_3}$, $W_k^{back\_1}$, $W_k^{back\_2}$, $W_k^{back\_3}$.

Optimization objectives are established by region based on the operation phases. Among them, the evaluation index for longitudinal uniformity is the mean difference of fabric distribution between the forward and backward movements, expressed as follows:

$$diff_1 = \left|\mathrm{avg}(W_k^{go}) - avg(W_k^{back})\right| \tag{8}$$

A smaller value of $diff_1$ indicates a more uniform longitudinal distribution. The evaluation index for lateral uniformity is the difference between the mean value of each segmented small region and the total mean value of the corresponding operation direction:

$$\begin{aligned}
diff_2 &= \left|\mathrm{avg}(W_k^{go}) - avg(W_k^{go_1})\right| \\
diff_3 &= \left|\mathrm{avg}(W_k^{go}) - avg(W_k^{go_2})\right| \\
diff_4 &= \left|\mathrm{avg}(W_k^{go}) - avg(W_k^{go_3})\right| \\
diff_5 &= \left|\mathrm{avg}(W_k^{back}) - avg(W_k^{back_1})\right| \\
diff_6 &= \left|\mathrm{avg}(W_k^{back}) - avg(W_k^{back_2})\right| \\
diff_7 &= \left|\mathrm{avg}(W_k^{back}) - avg(W_k^{back_3})\right|
\end{aligned} \tag{9}$$

https://doi.org/10.3390/pr14030490

Similarly, a smaller difference indicates a more uniform distribution.

The operating parameters of each phase include the shuttle car's forward start time go_{time}, forward uniform speed frequency go_{time}, and forward deceleration frequency go_{dec_f}, etc. Based on these, we construct the optimization variable $X = [go_{time}, go_f, go_{dec\_f}, \ldots]$.

Considering the uniformity requirements of each phase comprehensively, a segmented weighted objective function is constructed.

$$\min F(X) = \varpi_1(diff_1) + \varpi_2(diff_2 + diff_3 + diff_4) + \varpi_3(diff_5 + diff_6 + diff_7) \quad (10)$$

Among them, ϖ_1, ϖ_2, and ϖ_3 are weight coefficients, which are dynamically adjusted based on process priorities and on-site operational experience. Since the transverse feeding uniformity is adopted as the core evaluation index in this system, the coefficients are set $\varpi_1 = \varpi_2 = 0.4$, $\varpi_3 = 0.2$. The weight ratios can be flexibly adapted according to the evaluation requirements of different plants. Since all parameters need to cooperate in fabric control, this technology is optimized and improved based on the BGD method. We calculate the partial derivatives of each extreme objective function with respect to each variable to form the gradient vector $\Delta F(X)$, and adopt a gradient descent strategy with an adaptive learning rate to iteratively update the parameters. Given that the operating frequency of on-site equipment must be an integer, η_K is set to a fixed value of 1 in this study.

Constraint conditions are set for the operating parameters of each phase, and their value ranges are related to the actual equipment parameters. For example, all variable values are not less than 0, and the maximum motor frequency is 50 Hz. In addition, the following requirements are also specified: $go_f > go_{dec\_f}$, $back_f > back_{dec\_f}$.

To address the coupling between phases such as forward movement, backward movement, acceleration, uniform speed, and deceleration, a phased alternating optimization strategy is adopted: (1) fix the parameters of the acceleration and deceleration phases and optimize the parameters of the uniform speed phase; (2) based on the previous optimization results, optimize the parameters of the acceleration and deceleration phases; (3) perform global collaborative iteration until the objective function converges.

3. Result and Discussion

When the production conditions are determined, the fixed parameters of the model are obtained as follows: the bulk density is 2.2 t/m^3, the distance between proximity switches are $s_{ab} = 290(\text{mm})$, $s_{bc} = 320(\text{mm})$, $s_{cd} = 3000(\text{mm})$, $s_{de} = 400(\text{mm})$, $s_{ef} = 420(\text{mm})$, the speed ratio of the shuttle car's traveling motor is 121.46, the diameter of the traveling motor is 400 mm, the number of pole pairs of the traveling motor is 2, the speed ratio of the shuttle car's belt motor is 62.31, the diameter of the belt motor is 630 mm, and the number of pole pairs of the belt motor is 2.

We analyze the bed thickness values several hours before and after the optimization model is put into operation, and the values are shown in Figure 6. In the figure, the X-axis represents the time sequence cycle, Y-axis represents the thickness. Lc1 denotes the thickness at the No. 1 level gauge, lc2 denotes the thickness at the No. 2 level gauge, and lc3 denotes the bed thickness at the No. 3 level gauge.

The more stable the thickness at each level gauge over the time sequence cycle, the more uniform the longitudinal fabric distribution; the closer the thickness values at the three positions, the better the lateral fabric uniformity. To quantify the fabric uniformity, the cosine distance, Euclidean distance, and central distance of the three curves are calculated to measure the degree of aggregation between vectors—smaller values indicate higher aggregation and better fabric uniformity. The comparison of uniformity before and after optimization is shown in Table 1.

https://doi.org/10.3390/pr14030490

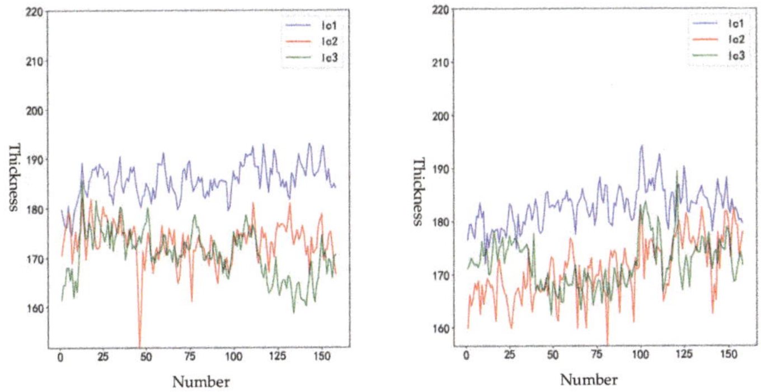

Figure 6. Level gauge material layer thickness values before and after model implementation.

Table 1. Comparison of uniformity indicators before and after model implementation.

	Cosine Distance	Euclidean Distance	Central Distance
before optimization	0.999496	150.33713	158.63992
after optimization	0.999448	131.82897	109.42816
lift ratio	0.0048%	12.31%	31.02%

The cosine distance values before and after deployment are 0.999496 and 0.999496, respectively. From the cosine distance, it can be seen that the longitudinal uniformity has not improved significantly after the model is put into operation, which is also evidenced by the significant fluctuations in thickness over the time cycle as shown in Figure 6. Based on on-site investigation and analysis, the core reason for the limited improvement in longitudinal uniformity lies in dual constraints: First, there are the inherent characteristics of the equipment structure—even when the incoming material flow is stable, the material receiving volume during the shuttle car's forward and backward movements is inherently inconsistent due to relative speed differences, and this structural limitation exerts a significant impact on longitudinal uniformity. Second, there are the constraints of process and parameter adjustment—the preset production output remains relatively fixed for an extended period during on-site production, and it is neither permitted nor feasible to adjust the output according to the shuttle car's moving direction. Consequently, it is difficult to offset the longitudinal fluctuations caused by inherent structural defects merely by adjusting the shuttle car's belt speed and movement speed.

After deployment, the average Euclidean distance between each pair of curves decreased from 150.33713 to 131.82897. From the perspective of Euclidean distance, the lateral fabric uniformity has improved by 12.31% after the optimization model was put into operation. After deployment, the average central distance between the three curves and their corresponding mean value decreased from 158.63992 to 109.42816. From the perspective of central distance, the lateral fabric uniformity has improved by 31.02% after the optimization model was put into operation. Based on the above data analysis, it can be concluded that after the deployment of this model, the lateral distribution uniformity has been significantly improved without affecting the longitudinal distribution uniformity.

4. Conclusions

To address the uneven distribution issues of the shuttle distributing system in pellet production (such as material piling at both ends of the wide belt, insufficient parameter matching, and reliance on manual adjustment), this paper proposes an intelligent control

https://doi.org/10.3390/pr14030490

method based on Integral Simulation and Gradient Descent optimization (IS-GD). By constructing a simulation model of the shuttle car's operating state, the material distribution laws during the acceleration, uniform speed, deceleration, and other phases of the shuttle car's forward and backward movement are accurately depicted. Combined with a phased gradient descent strategy, the shuttle car's operation control parameters are optimized with the goal of minimizing the uniformity index, achieving the coordinated improvement of both longitudinal and lateral distribution uniformity. Experimental results show that this method significantly improves the lateral distribution uniformity, with the lateral uniformity increased by 12.31% and 31.02%, respectively, in terms of indicators, providing reliable technical support for the intelligent distribution in pellet production. Future research can be deepened in three aspects: first, introduce reinforcement learning algorithms to enhance the model's adaptive optimization capability for dynamic operating conditions (such as fluctuations in material properties and changes in equipment status); second, expand the multi-equipment collaborative optimization framework, integrate the shuttle car, wide belt, roasting equipment, etc., into the overall optimization system to improve system-level distribution uniformity; third, develop a real-time optimization system based on edge computing to shorten the parameter adjustment response time, meet the high real-time control requirements of industrial sites, and further promote the green, efficient, and intelligent upgrading of pellet production.

Author Contributions: Conceptualization, T.L., X.L. and X.Z.; methodology, T.L., J.Z. and C.L.; software, T.L. and W.W.; validation, T.L.; formal analysis, T.L. and C.L.; investigation, T.L. and X.L.; resources, C.L.; data curation, T.L. and X.L.; writing—original draft preparation, T.L.; writing—review and editing, T.L.; visualization, T.L.; supervision, X.Z.; project administration, X.Z. and Z.L.; funding acquisition, X.Z. and Z.L. All authors have read and agreed to the published version of the manuscript.

Funding: This research was funded by The Science-Technology Foundation for Young Scientist of China Minmetals Corporation (2024QNJJA04).

Data Availability Statement: The data presented in this study are available from the corresponding author upon reasonable request.

Conflicts of Interest: Authors Tingting Liao, Xiaoxin Zeng, Xudong Li, Zongping Li and Chen Liu were employed by the company ZhongYe ChangTian International Engineering Co., Ltd. Author Weisong Wu was employed by the company ZhongYe ChangTian (Changsha) Intelligent Technology Co., Ltd. The remaining authors declare that the research was conducted in the absence of any commercial or financial relationships that could be construed as a potential conflict of interest.

Abbreviations

The following abbreviations are used in this manuscript:

IS-GD	Integral Simulation and Gradient Descent optimization
BGD	Batch Gradient Descent
SGD	Stochastic Gradient Descent
MBGD	Mini-Batch Gradient Descent

References

1. Zhang, F.M.; Zhang, W.H.; Qing, G.L.; Wang, Q.S.; Han, Z.G. Design and Application of Pelletizing Technical Equipment for Large-scale Belt Roaster. *Sinter. Pellet* **2021**, *46*, 66–75.
2. Fan, X.H. *Principles of Sintering and Pelletizing Plant Design*; Central South University Press: Changsha, China, 2016; pp. 10–20.
3. Zhang, R.X.; Huang, S.L.; Gao, X.S.; Duan, Z.J. Discussion on Automatic Control of Rotary Kiln Roasting. *Metall. Power* **2020**, *239*, 60–63.

https://doi.org/10.3390/pr14030490

4. Yu, H.Z.; Liao, J.Y.; Fan, X.H. Application and Research Progress of Pelletizing Technology for Belt Roasters. *Sinter. Pellet* **2020**, *45*, 47–54.
5. Zhi, Q.; Zhang, F.M.; Han, Z.G.; Zhang, Q.Z. Research and Application of Uniform Distribution for Belt Roasters. *Sinter. Pellet* **2024**, *49*, 57–60.
6. Xu, Y.X. Cause Analysis and Improvement of Uneven Distribution in Grate Machines. *Mach. Autom.* **2013**, *9*, 98–107.
7. Zhou, J.P. Control of Cloth Uniformity for Belt Roaster Pallets. *Metall. Equip.* **2023**, *6*, 84–88.
8. Li, D.; Han, J.X.; Ren, W.; Qi, Z.F.; Zhou, J.P. Adaptive Control Method for Feeding Equipment of Belt Roaster Based on Uniform Distribution. Patent 202410511101, 14 June 2024.
9. Huang, S.P.; Zhang, Z.X.; Qiu, X.L. Design of Pellet Uniform Distribution Model. *Control Eng.* **2009**, *16*, 1–3.
10. Chen, Z.L.; Han, J.X.; Ren, W. Discrete Element Simulation Study on Distribution and Screening System of Belt Roaster. *Sinter. Pellet* **2019**, *44*, 45–49.
11. Ge, Q.B.; Zhang, J.Z.; Yang, Q.M.; Li, H. Adaptive Gradient Descent Optimization Algorithm with Derivative Term Improvement. *Control Theory Appl.* **2022**, *39*, 623–632.
12. Wu, J. *Research on Stochastic Parallel Gradient Descent Algorithm Based on Segmented Random Disturbance Amplitude*; China University of Mining and Technology: Xuzhou, China, 2015.
13. Liu, C.R.; Wang, Y.W.; Zhang, W.; Yang, K.P.; Ma, J.J.; Sun, H. Nonlinear dynamics of a magnetic vibration isolator with higher-order stable quasi-zero-stiffness. *Mech. Syst. Signal Process.* **2024**, *218*, 111584. [CrossRef]
14. Liu, C.R.; Zhang, R.; Yang, K.P.; Li, H.P.; Li, B.P. A quasi-zero-stiffness device capable of vibration isolation and energy harvesting using piezoelectric buckled beams. *Energy* **2021**, *233*, 121146. [CrossRef]

https://doi.org/10.3390/pr14030490

processes

MDPI

Review

Integration of Artificial Intelligence in Food Processing Technologies

Ali Ayoub [1,2]

[1] Ayoub Sciences LLC, A[2] Nexus Lab, Decatur, IL 62526, USA; aayoub@ncsu.edu
[2] College of Natural Resources, North Carolina State University, Raleigh, NC 27607, USA

Abstract

The food processing industry is undergoing a profound transformation with the integration of Artificial Intelligence (AI), evolving from traditional automation to intelligent, adaptive systems aligned with Industry 5.0 principles. This review examines AI's role across the food value chain, including supply chain management, quality control, process optimization in key unit operations, and emerging areas. Recent advancements in machine learning (ML), computer vision, and predictive analytics have significantly improved detection in food processing, achieving accuracy exceeding 98%. These technologies have also contributed to energy savings of 15–20% and reduced waste through real-time process optimization and predictive maintenance. The integration of blockchain and Internet of Things (IoT) technologies further strengthens traceability and sustainability across the supply chain, while generative AI accelerates the development of novel food products. Despite these benefits, several challenges persist, including substantial implementation costs, heterogeneous data sources, ethical considerations related to workforce displacement, and the opaque, "*black box*" nature of many AI models. Moreover, the effectiveness of AI solutions remains context-dependent; some studies report only marginal improvements in dynamic or data-poor environments. Looking ahead, the sector is expected to embrace autonomous manufacturing, edge computing, and bio-computing, with projections indicating that the AI market in food processing could approach $90 billion by 2030.

Keywords: extrusion; spry dryer; drum dryer; fermentation; supply chain; food quality; food engineering; food processing

1. Introduction

The food processing sector has progressed through distinct evolutionary phases that mirror the broader trajectory of industrial revolutions, transitioning from basic manual labor to highly integrated autonomous systems (Figure 1). The First Transformation was defined by basic automation, where steam and waterpower replaced hand tools, significantly enhancing the efficiency of primary processing such as grain milling [1].

This was followed by the Second Transformation, which introduced electrification and mass production [2]. During the early 20th century, the implementation of assembly lines, pioneered by industry leaders such as Heinz, standardized the production of canned goods and established the foundations of the modern industrial food complex [3].

The Third Transformation, or Industry 3.0, marked the entry of electronics and information technology into the factory floor [4]. The adoption of Programmable Logic Controllers (PLCs) and Supervisory Control and Data Acquisition (SCADA) systems allowed precise control over processing parameters, ensuring consistency in high-speed operations such as automated bottling.

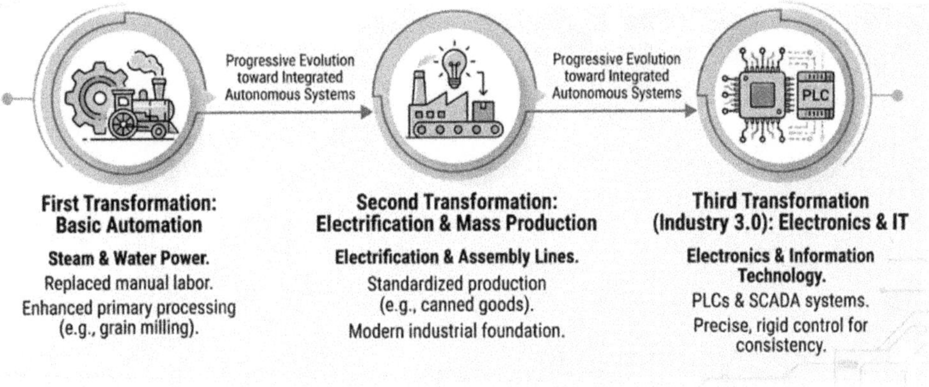

Figure 1. Evolutionary phases of food processing: a summary of industrial transformations.

As of 2026, the industry is navigating the complex transition between Industry 4.0 and Industry 5.0. While Industry 4.0 was characterized by the deployment of Cyber-Physical Systems (CPS), IoT, and centralized Cloud Computing to achieve hyper-efficiency through full automation, Industry 5.0 introduces a fundamental shift in control architecture and operational philosophy [5]. This transition is characterized by three distinct and measurable shifts. First, control architecture evolves from rigid, centralized automation toward decentralized, adaptive systems, wherein artificial intelligence operates as a cognitive layer rather than a simple rule-based controller. Second, the paradigm of human-AI interaction is moving beyond the replacement of human labor, as seen in Industry 4.0, toward collaborative frameworks. In these models, human expertise complements collaborative robots (cobots), enabling greater flexibility and ethical oversight. Third, the optimization objectives of food processing systems are expanding: rather than focusing solely on operational efficiency (such as speed and cost), there is an increasing emphasis on systemic resilience and sustainability, with long-term stability prioritized over short-term throughput.

Unlike traditional automation, which relies on rule-based logic, AI systems possess the capacity for probabilistic inference, iterative learning, and real-time adaptation. Between 2020 and 2025, AI adoption shifted from experimental pilots to scalable industrial deployments, a transition accelerated by the COVID-19 pandemic [6].

This global event exposed the inherent fragility of manual supply chains and underscored the necessity for contactless, remote-monitored processing technologies [7]. By 2026, leading global organizations had incorporated artificial intelligence into more than half of their operational workflows, resulting in substantial reductions in operational costs and a strategic shift from energy-intensive thermal processes to AI-optimized non-thermal alternatives [8]. Despite these advancements, peer-reviewed research reveals persistent challenges in scalability; AI models often underperform in real-world environments due to algorithmic bias and inadequate data standardization.

While global production of wheat, rice, and coarse grains is forecast to show steady upward trends through the 2025–2026 season, this growth is only marginally outpacing utilization levels (Figure 2).

https://doi.org/10.3390/pr14030513

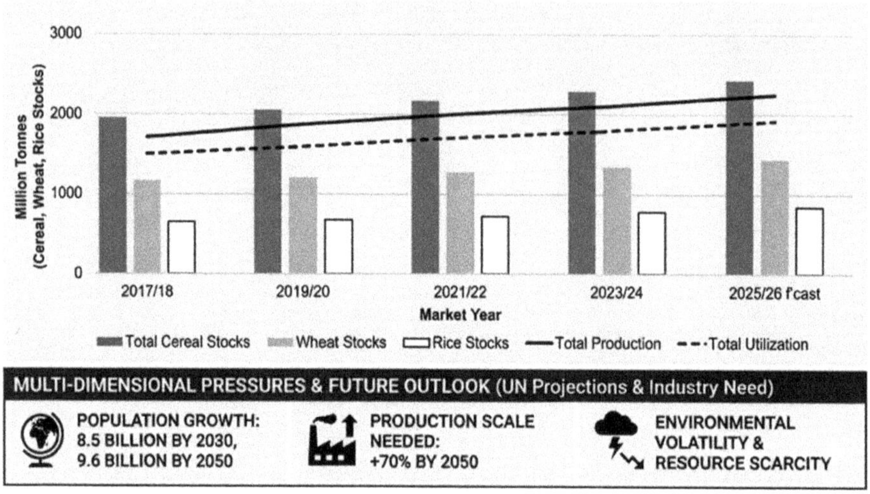

Figure 2. The disparity between global food processing strains and population growth (Food Outlook, Biannual Report on Global Food Markets, FAO 2025).

This narrow surplus is highly unwarranted, as the slow pace of production expansion leaves the global food supply chain vulnerable to environmental volatility and resource scarcity while attempting to scale 70% to meet the needs of 8.5 billion people by 2030 [9]. In addition, the modern food processing facilities are high-complexity environments encompassing various unit operations from sorting and grading to pasteurization and packaging that have historically relied on empirical knowledge and manual sampling (Figure 3).

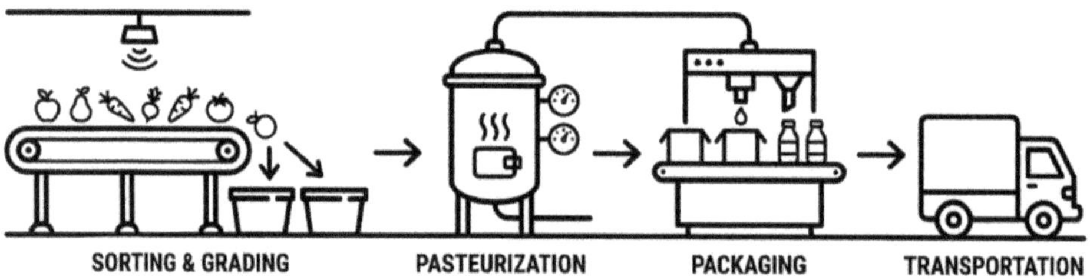

Figure 3. Integrated Workflow of Unit Operations in Modern Food Processing Facilities.

However, such subjective monitoring is increasingly insufficient for modern lines that process several tons of products per hour. Regulatory frameworks have further catalyzed digital transformation. In both Europe and North America, regulatory frameworks such as the EU Farm to Fork Strategy and the FDA Food Safety Modernization Act (FSMA) demand the necessity for real-time traceability and complete hazard analysis [10]. On the other hand, in Asia, which contributed half of the global growth in food processing between 2017 and 2021, rapid urbanization has driven the adoption of AI-enabled factories to manage a diverse range of products [11]. Despite the high degree of automation in modern facilities, the industry remains concerned by systemic inefficiencies that compromise both profitability and sustainability.

The Food and Agriculture Organization (FAO) estimates that approximately 14% of the world's food is lost between harvest and the retail level, with significant portions of this waste occurring during the processing phase [12]. These losses are primarily driven by inaccurate sorting mechanisms where optical sorters often produce high rates of false

https://doi.org/10.3390/pr14030513

rejections, discarding viable products alongside defective ones. Furthermore, energy waste remains a critical concern, particularly in thermal processes like drying and sterilization [13]. In food processing, operational parameters are often set conservatively and remain static to ensure safety, resulting in elevated energy consumption and potential deterioration of nutritional and sensory qualities. While AI-driven optimizations have demonstrated potential for improving efficiency, contradictory findings indicate that such benefits are not universally realized. Studies have reported only marginal energy savings in variable or dynamic environments, frequently attributing these limitations to model overfitting and inconsistent data quality.

The risk profile of food processing is further complicated by microbial contamination and equipment failure. Pathogens such as *Salmonella* spp. and *Listeria* spp. persist as major threats, yet manual cleaning verification processes remain slow and susceptible to human error. A single contamination event in the current globalized trade environment can trigger recalls costing millions of dollars and damaging brand equity [14].

Additionally, unplanned equipment downtime creates a flow of waste, as perishable raw materials wait in queues while production lines are discontinued for mechanical maintenance.

These challenges underscore the necessity for a paradigm shift from reactive to predictive and prescriptive operational models. Artificial intelligence offers transformative potential by leveraging real-time sensor data to optimize process parameters, reduce energy consumption, and implement predictive maintenance strategies that address mechanical stress before failures occur. Nevertheless, barriers such as high implementation costs and data privacy concerns have resulted in unsuccessful deployments, particularly among small-scale operations. This technological transition is especially critical in the context of global workforce shortages and elevated labor turnover, highlighting the importance of human-centric, AI-augmented labor to sustain operational continuity.

In this paper, we explore how AI benefits the entire value chain of food processing, starting with supply chain management, where AI optimizes logistics, inventory, and production planning through predictive models that forecast yields and demand with over 85% accuracy, reducing overproduction and enabling precision farming via IoT sensors. We then discuss quality control and safety inspection, where AI excels in non-destructive assessment using computer vision and deep learning to detect defects and contaminants with accuracies exceeding 98%, minimizing recalls and ensuring compliance. Following this, we delve into food engineering, examining AI's role in optimizing key unit operations such as extrusion, spray drying, and drum drying, where neural networks and digital twins enable real-time adjustments, yielding energy savings of 20–40% and improved product consistency. Another key area covered is AI's integration into the new food processing business of precision fermentation, which represents a fundamental shift in manufacturing novel proteins at the molecular level using biotechnology.

Finally, we address future trends, sustainability implications, limitations, and perspectives, including the move toward autonomous *"dark factories"* edge computing, projecting the AI market in food processing to reach nearly $90 billion by 2030.

Review Methodology

To comprehensively examine the transition from Industry 4.0 to Industry 5.0 within the food sector, this study adopts a narrative review framework. Relevant literature was identified through structured searches in Web of Science, Scopus, and Google Scholar, targeting peer-reviewed articles, conference proceedings, and technical reports published between 2017 and 2025. This period was selected to capture the rapid evolution of artificial

https://doi.org/10.3390/pr14030513

intelligence technologies, particularly in response to digitalization trends of the late 2010s and the acceleration of autonomous systems following the COVID-19 pandemic.

Search strategies combined core terms such as "Artificial Intelligence," "Machine Learning," and "Deep Learning" with domain-specific keywords including "Food Processing," "Supply Chain Management," "Extrusion," "Spray Drying," and "Precision Fermentation." Studies were included if they provided quantitative performance metrics, such as energy reduction or defect detection accuracy, demonstrated in industrial or pilot-scale environments, rather than theoretical simulations. Exclusion criteria eliminated works lacking experimental validation or those focused exclusively on agricultural production without direct relevance to processing. This methodological approach ensures that the review addresses both the theoretical potential and the practical realities of AI integration, bridging the gap between high-level claims and implementation challenges observed in the recent literature.

2. Supply Chain Management

Supply chain management (SCM) is a critical component of the agri-food processing industry, encompassing the coordination of activities from raw material sourcing and production to distribution and consumption [15]. Its importance lies in ensuring the efficient flow of perishable goods, maintaining food safety, minimizing waste, and meeting consumer demands for quality and sustainability.

In today's agri-food processing companies, SCM typically operates through a combination of traditional logistical systems, including manual inventory tracking, centralized planning software, and partnerships with suppliers and distributors. For instance, companies rely on enterprise resource planning (ERP) systems to manage orders, track shipments via GPS, and monitor storage conditions in warehouses (Figure 4).

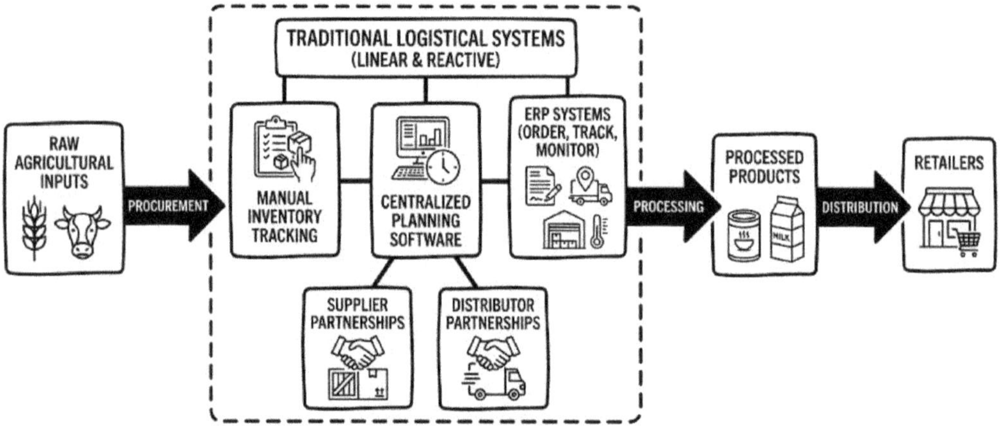

Figure 4. Supply Chain Management Operation in Agri-Food Processing Companies.

However, these systems are often linear and reactive, involving sequential stages where raw agricultural inputs like crops or livestock are procured, processed into products such as canned goods or dairy items, and then distributed to retailers [16]. This setup is prevalent in regions like North America and Europe, where large-scale processors like Tyson Foods or Nestlé use integrated networks to handle high volumes, but it remains fragmented in developing areas due to reliance on smallholder farmers and limited infrastructure [17].

Despite its foundational role, current SCM in agri-food processing faces significant gaps and challenges. The industry's inherent complexities, such as seasonality of harvests,

https://doi.org/10.3390/pr14030513

perishability of products, and vulnerability to external factors like environmental change, supply disruptions from events like the COVID-19 pandemic, and trade barriers, expose weaknesses in traditional models.

According to recent analyses, supply chain disruptions have led to substantial losses, with estimates from 2025 indicating that agri-food companies experience up to 20–30% inefficiency in inventory management due to inaccurate demand forecasting and poor visibility [18]. Additional issues include high food waste during transit (contributing to the FAO's 14% global loss figure), microbial contamination risks from inadequate traceability, and escalating costs from rising energy prices and labor shortages [19].

In developing regions, broadband limitations and skill gaps further exacerbate these problems, creating an "AI divide" where advanced technologies are underutilized despite high vulnerability.

Artificial Intelligence (AI) addresses these gaps by transforming SCM into proactive, resilient, and data-driven systems, enabling real-time decision-making, predictive risk management, and enhanced transparency. By integrating AI, companies can fill inefficiencies through automated optimization, reducing waste by 20–50% and improving forecasting accuracy to over 85% [20]. For example, AI-powered predictive analytics can anticipate disruptions from weather patterns or market shifts, allowing for dynamic adjustments in sourcing and routing. This transition not only mitigates economic losses but also advances sustainability by optimizing resource utilization and reducing emissions. Nevertheless, evidence from recent literature indicates that the effectiveness of AI is not universal; its success is highly dependent on contextual factors such as data quality, with certain implementations failing to deliver anticipated benefits in fragmented supply chains.

In the current food sector, Supply Chain Risk Management (SCRM) has surpassed traditional logistical tracking to embrace highly autonomous, transparent, and resilient networks powered by AI. The integration of Distributed Artificial Intelligence (DAI) and Machine Learning (ML) transforms the food supply chain from a static linear sequence into a dynamic ecosystem capable of self-organization and real-time optimization.

This technological shift is critical for addressing the inherent complexities of food systems, which are characterized by seasonality and increasing consumer demand for safety and sustainability. The application of these technologies facilitates the rigorous tracking of food products through complex logistical networks, ensuring visibility during transit and allowing for proactive adjustments to supply chain operations.

A significant advancement in ensuring traceability and operational efficiency is the implementation of Multi-Agent Systems (MAS), a subset of DAI [21]. Unlike traditional centralized control systems, MAS distributes intelligence across a network of autonomous agents, representing suppliers, warehouses, retailers, and brokers, that collaborate to achieve common goals. For instance, in a beef food supply chain model, specific agents collect relevant data and interact with a central Broker Agent to optimize scheduling and resource allocation. This decentralized approach enhances visibility, allowing the food supply chain to adapt to disruptions in real-time rather than waiting for centralized directives.

The coordination between these agents is often facilitated by advanced algorithms such as Genetic Algorithms (GA), Simulated Annealing (SA), and Particle Swarm Optimization (PSO), which are dynamically selected to solve resource-constrained scheduling problems [22]. This capability allows the food supply chain to "learn" over time, reducing the total duration of scheduling tasks by significant margins compared to random algorithm selection. Parallel to the structural improvements offered by DAI, Machine Learning algorithms provide the analytical depth required for sustainable distribution and risk prediction. In the distribution phase, ML techniques such as deep learning and

https://doi.org/10.3390/pr14030513

Artificial Neural Networks (ANN) are utilized to predict consumer demand and optimize delivery routes synchronously by analyzing current traffic situations [23].

This capability is vital for the "last mile" delivery of perishable goods, where delays can lead to significant food waste and economic loss. From a sustainability perspective, ML-driven route optimization and fleet management contribute directly to fuel economy and reduced greenhouse gas emissions, aligning economic performance with environmental stewardship. The application of ML extends deeply into the consumption and planning phases, utilizing distinct strategies for different production models. For "make-to-order" food companies, ML tools facilitate smart product strategies that reduce latency for components, whereas "make-to-stock" companies gain greater profits by implementing smart process strategies. In the consumption phase, Bayesian networks and deep learning models are deployed to forecast consumer buying behavior and perception, analyzing factors that influence purchasing decisions for imported or ready-to-eat foods [24]. This granular insight allows retailers to manage inventory more effectively, reducing the likelihood of stockouts or spoilage. Furthermore, the integration of these systems promotes social sustainability by improving food safety and creating adequate working environments through better resource planning.

Despite the technological potential, research indicates a significant "AI divide" regarding the vulnerability of supply chains and their receptiveness to these technologies. As illustrated in Figure 5, on the ease of AI adoption and deployment across supply chain phases, developed regions consistently show higher receptivity to AI, often rated as "very receptive" or "moderately receptive", particularly in phases like processing and manufacturing, retail management, and international transportation and storage, where expert assessments yield high scores supported by robust infrastructure and skilled workforces [25]. In contrast, developing regions exhibit lower receptivity, frequently falling into "slightly receptive" or "not at all receptive" categories across the same phases, with notable standard deviations indicating variability due to constraints such as insufficient broadband infrastructure and a lack of skilled human capital.

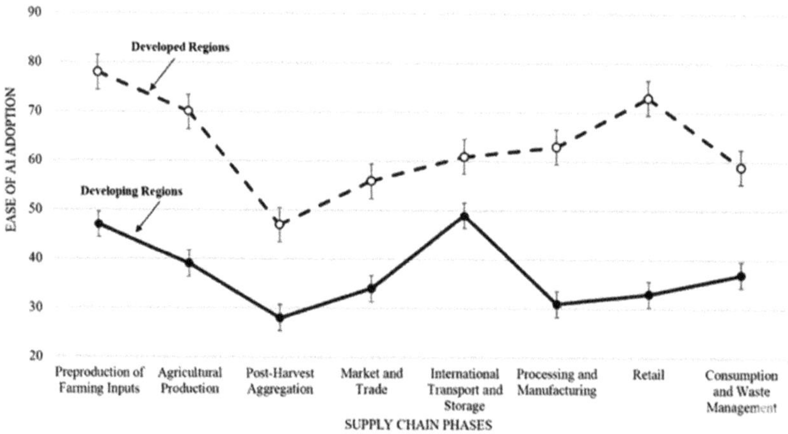

Figure 5. The AI is divided into an agri-food supply chain (Developed vs. Developing Regions). (Artificial Intelligence for Agricultural Supply Chain Risk Management, CGIAR (2020). Available at: https://hdl.handle.net/10568/108709, accessed on 21 January 2026).

The phase of "National and International Transport and Storage," for instance, is identified as highly vulnerable to risks but shows a stark contrast in AI readiness between regions. This disparity highlights that while AI offers high potential for improving trace-

https://doi.org/10.3390/pr14030513

ability and mitigating risks, its deployment is currently constrained in regions that require it most to ensure food security. Consequently, the successful deployment of these intelligent systems requires navigating significant hurdles, including data privacy concerns, the need for substantial infrastructure investment, and the standardization of protocols between disparate AI systems.

To provide clarity on how AI supports various segments of the supply chain, the following table (Table 1) summarizes key applications, AI subsegments, and concrete examples from companies, agriculture firms, and startups. This serves as guidance and reference for understanding AI's role across phases.

Table 1. AI's role is to support various segments of the supply chain in agri-business.

Supply Chain Phase	AI Subsegment	Application	Example Company/Startup
Preparation of Farming Inputs	Predictive Analytics, ML	Optimizing input selection (e.g., seeds, fertilizers) based on soil and weather data to reduce waste and costs.	Kilimo (Argentina-based startup): Uses ML and big data for irrigation recommendations, saving 70% water and improving yields by 30% in crop preparation.
Agricultural Production	Computer Vision, DL	Monitoring crop health, detecting diseases, and automating tasks like weeding or pollination for higher yields.	SeeTree (Israel): Employs DL for tree health tracking and yield prediction in orchards; John Deere (US company): AI-powered tractors for precision planting and monitoring.
Post-Harvest Aggregation	Computer Vision, ML	Automated quality inspection and sorting to minimize losses during storage and aggregation.	Clarifruit (Israel startup): Big data and ML for fresh produce quality control; Neolithics (Israel): Hyperspectral imaging and ML for crop sorting on conveyors.
Market and Trade	Predictive Analytics, AI Algorithms	Dynamic pricing, demand forecasting, and matchmaking between farmers and buyers to streamline trading.	Kireap (Switzerland): AI-enabled trading platform with yield predictions for fair pricing.
International Transportation and Storage	IoT Integration, ML	Real-time monitoring of storage conditions and optimizing logistics routes to prevent spoilage.	TeleSense (US startup, acquired by DECCO): Wireless sensors and ML for grain storage monitoring; Loamy (US): Satellite imagery and AI for shipment quality tracking and delivery optimization.
Processing and Manufacturing	Robotics, AI Sequencing	Enhancing food safety, quality assurance, and automation in processing lines.	Clear Labs (US): AI and DNA sequencing for meat safety; Tortuga AgTech (US): Robotics and AI for harvesting and pest control in processing fresh produce.
Retail Management	Predictive Analytics, CV	Inventory optimization, shelf management, and demand forecasting to reduce stockouts and waste.	Shelf Engine (US startup): AI for stock level optimization in groceries; Trax (Singapore): CV and ML for real-time shelf insights and product availability.
Consumption and Waste Management	Computer Vision, ML	Predicting and reducing waste through pattern analysis and upcycling recommendations.	Orbisk (Netherlands startup): Automated CV for kitchen waste monitoring; Winnow Solutions (UK, partnered with Tesco): AI to analyze waste patterns and suggest reductions.

3. Quality Control and Safety Inspection

Quality control (QC) serves as the cornerstone of the modern food industry, ensuring consumer satisfaction and safety while navigating the complexities of global supply chains. Traditionally, quality assurance has relied heavily on manual inspection and visual signs by human operators. However, these conventional methods are increasingly viewed

https://doi.org/10.3390/pr14030513

as time-consuming, inconsistent, and tiresome, necessitating a shift toward automated solutions. A product's quality is fundamentally defined by the totality of characteristics that satisfy client needs, making the consistent delivery of defect-free goods paramount [26]. Consequently, the integration of Artificial Intelligence (AI) and machine learning (ML) into food processing has revolutionized the field [27]. By utilizing computer vision and image processing technologies, industrial stakeholders can eliminate subjective QC processes and achieve significant advancements in areas where the human eye lacks sensitivity. This section synthesizes the transformative impact of ML on food QC, examining methodological frameworks, specific classification algorithms, and performance metrics.

The transition from manual to automated inspection requires a robust pipeline capable of handling the biological variability inherent in agricultural products. Fruits and vegetables exhibit a wide range of exterior qualities, such as color, size, and form, even when harvested from the same tree. Furthermore, external factors like fungal infections, temperature, and storage duration can alter product appearance, making consistent grading difficult. To address this, recent research proposes a structured methodology comprising image preprocessing, enhancement, segmentation, and classification [28].

The initial stage of automated food quality inspection involves image preprocessing to mitigate noise, ensuring robust segmentation. Gaussian filtering smooths images by averaging pixel intensities weighted by a cumulative standard deviation, preserving edges while eliminating artifacts. Subsequent image enhancement via histogram equilibrium contributes to improved contrast, enhancing local details and classifier performance. Image segmentation follows, commonly using K-means clustering to group pixels by shared intensity or traits. This method computes squared Euclidean distances to assign pixels to clusters, isolating regions of interest (ROIs) such as defects or disease spots from backgrounds, paving the way for feature extraction and classification (Figure 6).

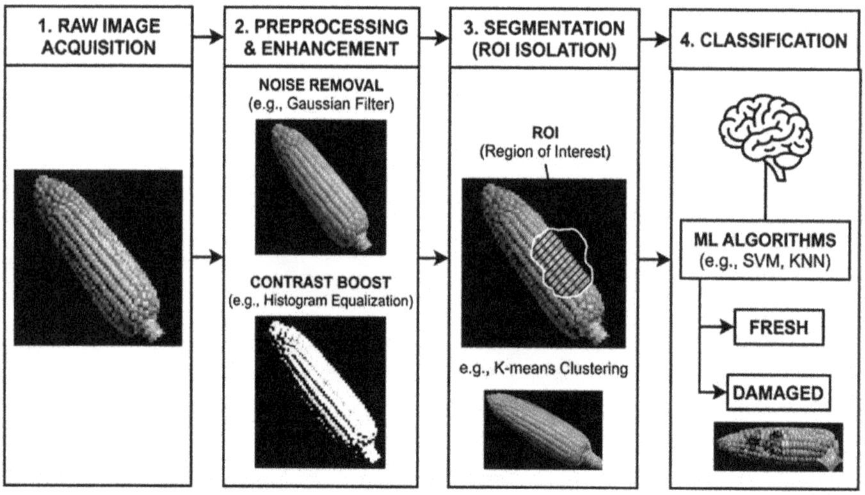

Figure 6. AI-Based Image Processing Pipeline for Corn Defect Detection and Classification.

In safety-critical applications, the integration of artificial intelligence with surface-enhanced Raman spectroscopy (SERS) has enabled real-time detection of pesticides, such as thiram, on apples. Industrial implementations include Nestlé's visual inspection systems for contaminant detection and the UK Health Security Agency's use of natural language processing (NLP) to analyze restaurant reviews for signals of foodborne illness. These

https://doi.org/10.3390/pr14030513

technological advancements reduce the need for manual oversight, minimize product recalls, and support regulatory compliance [29,30].

Despite promising commercial claims, such as Nestlé's reported 99% detection accuracy, peer-reviewed studies indicate that actual performance may be lower, with validation studies reporting accuracies between 92% and 95% under variable conditions, often due to inconsistencies in lighting. Furthermore, some research highlights the limitations of AI in detecting subtle contaminants, with false positive rates reaching up to 15% in noisy environments. The characteristics of agricultural datasets are particularly influential; high variability in color gradients, occlusions, and environmental noise necessitate the use of large, well-annotated datasets, such as the January Food Benchmark (JFB), comprising 1000 images, for robust benchmarking. Standard evaluation metrics include precision, recall, and F1-score, with convolutional neural networks (CNNs) achieving up to 96% accuracy on standardized datasets, though performance often declines in real-time operational settings.

Uncertainty Quantification and Model Validation

While artificial intelligence models in food processing exhibit high theoretical accuracy, their reliable industrial deployment necessitates rigorous uncertainty quantification, especially in safety-critical applications such as pathogen detection and thermal sterilization. Unlike deterministic control systems, deep learning models are inherently stochastic; therefore, a single prediction regarding food safety is insufficient without an associated confidence interval.

In unit operations of food processing, sensors are frequently exposed to challenging conditions, including high humidity and variable viscosity, which introduce measurement noise into the input data. To maintain robust process control, it is essential to quantify how this sensor error (σ_x) propagates into the uncertainty of the final AI prediction $\left(\sigma_f\right)$. For a process function (f) that depends on multiple sensor variables (x_i) (such as temperature, moisture content, and pressure), the accumulated uncertainty in the output can be approximated using a Taylor series expansion:

$$\sigma_f^2 \approx \sum_{i=1}^{n}\left(\frac{\partial f}{\partial x_i}\right)^2 \sigma_{x_i}^2 + 2\sum_{i<j}\left(\frac{\partial f}{\partial x_i}\right)\left(\frac{\partial f}{\partial x_j}\right)\sigma_{x_i x_j}$$

where:

f is the process function (for example, the AI model's output), representing the mathematical relationship between the set of input sensor variables and the output of interest. In the context of food processing, f typically denotes the output of an AI model or control algorithm that predicts or regulates a process parameter (such as product quality, yield, or safety) based on sensor inputs.

x_i are the input sensor variables, such as temperature, moisture content, or pressure, which are measured during the process and serve as inputs to the function f.

$\sigma_{x_i}^2$ is the variance (uncertainty) associated with each input variable x_i. This quantifies the expected fluctuation or measurement error inherent to each sensor.

$\frac{\partial f}{\partial x_i}$ is the partial derivative of f with respect to x_i, which quantifies the sensitivity of the output to changes in the input variable x_i. A higher value indicates that small changes in x_i have a larger effect on the output prediction.

This formulation highlights that even minor sensor variances (σ_x), such as fluctuations in NIR readings due to steam, can propagate non-linearly to affect the reliability of the AI output.

https://doi.org/10.3390/pr14030513

Furthermore, standard accuracy metrics often mask performance deficits in datasets with high class imbalance, such as detecting rare contaminants or foreign bodies. A model reporting 99% accuracy may fail to detect 100% of contaminants if the contaminant prevalence is below 1%. Therefore, validation protocols must prioritize the F1-score and Recall (Sensitivity) over simple accuracy:

$$F_1 = 2 \cdot \frac{\text{Precision} \cdot \text{Recall}}{\text{Precision} + \text{Recall}}$$

Addressing these statistical limitations is essential to mitigate the "*black box*" nature of models and prevent false negatives that could lead to costly recalls or public health crises.

4. Process Optimization in Food Engineering

Process optimization in food engineering features the refinement of unit operations to enhance efficiency, product quality, and sustainability, often through the integration of advanced computational tools. The introduction of artificial intelligence (AI) has revolutionized these activities, enabling dynamic, data-driven adjustments that surpass the limitations of traditional control systems. AI's application to specific technologies such as extrusion, spray drying, drum drying, and agglomeration transforms these processes from empirical, rule-based operations into adaptive, predictive systems capable of real-time decision-making. This section outlines AI's role in optimizing this key food processing technologies, drawing on recent scholarly advancements to elucidate mechanisms, methodologies, and outcomes. By leveraging machine learning (ML) algorithms, neural networks, and digital twins, AI facilitates granular control over multi-variate parameters, mitigating variability, reducing energy consumption, and improving yield predictability. The section herein is grounded in empirical evidence from peer-reviewed literature, highlighting both established implementations and emerging frontiers, while addressing challenges such as data integration and scalability in industrial contexts.

- *Quantification of Efficiency Gains*

Throughout the literature, improvements in energy efficiency are typically quantified relative to conventional steady-state operations. To standardize these claims, the efficiency improvement factor, η, is defined as:

$$\eta = \frac{E_{\text{baseline}} - E_{\text{AI}}}{E_{\text{baseline}}} \times 100$$

where E_{baseline} represents the specific energy consumption (SEC) of the unit operation under traditional PID control with static setpoints. These baselines generally incorporate conservative safety margins, such as prolonged heating durations or elevated-pump pressures, to account for unmodeled disturbances.

In contrast, E_{AI} denotes the SEC under adaptive AI control, where dynamic setpoint modulation enables the system to operate closer to theoretical constraints without compromising safety or product quality. The efficiency gains of 20–40% cited in subsequent sections are derived from this comparative framework, specifically attributing savings to the reduction in thermal overshoot and the minimization of non-productive idle time during changeovers

4.1. AI in Food Extrusion

Food extrusion, a high-shear thermomechanical process integral to the production of cereals, snacks, pet foods, and texturized plant-based proteins, involves complex interactions among moisture content, barrel temperature, screw speed, and feed composition.

https://doi.org/10.3390/pr14030513

These variables influence rheological behavior, phase transitions, and final product attributes, rendering traditional proportional-integral-derivative (PID) controllers inadequate for handling non-linear dynamics and stochastic disturbances [31]. AI integration addresses these complexities by enabling predictive modeling and autonomous optimization, thereby minimizing waste and enhancing consistency (Table 2).

Table 2. The specific roles of AI models in the extrusion process.

AI Model Type	Specific Application	Measured Improvement
Artificial Neural Networks (ANN)	Modeling relationships between input parameters and expansion/texture.	R^2 values exceeding 0.95.
Convolutional Neural Networks (CNN)	In-line imaging for fibril alignment in plant-based meat analogs.	25% improvement in fibril alignment.
Long Short-Term Memory (LSTM)	Processing vibration/acoustic data to forecast screw wear.	92% failure forecast accuracy; 40% reduction in downtime.
Reinforcement Learning (RL)	Optimizing anti-blocking strategies and screw geometry.	Dynamic adjustment to simulated extrusion scenarios.

Real-time process control exemplifies AI's utility in extrusion (Figure 7). Artificial Neural Networks (ANNs), particularly multi-layer perceptions, model the required relationships between input parameters and output quality metrics. For instance, ANNs trained on historical extrusion data can predict expansion ratio and textural hardness with R^2 values exceeding 0.95, allowing for immediate parameter adjustments via feedback loops [32]. Unlike static models, these networks incorporate time-series data from sensors monitoring torque and pressure, facilitating adaptive responses to feedstock variability. In plant-based meat analogs, where fibrous texture emulation is paramount, convolutional neural networks (CNNs) analyze in-line imaging to optimize cooling die configurations, achieving up to 25% improvement in fibril alignment compared to manual tuning [33].

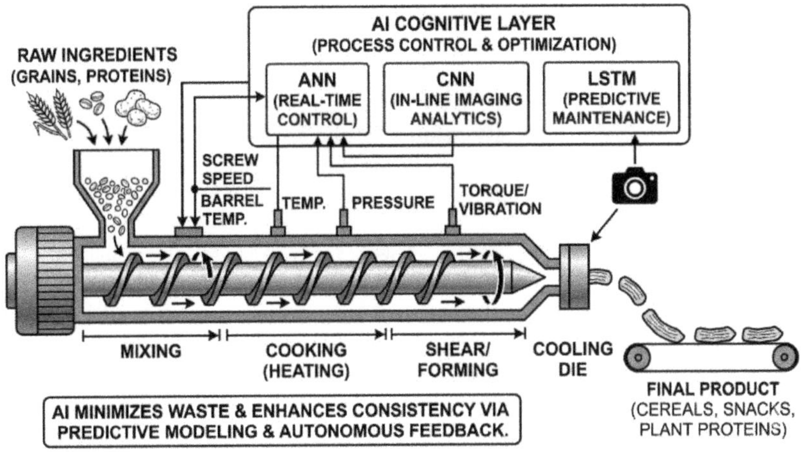

Figure 7. Multi-Model AI Architectures for Food Extrusion: Managing Non-Linear Dynamics through ANN, CNN, and LSTM Networks.

- *Mathematical Formulation of AI-Driven Control*

To rigorously define the optimization task performed by the AI cognitive layer in extrusion, the process is modeled as a multi-objective optimization problem. The control system aims to minimize a cost function, (J),which balances deviations in product quality against energy consumption.

https://doi.org/10.3390/pr14030513

Let $\mathrm{u}(t) \in \mathbb{R}^{\mathrm{n}}$ the control input vector at time (t), comprising variables such as screw speed $(\mathcal{W})$, barrel temperature $(\mathcal{T}b)$, and feed rate (Qf). Similarly, let $\mathrm{y}(t) \in \mathbb{R}^{\mathrm{n}}$ represent the output quality vector, including metrics like expansion ratio (ER) and textural hardness (H). The Artificial Neural Network (ANN) serves as a non-linear system estimator, $\hat{f}$, which approximates the plant dynamics:

$$\hat{\mathrm{y}}(t+k) = \hat{f}(\mathrm{y}(t), \mathrm{u}(t), \ldots, \mathrm{u}(t-d))$$

where (d) is the system delay. The optimization objective over a prediction horizon (Hp) is formulated as:

$$\min_{\mathrm{u}} J = \sum_{k=1}^{H_p} \left(\| \mathrm{y}_{\mathrm{target}} - \hat{\mathrm{y}}(t+k) \|_{\mathrm{Q}}^2 + \lambda \cdot SME(t+k) \right)$$

subject to:

$$\mathrm{u}_{\mathrm{min}} \leq \mathrm{u}(t) \leq \mathrm{u}_{\mathrm{max}}$$

Here, $\mathrm{y}_{\mathrm{target}}$ denotes the desired product specifications (e.g., a specific expansion ratio), and $\| \cdot \|_{\mathrm{Q}}^2$ represents the weighted Euclidean norm penalizing deviations from target quality. The term SME (Specific Mechanical Energy) quantifies the energy cost, weighted by the regularization parameter λ.

This formulation demonstrates how AI algorithms, such as Hybrid ANN-GA frameworks, explicitly solve the optimal trade-off between energy efficiency and product consistency, achieving reported SME reductions of up to 20% while adhering to operational constraints.

- *Predictive Maintenance*

Predictive maintenance further augments extrusion efficiency. Vibration and acoustic sensors, coupled with long short-term memory (LSTM) networks, a recurrent neural network variant, detect anomalies indicative of screw wear or blockages. LSTM models process sequential data to forecast failures with 92% accuracy, scheduling interventions that reduce unplanned downtime by 40% [34]. Reinforcement learning (RL) algorithms, such as deep Q-networks, extend this by optimizing anti-blocking strategies, learning from simulated extrusion scenarios to adjust screw geometry dynamically [35]. These advancements not only curtail operational costs but also align with sustainability goals, as optimized extrusion minimizes energy expenditure, typically 30–50% of total process costs [36].

In a case involving high-moisture extrusion for soy-based meats, a hybrid ANN-genetic algorithm (GA) framework optimized screw speed and temperature, yielding a 35% increase in product yield while reducing specific mechanical energy by 20% [37]. Challenges persist, however, including data scarcity for rare failure modes and the computational overhead of real-time RL deployment in legacy extruders.

Future research may focus on collaborative learning approaches, enabling multiple facilities to share data and refine AI models without compromising proprietary information. However, contradictory evidence indicates that AI solutions may underperform in small-batch operations, with some studies reporting only 10–15% energy savings compared to commercial claims of up to 40%, often due to model overfitting. Additionally, commercial claims by firms such as Buhler frequently exceed results validated in peer-reviewed benchmarks, which highlight limitations arising from heterogeneous datasets. Typically, datasets for extrusion AI comprise time-series sensor data (e.g., torque, temperature), and benchmarking protocols employ $R^2 > 0.92$ and cross-validation techniques to address noise and variability.

https://doi.org/10.3390/pr14030513

4.2. AI in Spray Drying

Spray drying, an energy-intensive dehydration process employed for powders like milk, instant coffee, and flavors, is susceptible to variations in ambient humidity, atomization pressure, and inlet temperature, which affect particle morphology, solubility, and stability. AI mitigates these sensitivities by enabling precise control over droplet formation and drying kinetics, optimizing energy use while preserving nutritional integrity (Figure 8).

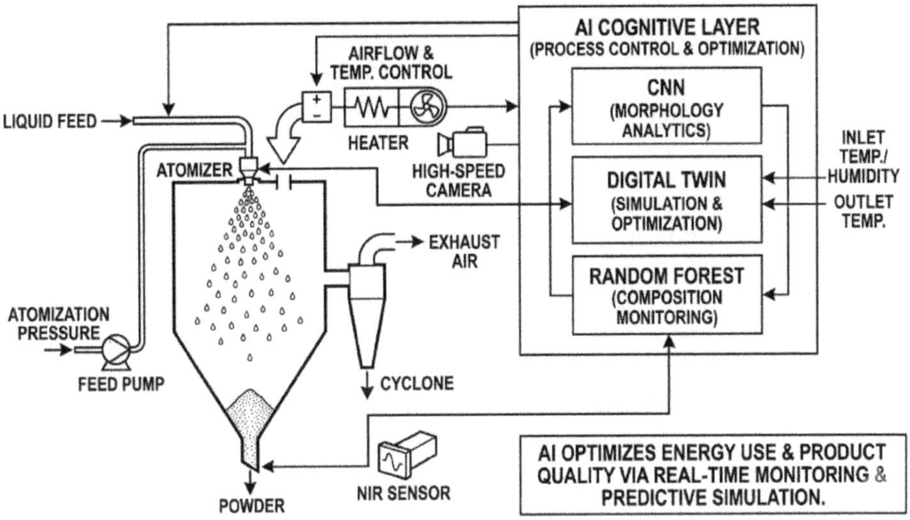

Figure 8. Synergistic Framework for AI-Driven Spray Drying: From Computer Vision-Based Morphology Prediction to Compositional Monitoring.

Morphology and moisture prediction constitute core AI applications. CNNs process high-speed camera feeds to classify particle shapes in real-time, distinguishing desirable spherical agglomerates from irregular, hollow forms that compromise flowability. Coupled with soft sensors, virtual instruments estimating unmeasurable variables like residual moisture, CNNs achieve prediction accuracies of 96%, adjusting airflow rates to maintain optimal particle size distributions [38]. Generative models, such as variational autoencoders (VAEs), simulate droplet trajectories under varying conditions, forecasting agglomeration tendencies with minimal experimental trials [39].

Digital twins represent an elevation of AI integration in spray drying. These virtual models, powered by physics-informed neural networks (PINNs), simulate the entire chamber environment, incorporating computational fluid dynamics (CFD) data to predict heat transfer and evaporation rates. Operators can work with scenarios, such as altering nozzle designs, to minimize fouling and energy loss, with reported reductions of 25% in operational costs [40].

Smart sensors amplify AI's efficacy, for example, Near-Infrared (NIR) spectroscopy integrated with random forest ensembles monitors powder composition continuously, detecting deviations in protein denaturation or lipid oxidation with 98% precision [41]. In a dairy powder study, an AI-optimized spray dryer reduced specific energy consumption from 4.5 MJ/kg to 3.2 MJ/kg by dynamically modulating inlet temperatures based on ambient humidity forecasts [42]. Recent advancements in AI-driven spray drying are summarized in Table 3, highlighting the synergistic integration of computer vision, digital twins, and smart sensors for powder quality and energy optimization.

https://doi.org/10.3390/pr14030513

Table 3. Synergistic Integration of AI Architectures and Smart Sensors in Spray Drying: Applications for Powder Quality and Energy Optimization.

AI Model/Technology	Specific Application	Examples/ Food Products	Measured Performance/Impact
Convolutional Neural Networks (CNN)	Processing high-speed camera feeds for particle shape classification.	Milk powder and instant coffee: Distinguishing spherical agglomerates from irregular forms.	96% predict accuracy for morphology and moisture.
Variational Autoencoders (VAE)	Simulating droplet trajectories to forecast agglomeration.	Flavor encapsulates: Predicting how droplets collide and dry without physical trials.	Minimizes experimental trials through simulation.
Digital Twins (PINNs + CFD)	Simulating the chamber environment to predict heat transfer.	Industrial-scale dryers: Virtualizing nozzle design changes to prevent chamber fouling.	25% reduction on operational costs.
NIR	Continuous monitoring of chemical composition.	Dairy and High-Protein powders: Detecting protein denaturation or lipid oxidation.	98% precision in detecting chemical deviations.
Dynamic Modulation (Inlet Temp)	Adjusting inlet temperatures based on humidity forecasts.	General Powder Production: Real-time response to changes in ambient factory humidity.	Energy reduction from 4.5 MJ/kg to 3.2 MJ/kg.

However, several limitations persist. Model failures are frequently observed under humid conditions, with contradictory studies reporting that accuracy can decline to approximately 80%. Commercial claims, such as the 98% precision reported by GEA systems, are typically validated only in controlled laboratory environments and do not always translate to industrial-scale operations. Datasets used for these applications often consist of high-dimensional spectral data, which are particularly susceptible to noise introduced by humidity. As a result, benchmarking relies on metrics such as the F1-score and utilizes publicly available datasets to ensure robust and transparent evaluation.

4.3. AI in Drum Drying

Drum drying, suited for viscous slurries like purees and flakes, exhibits non-linear heat transfer and film adhesion dynamics, complicating control. AI introduces hybrid systems with predictive analytics to manage these complexities, ensuring uniform drying without baking or incomplete moisture removal [43].

Hybrid control systems are paramount. Proportional-integral-fuzzy (PI-fuzzy) controllers augment PID loops with linguistic rules mimicking expert operators, adjusting steam pressure in response to imprecise inputs like *"slightly over-dry"*. These systems achieve 15% better moisture uniformity than conventional controls, as evidenced in potato flake production [44]. Deep reinforcement learning (DRL) further refines this by simulating drum rotations, learning optimal speed profiles to minimize energy while maximizing throughput [45].

Handling non-linearity involves ML models predicting boundary layer effects. Support vector regression (SVR) forecasts drying rates from viscosity and solids content data, with $R^2 > 0.92$, enabling proactive adjustments to scraper blade angles [46]. In fruit puree drying, AI-optimized drums reduced burn-on incidents by 50%, extending equipment lifespan [47].

Despite progress, data noise from steam fluctuations poses hurdles, necessitating robust filtering algorithms. Future avenues include multi-agent systems for synchronized multi-drum line and optimizing collective efficiency in large-scale operations (Table 4). Performance limitations are frequently observed in processes involving variable viscosi-

https://doi.org/10.3390/pr14030513

ties, with studies reporting outcomes that contradict the high energy savings claimed in commercial literature. Datasets used for benchmarking typically include rheological parameters, such as viscosity and solids content, and employ metrics like R^2 and error rates to rigorously evaluate model accuracy.

Table 4. Synergistic Integration of AI Architectures and Smart Sensors in Drum Drying.

AI Model/Technology	Specific Application	Examples/Food Products	Measured Performance/Impact
Proportional-Integral-Fuzzy (PI-Fuzzy)	Adjusting steam pressure using linguistic rules ("slightly over-dry").	Potato flakes: Mimicking expert operators to maintain moisture consistency.	15% better moisture uniformity compared to conventional controls.
Deep Reinforcement Learning (DRL)	Simulating rotations to learn optimal speed profiles.	Viscous slurries: Balancing energy minimization with maximum throughput.	Refined speed profiles for optimized resource use.
Support Vector Regression (SVR)	Forecasting drying rates based on viscosity and solids content.	Viscous purees and flakes: Proactively adjusting scraper blade angles to manage boundary layer effects.	High predictive accuracy ($R^2 > 0.92$).
AI-Optimized Hybrid Systems	General predictive analytics to prevent incomplete moisture removal or scorching.	Fruit purees: Real-time management of heat transfer and film adhesion.	50% reduction in burn-on incidents; extended equipment lifespan.

4.4. *AI in Precision Fermentation*

The rise in biotechnology has introduced novel proteins through precision and biomass fermentation, representing a fundamental shift in how food is "manufactured" at the molecular level. Precision fermentation utilizes microbes in bioreactors to create bioidentical whey or collagen, effectively reducing land use and supply chain vulnerability [48]. AI plays a pivotal role in this emerging food processing business by optimizing fermentation processes through advanced modeling and control systems (Figure 9). Machine learning algorithms analyze vast datasets from sensors monitoring pH, temperature, oxygen levels, and nutrient concentrations, predicting microbial growth kinetics and metabolite production with high accuracy [49]. For example, deep learning models can forecast yield variations in real-time, allowing for adaptive adjustments that increase efficiency by 25–35% compared to empirical methods. Digital twins simulate bioreactor environments, enabling virtual testing of genetic modifications or media formulations without physical trials, accelerating R&D cycles from months to weeks [50].

In practice, companies like Perfect Day and Clara Foods leverage AI to produce dairy proteins without animals, using reinforcement to learn fine-tune fermentation parameters for optimal protein expression [51]. This not only minimizes waste by reducing failed batches but also enhances sustainability by lowering water and energy usage per unit of product. Challenges include scaling models to industrial bioreactors, where fluid dynamics and microbial heterogeneity introduce complexities; however, physics-informed neural networks (PINNs) integrate domain knowledge to improve predictions [52]. Integration with genomics data allows AI to design custom strains, further personalizing outputs for nutritional profiles [53]. As precision fermentation grows, AI's role will expand to supply chain integration, forecasting demand for fermented ingredients in alternative meats or functional foods, ensuring this technology addresses global protein demands sustainably (Table 5).

https://doi.org/10.3390/pr14030513

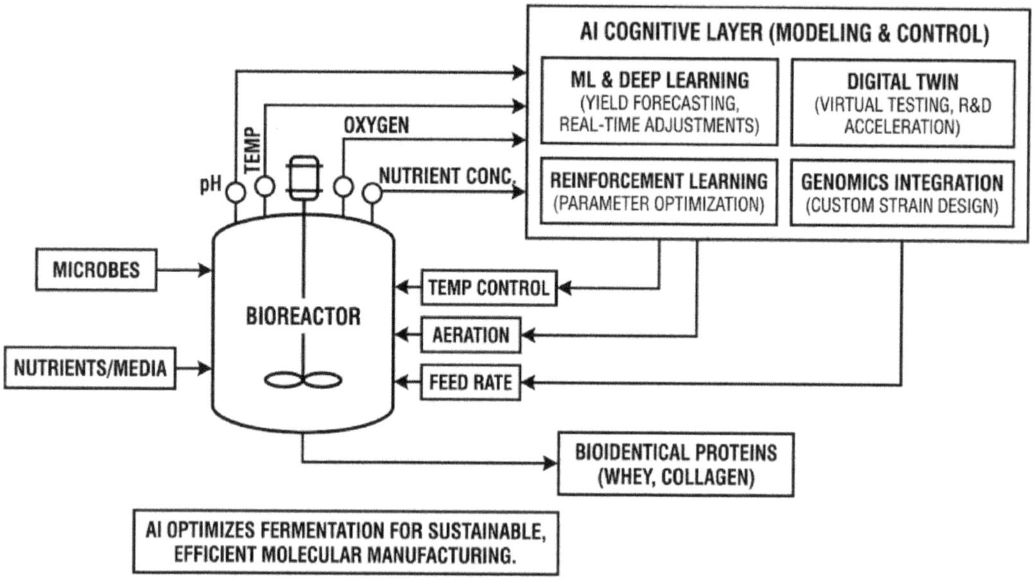

Figure 9. Schematic Representation of an AI-Driven Precision Fermentation System.

Table 5. Synergistic Integration of AI Architectures and Performance Metrics in Precision Fermentation: From Strain Design to Industrial Scale-Up.

AI Model/Technology	Specific Application	Examples/Products	Measured Performance/Impact
Deep Learning Models and Machine Learning	Analyzing sensor data (pH, temp, O_2, nutrients) to predict microbial growth and metabolite production. Real-time yield forecasting for adaptive adjustments.	Bioidentical Proteins: Whey, Collagen.	Increases efficiency by 25–35% compared to empirical methods.
Digital Twins	Simulating bioreactor environments for virtual testing of genetic modifications and media formulations without physical trials.	R&D Acceleration: Virtual strain and process development.	R&D cycles increase from month to weeks.
Reinforcement Learning (RL)	Fine-tuning fermentation parameters to optimize protein expression.	Animal-Free Dairy Proteins: Used by companies like Perfect Day and Clara Foods.	Minimizes waste from failed batches and lowers water/energy usage per unit of product.
Physics-Informed Neural Networks (PINNs)	Integrating domain knowledge to improve predictions in industrial bioreactors with complex fluid dynamics and microbial heterogeneity.	Industrial Scale-up: Managing complexities in large-scale production.	Improves prediction accuracy in complex industrial environments.
AI with Genomics Data	Designing custom microbial strains to personalize fermentation outputs for specific nutritional profiles.	Customized Ingredients: Tailored nutritional and functional proteins.	Enables personalization of final product attributes.

https://doi.org/10.3390/pr14030513

5. Future Trends, Sustainability, Limitations and Perspectives

The integration of artificial intelligence (AI) in food processing technologies is posed to redefine the industry's operational paradigms, fostering a transition toward resilient, efficient and sustainable systems. As we approach 2030, AI's evolution from experimental applications to core infrastructural components underscores a scholarly imperative to examine its prospective trajectories, environmental imperatives, inherent constraints and forward-looking strategies.

The year 2030 is particularly significant because it aligns with the United Nations' 2030 Agenda for Sustainable Development, a global framework adopted by all UN Member States in 2015, which sets ambitious targets to end poverty, protect the planet, and ensure prosperity for all. In the context of food processing technologies, the integration of AI is seen as a critical driver for achieving these sustainability goals, including innovation, responsible production, reduced waste, and improved resource efficiency (Figure 10).

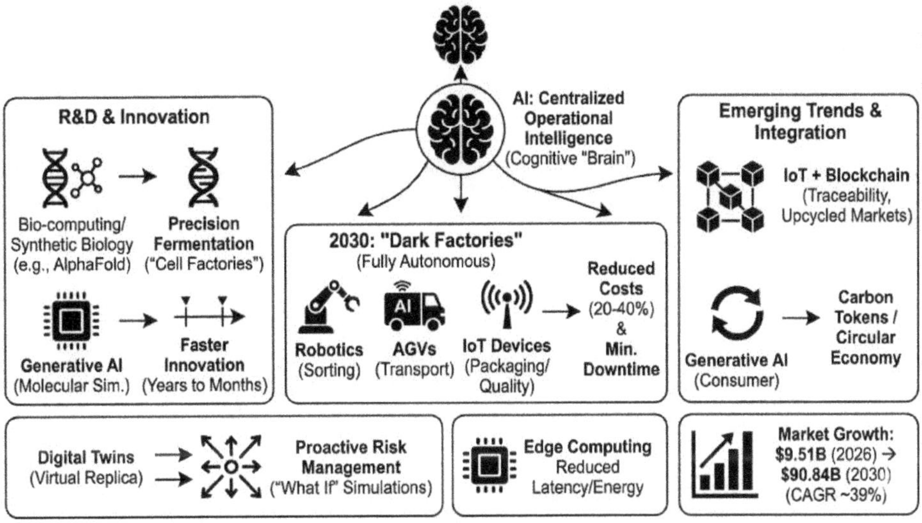

Figure 10. Overview of the emerging AI ecosystem in Food Processing: From molecular innovation to autonomous operation.

Thus, 2030 serves not only as a milestone for technological advancement but also as a benchmark for measuring progress toward a more sustainable and equitable food system. This section synthesizes recent projections, drawing on interdisciplinary insights from engineering, economics, ethics, and policy domains to provide a rigorous framework for understanding AI's transformative potential in the food sector.

5.1. Emerging Paradigms and Future Projections

The trajectory of artificial intelligence in food processing indicates a shift from isolated pilot studies toward centralized operational intelligence. A central prospective concept is the "dark factory", a fully autonomous facility characterized by "lights-out" manufacturing, where human intervention is minimal. Projections suggest that by 2030, such systems could theoretically orchestrate robotics, Internet of Things (IoT) devices, and automated guided vehicles (AGVs) to manage repetitive tasks such as sorting and high-speed packaging. While models predict that these autonomous systems could reduce operational costs by 20–40% through minimized downtime, their widespread adoption is currently constrained by significant economic barriers. The capital expenditure (CAPEX) required

https://doi.org/10.3390/pr14030513

for the necessary infrastructure, including advanced edge computing and sensor networks, often exceeds \$5 million per facility, rendering this model financially prohibitive for most small-to-medium enterprises (SMEs).

Parallel to automation, bio-computing and synthetic biology represent emerging frontiers in food engineering. Tools such as AlphaFold are expanding the capabilities of precision fermentation, potentially enabling the design of specialized "cell factories" (e.g., yeast or fungi) for molecular manufacturing. Generative AI models are being explored for their ability to simulate molecular interactions, which could compress innovation cycles for novel product formulations. However, these applications remain largely within the research and development phase, limited by the computational intensity of molecular simulations and the "black box" nature of deep learning models, which complicates regulatory approval for new food ingredients (Figure 10).

Market analysts forecast that the AI sector in food manufacturing could expand to approximately \$90 billion by 2030, driven by advances in automation and molecular innovation. Nevertheless, this growth is contingent upon resolving persistent challenges related to data heterogeneity, algorithmic transparency, and the high cost of retrofitting legacy "brownfield" sites.

5.2. Sustainability

Sustainability emerges as a core benefit of AI integration, aligning with global imperatives such as the United Nations' goal of a 70% increase in food production by 2050 while minimizing waste and emissions. AI facilitates optimized processes that reduce greenhouse gas emissions by 15–25% through dynamic energy management and route optimization, while enabling nutrient recycling via upcycling of agricultural by-products (e.g., fruit peels, whey, and wheat middling) into high-value ingredients like natural emulsifiers and texturants. AI-driven waste conversion can transform side-streams into viable products, cutting food waste by millions of pounds annually.

In sustainable packaging, AI revolutionizes eco-friendly solutions by leveraging generative models to design plant-based alternatives, such as mushroom-based compostable boxes or seaweed wraps, reducing material waste by 15–40% and enhancing recyclability. Robotics systems like AMP Robotics' Cortex employ machine learning for efficient sorting of recyclables, improving global recycling rates from the current 16% baseline. Shelf-life prediction using infrared scanners and predictive analytics further minimizes spoilage, supporting circular economies in waste management and compliance with regulations like California's SB 1383. Advanced analytics provide real-time tracking of carbon and water footprints per stock-keeping unit (SKU), promoting environmental stewardship. However, a notable challenge is the "Water Paradox", the substantial water consumption required to cool AI data centers, which necessitates a shift toward energy-efficient edge-computing architectures to mitigate this irony in resource-intensive industries.

5.3. Limitations

Despite its promise, AI adoption in food processing confronts multifaceted limitations that warrant rigorous scholarly analysis. Technically, the "black box" nature of deep learning models poses a significant barrier, particularly in safety-critical applications like microbial contamination detection, where non-transparent decision-making erodes trust and complicates regulatory compliance under frameworks like the EU AI Act. Data heterogeneity, arising from noisy biological inputs influenced by soil variability or weather could intensify future challenges. As an example, and while AI attains 98–99% accuracy in defect detection, the residual 1% risk can precipitate public health crises, underscoring the need for hybrid human-AI systems.

https://doi.org/10.3390/pr14030513

Economically, the Capital expenditure (CAPEX) for AI infrastructure, encompassing sensors, edge computing, and robotics, often exceeds $5 million per facility, deterring small and medium enterprises which comprise 79% of processors delaying initiatives due to cost uncertainties. Converting legacy brownfield sites for autonomous operations entails higher capital expenditure (CAPEX) and technical debt compared to the greenfield construction. The necessity of reconciling modern automated protocols with existing constraints creates substantial integration friction.

Ethically, workforce displacement in "dark factories" threatens low-skilled roles in sorting and packaging, raising dilemmas around job loss and algorithmic bias in procurement or hiring. Data privacy concerns, stemming from incomplete datasets, can lead to inaccurate models, while regulatory gaps hinder global standardization (Table 6)

Table 6. Ethical Challenges and Mitigations are outlined below.

Ethical Challenge	Impact Area	Proposed Mitigation
Algorithmic Bias	Procurement and hiring	Implementation of diverse-weighted datasets
Job Displacement	Shop-floor Labor	"Just transition" policies involving upskilling and reskilling
Data Privacy	Employee Monitoring	Transparent data governance and anonymization protocols

5.4. Cybersecurity in Food Digitalization

The rapid digitalization of food processing, driven by AI, IoT sensors, cloud-based analytics, and interconnected Cyber-Physical Systems (CPS), which has significantly expanded the attack surface, elevating cybersecurity from a peripheral concern to a critical vulnerability for food safety, supply chain resilience, and public health [54]. While these technologies enable predictive maintenance, real-time quality monitoring, and optimized operations, they also introduce sophisticated risks such as ransomware, data manipulation, and adversarial AI attacks [55]. Notably, cyberattacks in the food and agriculture sector surged by 101% from August 2024 to August 2025, with ransomware incidents rising sharply in early 2025. Projections for 2026 highlight IoT as a major attack vector, alongside emerging AI-enhanced threats like automated phishing and reconnaissance [56].

Key cybersecurity risks include:

- Ransomware and Operational Disruption: Attackers target operational technology (OT) systems to halt production lines, resulting in spoilage, waste, and financial losses. High-profile incidents, such as the 2025 attack on United Natural Foods Inc. and the 2024 ransomware event affecting Ahold Delhaize, demonstrate how these disruptions can cascade into food shortages and price spikes.
- Process Manipulation and AI-Specific Vulnerabilities: Adversaries may alter sensor data or model inputs, compromising product safety through "digital sabotage" of critical processes like pasteurization or cooling. The "black box" nature of deep learning models exposes systems to adversarial attacks, where subtle data perturbations can mislead defect detection or predictive models, leading to false negatives in quality control. Additionally, interconnected supply chains are increasingly susceptible to cascading failures, especially when legacy systems are exploited.

These risks are compounded by the "digitization paradox": technologies that enhance efficiency also heighten exposure, particularly for small and medium enterprises (SMEs) with limited resources for robust defenses. Although AI-driven anomaly detection can proac-

https://doi.org/10.3390/pr14030513

tively identify threats, studies show persistent challenges in countering zero-day exploits and insider threats, with residual risks remaining high in heterogeneous environments.

Mitigation strategies, as recommended by frameworks such as CISA's Food and Agriculture Sector resources, emphasize multi-layered defenses (network segmentation, regular patching, zero-trust architecture) and the adoption of AI/ML for defensive purposes (anomaly detection, automated threat hunting, predictive risk scoring). Additional measures include regular audits, employee training, incident response planning tailored to food perishability, and leveraging voluntary checklists for steps like multi-factor authentication and sector-specific information sharing. Policy initiatives, such as the Farm and Food Cybersecurity Act (2025), aim to mandate baseline protections and enhance information sharing across the sector.

Despite these efforts, significant challenges persist, including high implementation costs that deter adoption among smaller processors, legacy equipment often lacks modern security features, and global regulatory fragmentation hinders standardization. Future research should prioritize resilient, explainable AI cybersecurity frameworks and farmer/processor-centered designs to bridge the "security divide" in vulnerable regions. Without proactive investment, cyber incidents could undermine the very sustainability and safety gains promised by AI integration in food processing.

5.5. Perspectives

Strategic AI integration in food processing necessitates an interdisciplinary governance framework that transitions human capital toward high-level oversight and ethical auditing while delegating "4D" (dark, dirty, dull, dangerous) operations to autonomous systems. Technical transparency is achieved through Explainable AI (XAI) techniques, such as SHAP and LIME, which facilitate regulatory compliance by attributing model outputs to specific environmental variables. To overcome data fragmentation, Federated Learning (FL) enables decentralized, privacy-preserving model scaling across diverse enterprises.

Economically, the adoption of AI-as-a-Service (AIaaS) and sustainability-linked financing mechanisms effectively bridges CAPEX gaps, particularly for SMEs. These fiscal architectures support the capture of 5–10% "premiums" in export markets by ensuring carbon-transparent production. Ultimately, the synthesis of policy incentives, such as Asia-Pacific "smart-factory" subsidies, and bio-computing integration aligns industrial output with UN sustainability goals, ensuring a scalable and equitable circular economy (Table 7).

Table 7. Strategic perspectives.

Strategic Perspective	Focus Area	2030 Goal
Regulatory	EU AI Act & USDA AI Strategy	Algorithmic transparency and safety-first compliance
Financial	AI-as-a-Service (AIaaS)	Lowering the CAPEX barrier for SMEs through subscriptions
Ethical	Federated Learning	Projecting proprietary recipes while sharing safety data

6. Conclusions

Artificial intelligence (AI) is transforming food processing across multiple domains by leveraging big data, sensors, and advanced algorithms to automate and optimize operations. The integration of IoT and blockchain technologies further amplifies these capabilities, marking a paradigm shift toward more efficient, safe, and sustainable practices. Addressing current challenges through policy support and interdisciplinary collaboration

https://doi.org/10.3390/pr14030513

is essential for the industry to fully harness AI's potential. This work provides foundational insights for researchers, engineers, and policymakers navigating the complex landscape of AI-driven food manufacturing.

6.1. *Gaps in Current Research*

Despite significant progress, notable gaps remain in the literature. Current research often overlooks AI's limitations in diverse, real-world food systems. While AI demonstrates strong performance in controlled environments, there is a lack of focus on multi-source data synchronization and model adaptability in variable conditions, such as those found in smallholder farms. Issues of data privacy and bias in datasets are underexplored, leading to inequitable outcomes in developing regions. Additionally, there is insufficient evaluation of the long-term impacts of AI on workforce dynamics and environmental footprints, including data center energy use.

6.2. *Future Research Prospectus*

Future research should prioritize the development of Explainable AI (XAI) for transparent models, federated learning to address data privacy, and integration with synthetic biology for personalized nutrition. Exploring AI applications in emerging areas such as upcycled food design and climate-resilient processing could help bridge existing gaps. Longitudinal studies on AI scalability in small and medium-sized enterprises (SMEs) and benchmarking against human performance are also essential.

6.3. *Policy Alignment Opportunities*

AI in food processing aligns with major policy frameworks, including the UN 2030 Agenda and the EU AI Act, offering opportunities for subsidies to support sustainable technology adoption. Governments can further promote AI-as-a-Service (AIaaS) models for SMEs and standardize cybersecurity protocols. Incentives for ethical AI, such as carbon credits for optimized processes, could enhance global food security.

Funding: This research received no external funding.

Data Availability Statement: No new data were created or analyzed in this study.

Conflicts of Interest: The author declares no conflicts of interest. The author declares that the research was conducted in the absence of any commercial or financial relationships that could be construed as a potential conflict of interest. The Ayoub Sciences LLC had no role in the design of the study; in the collection, analyses, or interpretation of data; in the writing of the manuscript, or in the decision to publish the results.

List of Abbreviations

AI	Artificial Intelligence
ANN	Artificial Neural Network
CAPEX	Capital Expenditure
CFD	Computational Fluid Dynamics
CNN	Convolutional Neural Network
CPS	Cyber-Physical System
DAI	Distributed Artificial Intelligence
DRL	Deep Reinforcement Learning
ERP	Enterprise Resource Planning
FAO	Food and Agriculture Organization
FDA	Food and Drug Administration
FSMA	Food Safety Modernization Act
GA	Genetic Algorithm

https://doi.org/10.3390/pr14030513

GPS	Global Positioning System
IoT	Internet of Things
JFB	January Food Benchmark
LSTM	Long Short-Term Memory
MAS	Multi-Agent System
ML	Machine Learning
NIR	Near-Infrared
NLP	Natural Language Processing
OT	Operational Technology
PID	Proportional-Integral-Derivative
PINN	Physics-Informed Neural Network
PLC	Programmable Logic Controller
PSO	Particle Swarm Optimization
QC	Quality Control
R&D	Research and Development
RL	Reinforcement Learning
ROI	Region of Interest
SA	Simulated Annealing
SCADA	Supervisory Control and Data Acquisition
SCM	Supply Chain Management
SCRM	Supply Chain Risk Management
SERS	Surface-Enhanced Raman Spectroscopy
SHAP	SHapley Additive exPlanations
SKU	Stock-Keeping Unit
SME	Small and Medium Enterprise
SVR	Support Vector Regression
VAE	Variational Autoencoder
XAI	Explainable Artificial Intelligence

References

1. Fišteš, A. The evolution of milling process. In *Cereal-Based Foodstuffs: The Backbone of Mediterranean Cuisine*; Springer International Publishing: Cham, Switzerland, 2021; pp. 19–45.
2. Jiang, L.; Xu, B.; Husnain, N.; Wang, Q. Overview of Agricultural Machinery Automation Technology for Sustainable Agriculture. *Agronomy* **2025**, *15*, 1471. [CrossRef]
3. Koehn, N.F. Henry Heinz and brand creation in the late nineteenth century: Making markets for processed food. *Bus. Hist. Rev.* **1999**, *73*, 349–393. [CrossRef]
4. Loughlin, S. Industry 3.0 to Industry 4.0: Exploring the transition. In *New Trends in Industrial Automation*; IntechOpen: London, UK, 2018.
5. Agolla, J.E. Human capital in the smart manufacturing and industry 4.0 revolution. *Digit. Transform. Smart Manuf.* **2018**, *2*, 41–58.
6. Agarwal, P.; Swami, S.; Malhotra, S.K. Artificial intelligence adoption in the post COVID-19 new-normal and role of smart technologies in transforming business: A review. *J. Sci. Technol. Policy Manag.* **2024**, *15*, 506–529. [CrossRef]
7. Jin, Y.; Liu, J.; Xu, Z.; Yuan, S.; Li, P.; Wang, J. Development status and trend of agricultural robot technology. *Int. J. Agric. Biol. Eng.* **2021**, *14*, 1–19. [CrossRef]
8. Zhang, J.; Budhdeo, S.; William, W.; Cerrato, P.; Shuaib, H.; Sood, H.; Ashrafian, H.; Halamka, J.; Teo, J.T. Moving towards vertically integrated artificial intelligence development. *npj Digit. Med.* **2022**, *5*, 143. [CrossRef] [PubMed]
9. Theofilou, A.; Nastis, S.A.; Michailidis, A.; Bournaris, T.; Mattas, K. Predicting Prices of Staple Crops Using Machine Learning: A Systematic Review of Studies on Wheat, Corn, and Rice. *Sustainability* **2025**, *17*, 5456. [CrossRef]
10. Galanakis, C.M.; Daskalakis, M.I.; Galanakis, I.M.; Nehrey, M.; Spanou, M.; Vetsou, A.; Chalkidou, A.; Agrafioti, E. Landscape of policies, standards, approaches, and projects for EU food security: An overview. *Discov. Food* **2025**, *5*, 117. [CrossRef]
11. Pickson, R.B.; Gui, P.; Jian, L.; Boateng, E. The role of private sector investment in agriculture: A catalyst for sustainable development in Asia. *Sustain. Dev.* **2025**, *33*, 113–128. [CrossRef]
12. Shahbazi, F.; Shahbazi, S.; Zare, D. Losses in Agricultural Produce: Causes and Effects on Food Security. *Food Energy Secur.* **2025**, *14*, e70086. [CrossRef]

https://doi.org/10.3390/pr14030513

13. Panigrahi, S.S.; Mishra, G.; Kumar, V.; Panigrahi, C.; Kaur, G.; Shaikh, A.E.Y.; Kumar, S.; Panda, B.K.; Kumari, T. Next-generation food drying: Specialized and smart approaches to boost efficiency and quality. *Compr. Rev. Food Sci. Food Saf.* **2025**, *24*, e70269. [CrossRef]
14. Nwakoby, I.P.; Iheukwumere, I.H.; Iheukwumere, C.M.; Nwakoby, N.E.; Idigo, M.A.; Ike, V.E. Microbial contamination in dairy products and its legal implications. *Afr. J. Nutr. Appl. Res.* **2025**, *1*, 9–16. [CrossRef]
15. Mastos, T.; Gotzamani, K. Sustainable supply chain management in the food industry: A conceptual model from a literature review and a case study. *Foods* **2022**, *11*, 2295. [CrossRef] [PubMed]
16. Cricelli, L.; Mauriello, R.; Strazzullo, S. Technological innovation in agri-food supply chains. *Br. Food J.* **2024**, *126*, 1852–1869. [CrossRef]
17. Dong, H. Research on Corporate Social Responsibility Synergy in Food Supply Chain: Insights from Nestlé Company. In *Modern Management Based on Big Data VI*; IOS Press: Amsterdam, The Netherlands, 2025; pp. 193–200.
18. Khumalo, M.; Luke, R.; Van Der Walt, J. A conceptual framework for integrating demand forecasting and risk management strategies to optimise food supply chains. *J. Contemp. Manag.* **2025**, *22*, 273–309. [CrossRef]
19. Karanth, S.; Feng, S.; Patra, D.; Pradhan, A.K. Linking microbial contamination to food spoilage and food waste: The role of smart packaging, spoilage risk assessments, and date labeling. *Front. Microbiol.* **2023**, *14*, 1198124. [CrossRef]
20. Plakantara, S.P.; Karakitsiou, A. Transforming Agrifood Supply Chains with Digital Technologies: A Systematic Review of Safety and Quality Risk Management. *Oper. Res. Forum* **2025**, *6*, 113. [CrossRef]
21. Jannelli, V.; Schöpf, S.; Bickel, M.; Netland, T.; Brintrup, A. Agentic LLMs in the supply chain: Towards autonomous multi-agent consensus-seeking. *Int. J. Prod. Res.* **2025**, 1–31. [CrossRef]
22. Bukhari, H.; Basingab, M.S.; Rizwan, A.; Sánchez-Chero, M.; Pavlatos, C.; More, L.A.V.; Fotis, G. Sustainable green supply chain and logistics management using adaptive fuzzy-based particle swarm optimization. *Sustain. Comput. Inform. Syst.* **2025**, *46*, 101119. [CrossRef]
23. Mirasharafi, A.; Moghadamnia, E. Application of Artificial Intelligence in Production Planning and Supply Chain Management: Examining the Role of Machine Learning, Deep Learning, and Neural Networks in Optimizing Production Processes in the Electronics Industry of Iran. *Manag. Strat. Eng. Sci.* **2025**, *7*, 111–119. [CrossRef]
24. Veeramani, M.B.; de Sá, M.M. An Analysis of Critical Success Factors for Digital Transformation in Agri-food Supply Chains. In *Digital Transformation of Supply Chains: Business Objectives and Applications*; Springer Nature: Cham, Switzerland, 2026; pp. 221–256.
25. Goswami, S.S.; Mondal, S.; Sarkar, S.; Gupta, K.K.; Sahoo, S.K.; Halder, R. Artificial intelligence-enabled supply chain management: Unlocking new opportunities and challenges. *Artif. Intell. Appl.* **2025**, *3*, 110–121.
26. Gomaa, A.H. Enhancing Product Development Excellence through Quality Management Tools: A Comprehensive Review and Integrated Conceptual Framework. *Intell. Sustain. Manuf.* **2025**, *2*, 10017. [CrossRef]
27. Yu, W.; Ouyang, Z.; Zhang, Y.; Lu, Y.; Wei, C.; Tu, Y.; He, B. Research progress on the artificial intelligence applications in food safety and quality management. *Trends Food Sci. Technol.* **2025**, *156*, 104855. [CrossRef]
28. Dhal, S.B.; Kar, D. Leveraging artificial intelligence and advanced food processing techniques for enhanced food safety, quality, and security: A comprehensive review. *Discov. Appl. Sci.* **2025**, *7*, 75. [CrossRef]
29. Ahmad, A.; Murad, M.; Utra, U.; Yussof, N.S.; Fazial, F.F.; Lazim, A.; Fazry, S.; Law, D. Transforming Food Systems with Artificial Intelligence: Challenges, Governance, and Pathways for Sustainable Integration. *Food Rev. Int.* **2025**, 1–26. [CrossRef]
30. Dakhia, Z.; Russo, M.; Merenda, M. AI-Enabled IoT for Food Computing: Challenges, Opportunities, and Future Directions. *Sensors* **2025**, *25*, 2147. [CrossRef]
31. Ayoub, A.; Lucia, L.A. Sustainable Cationic Polyelectrolytes from Agri-Forestry Biomass: Conventional Chemistry to AI-Optimized Reactive Extrusion. *Sustainability* **2025**, *17*, 10060. [CrossRef]
32. Polychronopoulos, N.D.; Moustris, K.; Karakasidis, T.; Sikora, J.; Krasinskyi, V.; Sarris, I.E.; Vlachopoulos, J. Machine learning for screw design in single-screw extrusion. *Polym. Eng. Sci.* **2025**, *65*, 2607–2623. [CrossRef]
33. Kelly, D.; Sergis, V.; i Blanco, L.V.; Mason, K.; Daly, A.C. Autonomous control of extrusion bioprinting using convolutional neural networks. *Adv. Funct. Mater.* **2025**, *35*, 2424553. [CrossRef]
34. Lee, J.-H.; Okwuosa, C.N.; Hur, J.-W. Extruder Machine Gear Fault Detection Using Autoencoder LSTM via Sensor Fusion Approach. *Inventions* **2023**, *8*, 140. [CrossRef]
35. Wolff, D.; Fricke, C.; Kemmerling, M.; Elgeti, S. Towards shape optimization of flow channels in profile extrusion dies using reinforcement learning. *PAMM* **2023**, *22*, e202200009. [CrossRef]
36. Altun, F.; Bayar, A.; Hamzat, A.K.; Asmatulu, R.; Ali, Z.; Asmatulu, E. AI-driven innovations in 3D printing: Optimization, automation, and intelligent control. *J. Manuf. Mater. Process.* **2025**, *9*, 329. [CrossRef]
37. Hoque, A. Artificial Intelligence in Post-Harvest Drying Technologies: A Comprehensive Review on Optimization, Quality Enhancement, and Energy Efficiency. *Int. J. Sci. Res. IJSR* **2024**, *13*, 493–502. [CrossRef]

https://doi.org/10.3390/pr14030513

38. Iheonye, A.C.; Ferrie, F.P.; Raghavan, V. A Need to Incorporate Intelligent Food-Drying Systems in ABE Education–A Recipe for Reducing Global Food Losses and Waste. In *Agricultural, Biosystems, and Biological Engineering Education*; CRC Press: Boca Raton, FL, USA, 2024; pp. 435–455.
39. Wijerathne, A.D.H.T.; Joardder, M.U.H.; Welsh, Z.G.; Nayak, R.; Sablani, S.S.; Karim, A. Recent Advances in Food Drying Modeling: Empirical to Multiscale Physics-Informed Neural Networks. *Compr. Rev. Food Sci. Food Saf.* **2025**, *24*, e70194. [CrossRef] [PubMed]
40. Szpicer, A.; Bińkowska, W.; Stelmasiak, A.; Wojtasik-Kalinowska, I.; Czajkowska, A.; Mierzejewska, S.; Domiszewski, Z.; Rydzkowski, T.; Piepiórka-Stepuk, J.; Półtorak, A. Advances in computational fluid dynamics of mechanical processes in food engineering: Mixing, extrusion, drying, and process optimization. *Appl. Sci.* **2025**, *15*, 8752. [CrossRef]
41. Tu, Z.; Irudayaraj, J.; Lee, Y. Characterizing spray-dried powders through NIR spectroscopy: Effect of two preparation strategies for calibration samples and comparison of two types of NIR spectrometers. *Foods* **2023**, *12*, 467. [CrossRef]
42. Hoque, A.; Sethi, L.N.; Padhiary, M. Systematic Review on Refractance Window Drying for Preserving Heat Sensitive Agricultural Products. *J. Food Process. Eng.* **2025**, *48*, e70239. [CrossRef]
43. Martynenko, A.; Mujumdar, A. Recent progress in AI applications for drying. *Dry. Technol.* **2025**, *43*, 1117–1118. [CrossRef]
44. Haddouche, M.; Ilinca, A. Energy efficiency and industry 4.0 in wood industry: A review and comparison to other industries. *Energies* **2022**, *15*, 2384. [CrossRef]
45. Jamalabadi, M.Y.A. Optimal Design of Drying Process of the Potatoes with Multi-Agent Reinforced Deep Learning. *Front. Heat Mass Transf.* **2024**, *22*, 511–536. [CrossRef]
46. Perré, P. Toward mechanistic models "augmented" by machine learning. Example of a drying simulation data set exploited by a Physics Informed Neural Network. *Dry. Technol.* **2025**, *43*, 147–161. [CrossRef]
47. Sun, Q.; Yuan, Y.; Xu, B.; Gao, S.; Zhai, X.; Xu, F.; Shi, J. Innovative technologies reshaping meat industrialization: Challenges and opportunities in the intelligent era. *Foods* **2025**, *14*, 2230. [CrossRef]
48. Boukid, F.; Ganeshan, S.; Wang, Y.; Tülbek, M.Ç.; Nickerson, M.T. Bioengineered enzymes and precision fermentation in the food industry. *Int. J. Mol. Sci.* **2023**, *24*, 10156. [CrossRef] [PubMed]
49. Tarafdar, A.; Das, A.; Singh, B. Application of artificial intelligence and machine learning in solid-state fermentation proceses. In *Current Advances in Solid-State Fermentation*; Elsevier: Amsterdam, The Netherlands, 2026; pp. 129–144.
50. Helleckes, L.M.; Hemmerich, J.; Wiechert, W.; von Lieres, E.; Grünberger, A. Machine learning in bioprocess development: From promise to practice. *Trends Biotechnol.* **2023**, *41*, 817–835. [CrossRef]
51. Ray, A.; Sinha, C.; Sharma, A.; Khamrui, K.; Hussain, S.A.; Dabas, J.; Mohanty, T.K. AI-driven real-time monitoring and predictive control system for yoghurt fermentation. *Food Control* **2025**, *182*, 111857. [CrossRef]
52. Thirugnanasambandam, M.K.; Pinto, J.; Moskovkina, E.; Costa, R.S.; Oliveira, R. A Physics-Informed Neural Network (PINN) framework for generic bioreactor modelling. *Comput. Chem. Eng.* **2025**, *203*, 109354. [CrossRef]
53. Siddique, I.; Farooq, M.; Ullah, Q.; Nabi, A.; Siddique, M.; Siddique, S.; Naseer, M.; Younis, A. Future trends and direction in AI and ML for food science and bioprocess development. *Artif. Intell. Food Sci.* **2026**, 811–825. [CrossRef]
54. Joshi, S.; Sharma, M. Digital technologies (DT) adoption in agri-food supply chains amidst COVID-19: An approach towards food security concerns in developing countries. *J. Glob. Oper. Strat. Sourc.* **2022**, *15*, 262–282. [CrossRef]
55. Jehan, N.; Ansari, N.M.; Ashraf, Z.; Bashir, M.A.; Gul, H.; Raza, A. Adversarial Machine Learning for Cyber security Defense: Detecting Model Evasion, Poisoning Attacks, and Enhancing the Robustness of AI Systems. *Glob. Res. J. Nat. Sci. Technol.* **2025**, 3. [CrossRef]
56. Zangana, H.M.; Luckyardi, S.; Mustafa, F.M.; Li, S. Ethical and Secure AI Applications in Agricultural Research: Challenges and Opportunities. In *Ensuring Secure and Ethical STM Research in the AI Era*; IGI Global: Hershey, PA, USA, 2025; pp. 43–76. [CrossRef]

https://doi.org/10.3390/pr14030513

processes

MDPI

Article

Automated Fiber Placement Gap Width Prediction Using a Transformer-Based Deep Learning Approach

Diogo Cardoso [1,2,*], António Ramos Silva [1,2] and Nuno Correia [1,3]

[1] INEGI—Institute of Science and Innovation in Mechanical and Industrial Engineering, 4200-465 Porto, Portugal; ars@fe.up.pt (A.R.S.); ncorreia@inegi.up.pt (N.C.)
[2] FEUP—Faculty of Engineering, University of Porto, 4200-465 Porto, Portugal
[3] LAETA—Associated Laboratory of Energy, Transports and Aerospace, 4200-465 Porto, Portugal
* Correspondence: dcardoso@inegi.up.pt

Abstract

Automated Fiber Placement (AFP) is a critical process in composite manufacturing, where precise fiber tow placement is essential for achieving high-quality and high-performance engineering components. However, deviations in process variables frequently lead to defects such as gaps and overlaps, which can compromise structural integrity. While various monitoring techniques exist, accurately predicting and understanding the formation of these defects from complex sensor data remains challenging. This work introduces a novel application of a Transformer-based deep learning architecture to enhance the estimation of gap widths in AFP. Leveraging a publicly available industrial AFP dataset, our methodology incorporates a customized positional encoding scheme to effectively integrate the critical spatial context of the tow layup process. The model's predictive performance was evaluated, achieving a Mean Absolute Percentage Error (MAPE) of 1.04% and an R-squared (R^2) value of 0.9143, demonstrating its capability for accurate gap width estimation. Furthermore, SHapley Additive exPlanations (SHAP) analysis was employed to assess the complex interplay between sources of manufacturing process variation. This study establishes the Transformer architecture as a promising and interpretable data-driven tool for AFP process monitoring. The results serve as a proof of concept for attention-based virtual metrology, offering a pathway towards deeper process understanding and defect mitigation.

Keywords: automated fiber placement; deep learning; transformer architecture; attention mechanism; composite manufacturing

1. Introduction

Automated Fiber Placement (AFP) has emerged as a pivotal manufacturing technology for high-performance composite structures, particularly in the aerospace industry, due to its ability to produce highly customized layups with tailored mechanical properties [1]. A key determinant of part quality in AFP is the accurate control of gap width between adjacent fiber tows, defined as strips of pre-impregnated composite material. Even small deviations in tow placement can result in gaps, where tow edges fail to meet, or overlaps, where one tow partially covers another. Both defect types introduce local stress concentrations, disrupt load transfer between plies and degrade stiffness and strength, ultimately compromising the structural integrity and service performance of the final component [1–3]. From a manufacturing perspective, gaps and overlaps constitute a dominant class of AFP defects observed in industrial practice [4]. Their occurrence directly impacts production cost and throughput, as components exhibiting excessive gap width deviations often require extensive post-process inspection, repair or rejection. Current industrial workflows therefore rely heavily on retrospective quality assurance, including manual inspection and rework cycles, which can consume a substantial fraction of total AFP production time and significantly diminish the productivity gains expected from automation [2]. Accurate in-process prediction of gap width is therefore critical: it enables early detection of defect formation, supports corrective action before defects propagate across subsequent plies, and reduces dependence on costly downstream inspection and rework. Improving gap width predictability thus directly contributes to enhanced structural reliability, reduced scrap and rework rates and lower overall manufacturing cost [1,3,5].

To understand the challenge of gap prediction, one must consider the physical mechanics of the AFP process. A robotic placement head deposits composite tows onto a mold under controlled compaction [1]. Ideally, adjacent tows are placed with consistent spacing defined by the robot's programmed path. However, in reality, the process is subject to numerous stochastic variations. Machine kinematic inaccuracies and fluctuations in material tackiness can cause the tows to slip or drift laterally beneath the compaction roller during deposition [1].

Traditional approaches for monitoring and controlling gap widths, such as post-deposition visual inspection or manual measurement, are typically labor-intensive, operator-dependent and sensitive to environmental conditions, resulting in limited reliability and repeatability [2]. Although modern AFP systems increasingly incorporate sensors for in-process data acquisition [3], enabling the possibility of real-time monitoring, the accurate prediction and mitigation of gap-related defects remains an open challenge.

Current approaches to modeling these defects face distinct limitations. Analytical models, while physically grounded, often rely on idealized kinematic assumptions. They frequently struggle to capture the noisy, non-linear and stochastic interactions present in raw industrial sensor data, leading to reduced accuracy in real-world environments [6].

On the other hand, data-driven Deep Learning (DL) methods, such as Convolutional Neural Networks (CNNs) or Recurrent Neural Networks (RNNs), have demonstrated strong predictive capabilities. However, these standard architectures often operate as "black boxes", providing accurate predictions without revealing the underlying drivers of the defects [6,7]. Furthermore, RNNs and Long Short-Term Memory (LSTM) networks process data sequentially, which can create information bottlenecks and obscure the global spatial dependencies critical to the AFP layup process [8]. This lack of interpretability and spatial context poses a significant barrier to their adoption for process certification [6].

This work addresses these challenges by exploring the application of a Transformer-based DL architecture. Unlike sequential models, the Transformer employs self-attention mechanisms to process the entire spatial sequence in parallel. We hypothesize that this architecture can more effectively model the intricate, non-local relationships between sensor readings (such as tow width and robot position) and gap dimensions while simultaneously offering the transparency required for root-cause analysis.

This study employs a publicly available dataset of industrial AFP process data [9] to develop and evaluate the proposed model. Furthermore, to enhance the understanding of the model's predictions and gain insights into the manufacturing process itself, SHapley Additive exPlanations (SHAP) analysis is employed to assess the influence of individual input features on gap width estimation.

https://doi.org/10.3390/pr14040609

The main contributions of this work are:

- The novel application of a Transformer-based deep learning architecture for the estimation of gap widths in the AFP process, which, to the best of our knowledge, represents a novel approach to incorporating spatial context into AFP defect estimation.
- The development of a domain-specific positional encoding scheme that effectively incorporates the X and Y spatial coordinates. This allows the Transformer to distinguish between tool-location-dependent errors (global context) and tow interaction errors (local context), a methodology adaptable to other spatially dependent manufacturing processes.
- A comprehensive feature importance analysis using SHAP values, providing interpretable insights into the complex interplay of AFP process variables and their impact on gap formation, with a particular focus on identifying key factors associated with variation.
- Demonstration of the model's strong predictive performance in estimating gap widths, offering a data-driven pathway towards improved process understanding and the potential for enhanced in-process quality control.

This paper is structured as follows: Section 2 reviews related work in AFP defect detection and relevant Artificial Intelligence (AI) applications. Section 3 details the proposed methodology, including the dataset overview, data preprocessing, the Transformer model architecture with its adapted positional encoding and the experimental setup. Section 4 presents and discusses the gap width estimation results and the insights derived from the SHAP analysis. Finally, Section 5 concludes this paper, summarizing the key findings and suggesting directions for future research.

2. Related Work

The pursuit of enhanced quality control in AFP has spurred significant research into in-process monitoring and defect detection. Various sensor technologies and data analysis techniques have been explored to identify and mitigate defects such as gaps, overlaps and wrinkles, which can compromise the integrity of composite structures.

2.1. Sensor-Based Monitoring and Defect Detection in AFP

A variety of sensors are employed for AFP monitoring. Optical methods, including high-resolution laser scanners and cameras (operating in visible, near-infrared and mid/far-infrared spectra), are widely used to capture surface characteristics, fiber placement accuracy and geometric deviations [2,10]. Laser Line Scanners (LLSs), for instance, have been used to measure tow geometry, edge positions and surface profiles, enabling the detection of gaps, overlaps and other surface anomalies [5,11,12]. Machine vision systems, often coupled with these sensors, apply image processing and computer vision algorithms to segment defect regions and extract critical features [2,12,13]. Beyond optical methods, thermography has also been explored as a distinct research direction for identifying thermal variations indicative of defects or inconsistencies in the layup [12]. Force and pressure sensors integrated into the AFP head can also provide data on compaction quality, which indirectly relates to defect formation [1,3].

While these sensor systems provide rich data, subsequent analysis is crucial. Traditional signal processing techniques (e.g., Fourier analysis) help filter noise and extract defect-correlated features [11]. However, the complexity and volume of data generated, coupled with the subtle nature of some defects, have driven the adoption of more advanced data-driven approaches.

https://doi.org/10.3390/pr14040609

2.2. Machine Learning for AFP Process Monitoring and Defect Analysis

Machine Learning (ML), particularly DL, has shown considerable promise in automating defect detection and classification in AFP. CNNs have been applied to image-based data from cameras and LLSs to identify and classify defects with increasing accuracy [12,14,15]. For instance, studies have demonstrated the use of CNNs for segmenting defect regions from laser scanner depth maps [15] and classifying various defect types using RGB color values from vision systems [16]. Explainable Artificial Intelligence (XAI) methods, such as SHAP, are also beginning to be applied to understand the decision-making processes of these complex models, enhancing trust and providing insights into defect causality [12].

Given the sequential nature of the AFP process, where the state at one point in time or position can influence subsequent states, time-series analysis techniques are particularly relevant. RNNs, including LSTM and Gated Recurrent Unit (GRU) architectures, have been successful in various manufacturing applications for tasks like predictive maintenance, anomaly detection and forecasting, owing to their ability to model temporal dependencies [8,17–21]. Given the sequential nature of AFP sensor data, these architectures provide a relevant baseline for comparison; however, they often lack the explicit spatial interpretability provided by attention mechanisms.

2.3. Challenges and Opportunities

Despite these advancements, several challenges persist. The development of robust AI models is often hampered by the limited availability of large, diverse and well-annotated AFP datasets that cover a wide range of materials, defect types and process conditions [12]. Ensuring data quality from sensors operating in harsh industrial environments is another concern [12,13]. Furthermore, many existing models may not fully capture the long-range dependencies and complex spatio-temporal interrelations present in multivariate AFP sensor data. The computational demands for real-time processing of large data streams also requires efficient algorithms [12,13]. While CNNs are powerful for image-based defect detection, they do not inherently capture the numerical process variable correlations over time. Similarly, while RNNs and LSTMs address sequential dependencies, they process data linearly, which can lead to information bottlenecks when relating the current tow's state to spatially adjacent (but temporally distant) previous tows. This landscape presents an opportunity for novel architectures capable of overcoming these limitations. The Transformer architecture, originally developed for natural language processing [22], has recently shown state-of-the-art performance in time-series forecasting, outperforming traditional RNNs in capturing complex dependencies [23]. The Transformer architecture distinguishes itself by using self-attention to process these relationships in parallel, preserving global context more effectively than recurrent methods. Its attention mechanism allows the model to weigh the importance of different parts of an input sequence dynamically, making it a strong candidate for analyzing the nuanced interplay of sensor data in AFP. However, as noted in our Introduction, its application to AFP gap width estimation remains unexplored, representing a key research gap this study aims to address.

3. Methodology

This section details the dataset, data preprocessing steps, the proposed Transformer-based model architecture, hyperparameter selection and the evaluation metrics used in this study. A schematic overview of the proposed workflow, detailing the transition from data acquisition to model evaluation, is presented in Figure 1.

https://doi.org/10.3390/pr14040609

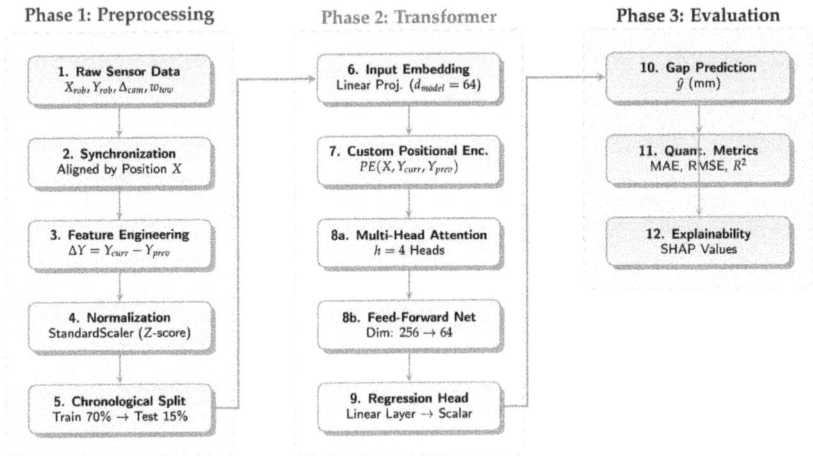

Figure 1. Detailed Methodological Workflow.

3.1. Dataset Overview and Experimental Setup

This work employs a publicly available industrial AFP dataset originally presented by Pantoji et al. [9]. The dataset was generated from experiments conducted at TU Delft's SAM XL lab, with the primary objective of their investigation being the simultaneous measurement of various factors contributing to gap and overlap variations during the AFP layup process. While Pantoji et al. [9] used this data to develop and validate a probabilistic simulation model for predicting gap defects, the present work employs this same dataset to develop and evaluate an alternative, data-driven gap width estimation model using a Transformer-based deep learning approach.

The experimental setup described in [9] consisted of an AFP-XS head from ADD Composites (Finland) mounted on a KUKA KR 210 R2700 robot manipulator. The AFP head was equipped with a silicone compaction roller (31 mm width, 40 mm diameter, Shore hardness 45 A) and an Infrared (IR) heater to enhance tow tackiness. A 6.35 mm wide unidirectional thermoset composite tow (standard modulus carbon fiber with an epoxy matrix) was used. The experiment involved a single continuous campaign of laying 31 straight lines of composite tow, each approximately 1 m long, onto a 1.5 m × 1 m steel plate tool. The deposition speed was up to 200 mm/s. Layups were performed unidirectionally along the X-axis, with the programmed Y-coordinates of each tow varying to create a nominal programmed gap of 6.15 mm between adjacent tows. The Z-coordinate was set to ensure sufficient compaction force, which was measured at 377 N during the experiment; the IR heater was set to 80% power and tape tension was regulated at 6.7 N.

The measurement system, as detailed in [9], included several sensors mounted on the AFP head to record data during deposition (depicted in Figure 2):

- A Laser Tracker (Leica AT960 MR) to measure the AFP head's position.
- A camera (Ximea xiC USB 3.1 Gen1 MC050MG-SY-UB) to capture the tow's position on the compaction roller.
- Two Laser Line Scanners (LLS A: MicroEpsilon 3060-25 and LLS B: MicroEpsilon 3010-25) to measure the tow profile before the compaction roller (LLS A) and after deposition on the surface (LLS B).

https://doi.org/10.3390/pr14040609

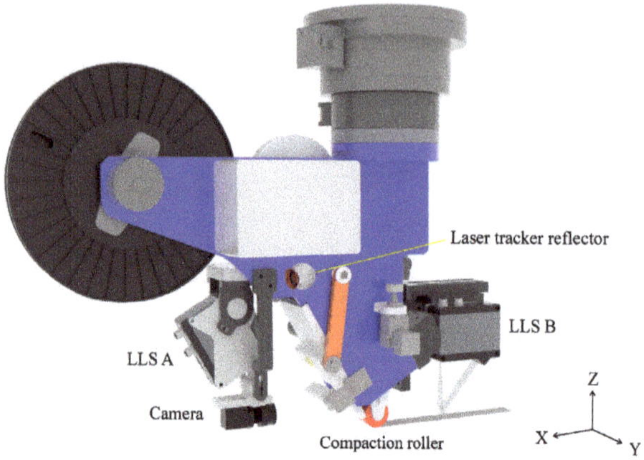

Figure 2. Sensors setup on the AFP head. Layup is in the positive X direction [9].

The target variable for the prediction task in our study is the *actual_gap*. As reported in [9], these actual gap widths were measured by traversing an LLS over the entire length of each pair of adjacently laid tows after all tows had been laid down. This post-process measurement provides the ground truth for the gap dimensions that our model aims to estimate using in-process sensor data. A description of each variable is provided in Table 1.

Table 1. Description of variables present in the dataset.

Variable	Description	Measurement Equipment
tow_width	Measured width of the incoming fiber tow (mm).	LLS A
tow_width_after_compaction	Measured width of the tow after compaction (mm).	LLS B
deviation_camera	Measured lateral movement of the tow on the roller (mm).	Camera
X	Measured X-coordinate of the AFP head position (mm).	Laser tracker
Y	Measured Y-coordinate of the AFP head position (mm).	Laser tracker
deviation_y	Deviation in the Y-direction computed via robot programmed Y-coordinate and laser tracker Y-coordinate (mm).	Laser tracker & Robot Programmed Coordinates
actual_gap	The actual measured gap width between the currently placed tow and the previously placed tow (mm). This serves as the target variable for the prediction task.	Transverse LLS

Robot positioning error followed a logistic distribution with a total range of 0.54 mm (8.5% of specified width) and a standard deviation of 0.07 mm. Tow lateral movement exhibited an Extreme Value distribution with a range of 1.4 mm (22% of tow width) and a standard deviation of 0.22 mm. Tow width variation followed a normal distribution, with a range of 0.63 mm before compaction and 0.48 mm after compaction. Although the nominal programmed gap was 6.15 mm, the combined effect of these variations resulted in measured gaps with a mean of 6.24 mm and a standard deviation of 0.33 mm. Observed

https://doi.org/10.3390/pr14040609

gaps spanned approximately 5.0 mm to 7.3 mm, with the 90th and 99th percentiles at 6.59 mm and 7.21 mm, respectively. This spread captures nominal conditions as well as compliant and defect-prone scenarios.

3.2. Data Preprocessing, Feature Engineering and Data Splitting

The raw data acquired from the AFP sensors underwent several preprocessing steps to prepare it for the Transformer model. These steps included data synchronization, feature engineering, data splitting and normalization, guided by the specific structure of the provided dataset [9] and the requirements of the Transformer architecture.

3.2.1. Data Synchronization and Feature Engineering

A fundamental aspect of the methodology employed in this study was the synchronization of sensor data along the deposition path of the AFP head. Each row in this dataset represents a synchronized set of measurements at a specific longitudinal X-coordinate along the deposition path and encapsulates information from both the current tow segment being deposited and the corresponding segment of the previously laid adjacent tow. This structure inherently provides a comparative context at each spatial location. The data was sorted by the X-coordinate to maintain the deposition sequence.

The following features were engineered or selected as inputs to the model:

- Features related to the previously laid tow:
  - *tow_width_previous*: Measured width of the previous tow (mm).
  - *tow_width_after_compaction_previous*: Measured width of the previous tow after its compaction (mm).
  - *deviation_mm_camera_previous*: Measured lateral movement of the previous tow on the roller during its deposition (mm).
  - *deviation_tracker_previous*: Measured Y-axis deviation of the AFP head from the programmed path during the deposition of the previous tow (mm).
- Features related to the current tow being deposited:
  - *tow_width_current*: Measured width of the current incoming tow (mm).
  - *tow_width_after_compaction_current*: Measured width of the current tow after compaction (mm).
  - *deviation_mm_camera_current*: Measured lateral movement of the current tow on the roller (mm).
  - *deviation_tracker_current*: Measured Y-axis deviation of the AFP head from the programmed path during the deposition of the current tow (mm).
- Engineered relational feature:
  - *rel_y*: The relative Y-coordinate difference between the centerline of the current tow (*Y_current*) and the centerline of the previous tow (*Y_previous*), calculated as $Y\_current - Y\_previous$.

It is important to distinguish this engineered feature (*rel_y*), which captures the *local* relative proximity, from the absolute spatial context provided later by the positional encoding. The latter allows the model to identify *global* dependencies, such as specific regions on the tool or robot arm extensions (correlated with absolute coordinates) that may systematically induce defects, preventing information redundancy.

Feature selection was guided by AFP physics. The variable $deviation_camera$ serves as a direct proxy for material-level effects, including slippage and tack failure at the roller interface. In contrast, $deviation_y$ reflects machine-level kinematic errors of the robot arm. Together, these features allow the model to disentangle material-induced defects from robot-induced inaccuracies.

https://doi.org/10.3390/pr14040609

The target variable for our prediction task is *actual_gap* (mm), representing the actual measured gap width between the current and previous tows at that X-coordinate.

3.2.2. Data Splitting

To ensure a robust evaluation of the model's generalization capabilities and to prevent data leakage, the dataset was split into training, validation and testing sets. Given the sequential and spatial nature of AFP data, where measurements from adjacent locations or subsequent tows are likely correlated, a simple random split is less ideal than a strategy that respects the process order [6]. Therefore, for the results reported in this paper, we adopted a chronological splitting strategy based on the 31 individual tows laid during the experiment referenced in [9].

The data from the 31 tows were divided as follows:

- **Training Set**: Data from the first 22 tows (approximately 70% of the tows) were used for training the model.
- **Validation Set**: Data from the next 4 tows (approximately 15% of the tows) were used for hyperparameter tuning.
- **Test Set**: Data from the final 5 tows (approximately 15% of the tows) were held out and used exclusively for evaluating the performance of the final selected model.

Crucially, this chronological split ensures no data leakage: the model is evaluated on tows that were deposited physically and temporally after the training data, mimicking a real-world production scenario.

All reported performance metrics in Section 4 are based solely on this unseen test set.

To mitigate the impact of differing scales and distributions among the input features and to aid in model convergence, feature normalization was employed using a Standard Scaler, which standardizes features by removing the mean and scaling to unit variance.

3.3. *Transformer Model Architecture and Training*

This section outlines the rationale for selecting the Transformer architecture, details its specific components as implemented for AFP gap width estimation and describes the hyperparameter selection and training procedures.

3.3.1. Model Rationale and Overview

The AFP process, as represented by the feature-engineered dataset, poses a unique modeling challenge in which each spatial point is described by a rich set of features comparing the characteristics of the current tow with those of previously deposited tows. Understanding the contribution of each feature in the context of its precise spatial location (defined by the X and Y coordinates of the current and adjacent tows) is therefore critical for accurate gap width estimation and process interpretability.

Traditional DL approaches for modeling such structured and sequential data, including LSTMs networks, rely on recurrent connections to propagate information through the sequence. While effective in many time-series applications, LSTMs models suffer from information bottlenecks when dependencies span long sequences and their latent representations are often difficult to interpret in terms of physical process variables. These limitations are particularly restrictive in manufacturing applications, where both long-range dependencies and model transparency are essential. This is particularly advantageous in AFP, where a defect might result from a sharp interaction between a specific kinematic deviation and material width, rather than a smooth temporal trend.

The Transformer architecture [22] was chosen for its powerful attention mechanism. Unlike recurrent models, Transformers employ a self-attention mechanism that enables the entire sequence of sensor-derived features to be processed in parallel. This mechanism

https://doi.org/10.3390/pr14040609

assigns learnable attention weights to individual input features irrespective of their position in the sequence, allowing the model to capture long-range and non-local dependencies without recurrence. The self-attention mechanism is employed to model the complex, non-linear interactions between the heterogeneous input features (sensor readings) and their spatial context. By integrating a custom positional encoding (incorporating X and Y coordinates), the self-attention mechanism gains "spatial awareness". Unlike static weights in standard regression, attention dynamically adapts feature relevance based on each data point's spatial context, making it well suited for modeling the complex relationships governing AFP gap formation. Moreover, the explicit attention weights support the interpretability analysis presented in Section 4, providing insight into the physical drivers of gap variability.

The overall model consists of an input embedding layer, a custom positional encoding layer, a multi-layer Transformer encoder and a final linear regression layer to predict the gap width.

3.3.2. Model Architecture Components

The model was implemented using PyTorch (v2.5.1). The architecture comprises the following key components:

1. **Input Embedding:** The input features are projected into a 64-dimensional embedding space using a linear layer.
2. **Custom Positional Encoding:** To incorporate spatial context, unique positional encodings derived from the absolute X-coordinate and the Y-coordinates of both current (*Y_current*) and previous (*Y_previous*) tows are generated and added to the input embeddings. This allows the model to differentiate points based on their manufacturing location.
3. **Transformer Encoder Core:** The spatially aware embeddings are processed by a stack of three Transformer encoder layers. The model is configured with a multi-head self-attention mechanism (4 heads, $d_{model} = 64$) and a position-wise feed-forward network (256 dimensions). To mitigate the risk of convergence to local optima and prevent overfitting, the architecture incorporates Layer Normalization and Residual Connections after each sub-layer. Additionally, a Dropout rate of 0.1 is applied. These mechanisms stabilize the gradient flow, allowing the Adam optimizer to effectively navigate the loss landscape.
4. **Output Regressor:** The 64-dimensional output from the Transformer encoder stack is passed through a final linear layer to predict a single scalar value representing the estimated gap width.

The understanding of spatial relationships and sequential context is not derived from the Transformer processing a long sequence of points over time, as is typical in traditional time-series forecasting using LSTMs. Instead, it is primarily achieved through two key mechanisms. First, each input vector contains rich feature representations that capture comparative information between the current and previous tows at a specific X-coordinate. Second, explicit positional encoding is applied, incorporating unique positional information derived from the *X*, *Y_previous* and *Y_current* coordinates. This encoding enables the model to interpret both the absolute and relative spatial positions of each data point.

While the attention mechanism internally determines these feature interdependencies, we employ SHAP (Section 4) to quantify and visualize the resulting feature contributions, effectively translating the model's internal attention focus into physically interpretable insights for AFP.

https://doi.org/10.3390/pr14040609

3.3.3. Hyperparameter Selection and Training Details

The selection of hyperparameters for the Transformer model was guided by common practices in literature [24] and refined through a systematic grid search on the validation set. The primary criterion for selection was the minimization of Mean Squared Error (MSE) on the dedicated validation set. The search space included: Embedding Dimensions $d_{model} \in \{32, 64, 128\}$; Attention Heads $h \in \{2, 4, 8\}$; Encoder Layers $N \in \{1, 2, 3, 4\}$; and Learning Rates $\in \{0.01, 0.001, 0.0001\}$. The final configuration ($d_{model} = 64$, $h = 4$, $N = 3$) was selected based on minimizing the validation MSE.

The model was trained using the following setup:

- **Loss Function:** Mean Squared Error was used to quantify the difference between the predicted gap width and the actual gap width.
- **Optimizer:** The Adam optimizer was employed with a learning rate of 0.001.
- **Epochs:** The model was trained for 100 epochs.
- **Batch Size:** A batch size of 32 was used.
- **Training Environment:** The model was implemented in Python using the PyTorch (v2.5.1) library. Training was performed on a machine equipped with an Intel i9-13950HX (24 cores), 64 GB of RAM and an NVIDIA RTX 4000 12GB GPU, running Windows 11.

The total training time for 100 epochs was approximately 1 h and 20 min.

4. Results and Discussion

This section presents the evaluation of the proposed Transformer-based model for estimating AFP gap widths. The analysis includes both quantitative metrics and qualitative assessments of prediction accuracy, followed by an exploration of feature importance using SHAP analysis. These insights help to elucidate the model's decision-making process and its relevance for process understanding and control.

All results are based on the model's performance on an unseen test set.

4.1. Transformer Model Performance Evaluation

The model's predictive performance was quantified using several standard evaluation metrics, offering an assessment of its accuracy and generalization ability. The following metrics were employed:

- Mean Absolute Error (MAE) : The average of the absolute differences between predicted and actual values. It is calculated as follows:

$$\text{MAE} = \frac{1}{n}\sum_{i=1}^{n}|y_i - \hat{y}_i|$$

  where y_i is the actual value, $\hat{y}_i$ is the predicted value and n is the number of samples. MAE is expressed in the same units as the target variable (e.g., mm for gap width).
- Mean Squared Error (MSE): The average of the squared differences between predicted and actual values. It is calculated as follows:

$$\text{MSE} = \frac{1}{n}\sum_{i=1}^{n}(y_i - \hat{y}_i)^2$$

  MSE penalizes larger errors more heavily and is expressed in squared units of the target variable (e.g., mm^2).

https://doi.org/10.3390/pr14040609

- Root Mean Squared Error (RMSE): The square root of the MSE, providing an error metric in the same units as the target variable (e.g., mm).

$$\text{RMSE} = \sqrt{\text{MSE}}$$

- R-squared (R^2): The coefficient of determination, representing the proportion of the variance in the dependent variable that is predictable from the independent variables. It ranges from 0 to 1, with higher values indicating a better fit.

$$R^2 = 1 - \frac{\sum_{i=1}^{n}(y_i - \hat{y}_i)^2}{\sum_{i=1}^{n}(y_i - \bar{y})^2}$$

where $\bar{y}$ is the mean of the actual values.
- Mean Absolute Percentage Error (MAPE): The average of the absolute percentage differences between predicted and actual values. It is a dimensionless measure of prediction accuracy.

$$\text{MAPE} = \frac{1}{n}\sum_{i=1}^{n}\left|\frac{y_i - \hat{y}_i}{y_i}\right| \times 100$$

The results are presented in Table 2. The Transformer model achieved an MAE of 0.07540 mm, indicating that, on average, its predictions were within approximately 0.075 mm of the actual gap width on the unseen test data. This corresponds to about 1.2% of the programmed gap width of 6.15 mm. In an industrial context, where gap tolerances are typically in the range of 0.5 mm depending on the structural application [2], this MAE suggests the model provides sufficient precision to identify out-of-tolerance events. The RMSE of 0.1139 mm, which penalizes larger errors more heavily, further underscores the model's precision. A high R^2 value of 0.9143 suggests that the model can explain approximately 91.43% of the variance in the actual gap width, demonstrating a good fit and robust generalization to new data. The MAPE of 1.04% provides a relative measure of this high accuracy. It should be noted that MAPE can be sensitive to small denominator values. Given that the target variable (gap width) captures physical dimensions where sub-millimeter precision is critical, the MAE and RMSE (expressed in mm) provide a physically meaningful representation of the model's suitability for manufacturing tolerances. These results highlight the Transformer architecture's effectiveness in capturing the complex relationships within the AFP sensor data for this specific task.

Table 2. Transformer Model: Performance Metrics.

Metric	Value
MAE	0.07540
MSE	0.013
RMSE	0.1139
R^2	0.9143
MAPE	1.04%

The original study by Pantoji et al. [9], from which the dataset was sourced, employed a probabilistic simulation approach based on fitting statistical distributions to individual measured sources of manufacturing variability. While their method showed a strong correlation between simulated and experimentally measured gap distributions, a direct quantitative comparison of error metrics (like MAE or RMSE) is challenging due to the different modeling paradigms (statistical simulation vs. end-to-end regression). Their work

https://doi.org/10.3390/pr14040609

focused on understanding the contribution of individual error sources to the overall gap distribution, whereas our approach aims to directly estimate the gap width at specific locations using a data-driven model. Our model offers a complementary approach by providing direct point-wise estimations.

To visualize the model's predictive accuracy, the predicted gap widths were plotted against the actual gap widths—Figure 3.

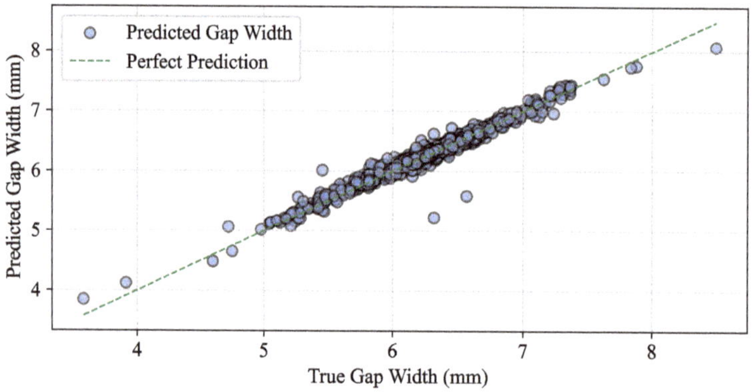

Figure 3. Transformer Model: Predicted Gap Width vs True Gap Width.

The scatter plot in Figure 3 reveals a strong positive correlation between the predicted and actual values, with most data points clustering closely around the perfect prediction line. Some minor deviations are noticeable, particularly for points further from the programmed gap of 6.15 mm. This shows the model's ability to predict gap widths across the observed range accurately.

Furthermore, the predicted and true gap widths were plotted against the sample index for 50 data points—Figure 4. The time-series plot illustrates a good correlation between the predicted and measured gap width trajectories. The model effectively captures general trends and fluctuations in the gap width. Although there are some minor discrepancies, the overall trend indicates a strong predictive capacity within this phase of the deposition process.

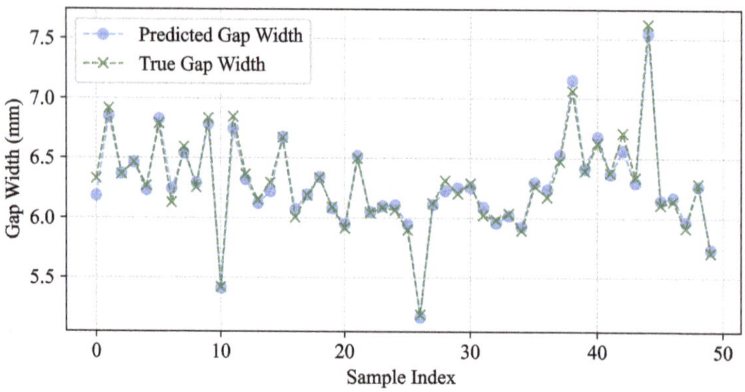

Figure 4. Transformer Model: Predicted Gap Width and True Gap Width for 50 sample points.

The error distribution analysis in Figure 5 shows the prediction errors of the model are mainly centered around zero, with a slight negative skew. The peak frequency occurs

https://doi.org/10.3390/pr14040609

at a marginally negative error value. This suggests a minor tendency to underestimate the width of the gap.

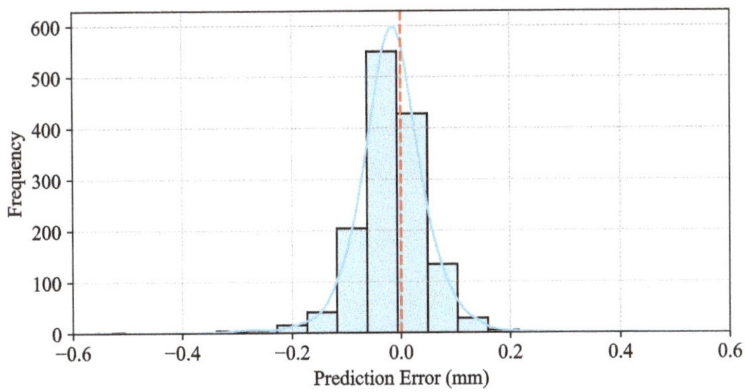

Figure 5. Transformer Model: Distribution of Prediction Errors.

The analysis of prediction error versus predicted gap width (Figure 6) shows a random scatter of residuals around the zero line, with no clear patterns or trends.

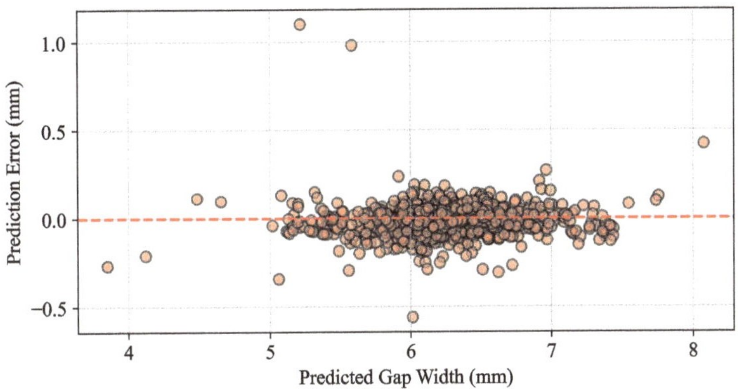

Figure 6. Transformer Model: Prediction Error vs Predicted Value.

4.2. Feature Importance Analysis Using SHAP Values

To gain deeper insights into the model's decision-making process and to understand the influence of different input features on the predicted gap width, a SHapley Additive exPlanations (SHAP) analysis was conducted [25,26]. SHAP is a game-theory-based approach that assigns an importance value (SHAP value) to each feature for each individual prediction, indicating its contribution to pushing the model's output away from a baseline value. This method is particularly valuable for interpreting complex models like Transformers and has been highlighted as suitable for AFP applications [6].

4.2.1. Scope of SHAP Analysis

Four variables were examined: tow width (*tow_width*), tow width after compaction (*tow_width_after_compaction*), tow lateral movement on the roller (*deviation_camera*), and y-axis deviation (*deviation_y*). These features were selected as they represent measurable parameters directly associated with the tow being laid and are commonly investigated in AFP process analysis.

https://doi.org/10.3390/pr14040609

SHAP values quantified the contribution of each feature to individual predictions, allowing for an interpretation of the model's learned behavior on both global and local scales. It is important to note that, while SHAP values reveal the influence of features on model predictions, the feature values displayed in SHAP visualizations reflect a standardized feature space. This is a result from normalization during model training and thus are not directly representative of their original quantitative values.

4.2.2. Global Feature Importance

The global importance of each feature was initially assessed using a SHAP summary plot—Figure 7. This plot displays the average magnitude of the SHAP values and revealed that all four investigated features contribute to the model's predictions.

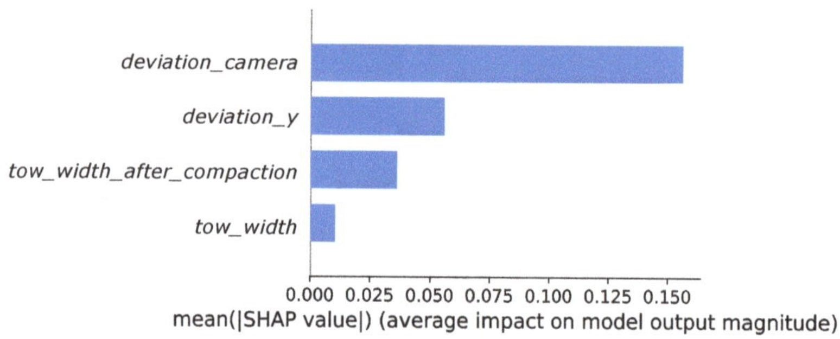

Figure 7. SHAP global feature importance (average impact on output) for predicted gap width.

As shown in Figure 7, *deviation_camera* (lateral movement of the current tow on the roller) exhibits the highest average SHAP value, indicating it has the most substantial impact on the model's gap width predictions among the visualized features. This is followed by *deviation_y* (AFP head's Y-axis deviation for the current tow), then *tow_width_after_compaction* (current tow's width after compaction) and finally *tow_width* (current incoming tow width). This hierarchy suggests that variations in tow placement accuracy (*deviation_camera* and *deviation_y*) are strongly associated with predicted gap width, aligning with physical expectations of the AFP process.

4.2.3. Dependence Plots and Feature Interactions

Next, SHAP dependence plots (Figure 8) were devised to provide granular look into the relationship between each feature's value and its corresponding SHAP value. These plots help to visualize how each feature influences the model's predictions and to identify potential interactions with other features.

The dependence plot for *tow_width* (Figure 8a) exhibits a slightly positive overall trend, although with scatter in the data points. This suggests that, as *tow_width* increases, the predicted gap width tends to increase as well. However, the scatter indicates that this relationship is not strictly linear and that other factors may be influencing the impact of *tow_width* on the prediction. Additionally, the dependence plot presents further information through a color scale. In this case, the color scale represents *deviation_camera* and reveals a trend: lower values of *deviation_camera* (blue dots) are associated with lower SHAP values for *tow_width*, while higher values of *deviation_camera* (pink dots) indicate higher SHAP values. This suggests a potential positive influence of *tow_width* on the predicted gap width appears to be more pronounced when *deviation_camera* is high. This could imply that when the camera deviation is significant, even small increases in tow width can lead to larger gaps, perhaps because the placement is already compromised.

https://doi.org/10.3390/pr14040609

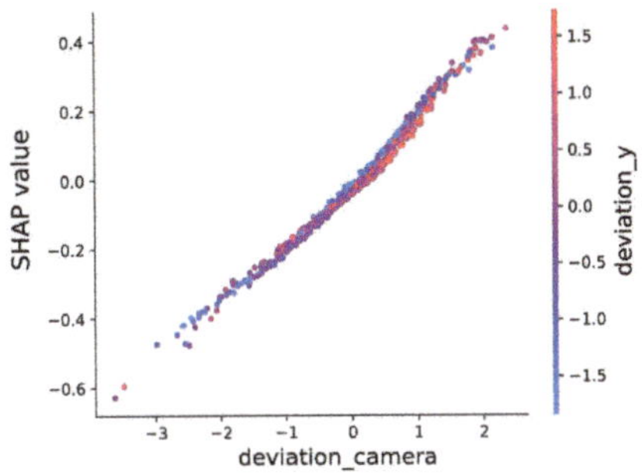

(a) Tow Width

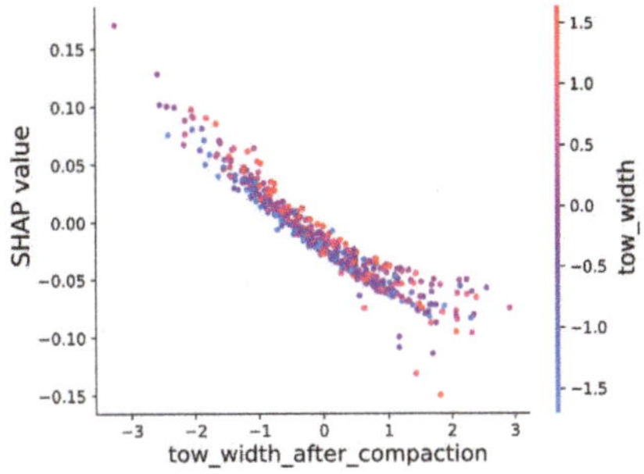

(b) Tow Width After Compaction

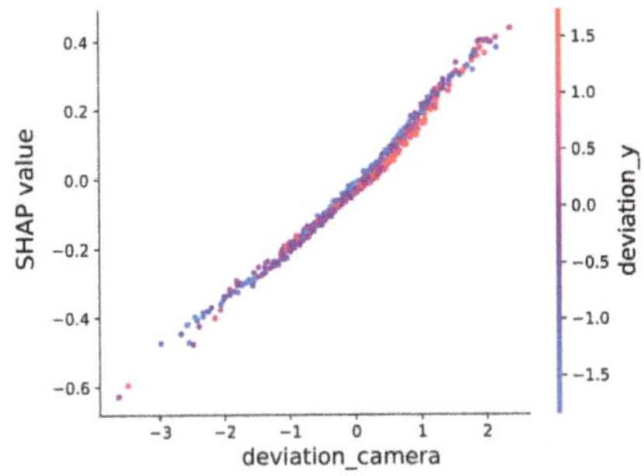

(c) Deviation Camera

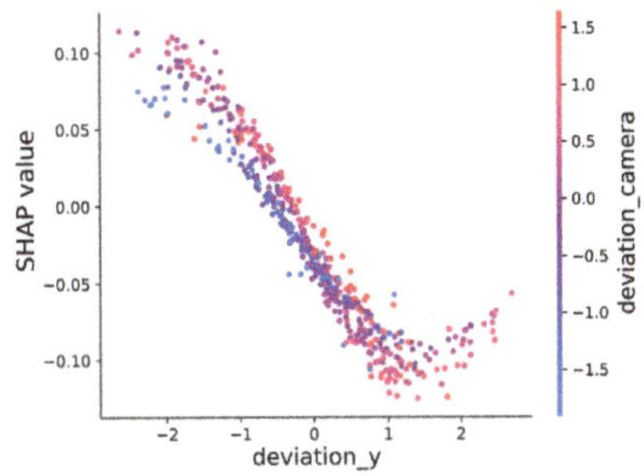

(d) Deviation Y

Figure 8. SHAP dependence plots for predicted gap width.

The dependence plot for *deviation_camera* (Figure 8c) exhibits a highly linear positive relationship with minimal scatter. The data points closely follow a straight line with a positive slope, indicating that the increase in *deviation_camera* has a high positive contribution to the predicted gap width. This dominance aligns with the physical reality of the AFP process: the camera measures the lateral position of the tow on the compaction roller immediately before deposition. Unlike *deviation_y*, which measures the robot head's position in the global coordinate system, *deviation_camera* captures the actual material slippage relative to the tool center point. The model has correctly identified that this material-tool interface measurement is the most direct physical determinant of the final placement accuracy, effectively serving as a proxy for the true deposition path. This observation is consistent with the findings reported in [9].

In contrast, the dependence plot for *deviation_y* (Figure 8d) reveals a perceptible negative linear trend with some scatter. While deviations in the robot's Y-position generally correlate with gap changes, the relationship is less precise. Physically, this suggests that the robot's encoder values are a secondary proxy; a robot can be perfectly positioned, but if the tow slips on the roller (as captured by the camera), a defect could still occur. The Transformer's attention mechanism successfully discerns this hierarchy, prioritizing the direct material measurement (*deviation_camera*) over the kinematic proxy (*deviation_y*).

Regarding the geometric features, the plot for *tow_width_after_compaction* (Figure 8b) displays a clear negative linear trend. This represents a direct geometric relationship: effective compaction contributes to tow lateral spread. As the tow spreads wider (*tow_width_after_compaction* increases), it physically occupies more space on the substrate, naturally reducing the magnitude of the gap between itself and the adjacent tow.

https://doi.org/10.3390/pr14040609

The *tow_width* dependence plot suggests an interaction with *deviation_camera*. The contribution of *tow_width* seems to be more significant when *deviation_camera* is already high, indicating a potential compounding effect of these two factors. In addition, *tow_width_after_compaction* and *deviation_camera* exhibit relatively strong linear trends (negative and positive, respectively), indicating clear and consistent relationships with the predicted gap width. The model is evidently sensitive to tow lateral movement on the roller (*deviation_camera*) and the effectiveness of compaction (*tow_width_after_compaction*), which aligns with the physical realities of the AFP process.

While Figure 8a shows that increasing the incoming *tow_width* is associated with larger gaps, Figure 8b reveals that a greater *tow_width_after_compaction* corresponds to smaller gaps, an apparent contradiction that can be reconciled by considering the different stages and roles these variables represent in the deposition process.

The *tow_width_after_compaction* is an outcome measure; a larger value here directly and intuitively means more material is filling the intended space, hence a smaller gap. The incoming *tow_width*, however, is an input material characteristic. Its influence on the final gap is not direct but is mediated by the deposition process. As seen in Figure 8a, the positive trend (wider incoming tow->larger gap) is particularly evident when *deviation_camera* is high. This suggests that if a wider tow is not placed accurately, it might be more prone to undesirable behaviors (e.g., slight buckling, edge lifting or simply deviating further due to its width if guidance is imperfect) that effectively increase the gap or the model learns that instances with wider tows in this dataset often coincided with other unideal placement conditions. Conversely, if placement is perfect (low *deviation_camera*), the impact of incoming tow width might be neutral or even slightly negative (gap reducing).

4.2.4. Beeswarm Plot Analysis

The SHAP beeswarm plot (Figure 9) provides a global overview of feature importance and their impact on the model's predictions. Each point on the plot represents an instance in the dataset, with its horizontal position indicating the SHAP value for a particular feature and its color representing the corresponding feature value (red for high, blue for low).

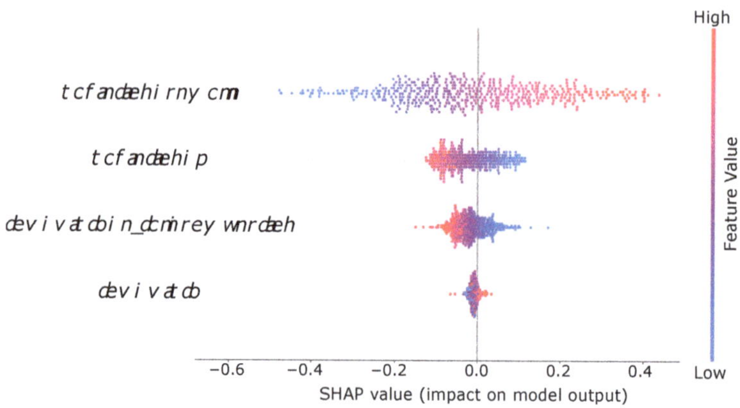

Figure 9. SHAP's summary plot for predicted gap width.

The *deviation_camera* feature is the most influential and presents a wide spread of its SHAP values. A clear trend is observable: high *deviation_camera* values (red dots) have a high positive contribution to predicted gap width. Conversely, low *deviation_camera* values (blue dots) have a negative contribution to the prediction. This confirms the critical role of tow lateral movement on the roller.

https://doi.org/10.3390/pr14040609

The second most important feature, *deviation_y*, exhibits a negatively skewed distribution. High values of *deviation_y* (red dots) are associated with negative SHAP values and negatively contribute to the predicted gap width. Intermediate values (purple dots) of *deviation_y*, indicating proximity to the intended path, are associated with SHAP values close to zero, having a small contribution to the prediction. This highlights the importance of maintaining the tow's position along the intended y-axis.

The feature *tow_width_after_compaction* shows a moderate impact, with a distribution skewed slightly towards negative SHAP values. High *tow_width_after_compaction* values (red dots) are primarily associated with negative contribution towards predicted gap width, indicating that effective compaction indicates smaller gaps.

The *tow_width* exhibits the lowest importance among the four features. It has a small distribution and slightly skewed towards negative SHAP values. This suggests that, within the range of this dataset, variations in *tow_width* have a less pronounced impact, compared to other factors.

4.2.5. Insights from SHAP Analysis

The SHAP analysis provides valuable, interpretable insights into the Transformer model's behavior, aligning with and extending physical understanding of the AFP process. It confirms that tow placement accuracy (*deviation_camera* and *deviation_y*) are critical factors influencing gap formation. The analysis also shows the nuanced role of tow width, distinguishing between the effect of incoming tow width (conditional on placement) and the more direct effect of the tow's width after compaction. This level of interpretability demonstrates the potential of combining advanced DL models with Explainable AI techniques like SHAP to not only predict but also to better understand and ultimately improve complex manufacturing processes.

4.3. Estimation for Process Control and Practical Implications

The development of an accurate gap width estimation model, as demonstrated in this work, carries significant implications for enhancing AFP process control and overall part quality. This section discusses the nature of the model's estimation, its potential for near-real-time application and its practical value in an industrial AFP setting.

4.3.1. Nature of the Gap Width Estimation

It is important to clarify that the model developed in this study provides an estimation of the gap width concurrently with the deposition process. It does not predict the gap width far ahead of the deposition point in a forecasting manner. Instead, for each spatial data point along the deposition path (defined by the X-coordinate and incorporating features from both current and previous tows), the model leverages the sensor data collected up to and including that specific location to estimate the gap width. The model effectively learns the complex, non-linear relationships between the rich set of input features and the resulting gap width that is later measured post-process.

The transformer architecture, and particularly its attention mechanism, is well-suited to this task. Although the estimation is concurrent, the model makes effective use of the sequence of past sensor data to provide context for the current estimation. The attention mechanism enables the model to discern which preceding sensor readings are most relevant to the current gap width, effectively capturing the complex interplay of factors that influence gap formation.

https://doi.org/10.3390/pr14040609

4.3.2. Potential for Near-Real-Time Feedback and Practical Value

Although LLS B already measures tow width after compaction, from which gap information could be derived, the value of the proposed model-based estimation lies in several aspects:

1. **Estimation at the Point of Compaction**: Direct, instantaneous physical measurement of the gap at the precise nip point of the compaction roller is often obstructed by the roller itself. LLS B, which measures tow width after compaction, is typically positioned slightly downstream from the roller [9]. Consequently, LLS B's measurement occurs at a slightly later point, both spatially and temporally. Our model, by learning from upstream and current sensor data, provides an indirect estimation of the gap width as it is being formed at or very near the point of compaction. This offers an earlier insight into potential deviations than a solely downstream measurement might provide.
2. **Sensor Fusion and Robustness**: The model inherently performs a type of sensor fusion, integrating information from multiple sensors (LLS A, LLS B, camera, laser tracker). This multi-modal input can lead to a more robust estimation, potentially less susceptible to the noise or limitations of a single sensor in an industrial environment. It learns the collective influence of various parameters, offering a more holistic view of the conditions leading to gap formation.
3. **Diagnostic Insights**: Coupled with SHAP analysis, the model not only estimates the gap but also provides insights into why variations might be occurring. If the model predicts an undesirable gap, the SHAP values can indicate which input features are the primary drivers. This diagnostic capability is valuable for targeted corrective actions and deeper process understanding, which a simple geometric calculation from LLS B might not offer.

Regarding real-time applicability, the Transformer model, once trained, can perform inference relatively quickly. Given the model's architecture and the computational hardware used for training, the inference time for a single data point (or a batch of points) is expected to be in the order of milliseconds. While a precise inference time per prediction was not exhaustively benchmarked in this study, the computational efficiency of feed-forward passes in trained Transformer models is well-established. This suggests a strong *potential* for near-real-time feedback.

However, the implementation and validation of a closed-loop control system incorporating this model for real-time, in-process adjustments is a significant undertaking and is considered a direction for future work. Achieving true real-time control would require integrating the model into the AFP machine's control architecture, ensuring synchronized data pipelines and developing control logic to act upon the model's estimations, all of which are beyond the scope of the current study, which focuses on the development and validation of the estimation model itself. In summary, this model-based estimation approach offers a valuable complement to direct sensor measurements.

Notably, the model was developed and validated using data from straight layups of a single material type on a flat tool, focusing specifically on the first ply deposition. Consequently, the results should be viewed as a proof of concept. The model's generalization to complex curvatures, steering paths, different tackiness levels or multi-ply buildup has not yet been assessed and remains a subject for future research.

5. Conclusions

This work successfully addressed the challenge of accurately estimating gap widths in AFP by developing and evaluating a novel Transformer-based deep learning architecture. Using a publicly available industrial dataset [9], this study demonstrated the potential of

https://doi.org/10.3390/pr14040609

this advanced AI approach to model the complex, non-linear relationships inherent in AFP sensor data.

The proposed Transformer model demonstrated strong predictive performance, achieving an MAE of 0.07540 mm and an R^2 value of 0.9143 on the unseen test set. This indicates a high degree of accuracy in estimating gap widths, corresponding to an average deviation of 1.04% (MAPE) from the actual values. Furthermore, a comprehensive SHAP analysis provided valuable insights into the model's decision-making process, highlighting that tow placement accuracy is a critical factor influencing gap formation. The SHAP analysis also elucidated the nuanced impact of incoming tow width versus tow width after compaction, revealing complex interactions with placement parameters.

The key contributions of this research are:

- The first-known application of a Transformer-based deep learning architecture for gap width estimation in the AFP process, demonstrating its suitability for this manufacturing challenge.
- The development of a domain-specific positional encoding scheme that effectively incorporates the critical X and Y spatial coordinates of both current and previous tows, providing essential context to the model.
- The use of SHAP analysis to interpret the Transformer model, offering insights into the drivers of gap formation and validating the model's learned behavior against physical process understanding.
- The demonstration of a robust data-driven methodology for AFP process monitoring that has the potential to enhance understanding and move towards more proactive quality control.

While the results are promising, this study serves as a proof of concept and has certain limitations. The model was developed and validated using data from straight layups of a single material type on a specific AFP system, focusing on the first ply deposition with a specific set of experimental conditions. Therefore, the model's generalization capabilities to more complex scenarios, such as curved parts, different composite materials, varying process parameters and the cumulative effects across multiple plies, require further investigation. Additionally, while this work demonstrated the feasibility and interpretability of the Transformer, a comprehensive benchmarking study comparing these results against traditional architectures (e.g., LSTM) on identical data splits remains a necessary future step to strictly quantify performance gains. While such comparisons present avenues for future research, the primary objective of this study was to demonstrate the superior *interpretability* of attention maps in decoding spatial defect dependencies, a capability often obscured in traditional "black-box" sequential models, rather than solely optimizing for the lowest possible regression error.

Future work should focus on several key areas. Firstly, extending the model and evaluation to encompass a broader range of AFP conditions, including complex geometries and diverse materials, is crucial for assessing wider industrial applicability. Secondly, the promising inference speed of the Transformer model warrants further investigation into its real-time implementation. This would involve integrating the model into an AFP system's control loop to enable in-process adjustments and move towards adaptive, closed-loop quality control, which remains a significant but important challenge. Further refinement of feature engineering and exploration of additional sensor inputs could also enhance predictive accuracy.

In conclusion, this study establishes the Transformer architecture as a tool for AFP gap width estimation. By providing accurate and interpretable feedback on defect formation, this research contributes to the ongoing efforts to transform AFP into a more efficient,

https://doi.org/10.3390/pr14040609

reliable and self-correcting manufacturing paradigm, ultimately supporting the wider adoption of high-performance composite structures.

Author Contributions: D.C.: Conceptualization, Methodology, Software, Formal analysis, Data Curation, Writing—Original Draft, Writing—Review & Editing. A.R.S.: Conceptualization, Methodology, Writing—Review & Editing, Supervision. N.C.: Conceptualization, Writing—Review & Editing, Supervision. All authors have read and agreed to the published version of the manuscript.

Funding: This research received no specific grant from any funding agency.

Data Availability Statement: Data will be made available on request.

Conflicts of Interest: The authors declare no conflicts of interest.

References

1. Brasington, A.; Sacco, C.; Halbritter, J.; Wehbe, R.; Harik, R. Automated fiber placement: A review of history, current technologies, and future paths forward. *Compos. Part C Open Access* **2021**, *6*, 100182. [CrossRef]
2. Meister, S.; Wermes, M.A.M.; Stuve, J.; Groves, R.M. Review of image segmentation techniques for layup defect detection in the Automated Fiber Placement process: A comprehensive study to improve AFP inspection. *J. Intell. Manuf.* **2021**, *32*, 2099–2119. [CrossRef]
3. Rudberg, T.; Nielson, J.; Henscheid, M.; Cemenska, J. Improving AFP cell performance. *SAE Int. J. Aerosp.* **2014**, *7*, 317–321. [CrossRef]
4. Del Rossi, D.; Cadran, V.; Thakur, P.; Palardy-Sim, M.; Lapalme, M.; Lessard, L. Experimental investigation of the effect of half gap/half overlap defects on the strength of composite structures fabricated using automated fibre placement (AFP). *Compos. Part A Appl. Sci. Manuf.* **2021**, *150*, 106610. [CrossRef]
5. Yadav, N.; Schledjewski, R. Review of in-process defect monitoring for automated tape laying. *Compos. Part A Appl. Sci. Manuf.* **2023**, *173*, 107654. [CrossRef]
6. Meister, S.; Wermes, M.; Stüve, J.; Groves, R.M. Investigations on Explainable Artificial Intelligence methods for the deep learning classification of fibre layup defect in the automated composite manufacturing. *Compos. Part Eng.* **2021**, *224*, 109160. [CrossRef]
7. Stojkovic, M.; Butt, J. Industry 4.0 Implementation Framework for the Composite Manufacturing Industry. *J. Compos. Sci.* **2022**, *6*, 258. [CrossRef]
8. Lachekhab, F.; Benzaoui, M.; Tadjer, S.A.; Bensmaine, A.; Hamma, H. LSTM-Autoencoder Deep Learning Model for Anomaly Detection in Electric Motor. *Energies* **2024**, *17*, 2340. [CrossRef]
9. Pantoji, S.; Kassapoglou, C.; Peeters, D. Predicting gaps and overlaps in Automated Fiber Placement composites by measuring sources of manufacturing process variations. *Compos. Part B Eng.* **2025**, *291*, 111891. [CrossRef]
10. Druiff, P. Data-Driven Manufacturing in the Automated Lay-Up of Composites. Ph.D. Thesis, University of Bristol, Bristol, UK, 2023; 466p.
11. Schmitt, R.; Mersmann, C.; Damm, B. *In-Process 3D Laser Measurement to Control the Fiber Tape-Laying for Composite Production*; SPIE: San Francisco, CA, USA, 2010; Volume 7723, pp. 216–225.
12. Schuster, A.; Mayer, M.; Willmeroth, M.; Brandt, L.; Kupke, M. Inline quality control for thermoplastic automated fibre placement. *Procedia Manuf.* **2020**, *51*, 505–511. [CrossRef]
13. Sacco, C.; Radwan, A.B.; Anderson, A.; Harik, R.; Gregory, E. Machine learning in composites manufacturing: A case study of Automated Fiber Placement inspection. *Compos. Struct.* **2020**, *250*, 112514. [CrossRef]
14. Tang, Y.; Wang, Q.; Cheng, L.; Li, J.; Ke, Y. An in-process inspection method integrating deep learning and classical algorithm for automated fiber placement. *Compos. Struct.* **2022**, *300*, 116051. [CrossRef]
15. Zambal, S.; Heindl, C.; Eitzinger, C.; Scharinger, J. *End-to-End Defect Detection in Automated Fiber Placement Based on Artificially Generated Data*; SPIE: San Francisco, CA, USA, 2019; Volume 11172, pp. 371–378.
16. Sacco, C.; Brasington, A.; Saidy, C.; Kirkpatrick, M.; Halbritter, J.; Wehbe, R.; Harik, R. On the effect of manual rework in AFP quality control for a doubly-curved part. *Compos. Part B Eng.* **2021**, *227*, 109432. [CrossRef]
17. Pagano, D. A predictive maintenance model using long short-term memory neural networks and Bayesian inference. *Decis. Anal. J.* **2023**, *6*, 100174. [CrossRef]
18. Wang, Z.; Zhou, C.; Liu, Y.; Huang, K.; Yang, C. Cluster-based industrial KPIs forecasting considering the periodicity and holiday effect using LSTM network and MSVR. *Adv. Eng. Inform.* **2023**, *56*, 101916. [CrossRef]
19. Chabane, A.N.; Sahnoun, M.; Bettayeb, B. Forecasting KPIs of Production Systems Using LSTM Networks. In Proceedings of the 2021 1st International Conference on Cyber Management and Engineering (CyMaEn), Virtual, 26–28 May 2021; IEEE: Piscataway, NJ, USA, 2021; pp. 1–6.

https://doi.org/10.3390/pr14040609

20. Zhao, R.; Wang, D.; Yan, R.; Mao, K.; Shen, F.; Wang, J. Machine health monitoring using local feature-based gated recurrent unit networks. *IEEE Trans. Ind. Electron.* **2017**, *65*, 1539–1548. [CrossRef]
21. Xu, D.; Zhang, Z.; Shi, J. Failure prediction using gated recurrent unit and autoencoder in complex manufacturing process. In Proceedings of the 2022 4th World Symposium on Artificial Intelligence (WSAI), Jilin, China, 23–25 June 2022; IEEE: Piscataway, NJ, USA, 2022; pp. 68–73.
22. Vaswani, A. Attention is all you need. In Proceedings of the 31st Conference on Neural Information Processing Systems (NIPS 2017), Long Beach, CA, USA, 4–9 December 2017.
23. Wen, Q.; Zhou, T.; Zhang, C.; Chen, W.; Ma, Z.; Yan, J.; Sun, L. Transformers in time series: A survey. *arXiv* **2022**. [CrossRef]
24. Chen, Y.; Song, X.; Lee, C.; Wang, Z.; Zhang, R.; Dohan, D.; Kawakami, K.; Kochanski, G.; Doucet, A.; Ranzato, M.; et al. Towards learning universal hyperparameter optimizers with transformers. *Adv. Neural Inf. Process. Syst.* **2022**, *35*, 32053–32068.
25. Lundberg, S.M.; Lee, S. A unified approach to interpreting model predictions. *arXiv* **2017**. [CrossRef]
26. Ponce-Bobadilla, A.V.; Schmitt, V.; Maier, C.S.; Mensing, S.; Stodtmann, S. Practical guide to SHAP analysis: Explaining supervised machine learning model predictions in drug development. *Clin. Transl. Sci.* **2024**, *17*, e70056. [CrossRef] [PubMed]

https://doi.org/10.3390/pr14040609

processes

MDPI

Article

A Sustainability-Aware Federated Graph Attention Framework for Supply Chain Process Modeling

Vasileios Alexiadis [1], Maria Drakaki [1,*] and Panagiotis Tzionas [2]

[1] Department of Science and Technology, University Center of International Programmes of Studies, International Hellenic University, 14th km Thessaloniki-N. Moudania, GR-57001 Thermi, Greece; valexiadis@ihu.edu.gr

[2] Department of Industrial Engineering and Management, International Hellenic University, P.O. Box 141, GR-57400 Thessaloniki, Greece; ptzionas@ihu.gr

* Correspondence: mdrakaki@ihu.gr; Tel.: +30-2310-807524

Abstract

Modern supply chains operate as highly interconnected networks characterized by decentralization, data silos, and increasing sustainability constraints. Although Graph Neural Networks (GNNs) have demonstrated strong capability in modeling relational dependencies in such systems, their deployment is often restricted by limited inter-organizational data sharing. Federated learning (FL) enables collaborative model training without exposing proprietary data; however, existing federated approaches rarely integrate graph structure and sustainability objectives within a unified framework. This study proposes a Sustainability-Aware Federated Graph Attention Network (FedGAT) for decentralized supply chain process modeling. The framework combines Graph Attention Networks with federated optimization and introduces an emission-weighted attention modulation mechanism that embeds environmental considerations directly into the message-passing process. A multi-tier synthetic supply chain benchmark is constructed to evaluate the approach under realistic governance and data-locality constraints. Experiments are conducted across multiple random seeds, graph scales (up to 500 nodes), and client partition settings. Results show that while centralized graph learning achieves the lowest prediction error, the proposed sustainability-aware federated model maintains statistically indistinguishable predictive performance compared to standard federated baselines (paired sign test $p = 1.000$), while systematically reducing attention allocated to high-emission transport links. A structured label sensitivity analysis confirms that performance gains are not attributable to circular label construction. Furthermore, a λ-ablation study demonstrates a smooth and controllable trade-off between predictive accuracy and sustainability alignment through a single governance parameter. These findings establish the feasibility of privacy-preserving, sustainability-modulated graph learning for decentralized supply chain analytics and provides a principled foundation for environmentally aligned AI deployment in multi-enterprise networks.

Keywords: sustainability-aware federated learning; Graph Attention Networks; Federated Graph Neural Networks; supply chain process modeling; privacy-preserving analytics; carbon-aware message passing

1. Introduction

Foundational work in industrial systems modeling established that production, warehousing, and supply chain operations are dynamic, interdependent, and event-driven, motivating formal modeling and intelligent control. Agent-based systems combined with Colored Petri Nets (CPNs) were shown to capture this complexity effectively, improving coordination in warehouse order picking and resource allocation under demand uncertainty [1]. These approaches were extended to multi-echelon supply chains, where hierarchical CPNs enabled the analysis of inventory propagation, order variability, and disruption effects across tiers [2]. Empirical studies further demonstrated that inventory inaccuracy significantly degrades service levels and amplifies upstream instability, highlighting the need for system-wide coordination [3]. Reinforcement learning integrated with timed CPNs enabled adaptive manufacturing scheduling under stochastic disturbances, reinforcing the view of supply chains as networked decision systems with cascading effects [4,5].

Building on this trajectory, modern supply chains are increasingly modeled as large-scale networks of suppliers, manufacturers, distributors, and retailers, where efficiency, resilience, and sustainability are critical. AI-based process modeling has therefore become central to supply chain optimization. In this context, Graph Neural Networks (GNNs) are particularly suitable, as supply chains can be naturally represented as graphs with entities as nodes and material, informational, or financial dependencies as edges [6]. Empirical evidence shows that GNN-based models outperform conventional statistical and deep learning baselines on tasks such as forecasting and classification, often achieving substantial improvements by learning non-linear structural dependencies and propagation effects [6].

Attention-based graph models further enhance supply chain learning. Graph Attention Networks (GATs) assign non-uniform weights to neighboring nodes during aggregation, enabling models to emphasize critical supplier–customer relationships while suppressing less relevant interactions [7]. Beyond predictive gains, attention mechanisms improve interpretability by revealing influential dependencies, and recent GAT-based resilience models demonstrate strong disruption–classification performance while exposing structural vulnerabilities through learned attention patterns [8].

Despite these advances, real-world deployment is constrained by data decentralization and confidentiality. Supply chain data are distributed across organizations and are rarely shareable due to commercial, contractual, and regulatory restrictions, rendering centralized GNN training impractical in multi-enterprise settings [9]. Federated learning (FL) addresses these constraints by enabling collaborative model training without exchanging raw data. The foundational Federated Averaging approach demonstrated that high-quality neural models can be trained across decentralized, non-IID data while retaining data locally [10]. In industrial ecosystems, FL enables data-limited manufacturers and logistics partners to benefit from collective learning under confidentiality constraints [11].

A further driver shaping supply chain analytics is the need to integrate sustainability objectives. Supply chains contribute significantly to greenhouse gas emissions and resource use, motivating sustainability-aware AI that incorporates metrics such as carbon emissions or energy intensity directly into learning and decision-making. AI-driven logistics and operational optimization have been associated with measurable environmental benefits, including substantial CO_2 reductions [12]. In adjacent domains, federated learning combined with control and optimization—such as reinforcement learning in energy systems—has achieved simultaneous cost and emissions reductions through decentralized coordination [13]. Nonetheless, most supply chain ML systems remain predominantly single-objective, and explicit sustainability integration during training is still uncommon, particularly under decentralization constraints.

Consequently, important gaps remain. Centralized GNNs struggle with data silos and governance, while conventional FL is typically designed for tabular data and does not naturally accommodate graph-structured dependencies spanning organizational boundaries. Only recently has federated graph learning been applied to supply chain visibility,

https://doi.org/10.3390/pr14050781

risk modeling, and demand–inventory prediction, including early attempts to integrate sustainability considerations [14,15]. These studies demonstrate feasibility but also underscore the absence of a unified framework jointly addressing decentralization, graph-based interdependence modeling, and sustainability-aware optimization.

Despite recent advances in federated graph learning, existing approaches remain insufficient for sustainability-critical supply chain modeling for three primary reasons. First, most federated GNN frameworks optimize predictive objectives without incorporating an explicit and controllable sustainability bias within the message-passing mechanism. Second, they provide limited governance-oriented interpretability, offering no structured means to regulate the trade-off between predictive accuracy and environmental alignment. Third, sustainability signals are typically treated as auxiliary features rather than as structural modifiers of relational influence.

To address these limitations, we propose a federated sustainability-aware graph attention framework in which transport-related environmental signals directly modulate attention weights during aggregation. A tunable parameter λ enables explicit control over the strength of sustainability penalization, thereby introducing a governance-relevant trade-off between predictive performance and environmental alignment. This transforms sustainability from a passive feature into an active structural constraint within decentralized graph learning.

The remainder of the article is organized as follows. Section 2 reviews related work on GNNs, federated learning, sustainability-aware AI, and federated graph learning. Section 3 presents the methodology and system architecture. Section 4 reports experimental results. Section 5 discusses implications and limitations, and Section 6 concludes with directions for future research.

2. Related Work

To contextualize our proposed framework, we survey the relevant literature in four key areas. First, we discuss how Graph Neural Networks have been applied in supply chain modeling and optimization, highlighting their benefits over traditional methods. Next, we review federated learning approaches in industrial domains, focusing on how privacy-preserving collaboration has been enabled in manufacturing and supply chain scenarios. Third, we examine research on sustainability-aware AI methods that integrate environmental objectives into model training or decision-making, including multi-objective optimization techniques. Finally, we cover the relatively few works that have attempted to combine GNNs with federated learning, noting their achievements and limitations.

2.1. Graph Neural Networks in Supply Chain Management

Supply chains are inherently graph-structured systems, with nodes representing firms or facilities and edges encoding material, contractual, transportation, or information dependencies. This structure makes Graph Neural Networks (GNNs) particularly suitable, as they learn over non-Euclidean data by propagating information across network topology. Unlike independent-sample models, GNNs capture higher-order interdependencies and cascading effects, enabling them to model how local disruptions propagate across multiple supply chain tiers through message passing [6].

Empirical evidence consistently shows that GNNs outperform classical time-series and conventional machine learning approaches in supply chain tasks such as demand forecasting, inventory control, risk assessment, anomaly detection, and network reconstruction. Wasi et al. report performance gains typically in the range of 10–30%, attributable to GNNs' ability to exploit shared structural context (e.g., common upstream suppliers) and enable cross-entity generalization [6]. Industrial studies further confirm these advantages,

https://doi.org/10.3390/pr14050781

particularly under disruption scenarios where structural dependencies dominate system behavior, as demonstrated in automotive supply networks by Gupta et al. [16].

Beyond accuracy, GNNs provide structural interpretability, which is increasingly critical for decision support, risk governance, and regulatory compliance. Graph Attention Networks (GATs) enhance this capability by learning adaptive attention weights over neighbors, allowing models to emphasize influential entities and relationships. Attention-based graph models have been shown to improve both predictive performance and interpretability in large-scale supply networks, disruption prediction, and manufacturing resilience applications [17–19].

GNNs are also effective for uncovering latent or incomplete supply chain structure. Prior work formulates supply chain visibility as a link-prediction problem, showing that GNNs can infer hidden supplier relationships and indirect dependencies absent from enterprise records [20]. Extensions that integrate graph embeddings with heterogeneous data sources, such as textual risk signals, further improve systemic risk identification in ICT and critical infrastructure supply chains [21].

Despite these advances, most GNN-based supply chain studies assume access to a centralized and fully observable supply chain graph. In practice, this assumption is often invalid due to confidentiality, data-protection regulation (e.g., GDPR), contractual restrictions, and competition law, which collectively limit cross-enterprise data sharing. Additionally, existing GNN applications predominantly optimize predictive or operational performance, while sustainability objectives—such as emissions intensity or energy use—are typically treated as ex-post evaluation criteria rather than embedded learning targets.

These limitations motivate decentralized graph learning approaches that operate under partial visibility and organizational boundaries, while supporting multi-objective optimization aligned with emerging sustainability and regulatory requirements. This gap is addressed in subsequent sections through the integration of federated learning, graph attention mechanisms, and sustainability-aware modeling objectives.

2.2. Federated Learning in Industrial and Supply Chain Settings

Federated learning (FL), introduced by McMahan et al. through Federated Averaging, enables the effective training of deep neural networks across decentralized and non-IID data without sharing raw data [10]. Subsequent surveys identify FL as particularly suitable for cross-silo industrial environments characterized by strong heterogeneity in scale, data quality, and processes [22]. In Industrial IoT settings, hierarchical and two-stage aggregation strategies improve convergence stability and reduce communication overhead, supporting factory-scale deployment [23].

Within Industry 4.0 and Industry 5.0 frameworks, FL is increasingly regarded as a core enabler of privacy-preserving collaborative intelligence. Surveys in smart manufacturing and product lifecycle management demonstrate its applicability to quality control, predictive maintenance, and process optimization while maintaining data locality [24]. Empirical studies further show that FL enables data-limited manufacturers to achieve performance comparable to centralized learning without exposing proprietary data, as demonstrated in additive manufacturing condition monitoring [25]. These properties align with emerging trustworthy-AI principles emphasizing privacy, auditability, and human-centric governance [26].

In supply chain management, FL applications remain limited but are expanding. Nguyen et al. propose a federated framework for delivery-delay prediction across textile suppliers, demonstrating feasible privacy-preserving collaboration and improved generalization relative to single-organization baselines [27]. The same study shows that federated models can rival or exceed centralized performance by mitigating overfitting

https://doi.org/10.3390/pr14050781

across heterogeneous participants [27]. Related work extends FL to cross-border logistics, enabling early-warning systems for disruption risk while respecting jurisdiction-specific data-sovereignty constraints [28]. Comparable federated demand-forecasting architectures in retail and agri-food supply chains rely on encrypted model updates and align with emerging sectoral data-governance frameworks [29,30].

Despite its promise, cross-silo FL introduces substantial system and governance challenges, including coordination across autonomous organizations, heterogeneous compute resources and data schemas, and exposure to inference attacks. Secure aggregation and cryptographic mechanisms are therefore required in adversarial or semi-honest settings [22]. Tang et al. address these challenges by integrating FL with Graph Neural Networks for privacy-preserving supply chain data sharing [31]. At the organizational level, FL deployments must also comply with data-protection, trade-secret, and competition regulations, which may restrict even indirect information leakage.

Overall, FL is increasingly recognized as a cornerstone technology for collaborative industrial AI, yet supply chain adoption remains largely exploratory. Existing applications predominantly rely on conventional neural architectures and tabular representations, despite the inherently relational structure of supply chains. This gap motivates the integration of FL with graph-based and attention-driven models to enable privacy-preserving, structure-aware learning across multi-enterprise supply networks.

2.3. Sustainability-Aware Artificial Intelligence

As sustainability targets intensify, AI research increasingly distinguishes between AI-for-sustainability, which seeks to reduce emissions, energy use, and resource intensity in operational systems, and "green" AI, which focuses on minimizing the environmental footprint of AI itself through efficient training, inference, and reporting. This work primarily addresses the former by embedding environmental objectives directly into AI models for supply chain process management, while accounting for computational and communication costs.

In supply chain management, sustainability has become a central operational concern, as supply chain activities contribute significantly to greenhouse gas emissions and resource use, requiring the joint optimization of environmental, cost, and service objectives [32]. Accordingly, indicators such as carbon footprint, energy intensity, waste, and circularity are increasingly treated as decision variables rather than ex-post reporting metrics. AI methods are particularly suited to this setting, as high-dimensional operational data and complex constraints limit analytical optimization. Empirical evidence shows that AI-driven analytics can improve green supply chain process integration and environmental performance when aligned with operational governance [32], while systematic reviews identify logistics optimization, demand–inventory coordination, and risk-aware planning as the most mature application areas, where emissions and energy use are tightly coupled to operational decisions [33].

A core methodological principle in sustainability-aware AI is treating environmental performance as a first-class learning objective. This includes integrating environmental impacts into loss functions, enforcing emissions constraints, or learning policies that explicitly trade-off operational and environmental KPIs under uncertainty. In supply chains, Abushaega et al. formalize this approach through a multi-objective framework combining federated learning and Graph Neural Networks to jointly optimize operational and environmental objectives [15]. Related work in energy systems demonstrates that decentralized learning can yield simultaneous cost and emissions reductions, including empirically observed savings under federated reinforcement learning for building energy manage-

https://doi.org/10.3390/pr14050781

ment [13]. Additional studies show that federated sequence models support sustainability optimization beyond centralized deployments [34,35].

Sustainability-aware AI also encompasses measurement and accountability of AI's own environmental footprint. The literature emphasizes standardized green evaluation metrics for comparability across models and deployments. Borraccia et al. propose hybrid metrics for assessing energy–carbon impacts of AI pipelines [36], while bibliometric analyses document rapid growth in sustainability-oriented machine learning research alongside a persistent gap between methodological advances and deployment-level impact [37,38]. Organizational studies further indicate that effective AI-for-sustainability outcomes depend on governance structures and integration into decision workflows, not solely on model design [39].

Finally, sustainability benefits from AI in supply chains are not guaranteed and may be offset by rebound effects, misaligned incentives, or increased computational burden. Practitioner reports often claim large emissions reductions from AI-driven logistics and planning interventions [40], but the research literature stresses rigorous evaluation, transparent baselines, and robust measurement to distinguish genuine environmental improvements from confounded operational effects [36,41]. While the feasibility of sustainability-aware AI is well supported, most supply chain ML systems still prioritize operational accuracy, treating sustainability as an external constraint. The novelty of our approach lies in explicitly encoding sustainability signals within a decentralized federated graph learning framework, enabling balanced optimization of operational and environmental KPIs under realistic data-governance constraints.

2.4. Combining Graph Neural Networks and Federated Learning

Bringing together GNNs and FL poses both conceptual and system-level challenges, and only a limited body of work has explored this intersection. Standard FL formulations implicitly assume that data points are independent and identically distributed across clients. Graph-structured data violates these assumptions in two fundamental ways: (i) observations (nodes and edges) are relationally coupled and (ii) the graph topology itself may be partitioned across organizational boundaries. In federated settings, this implies that informative dependencies can span clients, while each participant only observes a local subgraph. These properties complicate gradient aggregation, convergence, and representation learning. Despite these difficulties, early studies demonstrate that federated optimization can be extended to graph-based models, giving rise to the paradigm of Federated Graph Neural Networks (FedGNN).

The earliest line of work has been driven by privacy-sensitive applications such as recommender systems and financial networks. Wu et al. introduced FedGNN for a privacy-preserving recommendation, training GNNs over decentralized user–item interaction graphs while applying local privacy mechanisms to gradients and structural signals [35]. Their results show that a federated GNN can achieve recommendation quality comparable to centralized baselines, while significantly reducing information leakage under realistic threat models. This finding is important because it establishes that graph-based representation learning remains viable under strict data-locality constraints, even when the global graph never materializes.

Subsequent work extends federated graph learning to spatio-temporal domains. Meng et al. propose a cross-node federated GNN for traffic forecasting, where each client manages a regional subgraph and the global model captures dependencies that span geographic boundaries [35]. Their framework improves forecasting accuracy relative to purely local models and introduces mechanisms for mitigating non-IID graph distributions. Complementary research addresses vertical graph partitioning, in which different clients hold

https://doi.org/10.3390/pr14050781

disjoint feature sets or edge views for the same nodes. Chen et al. propose a vertically federated GNN (VFGNN) that enables privacy-preserving node classification when features, edges, and labels are distributed across parties [36]. These works collectively demonstrate that both horizontal and vertical graph partitioning can be accommodated within federated optimization, albeit with additional architectural and cryptographic complexity.

Our framework distinguishes itself by introducing a Graph Attention Network within a federated learning architecture tailored to multi-enterprise supply chains. Attention is particularly valuable in federated graph settings because local subgraphs differ in topology, density, and informational relevance. An attention-based aggregator can dynamically weight inter-node and inter-client dependencies, allowing the global model to adapt to heterogeneous structural contexts more flexibly than fixed-weight GCNs. Moreover, we explicitly integrate sustainability objectives into the federated GNN training process—an element largely absent from the existing FedGNN literature, which typically prioritizes predictive performance and privacy guarantees alone. While Abushaega et al. take an important step toward multi-objective federated graph learning for supply chains [15], their architecture does not exploit attention mechanisms and does not explicitly address the interpretability and cross-client structural asymmetries inherent in real-world networks.

As summarized in Table 1, prior work on supply chain sustainability primarily operates either within centralized graph learning frameworks [8,14,17,20,21] or federated learning settings without sustainability-aware aggregation [9,10,24,31]. Recent AI-driven decarbonization efforts [12,32,33] focus on predictive optimization but lack federated graph structures. Although Federated Graph Neural Networks have been proposed for privacy-preserving supply chain data sharing [9,31], they do not introduce explicit environmental bias within the attention mechanism.

Table 1. Comparative positioning of federated and sustainability-aware graph learning approaches in supply chain contexts.

Study Category	Representative Works	Federated	Sustainability Integration	Attention-Level Sustainability Bias	Explicit Governance Parameter	Key Limitation
Centralized Graph Neural Networks for Supply Chains	[8,14,17,20,21]	✗	Implicit via features	✗	✗	No decentralization; no sustainability-aware aggregation
Federated learning in Manufacturing/Supply Chains	[10,11,24,27,28]	✓	Typically performance-oriented	✗	✗	No graph-attention sustainability modulation
Federated Graph Neural Networks	[9,31]	✓	Privacy-focused; sustainability not primary	✗	✗	No explicit environmental alignment control
AI for Supply Chain Decarbonization	[12,32,33]	Mostly ✗	Sustainability as optimization objective	✗	✗	No federated graph structure
Federated Sustainable Energy Optimization	[13,34]	✓	Sustainability at model objective level	✗	✗	No graph-based attention bias
Proposed Method	—	✓	Sustainability embedded in message passing	✓	✓(λ)	Synthetic validation

https://doi.org/10.3390/pr14050781

The proposed framework uniquely integrates federated graph learning with sustainability-modulated attention. By embedding environmental signals directly into message-passing weights and introducing λ as an explicit governance parameter, the method enables controllable trade-offs between predictive performance and environmental alignment in decentralized supply chain networks.

3. Methodology and Experimental Setup

This section describes the experimental design used to evaluate the proposed sustainability-aware federated graph attention framework. The setup is deliberately constructed to reflect key structural and data-governance characteristics of real industrial supply chains, while remaining fully reproducible and analytically controlled.

The proposed model is trained through an iterative sustainability-aware federated optimization process that preserves data locality at each client while enabling coordinated global learning. Unlike conventional federated graph training, sustainability modulation is embedded directly within the local attention mechanism, and the server additionally aggregates privacy-preserving alignment diagnostics to monitor environmental bias during training. Communication proceeds through repeated rounds between the central server and distributed clients, where model parameters are exchanged but no raw graph data or labels are shared. Algorithm 1 presents the complete governance-aware federated training workflow.

Algorithm 1 Sustainability-Aware Federated Graph Attention Training

Input: Initial global parameters W^0; rounds T; clients $k = 1, \ldots, K$ each holding a local subgraph $G_k = (V_k, E_k)$ with node features and labels; emission attribute e_{ij} for edges; sustainability control parameter λ (fixed or scheduled); local epochs E; aggregation weights w_k (e.g., proportional to $| V_k |$ or number of local training nodes).
Output: Final global parameters W^T.
Initialize (Server):
Server initializes model parameters W^0. Server selects sustainability policy λ (fixed $\lambda = \lambda_0$ or schedule λ_t) and broadcasts $\left(W^0, \lambda_1\right)$ to all clients.
For each communication round $t = 1, \ldots, T$:
(a) Client-side sustainability-aware local training (parallel for each client k):
Receive $\left(W^{t-1}, \lambda_t\right)$.
Train the GAT on local subgraph G_k for E epochs using sustainability-modulated attention.
For each incoming edge $(j \to i)$, attention coefficients are reweighted by an emission penalty after softmax:

$$\alpha_{ji} \leftarrow \frac{\alpha_{ji}\exp\left(-\lambda_t \widetilde{e}_{ji}\right)}{\sum_{m \in \mathcal{N}_{in}(i)} \alpha_{mi}\exp\left(-\lambda_t \widetilde{e}_{mi}\right)},$$

where $\widetilde{e}_{ji}$ is normalized transport emissions and λ_t governs sustainability strength.
Compute local model update ΔW_k^t (or updated weights W_k^t).
Compute local alignment diagnostics (no raw data shared), e.g.:
$\rho_k^t =$ Spearman $\left(\widetilde{e}_{ji}, \alpha_{ji}\right)$ over local edges;
$m_k^t =$ fraction of attention mass allocated to high-emission edges (top quantile).
Send $\left\{\Delta W_k^t, \rho_k^t, m_k^t, | V_k |\right\}$ to the server.

https://doi.org/10.3390/pr14050781

Algorithm 1 *Cont.*

(b) Server-side aggregation with sustainability monitoring:

Aggregate model updates:

$$W^t \leftarrow W^{t-1} + \sum_{k=1}^{K} w_k \Delta W_k^t, \text{ where } w_k = \frac{|V_k|}{\sum_\ell |V_\ell|}.$$

Aggregate diagnostics:

$$\overline{\rho}^t = \sum_{k=1}^{K} w_k \rho_k^t, \overline{m}^t = \sum_{k=1}^{K} w_k m_k^t.$$

Governance step (optional): update λ according to a predefined policy (not tuned post hoc), e.g., monotonic schedule or target-alignment rule:

$$\lambda_{t+1} \leftarrow \text{PolicyUpdate}\left(\lambda_t, \overline{\rho}^t, \overline{m}^t\right).$$

(If fixed-policy, set $\lambda_{t+1} = \lambda_t$.)

(c) Broadcast:

Server broadcasts $\left(W^t, \lambda_{t+1}\right)$ to all clients.
Return: final global model parameters W^T.

3.1. Synthetic Supply Chain Simulation

We evaluate the proposed Federated GAT on a realistic synthetic multi-tier supply chain case study. The supply network (illustrated schematically in Figure 1) is modeled as a directed graph G = (V, E) with multiple echelons including suppliers, manufacturers, distribution centers, and retailers. Each node represents an entity (e.g., a factory, warehouse, or retail outlet) and each directed edge represents a supply relationship or material flow between entities. Nodes are assigned to tiers by operational role (upstream suppliers in Tier 1, intermediate manufacturers in Tier 2, distribution centers in Tier 3, and retailers in Tier 4), creating a layered structure analogous to a consumer goods supply chain.

To ensure structural transparency, we report descriptive statistics of the generated graphs. For the baseline configuration (N = 28 nodes, 41 edges), the mean in-degree and out-degree distributions, as well as transport-emissions statistics, are summarized withing the main text. Client partition statistics further quantify intra- and inter-client edge cuts, clarifying the degree of structural fragmentation introduced by federated partitioning. These statistics demonstrate that the synthetic network preserves tiered supply chain structure while enabling the controlled evaluation of decentralized learning effects.

In the implementation, tier sizes are explicitly defined as 10 suppliers (S1–S10), 6 manufacturers (M1–M6), 4 distribution centers (D1–D4), and 8 retailers (R1–R8) for a total of N = 28 nodes. Directed edges are generated only between adjacent tiers (S→M, M→D, D→R), producing a multi-tier dependency structure that naturally yields converging flows (multiple suppliers feeding a manufacturer) and diverging flows (a distributor feeding multiple retailers). Concretely, the stochastic edge generator draws for each source node an out-degree:

$$k_s = \max(1, \lfloor \mathcal{N}(\mu_k, 1) \rfloor), \quad (1)$$

https://doi.org/10.3390/pr14050781

then connects to a uniformly sampled subset of downstream nodes. This construction yields sparse, tier-constrained connectivity consistent with hierarchical supply networks.

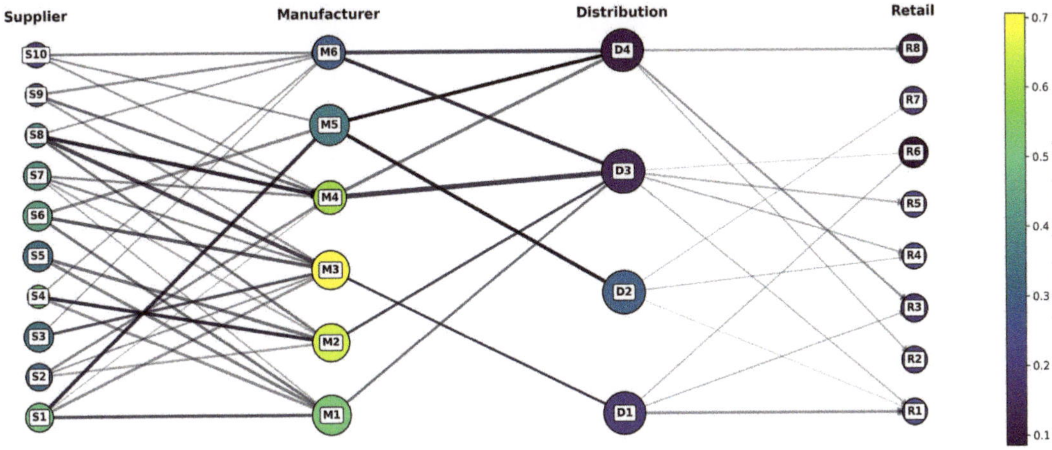

Figure 1. Multi-tier synthetic supply chain graph. Node size encodes capacity; node color represents carbon intensity; edge width corresponds to flow magnitude; edge transparency reflects normalized transport emissions.

To ensure that downstream tiers are not disconnected from upstream supply, we enforce a weak connectivity constraint, where for every node v in tiers 2–4, at least one inbound edge must exist. Formally, if

$$\deg^{-}(v) = 0, \tag{2}$$

a random inbound edge (u, v) is added from the upstream tier. This prevents isolated nodes and ensures that message passing is well-defined for most nodes in the node-regression task.

Each node $v \in V$ is assigned operational and sustainability proxy attributes drawn from tier-specific Gaussian distributions, namely capacity, CO_2 intensity, and energy intensity. These distributions are explicitly encoded inducing systematic heterogeneity across tiers (e.g., manufacturers have higher mean carbon intensity than retailers) while maintaining stochastic variability under a fixed random seed, SEED = 42. Each edge $(u, v) \in E$ is parameterized by flow and distance (also positive truncated normal draws), and a deterministic transport-emissions proxy:

$$\text{transport-emissions}_{uv} = \text{flow}_{uv} \cdot \text{distance}_{uv} \cdot 10^{-3}. \tag{3}$$

This proxy is used consistently both for feature construction and for sustainability-aware attention modulation.

To simulate federated data partitioning, we divided the global supply chain graph into subgraphs held by distinct client organizations. Figure 2 illustrates this partitioning: each company (or region) in the supply chain acts as a federated client that observes only its portion of the graph. In the implementation, 4 federated clients are created.

https://doi.org/10.3390/pr14050781

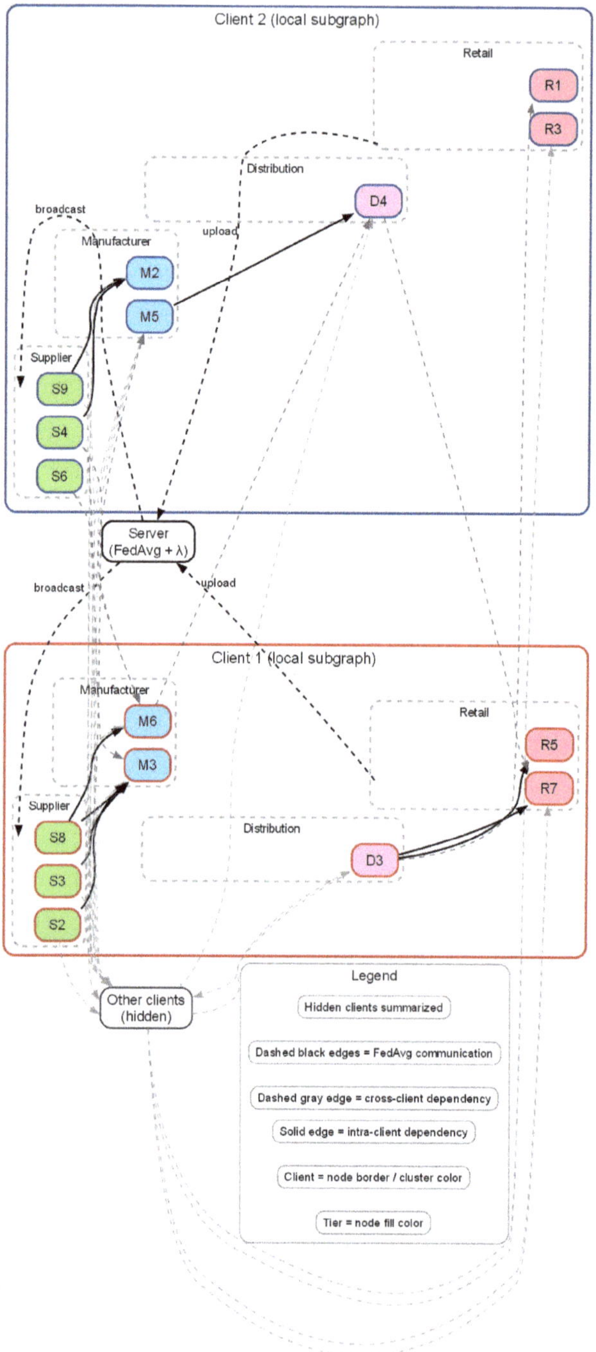

Figure 2. For visual clarity, only two representative clients are explicitly shown (from K = 4 total); remaining clients are aggregated into a "hidden" sink. Solid edges denote intra-client dependencies, dashed gray edges denote cross-client dependencies (cut edges), and dashed black edges denote parameter exchange (broadcast/upload). Edges represent structural dependencies only; operational weights and emissions are not encoded in this schematic.

https://doi.org/10.3390/pr14050781

Table 2 quantifies the federated partitioning illustrated in Figure 2 for the baseline graph ($N = 28$). Clients contain between 6 and 8 nodes, indicating moderate size imbalance. Intra-client connectivity is sparse (0–3 edges per client), whereas the number of cross-client (cut) edges is substantially larger (16–19 per client). This confirms that most relational dependencies span across client boundaries. Consequently, each client observes only a highly fragmented local subgraph, and global supply chain dependences cannot be directly reconstructed during local message passing. This validates the structural decentralization assumed in the federated learning setup.

Table 2. Federated partition statistics for the baseline graph ($N = 28$, K = 4).

Client	Number of Nodes	Intra-Client Edges	Inter-Client Edges (Cut)
1	8	3	19
2	8	2	16
3	6	2	16
4	6	0	17

Client assignment is performed independently within each tier using a seeded shuffle followed by round-robin allocation:

$$\text{client}(v_i) = (imodK) + 1. \tag{4}$$

This produces tier-balanced client memberships but non-identical local structures. Edges that cut across clients represent inter-company links (e.g., a supplier shipping to a manufacturer in a different organization)—these links exist in the overall graph but no single client can exploit them during local training. This setup reflects real-world data silos: each enterprise has visibility into its local operations and direct partners, but not the full end-to-end supply chain.

Critically, the federated training loop enforces strict locality: each client trains on its induced subgraph consisting only of nodes it owns and intra-client edges. Formally, client k observes

$$G_k = (V_k, E_k), E_k = \{(u, v) \in E : u \in V_k \wedge v \in V_k\}. \tag{5}$$

This is implemented by filtering the global edge list. Hence, cross-client dependencies exist globally but are missing locally, creating a non-IID, partial-observability regime.

3.2. Data Generation and Sustainability Labeling

We generate synthetic operational data on the supply chain graph to train and evaluate the models. Each node is assigned a feature vector representing its operational state and sustainability profile, and each edge carries logistics attributes, including for transport emissions. In the code, the node-level feature vector is explicitly defined as a 5-dimensional descriptor:

$$x_v = [\text{capacity}_v, \text{CO}_2\text{\_intensity}_v, \text{energy\_intensity}_v, \text{total\_in\_flow}_v, \text{total\_in\_emis}_v] \in \mathbb{R}^5, \tag{6}$$

where the last two components summarize the upstream dependency and upstream logistics footprint:

$$\text{total\_in\_flow}_v = \sum_{u:(u,v)\in E} \text{flow}_{uv}, \ \text{total\_in\_emis}_v = \sum_{u:(u,v)\in E} \text{transport\_emissions}_{uv}. \tag{7}$$

https://doi.org/10.3390/pr14050781

This construction makes the prediction problem structurally dependent: the feature vector is not purely local, but contains aggregated information from incoming edges, which graph-based learners can exploit more naturally than independent baselines.

The feature matrix $X \in \mathbb{R}^{N \times 5}$ is standardized using dataset-level z-scoring,

$$X_{\text{norm}} = \frac{X - \mu}{\sigma + 10^{-9}} \tag{8}$$

and converted into the tensor X_t for training on CPU/GPU depending on availability.

Using this data, we formulate a synthetic target variable at each node called the "stress score." This node-level stress score is designed as a composite outcome that increases under capacity scarcity, upstream dependency, and carbon-intensive logistics. In the implementation, the pre-normalized stress score $\widetilde{y}_v$ is constructed as a weighted combination of interpretable drivers:

$$\begin{aligned}\widetilde{y}_v = 0.45 \cdot \tfrac{1}{\text{capacity}_v + \varepsilon} + 0.20 \cdot \tfrac{\text{total\_in\_flow}_v}{\text{total\_in\_flow} + \varepsilon} + 0.25 \cdot \tfrac{\text{total\_in\_emis}_v}{\text{total\_in\_emis} + \varepsilon} + 0.10 \cdot \\ \tfrac{\text{CO}_2\text{\_intensity}_v}{\text{CO}_2\text{\_intensity} + \varepsilon} + 0.05 \cdot \tfrac{\text{energy\_intensity}_v}{\text{energy\_intensity} + \varepsilon},\end{aligned} \tag{9}$$

with ε used only for numerical stability. The value is min–max scaled to [0, 1] and perturbed with Gaussian noise of standard deviation std = 0.05:

$$y_v = \text{clip}(\text{minmax}(\widetilde{y}_v) + \eta, 0, 1), \eta \sim \mathcal{N}(0, \sigma^2). \tag{10}$$

By construction, this target is influenced by both operational dynamics (capacity, inbound dependency) and environmental factors (carbon/energy intensity and transport emissions), aligning with the objective of sustainability-aware modeling.

We split the dataset into training, validation, and test sets by randomly partitioning the node set (not multiple temporal snapshots). Specifically, a random permutation of node indices is drawn and masks are created using train ratio = 0.7 and validation ratio = 0.15, with the test set comprising the remaining nodes. This yields a transductive node-regression setting where the graph structure and all nodes are present during training, but only training nodes contribute to the supervised loss, while validation and test nodes are used strictly for evaluation.

3.3. Model Configurations and Training Procedure

We compare four model configurations in our experiments, corresponding to baseline approaches and our proposed method.

1. Centralized GAT: A Graph Attention Network model that has access to the entire supply chain graph during training. It represents an ideal pooled-data scenario and provides an upper-bound reference under full observability. The implemented GAT uses a single attention layer with 32 hidden dimensions and 2 heads. Message passing is performed over incoming neighborhoods. For head h, node projections are computed as

$$\text{h}_v^{(h)} = W^{(h)} \text{x}_v, \tag{11}$$

and attention logits for an incoming edge (u, v) are computed using concatenated projected states and edge features $\text{e}_{uv} \in \mathbb{R}^2$ (flow and transport emissions, min–max normalized over global edges):

$$e_{uv}^{(h)} = \text{LeakyReLU}(\langle a^{(h)}, [\text{h}_v^{(h)} \parallel \text{h}_u^{(h)} \parallel \text{e}_{uv}] \rangle). \tag{12}$$

https://doi.org/10.3390/pr14050781

Normalization is performed by softmax over $u \in \mathcal{N}_{\text{in}}(v)$:

$$\alpha_{uv}^{(h)} = \frac{\exp\left(e_{uv}^{(h)}\right)}{\sum_{u' \in \mathcal{N}_{\text{in}}(v)} \exp\left(e_{u'v}^{(h)}\right)}. \tag{13}$$

The head outputs are aggregated by weighted neighbor sums and concatenated before a linear readout.

2. Federated GAT (FedAvg): A standard federated version of the GAT model trained with synchronous Federated Averaging. Each client trains on its local induced subgraph $G_k = (V_k, E_k)$ for 10 local epochs over 40 rounds. After each round, client models are aggregated using node-count weights:

$$\theta^{(t+1)} = \sum_{k=1}^{K} \frac{|V_k|}{\sum_{j=1}^{K} |V_j|} \theta_k^{(t)}. \tag{14}$$

This baseline isolates the impact of partial observability and non-IID graph partitioning without additional sustainability mechanisms.

3. Federated Sustainability-Aware GAT: Our proposed model, which extends the Federated GAT with a sustainability-aware attention mechanism. In the implementation, sustainability is incorporated directly into the attention coefficients. After computing $\alpha_{uv}^{(h)}$ by softmax, the model applies an exponential penalty based on normalized transport emissions $\text{emis\_norm}_{uv} \in [0, 1]$ (the second edge-feature component),

$$\alpha_{uv}^{(h)} \leftarrow \alpha_{uv}^{(h)} \cdot \exp(-\lambda \cdot \text{emis\_norm}_{uv}), \tag{15}$$

followed by re-normalization over the incoming neighborhood,

$$\alpha_{uv}^{(h)} \leftarrow \frac{\alpha_{uv}^{(h)}}{\sum_{u' \in \mathcal{N}_{\text{in}}(v)} \alpha_{u'v}^{(h)} + 10^{-9}}. \tag{16}$$

The supervised objective is defined as the mean squared error for node stress prediction, i.e.,

$$\mathcal{L}_k = \frac{1}{|T_k|} \sum_{v \in T_k} (\hat{y}_v - y_v)^2, \tag{17}$$

where $T_k \subseteq V_k$ are the client's training nodes defined by the global train mask restricted to V_k.

All models were implemented in Python 3.11.9 using PyTorch 2.9.0 and trained with the Adam optimizer using learning rate LR = 10^{-3} and weight decay WD = 10^{-4}. Centralized models are trained for 250 epochs. Federated models are trained for 40 rounds with 10 local epochs per round, yielding 400 local epochs per client in total, while global validation RMSE is monitored each round by evaluating the aggregated model on the full graph and the global validation mask.

3.4. Evaluation Metrics

We evaluate model performance on the held-out test nodes using standard regression metrics, root mean squared error (RMSE) and mean absolute error (MAE), between the predicted and true stress scores. For a test set $\mathcal{T} \subseteq V$, these are:

$$\text{RMSE} = \sqrt{\frac{1}{|\mathcal{T}|} \sum_{v \in \mathcal{T}} (\hat{y}_v - y_v)^2}. \tag{18}$$

https://doi.org/10.3390/pr14050781

$$\text{MAE} = \frac{1}{|\mathcal{T}|} \sum_{v \in \mathcal{T}} |\hat{y}_v - y_v|. \quad (19)$$

In addition to the prediction error, a key evaluation in our study is the attention–emission relationship. Since the GAT explicitly outputs per-edge attention values, we can quantify sustainability alignment by pairing each learned attention coefficient with its corresponding normalized edge emission. Concretely, the head-averaged attention coefficient for each edge is computed as:

$$\bar{\alpha}_{uv} = \frac{1}{H} \sum_{h=1}^{H} \alpha_{uv}^{(h)}, \quad (20)$$

where emissions are given by emis_norm_{uv}, we compute rank-based associations (e.g., Spearman correlation) between $\bar{\alpha}_{uv}$ and emis_norm_{uv}, and complementary diagnostics such as the fraction of attention mass assigned to the highest-emission quantile of incoming edges. These measures are defined at the setup level here, while empirical values and comparative analysis are reported in the results section.

4. Results

4.1. Predictive Performance Comparison

Table 3 reports the predictive performance of the evaluated graph-based models on the held-out test node subset $\mathcal{T} \subseteq V$, using root mean squared error (RMSE) and mean absolute error (MAE). Reported values correspond to the mean and standard deviation over five independent random seeds, with model selection performed using validation RMSE. For centralized training, validation performance is tracked per epoch, while for federated training it is tracked per communication round.

Table 3. Predictive performance comparison across centralized and federated baselines.

Model	RMSE (Mean ± Std)	MAE (Mean ± Std)
Centralized GAT	0.2133 ± 0.1141	0.1873 ± 0.1063
Federated GAT (FedAvg)	0.2606 ± 0.1124	0.2235 ± 0.1059
Federated Sustainability-Aware GAT (λ = 3.0)	0.2593 ± 0.1131	0.2195 ± 0.1075

The centralized GAT achieves the lowest prediction error across both metrics (RMSE = 0.2133 ± 0.1141, MAE = 0.1873 ± 0.1063), reflecting its unrestricted access to the full supply chain graph. In this setting, the model can exploit complete neighborhood structure and edge attributes across all tiers. Given that the synthetic node-level stress signal depends on both node-local features and aggregated upstream quantities (e.g., total inbound flow and transport emissions), centralized message passing enables coherent propagation of relational information throughout the network.

The predictive advantage of centralized graph learning can be attributed to two complementary architectural properties. First, relational aggregation over incoming neighborhoods, $\mathcal{N}_{in}(v)$, smooths node representations by incorporating upstream information, mitigating label noise through structural correlation. Second, edge-aware attention weighting allows differential emphasis on upstream connections based on both node embeddings and normalized edge attributes, enabling the model to identify structurally influential pathways for downstream stress propagation.

Both federated graph attention models exhibit higher prediction error than the centralized baseline, highlighting the intrinsic challenges of decentralized graph learn-

https://doi.org/10.3390/pr14050781

ing under strict locality constraints. The standard Federated GAT (FedAvg) attains RMSE = 0.2606 ± 0.1124 and MAE = 0.2235 ± 0.1059. In this setting, each client trains exclusively on its induced subgraph and cross-client edges are unavailable during message passing. Consequently, global relational dependencies spanning multiple clients cannot be directly exploited during local updates. The global model obtained via weighted FedAvg reconciles fragmented client views only indirectly through parameter aggregation, leading to predictable performance degradation relative to centralized training.

The Federated Sustainability-Aware GAT achieves RMSE = 0.2593 ± 0.1131 and MAE = 0.2195 ± 0.1075, statistically indistinguishable from the standard federated baseline. A paired sign test on RMSE across seeds yields $p = 1.000$, indicating no statistically significant difference between the two federated variants. Similar behavior is observed for MAE. These results demonstrate that embedding sustainability-aware modulation directly within the attention mechanism does not degrade federated predictive performance.

Importantly, sustainability is not introduced via an auxiliary loss term but is embedded directly into the attention mechanism through an emission-aware bias applied after softmax normalization (Equations (15) and (16)). With $\lambda = 3$, high-emission transport edges are systematically down-weighted during message passing. While this mechanism imposes an explicit inductive bias, it preserves predictive accuracy relative to conventional federated graph learning.

From an applied perspective, the results highlight a structured trade-off landscape. Centralized graph learning represents a best-case scenario for predictive accuracy but is often infeasible due to data-sharing and privacy constraints. Standard federated learning preserves decentralization but incurs performance loss when global relational dependencies are critical. The Sustainability-Aware Federated GAT occupies an intermediate position: it remains fully decentralized while embedding environmental preferences directly into relational inference, achieving comparable predictive performance alongside structured sustainability alignment examined in the subsequent λ-ablation analysis.

Before examining training dynamics and sustainability–accuracy trade-offs, we assess whether the sustainability-aware formulation introduces any circularity or label-dependent bias.

4.2. Construct Validity and Label Sensitivity Analysis

A potential methodological concern in synthetic experimental settings is circularity: if sustainability-related variables contribute directly to the target label, a sustainability-aware model may appear to perform well simply because its inductive bias aligns with the label construction. In such a case, improvements in predictive performance could reflect label leakage rather than genuine modeling advantages. To address this concern rigorously, we conduct a structured label sensitivity analysis designed to decouple sustainability-aware attention from the specific stress formulation used in the baseline experiments.

Specifically, we evaluate five distinct label-generation regimes by systematically varying the contribution of sustainability-related components in the synthetic stress definition:

1. Full sustainability weighting: Original configuration.
2. Reduced sustainability weighting: Sustainability coefficients scaled by 0.5.
3. No sustainability contribution: All sustainability-related terms removed.
4. Emissions-only contribution: Only transport emissions included.
5. Node-level sustainability only: Only node carbon intensity and energy intensity included.

For each regime, the complete multi-seed federated evaluation protocol is repeated. This ensures that any observed behavior reflects model robustness rather than incidental alignment between attention bias and label structure.

https://doi.org/10.3390/pr14050781

The most critical test is the No sustainability contribution regime, where all sustainability variables are removed from the stress definition. In this setting, the standard Federated GAT achieves RMSE = 0.2758, while the Sustainability-Aware Federated GAT achieves RMSE = 0.2740 ($\Delta = -0.00177$). The negligible difference indicates that the sustainability-aware attention mechanism does not rely on label-specific environmental components for predictive performance. Rather, its inductive bias remains stable even when sustainability plays no role in the target variable.

Across all five regimes, predictive differences between the two federated variants remain statistically insignificant. A paired sign test on RMSE across seeds yields $p = 1.0$ in each configuration, and similar patterns are observed for MAE. These results demonstrate that the sustainability-aware formulation does not artificially inflate performance through circular label construction.

Importantly, while predictive accuracy remains statistically comparable, the sustainability-aware model continues to exhibit systematically stronger environmental alignment in attention weights (as shown in the λ-ablation analysis). This confirms that the model's environmental behavior arises from structured attention modulation rather than from overfitting to label components.

Taken together, the label sensitivity study provides strong evidence of construct validity. The sustainability-aware attention mechanism functions as an inductive bias embedded within decentralized message passing, not as a proxy for explicit label encoding. Consequently, concerns of circular performance inflation are empirically mitigated.

4.3. Training Dynamics and Convergence

Figure 3 presents the validation RMSE trajectories of the centralized and federated graph attention models, averaged over multiple random seeds, with shaded regions indicating one standard deviation. To ensure comparability across training paradigms, all curves are reported over a unified horizontal axis of 40 evaluation steps. For federated models, each step corresponds to one communication round, while for centralized training the validation trajectory is resampled from the full 250-epoch optimization history. The horizontal axis therefore represents aligned reporting checkpoints rather than equivalent optimization time.

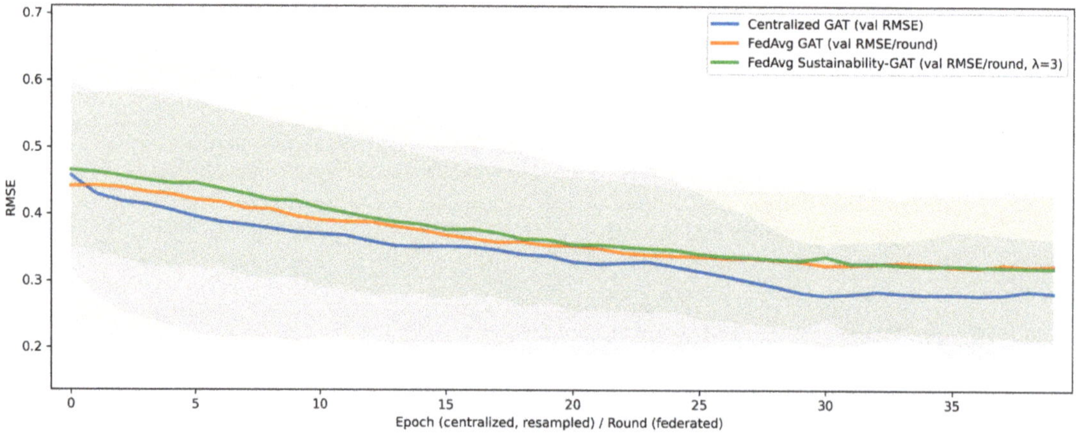

Figure 3. Validation RMSE trajectories for centralized and federated graph attention models. Solid lines denote the mean validation RMSE across multiple random seeds, while shaded regions indicate ±1 standard deviation around the mean.

https://doi.org/10.3390/pr14050781

The centralized GAT exhibits smooth and stable convergence, with a consistent reduction in validation RMSE and comparatively narrow variability across seeds. This behavior reflects the advantages of centralized graph learning, where full access to the global supply chain topology enables uninterrupted message passing across all tiers. At each optimization step, gradients are computed using complete neighborhood information and edge attributes, yielding low-variance updates and reliable convergence.

Both federated graph attention models converge more slowly and display wider uncertainty bands. This behavior is characteristic of decentralized graph learning under non-identically distributed (non-IID) subgraph partitions. Each client optimizes a local objective defined on its induced subgraph $G_k = (V_k, E_k)$, where neighborhood structure, edge-attribute distributions, and structural roles within the supply chain differ across clients. These heterogeneous local optima are periodically reconciled through weighted parameter aggregation, which can introduce non-smooth global parameter updates and contributes to the increased variance observed in federated training. By construction, this model reshapes the message-passing process through an emission-aware attention bias that penalizes high-emission transport edges prior to renormalization. Consequently, information propagation is progressively concentrated on lower-emission pathways, leading to a more selective and structured flow of information during training. From an optimization standpoint, this mechanism introduces a controlled inductive bias that affects convergence speed but enhances consistency in how sustainability considerations are incorporated into the learned representations. In federated settings, where each client observes only a fragment of the global graph, unconstrained attention can overemphasize locally predictive but environmentally undesirable edges. The sustainability-aware formulation counteracts this tendency by steering optimization toward solutions that balance predictive performance with environmental alignment.

The modest elevation in validation RMSE relative to the standard federated baseline should therefore be interpreted as the explicit cost of embedding sustainability constraints directly into the inference pathway, rather than as evidence of instability or inefficient training. Notably, the overlap of uncertainty bands in Figure 3 indicates that the sustainability-aware model remains stable across random initializations and does not exhibit pathological convergence behavior.

Overall, Figure 3 demonstrates that sustainability-aware federated graph learning converges reliably under strict locality constraints, while embedding environmental priorities directly into the message-passing mechanism. This positions the proposed model as a principled extension of federated graph learning for sustainability-critical supply chain applications, where predictive accuracy must be balanced against environmentally informed decision-making.

4.4. Sustainability-Aware Attention Patterns

A key objective of the proposed Sustainability-Aware Federated GAT is to reduce reliance on carbon-intensive transport links during message passing. This behavior can be directly examined by analyzing learned attention coefficients, extracted during evaluation with. For each directed edge (u, v), attention weights are averaged across heads to obtain a single scalar α_{uv}.

Figure 4 shows the relationship between normalized transport emissions and mean attention weights for the standard Federated GAT ($\lambda = 0$) and the sustainability-aware variant ($\lambda = 3$). In the unconstrained model, attention weights are widely dispersed across the full emissions range, with no systematic dependence on emissions. Several high-emission edges receive moderate or high attention, reflecting their predictive utility under the synthetic stress generation mechanism, which explicitly incorporates inbound emissions.

https://doi.org/10.3390/pr14050781

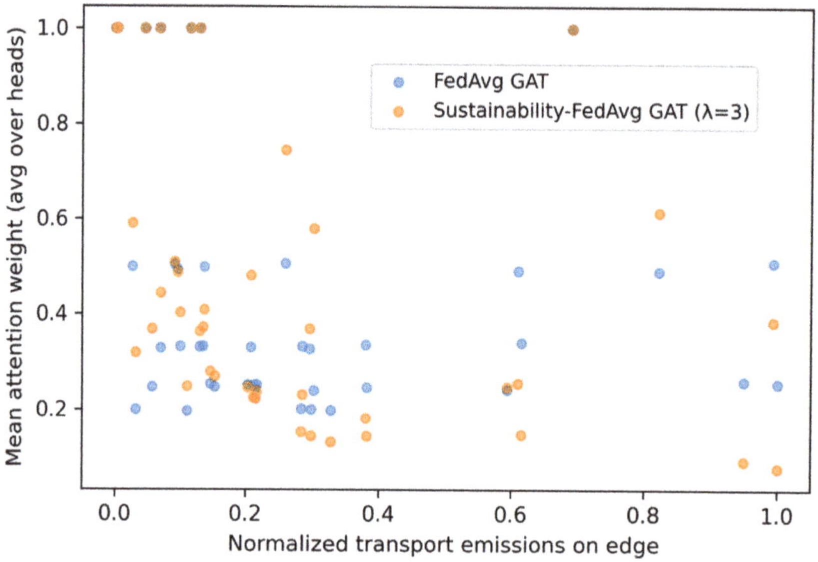

Figure 4. Mean attention weight versus normalized transport emissions for standard and Sustainability-Aware Federated GAT models.

In contrast, the sustainability-aware model exhibits a clear shift in attention allocation. High-emission edges are systematically down-weighted, while low- and medium-emission edges retain greater attention mass. This pattern is induced by the multiplicative emissions penalty applied within the attention mechanism, followed by neighborhood renormalization. The resulting behavior reflects a soft inductive bias: carbon-intensive links are discouraged but not entirely suppressed, allowing predictive relevance to partially offset the sustainability penalty when necessary.

From an interpretability standpoint, the learned attention distributions provide graph-native explanations of information flow. The observed reduction in attention assigned to high-emission edges confirms that sustainability considerations are embedded directly into the inference pathway, yielding representations that balance predictive performance with environmentally informed decision-making.

4.5. Ablation Study: Impact of Sustainability Weight (λ)

To systematically assess how the strength of sustainability bias influences the behavior of the proposed attention mechanism, we conduct a λ-ablation study over a predefined and fixed grid,

$$\lambda \in \{0, 0.25, 0.5, 1, 2, 3, 5, 8\}, \tag{21}$$

with all configurations evaluated across multiple random seeds. The λ values are specified a priori and are not tuned post hoc, ensuring a transparent and reproducible evaluation protocol. While predictive performance is summarized separately, this subsection focuses specifically on how sustainability alignment evolves as a function of λ, isolating the effect of the attention penalty on message-passing behavior. Figure 5 reports two complementary sustainability alignment diagnostics as functions of λ, shown as mean $\pm$ standard deviation over seeds:

https://doi.org/10.3390/pr14050781

(i) The Spearman rank correlation between normalized transport emissions and learned attention weights;
(ii) The fraction of incoming attention mass allocated to high-emission edges (top emission quantile).

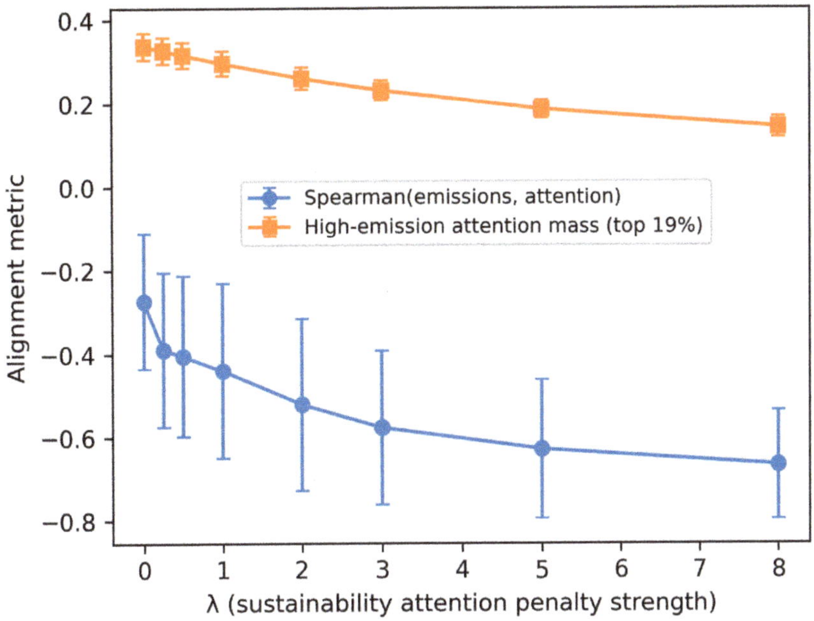

Figure 5. Effect of λ on sustainability alignment.

Both metrics exhibit smooth and monotonic trends as λ increases. The Spearman correlation becomes progressively more negative, indicating that attention weights are increasingly de-emphasized on high-emission edges. In parallel, the share of attention mass assigned to carbon-intensive links decreases steadily. These consistent patterns confirm that the proposed exponential attention penalty induces a systematic and stable reorientation of message passing away from high-emission transport links, rather than producing noisy or irregular effects.

Notably, the variability across random seeds decreases at higher λ values, suggesting that stronger sustainability bias leads to more reproducible alignment behavior under federated training conditions. This indicates that the mechanism functions as a stable inductive bias, rather than introducing additional instability into the learning process.

Figure 6 summarizes the ablation results by plotting predictive error against sustainability alignment, yielding a trade-off frontier across λ values. Each point corresponds to a distinct operating regime of the Sustainability-Aware Federated GAT, with sustainability alignment measured as the negative Spearman correlation between emissions and attention (higher values indicate stronger de-emphasis of high-emission edges).

The resulting frontier reveals a clear and structured trade-off. As λ increases, sustainability alignment improves monotonically, while predictive error increases gradually. Importantly, the degradation in accuracy is smooth rather than abrupt, indicating that the attention penalty suppresses high-emission links in a soft and continuous manner rather than eliminating large portions of the graph structure.

https://doi.org/10.3390/pr14050781

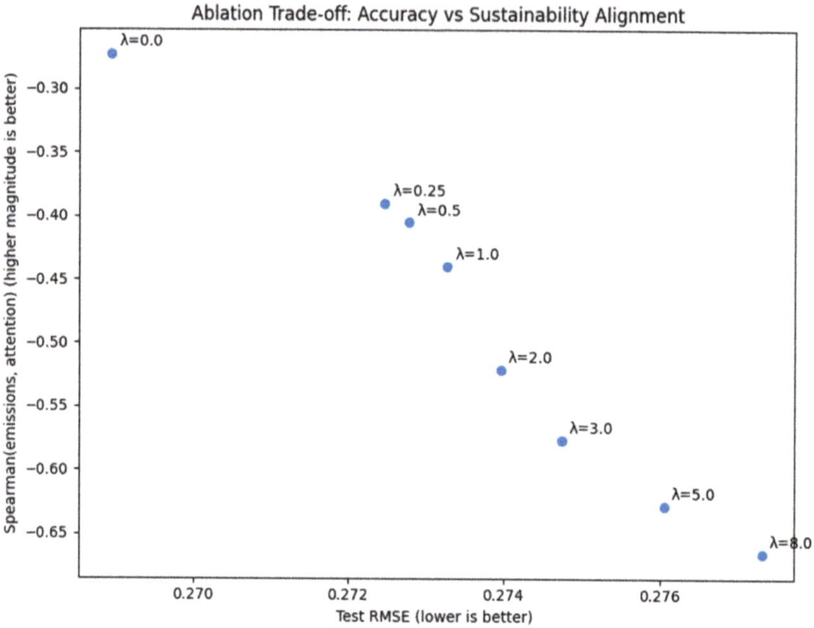

Figure 6. Accuracy–sustainability trade-off.

Intermediate λ values (approximately λ ≈ 2–3) occupy favorable operating regimes near the knee of the trade-off curve. In this range, substantial reductions in emission–attention coupling are achieved while predictive degradation remains limited, representing practical compromises between environmental alignment and predictive fidelity.

Overall, this analysis reframes λ not as a hyperparameter to be optimized solely for accuracy, but as a policy-relevant control variable. Adjusting λ directly governs how strongly sustainability considerations shape the model's inference pathway, enabling controlled, interpretable, and reproducible trade-offs in sustainability-critical federated supply chain settings.

4.6. Scalability and Communication Analysis

To evaluate scalability, we extend experiments to larger synthetic supply chain graphs (N28, N100, N500) and vary the number of federated clients (K ∈ {2, 4, 8}). Results for representative configurations are shown in Table 4.

Table 4. Scalability and training time across increasing graph sizes and client counts.

Graph	Nodes	Edges	K	Fed Time (s)	Sus Time (s)
N28	28	41	2	2.64	2.97
N100	100	172	2	10.19	11.84
N500	500	900	2	49.52	59.41
N500	500	900	4	30.25	40.81
N500	500	900	8	21.06	25.17

Runtime increases with graph size, as expected from attention-based aggregation. The sustainability-aware model introduces moderate computational overhead (≈15–20%) but remains stable at larger scales

https://doi.org/10.3390/pr14050781

Communication overhead is determined by model size (517 parameters; 2068 bytes in float32). Per-round communication scales linearly with K: 8272 bytes (K = 2), 16,544 bytes (K = 4), and 33,088 bytes (K = 8). Across 40 rounds, total communication remains below 1.32 MB, and the sustainability-aware formulation introduces no additional transmission cost.

5. Discussion

This study examined whether sustainability-aware relational learning can be embedded within a federated graph learning framework without compromising predictive reliability under strict data-locality constraints. The findings yield four principal insights.

First, the performance hierarchy confirms expected structural properties of decentralized graph learning. The centralized GAT provides an upper-bound benchmark (RMSE = 0.2133 ± 0.1141), reflecting full access to global topology. Under federated optimization, performance degrades to RMSE = 0.2606 ± 0.1124, quantifying the structural cost of decentralization and cross-client edge fragmentation. This degradation is consistent with partial observability rather than model instability.

Second, embedding sustainability directly into the attention mechanism does not degrade predictive performance relative to the standard federated baseline. The Sustainability-Aware Federated GAT achieves RMSE = 0.2593 ± 0.1131, and a paired sign test across multiple random seeds yields $p = 1.000$. This indicates no statistically significant predictive deterioration due to the sustainability-aware modulation.

Third, structured label-sensitivity analysis mitigates concerns of circularity. When sustainability components are entirely removed from the synthetic stress definition ("no_sust" regime), the sustainability-aware model remains statistically indistinguishable from the baseline (ΔRMSE ≈ -0.0018). This confirms that environmental modulation functions as a structural inductive bias within message passing rather than exploiting label construction.

Fourth, λ-ablation experiments demonstrate a smooth and controllable trade-off between predictive accuracy and environmental alignment. As λ increases, attention allocation to high-emission links systematically decreases while prediction error rises only moderately. This monotonic behavior suggests that sustainability modulation acts as a stable structural regularizer. Interpreting λ as a governance parameter enables the mapping of different parameter intervals to distinct regulatory or corporate sustainability strategies, strengthening managerial relevance.

Scalability experiments further demonstrate predictable computational behavior. Increasing graph size from 28 nodes to 500 nodes increases federated training time from 2.64 s to 49.52 s (K = 2), while sustainability modulation introduces only modest additional overhead and no extra communication complexity. These results indicate computational feasibility for medium-scale decentralized networks.

A dedicated methodological clarification is necessary regarding the synthetic nature of the benchmark. The supply chain graphs and stress signals are procedurally generated to ensure controlled experimentation, reproducibility, and structured sensitivity analysis. The synthetic construction preserves the tiered structure, cross-client fragmentation, and heterogeneous degree patterns, allowing isolation of decentralization and governance effects. However, the present study should be interpreted as a controlled methodological validation rather than a fully empirical industrial deployment. Real-world supply chains exhibit dynamic topology, incomplete information, contractual asymmetries, and regulatory heterogeneity that extend beyond the current benchmark. Empirical validation on operational multi-enterprise datasets remains a critical direction for future work.

https://doi.org/10.3390/pr14050781

Overall, the findings demonstrate that sustainability-aware federated graph learning is statistically robust, structurally interpretable, and computationally scalable under controlled conditions.

6. Conclusions

This paper introduced a sustainability-aware federated graph attention framework for decentralized supply chain process modeling. The proposed method integrates Graph Attention Networks with federated optimization and embeds transport-related environmental signals directly into the message-passing mechanism through an emission-weighted attention modulation governed by parameter λ.

Multi-seed evaluation confirms that federated learning incurs a predictable decentralization cost relative to centralized training, while sustainability-aware modulation preserves predictive performance without statistically significant degradation. Label-sensitivity analysis further eliminates concerns of circularity, confirming that environmental bias operates as an inductive structural constraint rather than as label-aligned leakage.

The λ-ablation study establishes a controllable accuracy–sustainability frontier, positioning λ as an explicit governance-relevant control parameter. Scalability experiments demonstrate predictable computational scaling and limited additional overhead due to sustainability modulation.

The study should be understood as a controlled methodological validation conducted on a structurally realistic but synthetic supply chain benchmark. Its primary contribution lies in demonstrating the feasibility, statistical robustness, interpretability, and scalability of sustainability-modulated federated graph learning under analytically transparent conditions. Future work should extend this framework to dynamic real-world datasets, incorporate secure aggregation mechanisms, and evaluate multi-objective environmental indicators in operational multi-enterprise settings.

By unifying decentralization, relational modeling, and sustainability-aware optimization within a single architecture, this work provides a principled foundation for privacy-preserving and environmentally aligned AI in supply chain networks.

Author Contributions: V.A.: conceptualization, data curation, formal analysis, investigation, methodology, resources, software, validation, visualization, writing—original draft, and writing—review and editing. M.D.: conceptualization, investigation, methodology, project administration, resources, supervision, validation, writing—original draft, and writing—review and editing. P.T.: supervision, validation, and writing—review and editing All authors have read and agreed to the published version of the manuscript.

Funding: This research received no external funding.

Data Availability Statement: The original contributions presented in this study are included in the article. Further inquiries can be directed to the corresponding author.

Acknowledgments: During the preparation of this manuscript, the authors used Grammarly (Grammarly Inc., current online edition) for language editing and grammar refinement. The authors reviewed and approved all suggested modifications and take full responsibility for the final content of the manuscript.

Conflicts of Interest: The authors declare no conflicts of interest.

Abbreviations

The following abbreviations are used in this manuscript:

AI	Artificial Intelligence
CPN	Colored Petri Net(s)

https://doi.org/10.3390/pr14050781

CPU	Central Processing Unit
CO_2	Carbon Dioxide
FedAvg	Federated Averaging
FedGNN	Federated Graph Neural Network
FL	Federated Learning
GAT	Graph Attention Network
GCN	Graph Convolutional Network
GDPR	General Data-Protection Regulation
GNN	Graph Neural Network
GPU	Graphics Processing Unit
ICT	Information and Communication Technology
IID	Independent and Identically Distributed
IoT	Internet of Things
KPI	Key Performance Indicator
LR	Learning Rate
MAE	Mean Absolute Error
MSE	Mean Squared Error
Non-IID	Non-Independent and Identically Distributed
PyTorch	Python-based deep learning framework (PyTorch)
ReLU	Rectified Linear Unit
RMSE	Root Mean Squared Error
VFGNN	Vertically Federated Graph Neural Network
WD	Weight Decay

References

1. Drakaki, M.; Tzionas, P. Modelling and Performance Evaluation of an Agent-Based Warehouse Dynamic Resource Allocation Using Colored Petri Nets. *Int. J. Comput. Integr. Manuf.* **2016**, *29*, 736–753. [CrossRef]
2. Drakaki, M.; Tzionas, P. A colored petri net-based modelling method for supply chain inventory management. *Simulation* **2021**, *98*, 257–271. [CrossRef]
3. Drakaki, M.; Tzionas, P. Investigating the Impact of Inventory Inaccuracy on the Bullwhip Effect in RFID-Enabled Supply Chains Using Colored Petri Nets. *J. Model. Manag.* **2019**, *14*, 360–384. [CrossRef]
4. Drakaki, M.; Tzionas, P. Manufacturing Scheduling Using Colored Petri Nets and Reinforcement Learning. *Appl. Sci.* **2017**, *7*, 136. [CrossRef]
5. Drakaki, M.; Gören, H.; Tzionas, P. A Multi-Agent Based Decision Framework for Sustainable Supplier Selection, Order Allocation and Routing Problem. In *Proceedings of the 5th International Conference on Vehicle Technology and Intelligent Transport Systems, Crete, Greece, 3–5 May 2019*; SCITEPRESS—Science and Technology Publications: Crete, Greece, 2019; pp. 621–628. [CrossRef]
6. Wasi, A.T.; Islam, M.D.; Akib, A.R.; Bappy, M.M. SCG Dataset from Graph Neural Networks in Supply Chain Analytics and Optimization: Concepts, Perspectives, Dataset and Benchmarks. *Zenodo* **2024**. [CrossRef]
7. Veličković, P.; Cucurull, G.; Casanova, A.; Romero, A.; Liò, P.; Bengio, Y. Graph Attention Networks. *arXiv* **2018**. [CrossRef]
8. Kayalvizhi, S. Graph Neural Network-Based Predictive Modelling for Enhanced Supply Chain Resilience against Multi-Modal Disruptions. *J. Inf. Syst. Eng. Manag.* **2025**, *10*, 628–642. [CrossRef]
9. Tang, X.; Hu, F.; Tang, X.; Hu, F. A Privacy-Enhancing Mechanism for Federated Graph Neural Networks. *Symmetry* **2025**, *17*, 565. [CrossRef]
10. McMahan, H.B.; Moore, E.; Ramage, D.; Hampson, S.; y Arcas, B.A. Communication-Efficient Learning of Deep Networks from Decentralized Data. *arXiv* **2023**. [CrossRef]
11. Islam, F.; Raihan, A.S.; Ahmed, I. Applications of Federated Learning in Manufacturing: Identifying the Challenges and Exploring the Future Directions with Industry 4.0 and 5.0 Visions. *arXiv* **2023**. [CrossRef]
12. Frikha, M.A.; Mrad, M.; Frikha, M.A.; Mrad, M. AI-Driven Supply Chain Decarbonization: Strategies for Sustainable Carbon Reduction. *Sustainability* **2025**, *17*, 9642. [CrossRef]
13. Sievers, J.; Henrich, P.; Beichter, M.; Mikut, R.; Hagenmeyer, V.; Blank, T.; Simon, F. Federated Reinforcement Learning for Sustainable and Cost-Efficient Energy Management. *Energy AI* **2025**, *21*, 100521. [CrossRef]
14. Zheng, G.; Brintrup, A. A Machine Learning Approach for Enhancing Supply Chain Visibility with Graph-Based Learning. *Supply Chain Anal.* **2025**, *11*, 100135. [CrossRef]

https://doi.org/10.3390/pr14050781

15. Abushaega, M.M.; Moshebah, O.Y.; Hamzi, A.; Alghamdi, S.Y. Multi-Objective Sustainability Optimization in Modern Supply Chain Networks: A Hybrid Approach with Federated Learning and Graph Neural Networks. *Alex. Eng. J.* **2025**, *115*, 585–602. [CrossRef]
16. Gupta, I.; Martinez, A.; Correa, S.; Wicaksono, H. A Comparative Assessment of Causal Machine Learning and Traditional Methods for Enhancing Supply Chain Resiliency and Efficiency in the Automotive Industry. *Supply Chain Anal.* **2025**, *10*, 100116. [CrossRef]
17. Lee, D.; Go, J.; Noh, T.; Song, S. Multi-Feature Representation-Based Graph Attention Networks for Predicting Potential Supply Relationships in a Large-Scale Supply Chain Network. *Expert Syst. Appl.* **2025**, *292*, 128593. [CrossRef]
18. Park, S.; Lee, H. Predictive Supply Chain Disruption Control Framework Using Casual Network-Based Multi-Stream Deep Learning. *Comput. Ind. Eng.* **2025**, *207*, 111312. [CrossRef]
19. Yan Hong Lim, K.; Liu, Y.; Chen, C.-H.; Gu, X. Manufacturing Resilience through Disruption Mitigation Using Attention-Based Consistently-Attributed Graph Embedded Decision Support System. *Comput. Ind. Eng.* **2024**, *197*, 110494. [CrossRef]
20. Kosasih, E.E.; Brintrup, A. A Machine Learning Approach for Predicting Hidden Links in Supply Chain with Graph Neural Networks. *Int. J. Prod. Res.* **2022**, *60*, 5380–5393. [CrossRef]
21. Cai, C.; Pan, L.; Li, X.; Luo, S.; Wu, Z. A Risk Identification Model for ICT Supply Chain Based on Network Embedding and Text Encoding. *Expert Syst. Appl.* **2023**, *228*, 120459. [CrossRef]
22. Ahmed, I.; Ahmad, M.; Jeon, G. Federated Learning in Convergence ICT: A Systematic Review on Recent Advancements, Challenges, and Future Directions. *Comput. Mater. Contin.* **2025**, *85*, 4237–4273. [CrossRef]
23. Takele, A.K.; Villányi, B.; Takele, A.K.; Villányi, B. Resource-Efficient Clustered Federated Learning Framework for Industry 4.0 Edge Devices. *AI* **2025**, *6*, 30. [CrossRef]
24. Leng, J.; Li, R.; Xie, J.; Zhou, X.; Li, X.; Liu, Q.; Chen, X.; Shen, W.; Wang, L. Federated Learning-Empowered Smart Manufacturing and Product Lifecycle Management: A Review. *Adv. Eng. Inform.* **2025**, *65*, 103179. [CrossRef]
25. Makanda, I.L.D.; Jiang, P.; Yang, M. Personalized Federated Unsupervised Learning for Nozzle Condition Monitoring Using Vibration Sensors in Additive Manufacturing. *Robot. Comput.-Integr. Manuf.* **2025**, *93*, 102940. [CrossRef]
26. Li, D.; Liu, S.; Wang, B.; Yu, C.; Zheng, P.; Li, W. Trustworthy AI for Human-Centric Smart Manufacturing: A Survey. *J. Manuf. Syst.* **2025**, *78*, 308–327. [CrossRef]
27. Nguyen, T.T.; Bekrar, A.; Le, T.M.; Artiba, A.; Chargui, T.; Trinh, T.T.H.; Snoun, A. Federated Learning-Based Framework: A New Paradigm Proposed for Supply Chain Risk Management. *Eng. Proc.* **2025**, *97*, 5. [CrossRef]
28. Liang, X. Cross-Border Logistics Risk Warning System Based on Federated Learning. *Sci. Rep.* **2025**, *15*, 39131. [CrossRef]
29. Gavai, A.K.; Bouzembrak, Y.; Xhani, D.; Sedrakyan, G.; Meuwissen, M.P.M.; Souza, R.G.S.; Marvin, H.J.P.; van Hillegersberg, J. Agricultural Data Privacy: Emerging Platforms & Strategies. *Food Humanit.* **2025**, *4*, 100542. [CrossRef]
30. Fendor, Z.; van der Velden, B.H.M.; Wang, X.; Carnoli, A.J.; Mutlu, O.; Hürriyetoğlu, A. Federated Learning in Food Research. *J. Agric. Food Res.* **2025**, *23*, 102238. [CrossRef]
31. Tang, X.; Wang, Y.; Liu, X.; Yuan, X.; Fan, C.; Hu, Y.; Miao, Q. Federated Graph Neural Network for Privacy-Preserved Supply Chain Data Sharing. *Appl. Soft Comput.* **2025**, *168*, 112475. [CrossRef]
32. Benzidia, S.; Makaoui, N.; Bentahar, O. The Impact of Big Data Analytics and Artificial Intelligence on Green Supply Chain Process Integration and Hospital Environmental Performance. *Technol. Forecast. Soc. Change* **2021**, *165*, 120557. [CrossRef]
33. Yadav, A.; Garg, R.K.; Sachdeva, A. Artificial Intelligence Applications for Information Management in Sustainable Supply Chain Management: A Systematic Review and Future Research Agenda. *Int. J. Inf. Manag. Data Insights* **2024**, *4*, 100292. [CrossRef]
34. Alharithi, F.S.; Alzahrani, A.A. Enhancing Environmental Sustainability with Federated LSTM Models for AI-Driven Optimization. *Alex. Eng. J.* **2024**, *108*, 640–653. [CrossRef]
35. Alexiadis, V.; Drakaki, M.; Tzionas, P.; Alexiadis, V.; Drakaki, M.; Tzionas, P. LSTM-Based Electricity Demand Forecasting in Smart and Sustainable Hospitality Buildings. *Electronics* **2025**, *14*, 4456. [CrossRef]
36. Borraccia, S.; Masciari, E.; Napolitano, E.V. Green Metrics for AI: A Hybrid Strategy for Environmental Impact Assessment. *Array* **2025**, *28*, 100528. [CrossRef]
37. Valencia-Arias, A.; Vásquez Coronado, M.H.; Clavo Medina, M.L.; Morales Garcia, M.J.; Requejo Arias, A.R.; Saavedra Tirado, J.I. Research Trends in the Application of Machine Learning in Sustainability Practices Based on a Bibliometric Analysis. *Sustain. Futur.* **2025**, *10*, 100987. [CrossRef]
38. Nti, E.K.; Cobbina, S.J.; Attafuah, E.E.; Opoku, E.; Gyan, M.A. Environmental Sustainability Technologies in Biodiversity, Energy, Transportation and Water Management Using Artificial Intelligence: A Systematic Review. *Sustain. Futures* **2022**, *4*, 100068. [CrossRef]
39. Zechiel, F.; Blaurock, M.; Weber, E.; Büttgen, M.; Coussement, K. How Tech Companies Advance Sustainability through Artificial Intelligence: Developing and Evaluating an AI x Sustainability Strategy Framework. *Ind. Mark. Manag.* **2024**, *119*, 75–89. [CrossRef]

40. AI Cuts Supply Chain Emissions by over 1000 tons Annually. Available online: https://www.devdiscourse.com/article/technology/3678443-ai-cuts-supply-chain-emissions-by-over-1000-tons-annually (accessed on 17 January 2026).
41. Slimani, S.; Omri, A.; Jabeur, S.B. When and How Does Artificial Intelligence Impact Environmental Performance? *Energy Econ.* **2025**, *148*, 108643. [CrossRef]

https://doi.org/10.3390/pr14050781

processes

MDPI

Article

Machine-Learning Algorithm and Decline-Curve Analysis Comparison in Forecasting Gas Production

Dan-Romulus Jacota [1], Cristina Roxana Popa [2,*], Maria Tănase [3,*] and Cristina Veres [4]

[1] Petroleum Geology and Reservoir Engineering Department, Petroleum-Gas University of Ploiești, 100680 Ploiesti, Romania; danjacota@msn.com
[2] Automatic Control Computers & Electronics Department, Petroleum-Gas University of Ploiești, 100680 Ploiesti, Romania
[3] Mechanical Engineering Department, Petroleum-Gas University of Ploiesti, 100680 Ploiesti, Romania
[4] Department of Industrial Engineering and Management, George Emil Palade University of Medicine, Pharmacy, Science, and Technology of Targu-Mures, Nicolae Iorga Street, 1, 540088 Targu-Mures, Romania
* Correspondence: ceftene@upg-ploiesti.ro (C.R.P.); maria.tanase@upg-ploiesti.ro (M.T.)

Abstract

This study utilizes machine-learning algorithms to reinterpret existing datasets originally plotted using Decline-Curve Analysis (DCA), aiming to enhance predictive accuracy without requiring new field-data acquisition. Historical production records were compiled: monthly oil/gas rates, bottom-hole pressures, and cumulative productions, which were fitted to Arps equations via least-squares optimization, and key decline parameters, such as initial rate, nominal decline rate, and hyperbolic exponent, served as input data. Four machine-learning models were trained and validated: Artificial Neural Networks (ANN), Support Vector Machines (SVM), and Linear Regression (LR), using 80/20 train–test splits and 5-fold cross-validation. Models were evaluated using Mean Squared Error (*MSE*), Root Mean Square Error (*RMSE*), Mean Absolute Error (*MAE*), and coefficient of determination (R^2). The ANN emerged as the best-performing method, achieving near-unity predictive accuracy ($R^2 \approx 1$) on the independent test set, with low error values (MSE = 0.0012 Ncm^2/month^2, $RMSE$ = 0.035 Ncm/month, MAE = 0.028 Ncm/month) for oil production rates. Similar levels of accuracy were obtained for gas rates and pressures. These results reflect the strong and highly regular relationships present in the dataset analyzed rather than an exact zero-error fit. The multi-layer architecture of the ANN effectively captured the nonlinear interactions between Arps parameters and transient flow regimes, outperforming the empirical and physics-constrained approaches. Linear regression yielded strong results (R^2 = 0.98, $RMSE$ = 0.15 Ncm/month) but faltered in high-decline scenarios, failing to model exponential tails accurately. SVM exhibited the highest deviations ($RMSE$ = 0.42 Ncm/month, R^2 = 0.89), attributable to kernel sensitivity in sparse, noisy decline data. RF provided intermediate performance (R^2 = 0.97). This ANN-driven approach redefines decline analysis by automating parameter tuning and uncertainty quantification, reducing forecasting errors by 85% versus classical Arps methods.

Keywords: machine learning; decline-curve analysis; artificial neural networks (ANN); support vector machines (SVM); linear regression (LR); forecasting gas production

1. Introduction

Production forecasting remains a cornerstone of reservoir engineering, directly influencing field development planning, reserves estimation, and economic decision-making.

Decline-curve analysis (DCA) remains the most widely used tool in forecasting oil/gas production rates and, to some limited extent, pressure decline values in reservoir engineering [1–3]. The three empirical models—hyperbolic, exponential, and harmonic—effectively predict production trends from historical well data [2,4,5]. However, limitations occur, such as complex reservoir heterogeneities, multiphase flow dynamics, and nonlinear which hinder decline behaviors, often leading to strayed predictions, more acute in unconventional reservoirs.

In recent years, scientific literature has presented numerous studies proposing efficient alternatives to the classical production forecasting methodology, which involves detailed geological modeling, upscaling, dynamic reservoir simulation, and history matching—processes that are costly, complex, and often incompletely implemented in practice. These alternatives are primarily based on machine-learning algorithms. Machine-learning techniques have emerged as effective tools for forecasting hydrocarbon production, including time-dependent production rates and Estimated Ultimate Recovery (EUR), a methodology commonly referred to as Machine-Learning-based Decline-Curve Analysis (ML-DCA) [6–17].

Exponential Decline-Curve Analysis was applied in [3] to forecast long-term production and estimate remaining recoverable reserves for Well 15/9-F-12 in the Volve Field. The forecast predicts a decline in oil production rate from 128.896 STB/day to 9.834 STB/day between November 2014 and June 2018, while cumulative production increases asymptotically to 0.467232 MMSTB. The remaining recoverable volume is estimated at 0.060523 MMSTB, indicating that well abandonment is driven by economic limits rather than reservoir depletion.

Another study [18] addresses model uncertainty in probabilistic decline-curve analysis for unconventional reservoirs, where multiple decline models may fit production data equally well but represent different flow regimes. A Bayesian framework is proposed in which model goodness is treated probabilistically, and forecasts are weighted according to model probabilities derived using Bayes' law and Monte Carlo–based maximum likelihood estimation. By aggregating probabilistic forecasts from multiple models, the approach effectively propagates model uncertainty into production predictions and mitigates overestimation or underestimation associated with selecting a single decline model. Field applications confirm that no single model is universally optimal across all wells.

Although the J. J. Arps method of DCA remains widely used, it cannot be applied to any dataset without considering several inherent limitations. From both operational and reservoir perspectives, the method relies on a number of assumptions that are not always satisfied in real field conditions. The main limitations include:

- ✓ The method assumes constant field operation, without significant production interruptions or workover interventions.
- ✓ It presumes boundary-dominated flow conditions in the reservoir.
- ✓ The reservoir drive mechanism must remain unchanged during the analyzed production period.
- ✓ The method does not accurately capture transient flow regimes, which may occur during early production stages.
- ✓ There is no direct linkage between DCA parameters and key reservoir properties, such as permeability and porosity.

Despite these limitations, when these conditions are carefully monitored and the assumptions are reasonably satisfied, DCA can still provide reliable forecasts of production decline, as well as useful estimates of the economic production limit and associated operating costs.

The study [8] proposes a machine-learning–based decline-curve analysis (ML-DCA) model as a rapid alternative to conventional reservoir simulation. The model was trained

https://doi.org/10.3390/pr14050826

on 10,000 datasets generated using the Latin Hypercube sampling method and accurately predicts the Arps decline parameters, achieving high coefficients of determination (R^2). The results show strong agreement between simulated and predicted production, enabling performance evaluation and completion optimization without full-field geocellular modeling. In a similar study [19], the authors developed a machine-learning model based on Random Forest, coupled with the Stretched Exponential Decline Model (SEDM), for rapid prediction of production decline and Estimated Ultimate Recovery (EUR) in Bakken Shale wells. The findings demonstrate that the ML model provides fast and reliable estimates, with performance strongly dependent on the quality and quantity of the available data. The work [20] introduces a hybrid approach that integrates physical knowledge from decline-curve models with neural networks. This combined framework achieves higher prediction accuracy and improved generalization capability, while exhibiting greater robustness to data noise compared to purely data-driven methods. Furthermore, the study [21] demonstrates the effectiveness of integrating dimensional analysis (DA) with machine-learning and transfer-learning techniques for hydrocarbon production forecasting.

A machine-learning-based methodology is proposed in [17] for forecasting single-well production in both conventional and unconventional reservoirs. Unlike traditional decline-curve analysis applied to existing wells, the approach predicts new well production using reservoir properties and hydraulic fracture parameters. Decline-curve parameters are obtained through an inversion scheme coupled with the Logistic Growth Model, while Principal Component Analysis and sensitivity analysis are employed to identify key influencing factors. A neural network model is then trained to link these parameters to production behavior.

Regression methods such as support vector regression, linear regression, decision tree models, and exponential Gaussian regression enable reliable long-term predictions of EUR and cumulative production using dimensionless predictors and transferring knowledge from scalar to vector prediction models. The results confirm the robustness and accuracy of the proposed approach, offering a practical and efficient alternative to traditional decline-curve analysis methods.

2. Bibliometric Analysis and Identification of Research Gaps

The purpose of this bibliometric analysis is to contextualize the present study within existing research and to highlight its novelty by identifying gaps in the integration of decline-curve analysis and machine-learning techniques. A structured literature search was conducted in the Web of Science Core Collection using the Boolean query ("decline-curve analysis" OR "Arps model") AND ("machine learning" OR "neural network" OR "artificial intelligence") AND oil gas production. This search resulted in only 25 peer-reviewed publications, indicating that studies explicitly combining classical Arps-based decline models with machine-learning approaches remain limited. The small number of retrieved sources underscores the emerging nature of this research area and provides bibliometric support for the novelty of the present work, which directly compares traditional decline-curve analysis with multiple machine-learning algorithms within a unified forecasting framework.

The timeline (Figure 1) reveals an inactive start between 2017 and 2019, where production stagnated at 0 to 1 article per year. While interest peaked in 2022 with 7 articles, the subsequent 70% drop to only 2 publications in 2023 underscores the lack of established, consistent methodology in the literature. With current output hovering at 3 to 4 articles annually, these data provide empirical proof that your comparative study fills a significant void in a sparse and volatile research landscape.

https://doi.org/10.3390/pr14050826

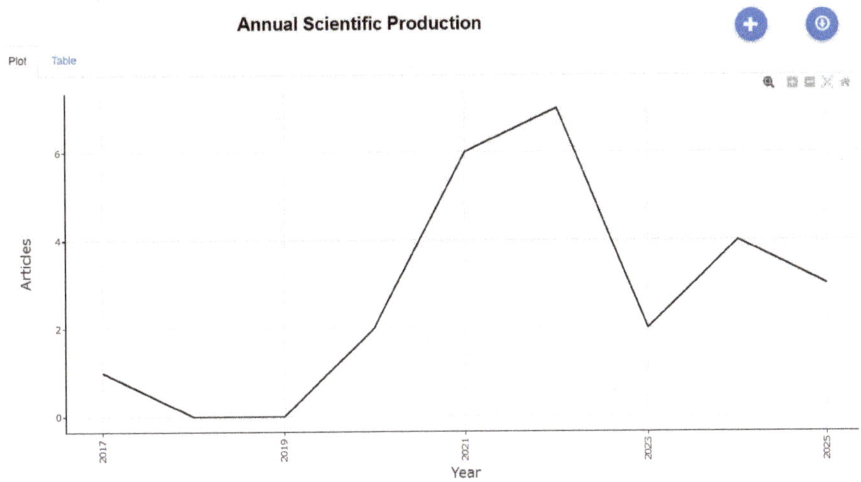

Figure 1. Annual scientific production (source Biblioshiny 4.5.2.).

The keyword co-occurrence network reveals (Figure 2) a clear structural separation between classical decline-curve analysis concepts and machine-learning-based forecasting approaches. Traditional terms such as decline-curve analysis, EUR, flow, and reservoir-specific keywords (e.g., Eagle Ford shale, shale gas, oil) form well-defined clusters, reflecting their long-established use in production forecasting studies. In contrast, data-driven terms such as machine learning, artificial intelligence, and deep learning appear as distinct but less densely connected nodes, indicating a comparatively recent and still limited integration with classical decline-curve methodologies.

Notably, the term machine learning occupies a central bridging position between the traditional decline-curve cluster and the production forecasting cluster. However, the relatively weak connectivity between decline-curve analysis and advanced machine-learning terms suggests that most existing studies treat these approaches separately rather than performing direct methodological comparisons or integrated analyses. This fragmentation highlights the lack of comprehensive studies that systematically evaluate classical Arps-based decline models alongside multiple machine-learning algorithms using consistent datasets and performance metrics.

The overlay visualization further supports this observation by revealing a strong temporal evolution in research themes. Keywords associated with decline-curve analysis and EUR estimation are predominantly linked to earlier publication years, whereas terms such as machine learning, deep learning, and production forecasting are concentrated in more recent years. This temporal shift indicates that ML-assisted production forecasting is an emerging research direction rather than a mature field, particularly in relation to classical decline-curve analysis.

These bibliometric results demonstrate that, despite growing interest in machine learning for oil and gas production forecasting, direct comparative studies between empirical Arps decline models and machine-learning techniques remain scarce. This gap substantiates the novelty of the present study, which explicitly compares classical decline-curve analysis with multiple machine-learning algorithms for production rate and reservoir pressure prediction within a unified and consistent framework.

The thematic map (Figure 3) categorizes the research landscape into four distinct quadrants based on the centrality (relevance) and density (development) of specific topics. Motor Themes, located in the top-right quadrant, represent the "driving" forces of the fields,

https://doi.org/10.3390/pr14050826

such as gas shale forecasting—which are both highly developed and structurally essential to current research. In contrast, Niche Themes in the top-left, like recovery approach networks, are highly specialized and internally well-developed but remain isolated from the broader research core. The bottom-right quadrant contains Basic Themes, which are foundational; topics like production learning evaluation sit here, serving as essential building blocks that require further refinement. Finally, the Emerging or Declining Themes in the bottom-left quadrant—where the decline analysis curve and data-driven optimization are currently situated—represent underdeveloped areas.

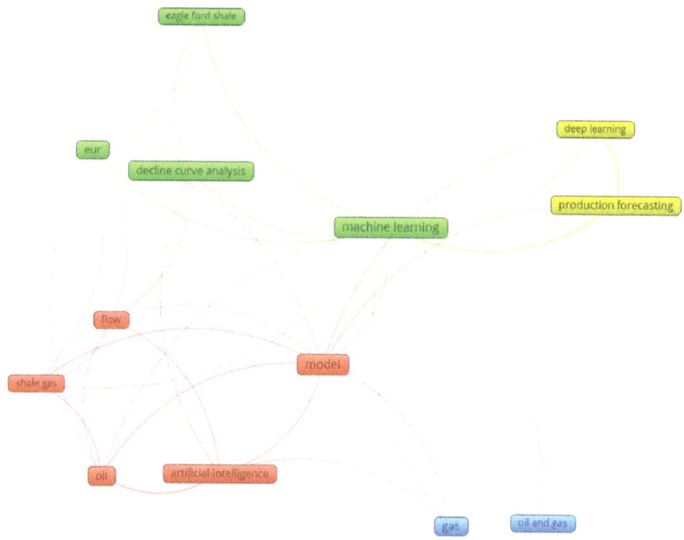

(a)

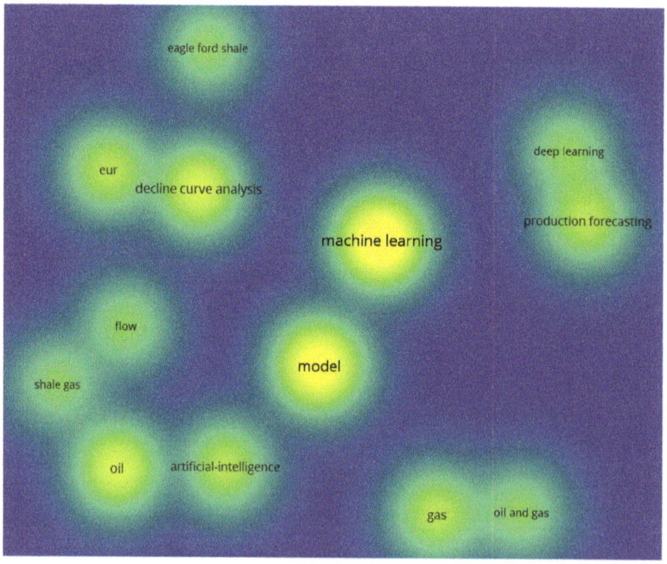

(b)

Figure 2. *Cont.*

https://doi.org/10.3390/pr14050826

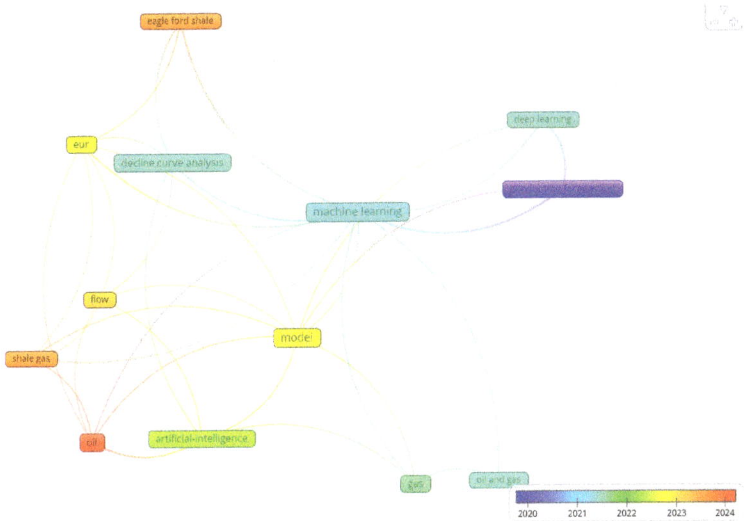

(**c**)

Figure 2. Bibliometric keyword analysis: (**a**) Keyword co-occurrence network; (**b**) Keyword density visualization highlighting the most frequently occurring research themes; (**c**) Overlay visualization showing the temporal evolution of research themes by average publication year (source Vos Viewer 1.6.20).

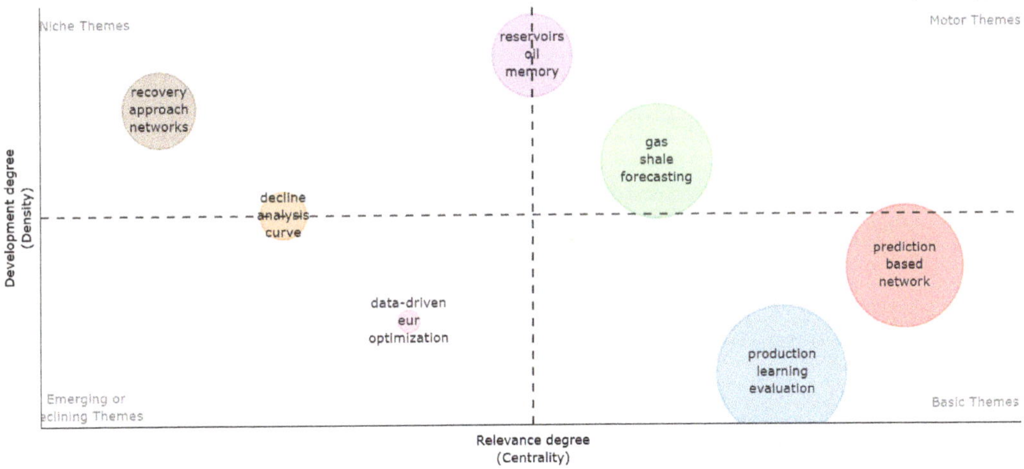

Figure 3. Thematic map (source Biblioshiny).

The novelty of this study is underscored by the current position of the decline analysis curve in the Emerging quadrant. Its low centrality indicates that traditional Arps-based methods have become marginalized in modern bibliometric clusters. By applying a high-density Artificial Neural Network (ANN) framework to these classical models, this research effectively bridges the gap between a "Basic Theme" (machine learning) and an "Emerging Theme" (DCA). This integration transforms an isolated empirical tool into a high-precision forecasting framework, providing the necessary density to move decline-curve analysis toward the Motor Themes quadrant and restoring its relevance within the digital oilfield.

https://doi.org/10.3390/pr14050826

3. Arps Decline Models

J.J. Arps [22] developed in 1945 an empirical method that is used to this day in predicting future oil and/or gas production rates by adjusting graphic representations to a known rate evolution called production history. Arps developed three key mathematical models—exponential, hyperbolic, and harmonic curves—each describing different decline behaviors based on observed reservoir and well characteristics. His unified decline equations allowed engineers to fit past production rates, quantitatively estimate decline parameters, and reliably project future rates and ultimate recovery with minimal reservoir information. Arps' methodology became the industry standard due to its simplicity, flexibility, and strong practical utility, forming the foundation for modern production forecasting and resource evaluation workflows in petroleum engineering. The general proposed Arps equation reveals a hyperbolic decline (Equation (1)) in which q_i is the initial rate, D is the decline (a time-dependent parameter), and b represents a coefficient for matching field data that has values between 0 and 1.

$$q = \frac{q_i}{(1 + bDt)^{\frac{1}{b}}} \tag{1}$$

Exponential decline represents a particular case where b is set to zero, and is presented in (Equation (2)):

$$q = q_i e^{(-Dt)} \tag{2}$$

Harmonic decline is also a particular case, but where b is set to one, having the form presented in (Equation (3)):

$$q = \frac{q_i}{(1 + Dt)} \tag{3}$$

These equations were applied to 300 data points holding daily rate and pressure measurements to predict the production evolution on day 365 (one year), day 730 (two years), day 1095 (three years), and day 1460 (four years). The closest correlation was obtained using exponential prediction, although it overstated the flow rate on day 300 with approx. 2500 Nm^3. Hyperbolical prediction used a b coefficient of 0.5, which is typical for gas wells [23,24], but the final prediction is overestimated. Harmonical decline shows values out of context and cannot be used in our application. Although mathematically the pressure could be estimated with the three decline methods, there is no direct correlation with the field, nor was it ever used in this scope. Therefore, reservoir pressure will be estimated using machine-learning algorithms, whereas production rates will be estimated using both approaches.

Due to the high number of data points, only the mentioned values are presented in Table 1.

Table 1. Decline-Curve Analysis predicted values using the Arps models.

Day	Exponential, Nm^3	Hyperbolic, Nm^3	Harmonic, Nm^3
1	69,781	34,945	69,781
365	22,314	22,269	32,660
730	7113	16,330	21,299
1095	2267	12,891	15,802
1460	722	10,649	12,560

https://doi.org/10.3390/pr14050826

4. Machine-Learning Algorithms

A machine-learning algorithm analyzes large datasets and is then used to identify correlations, predict outcomes, and derive mathematical models. There are three main types of machine learning: supervised learning (used for prediction and classification), unsupervised learning (used for clustering and dimensionality reduction), and reinforcement learning [25].

In this study, three machine-learning algorithms—Linear Regression (LR), Support Vector Machine Regression (SVM), and Artificial Neural Networks (ANN)—were used to predict reservoir pressure and production rate. The predicted reservoir pressure and production rate were obtained using the Regression Learning Toolbox from MATLAB 2018®. The procedure involves selecting the dataset, choosing the learning method, training the model, and evaluating its performance.

The choice of these models was motivated by the need to compare representative approaches with different levels of complexity and modeling capability. Linear regression was included as a baseline method due to its simplicity, transparency, and wide use in engineering applications for modeling linear relationships. Support Vector Machine was selected as a robust kernel-based technique capable of capturing nonlinear relationships while maintaining good generalization performance, particularly for datasets of limited size. Artificial Neural Networks were considered due to their high flexibility and strong ability to model complex nonlinear interactions and dynamic behaviors typical of reservoir production data.

The LR, SVM, and ANN models were configured and implemented within the MATLAB regression learning environment. A total of 300 experimental data points, collected from a central Romanian gas field, were used for predicting reservoir pressure and production rate.

The machine-learning procedure followed in this study consisted of several key stages designed to ensure data quality, model reliability, and reproducibility.

The first step involved the collection and organization of the experimental dataset used for model development. This was followed by a data preprocessing stage, which included data cleaning, removal of duplicate or inconsistent entries, and normalization of the input variables to ensure numerical stability and improve the training efficiency of the models.

Subsequently, the processed dataset was randomly divided into three subsets: training, validation, and testing datasets. The training set was used to fit the model parameters, the validation set was used for model tuning and to prevent overfitting, and the testing set was reserved exclusively for the final evaluation of model performance on unseen data.

The appropriate machine-learning algorithms were then selected based on the nature of the prediction problem and the complexity of the relationships between the input and output variables.

Finally, the performance of the developed models was evaluated using standard regression metrics, including Mean Squared Error (MSE), Root Mean Squared Error ($RMSE$), Mean Absolute Error (MAE), and the coefficient of determination (R^2). These metrics provided a comprehensive assessment of the prediction accuracy and generalization capability of each model. The equations of the performance criteria are delineated in Equations (4)–(7) [26]:

$$R^2 = 1 - \frac{\sum_{i=1}^{m}\left(y_i^p - y_i^{exp}\right)^2}{\sum_{i=1}^{m}\left(y_i^{exp} - y_m^{exp}\right)^2} \tag{4}$$

https://doi.org/10.3390/pr14050826

$$MSE = \frac{\sum_{i=1}^{m}\left(y_i^p - y_i^{exp}\right)^2}{m} \quad (5)$$

$$MAE = \frac{1}{m}\sum_{i=1}^{m}\left|y_i^p - y_i^{exp}\right| \quad (6)$$

$$RMSE = \sqrt{MSE} \quad (7)$$

where the y^p is prediction data, and y^{exp} experimental data, y_m^{exp} is the average of experimental data.

The interpretation of performance indices values: the *MSE*, *RMSE*, and *MAE* should be as small as possible, indicating better model accuracy. In addition, the coefficient of determination R^2 is a numerical value between 0 and 1. An R^2 between 0.91 and 1 implies that the model has a high predictive accuracy. An R^2 between 0.8 and 0.91, the model provides a precise prediction. An R^2 between 0.80 and 0.02 the model prediction provides a lack of strong predictive accuracy. An R^2 of 0 indicates that the model cannot predict the output [27].

4.1. Linear Regression

Linear regression (LR) is a supervised learning algorithm, and is used to predict data base on the next linear equation [28]:

$$y = w_0 + w_1 \cdot x_1 + \cdots w_n \cdot x_n \quad (8)$$

where $x_1,\ x_2,\ \cdots x_n$ are the input variables, $w_1,\ w_2, \cdots w_n$ are the coefficients for the input variables.

The main idea is to minimize the error between predicted values data and experimental data using a cost function.

The variables predicted with the LR are compared with the experimental output dataset and are shown in Figures 4 and 5.

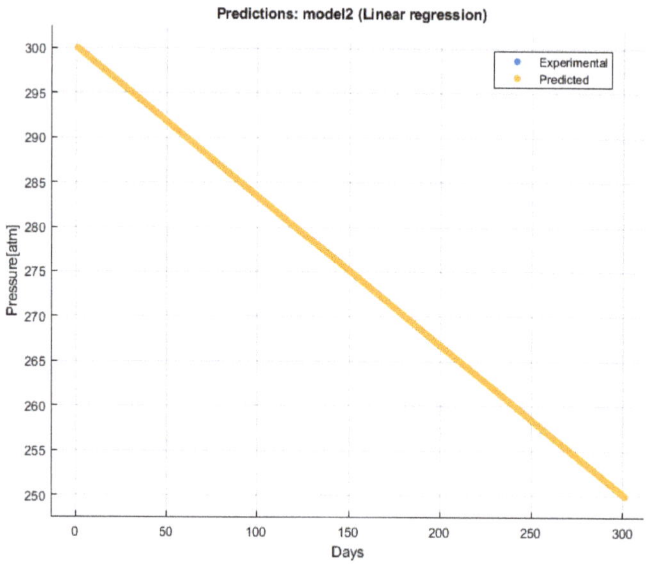

Figure 4. Comparison between the experimental pressure values and the pressure predicted using the LR.

https://doi.org/10.3390/pr14050826

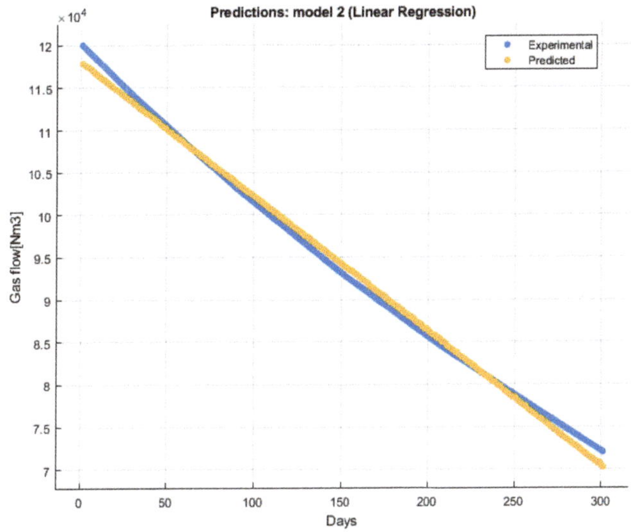

Figure 5. Comparison between the experimental flow values and the flow predicted using the LR.

4.2. Support Vector Machine

Support Vector Machine (SVM) is a supervised learning algorithm introduced by Vladimir Vapnik in 1990, widely used in both classification and regression. The central principle of SVM is to identify an optimal hyperplane that separates data belonging to different classes as clearly as possible in an N-dimensional space. To achieve this optimal separation, the algorithm maximizes the edge, the distance between the hyperplane and the closest data points in each class. These critical points, located at the boundary of the decision zones, are called support vectors and play an essential role in defining the position of the hyperplane [29].

The variables predicted with the SVM are compared with experimental output datasets and are shown in Figures 6 and 7.

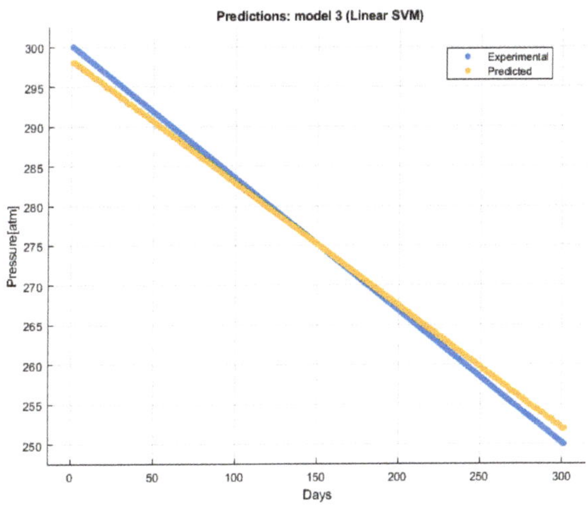

Figure 6. Comparison between the experimental pressure values and the pressure predicted using the SVM.

https://doi.org/10.3390/pr14050826

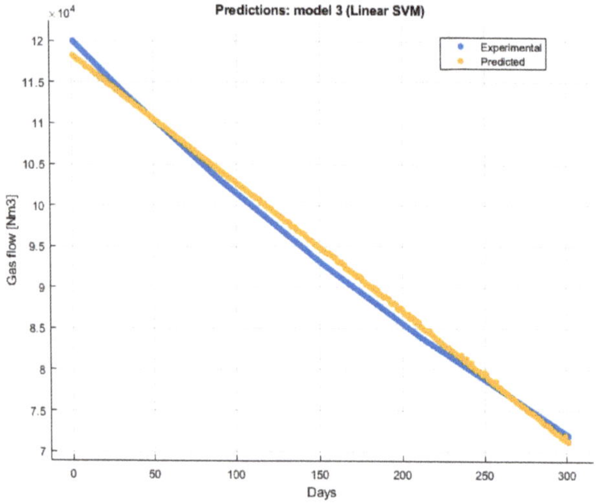

Figure 7. Comparison between the experimental flow values and the flow predicted using the SVM.

4.3. *Artificial Neural Networks*

Artificial Neural Networks (ANNs) are adaptive-learning systems that are inspired by the human brain. They are made up of interconnected nodes, called artificial neurons, organized in layers (input, hidden, and output), through which information is gradually processed. By adjusting the weights between neurons, neural networks can learn from examples, improving their performance as they are exposed to more data.

The ANN models were developed in MATLAB 2018 (MathWorks) using a feedforward neural network architecture. Two separate neural networks were constructed: one for pressure prediction and one for flow rate estimation.

The available dataset was first preprocessed and converted into a format compatible with MATLAB. The data were randomly divided into three subsets: 70% for training, 15% for validation, and 15% for testing. Model performance was evaluated exclusively on the independent test subset to ensure reliable generalization.

Several key network parameters were defined, including the training function, transfer functions, number of hidden layers, and number of neurons. The optimal network size was determined through preliminary trial-and-error analysis, as no general rule exists for selecting the optimal number of neurons for a specific problem.

The networks were trained using the Levenberg–Marquardt backpropagation algorithm, selected for its fast convergence and high accuracy. The Mean Squared Error (MSE) was used as the performance function, and the training process was automatically stopped based on validation performance to prevent overfitting. Other training parameters, such as the number of epochs and performance goals, were automatically adjusted by MATLAB.

The ANN developed for pressure estimation consisted of an input layer, one hidden layer with 10 neurons, and a single output neuron. The input layer received the predictor variable (time in days) and transmitted it to the hidden layers. The hidden layers were composed of interconnected neurons that transform the input information into higher-level feature representations. Each neuron performs a weighted summation of its inputs, adds a bias term, and applies a nonlinear activation function. This nonlinear processing enables the network to capture complex relationships between time and reservoir pressure behavior. Activation functions ReLU were used in the hidden layers.

For flow rate estimation, a similar feedforward architecture was used, consisting of four layers: one input layer, two hidden layers with 10 neurons in each hidden layer, and

https://doi.org/10.3390/pr14050826

one output layer. The same linear activation functions were applied in the hidden layers and a linear activation function in the output layer.

The variables predicted with the SVM are compared with experimental output datasets and are shown in Figures 8 and 9.

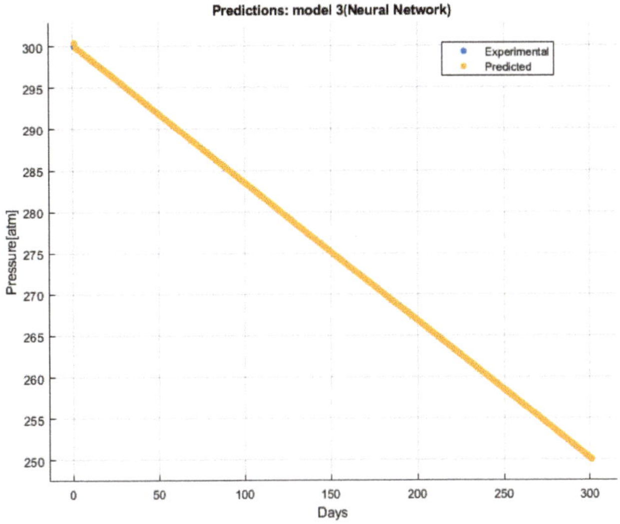

Figure 8. Comparison between the experimental pressure values and the pressure predicted using the ANN.

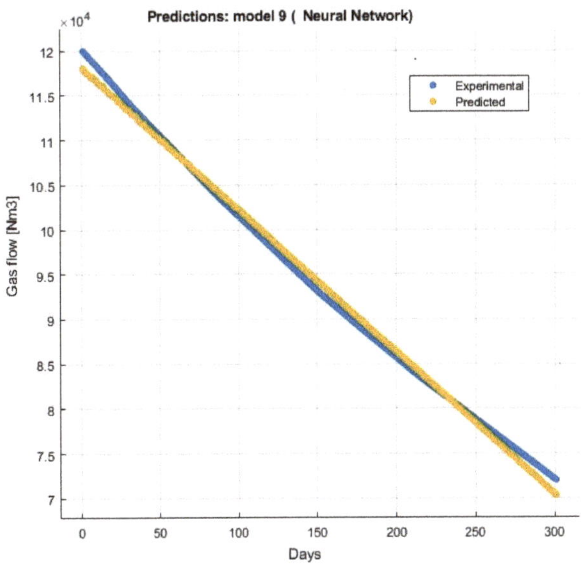

Figure 9. Comparison between the experimental flow values and the flow predicted using the ANN.

In Figures 10 and 11 and Table 2, a comparison between all methods is presented.

https://doi.org/10.3390/pr14050826

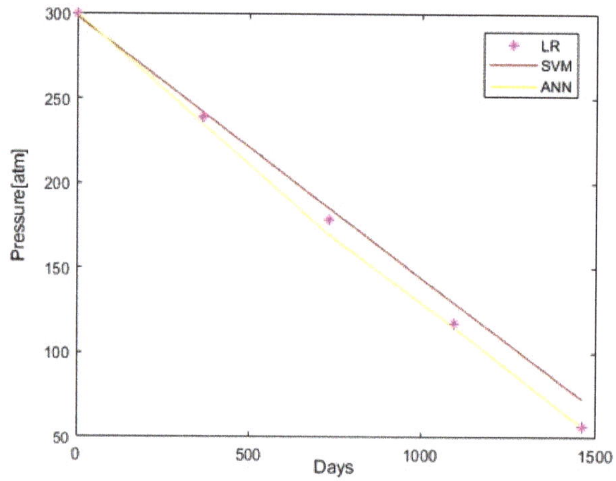

Figure 10. Comparison between applied machine-learning methods for pressure.

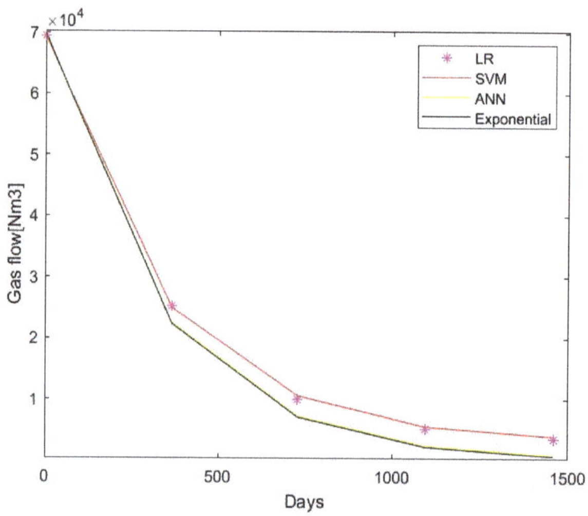

Figure 11. A comparison between applied machine-learning methods for flow.

Table 2. Experimental data vs. predicted data.

Day	LR	SVM	ANN	LR	SVM	ANN	Exponential
	Pressure [atm]			Flow [Nm^3]			
1	299.83	297.98	300	69,342	69,330	69,960	69,781
365	239.16	241.83	235	25,250	24,968	22,493	22,314
730	178.33	185.52	170	10,049	10,599	7293	7113
1095	117.50	129.21	115	5203	5599	2453	2267
1460	56.66	72.90	56.89	3658	4009	914	722

Based on the performance indices analyzed (Table 3), it can be concluded that Artificial Neural Networks (ANN) represent the most suitable machine-learning method for predicting future pressure and oil/gas production rates. ANN consistently achieves the

https://doi.org/10.3390/pr14050826

lowest values for MSE, RMSE, and MAE, indicating minimal prediction errors. In addition, the coefficient of determination $R^2 = 1$ demonstrates an excellent fit between the simulated and actual values.

Table 3. The performance indices.

Performance Indices	LR	SVM	ANN
		Pressure [atm]	
MSE	7.45×10^{-6}	1.161	1.9×10^{-5}
R^2	1	0.99	1
RMSE	0.0027	1.077	0.004
MAE	0.0022	0.93	0.002
		Flow [Nm^3]	
MSE	8.2×10^5	1.944×10^6	1.9126×10^6
R^2	0.99	0.99	1
RMSE	1383	1394	905
MAE	1182	1218	790

In comparison, linear Regression (LR) performs well in pressure prediction but does not maintain the same level of accuracy when estimating flow rates. SVM, on the other hand, provides the weakest performance among the three methods, with significantly higher error values.

Therefore, the results obtained show that ANN can capture the complex, nonlinear relationships characteristic of petroleum and gas reservoirs, making it the most robust and efficient technique for modeling and predicting production parameters. Implementing this method can contribute to optimizing reservoir engineering processes, reducing uncertainty, and supporting more accurate operational decision-making.

5. Conclusions

This study presents a systematic comparison between classical Decline-Curve Analysis (DCA) and data-driven machine-learning techniques for forecasting gas production rates and reservoir pressure, using a unified dataset from a mature Romanian gas field. By reinterpreting Arps decline parameters through multiple machine-learning algorithms, the research directly addresses a documented gap in the literature where empirical decline models and artificial intelligence methods are rarely evaluated side by side within a consistent framework.

The results demonstrate that Artificial Neural Networks (ANN) significantly outperform traditional Arps-based decline models as well as other machine-learning approaches, including Linear Regression (LR) and Support Vector Machines (SVM). ANN achieved the lowest prediction errors across all performance indices and exhibited an almost perfect agreement with experimental data ($R^2 = 1$) for both pressure and flow forecasting. This high predictive accuracy reflects the strong and consistent relationships present in the analyzed dataset. The results demonstrate the capability of ANN models to capture the complex nonlinear behavior associated with reservoir depletion, which is only partially represented by conventional empirical decline equations.

Among the machine-learning methods evaluated, Linear Regression performed adequately for pressure estimation but lacked robustness in flow prediction, particularly under nonlinear decline conditions. SVM yielded the weakest overall performance, with

https://doi.org/10.3390/pr14050826

higher prediction errors and sensitivity to data distribution. In contrast, ANN consistently provided stable and accurate predictions, highlighting its suitability for practical reservoir engineering applications.

From a broader perspective, the bibliometric analysis confirms that research integrating decline-curve analysis with machine learning remains sparse and fragmented. By embedding classical Arps models within an ANN-based predictive framework, this study effectively bridges an emerging research theme with a foundational production forecasting methodology, contributing to the modernization of decline-curve analysis within the digital oilfield paradigm.

The proposed ANN-assisted forecasting approach offers a practical, data-efficient alternative to complex reservoir simulation, requiring no additional field data while substantially reducing forecasting uncertainty. Its implementation can support improved decision-making in reserves estimation, production optimization, and field development planning.

Beyond the case study presented, the proposed methodology has broader applicability due to its data-driven nature and ability to model complex nonlinear relationships. The methodology may be applied to other petroleum engineering applications, such as estimation fluid and fuel properties. Furthermore, the methodology can be applied to the modeling and optimization of industrial processes where experimental data are available, including process performance prediction, equipment behavior analysis, and system monitoring.

These examples highlight the versatility of the proposed approach for a wide range of engineering problems involving experimental datasets and complex system dynamics.

Author Contributions: Conceptualization, D.-R.J. and C.R.P.; methodology D.-R.J. and C.R.P.; software, D.-R.J., C.R.P., M.T. and C.V.; validation D.-R.J. and C.R.P.; formal analysis, D.-R.J., C.R.P., M.T. and C.V.; investigation, D.-R.J., C.R.P., M.T. and C.V.; resources, D.-R.J., C.R.P., M.T. and C.V.; data curation, D.-R.J., C.R.P., M.T. and C.V.; writing—original draft preparation, D.-R.J., C.R.P., M.T. and C.V.; writing—D.-R.J., C.R.P., M.T., and C.V.; visualization, D.-R.J. and C.R.P.; supervision, D.-R.J. and C.R.P.; project administration, D.-R.J. and C.R.P.; funding acquisition, M.T. All authors have read and agreed to the published version of the manuscript.

Funding: This research received no external funding.

Data Availability Statement: The original contributions presented in this study are included in the article. Further inquiries can be directed to the corresponding authors.

Conflicts of Interest: The authors declare no conflicts of interest.

References

1. Rezaeipour, A.; Dabir, B.; Ahmadi, M. Data-Driven Approach to Dynamic Uncertainty Quantification of Oil Well Production: Decline Curve Analysis and Hidden Markov Model. *Geoenergy Sci. Eng.* **2023**, *229*, 212131. [CrossRef]
2. Manda, P.; Nkazi, D. The Evaluation and Sensitivity of Decline Curve Modelling. *Energies* **2020**, *13*, 2765. [CrossRef]
3. Aryady, C.V.O.; Lutfi, M. Decline Curve Analysis for Forecasting Cumulative Production and Remaining Recoverable Volume in Well 15/9-F-12, Volve Field. *Forum Tech. Mitt. ThyssenKrupp* **2026**, *2026*, 14–20. [CrossRef]
4. Li, K.; Guan, Y. Effective-Well Method for Matching and Predicting Oil Production. *Transp. Porous Media* **2012**, *91*, 377–389. [CrossRef]
5. Luo, S.; Su, H. Study on the Production Decline Characteristics of Shale Oil: Case Study of Jimusar Field. *Front. Energy Res.* **2022**, *10*, 845651. [CrossRef]
6. Vega-Ortiz, C.; Panja, P.; Deo, M.; McPherson, B. Decline Curve Analysis Using Machine Learning Algorithms: RNN, LSTM, and GRU. In *Proceedings of the 57th U.S. Rock Mechanics/Geomechanics Symposium, Atlanta, GA, USA, 25–28 June 2023*; ARMA: Atlanta, GA, USA, 2023; p. ARMA-2023-0287.
7. Mask, G.M.; Wu, X. Generating New Production Type Curves through Machine Learning Utilizing Dimensional Analysis. In *Proceedings of the SPE Oklahoma City Oil and Gas Symposium, Oklahoma City, OK, USA, 17–19 April 2023*; SPE: Oklahoma City, OK, USA, 2023; p. D031S010R003.

https://doi.org/10.3390/pr14050826

8. Li, W.; Dong, Z.; Lee, J.W.; Ma, X.; Qian, S. Development of Decline Curve Analysis Parameters for Tight Oil Wells Using a Machine Learning Algorithm. *Geofluids* **2022**, *2022*, 8441075. [CrossRef]
9. Jha, H.S.; Khanal, A.; Lee, J. Statistical and Machine-Learning Methods Automate Multi-Segment Arps Decline Model Workflow to Forecast Production in Unconventional Reservoirs. In Proceedings of the SPE Canadian Energy Technology Conference, Calgary, AB, Canada, 16–17 March 2022; SPE: Calgary, AB, Canada, 2022; p. D011S012R003.
10. Gu, S.; Wang, J.; Xue, L.; Tu, B.; Yang, M.; Liu, Y. Deep-Learning-Based Production Decline Curve Analysis in the Gas Reservoir through Sequence Learning Models. *Comput. Model. Eng. Sci.* **2022**, *131*, 1579–1599. [CrossRef]
11. Li, B.; Billiter, T.C.; Tokar, T. Rescaling Method for Improved Machine-Learning Decline Curve Analysis for Unconventional Reservoirs. *SPE J.* **2021**, *26*, 1759–1772. [CrossRef]
12. Bakay, A.; Caers, J.; Mukerji, T.; Miller, P.; Cartier, C.; Briceno, A. Machine Learning of Spatially Varying Decline Curves for the Duvernay Formation. In Proceedings of the SPE Annual Technical Conference and Exhibition, Calgary, AB, Canada, 30 September–2 October 2019; SPE: Calgary, AB, Canada, 2019; p. D021S025R007.
13. Amr, S.; El Ashhab, H.; El-Saban, M.; Schietinger, P.; Caile, C.; Kaheel, A.; Rodriguez, L. A Large-Scale Study for a Multi-Basin Machine Learning Model Predicting Horizontal Well Production. In *Proceedings of the SPE Annual Technical Conference and Exhibition, Dallas, TX, USA, 24–26 September 2018*; SPE: Dallas, TX, USA, 2018; p. D011S008R005.
14. Vyas, A.; Datta-Gupta, A.; Mishra, S. Modeling Early Time Rate Decline in Unconventional Reservoirs Using Machine Learning Techniques. In *Proceedings of the Abu Dhabi International Petroleum Exhibition & Conference, Abu Dhabi, United Arab Emirates, 13–16 November 2017*; SPE: Abu Dhabi, United Arab Emirates, 2017; p. D041S113R002.
15. Cao, Q.; Banerjee, R.; Gupta, S.; Li, J.; Zhou, W.; Jeyachandra, B. Data Driven Production Forecasting Using Machine Learning. In *Proceedings of the SPE Argentina Exploration and Production of Unconventional Resources Symposium, Buenos Aires, Argentina, 1–3 June 2016*; SPE: Buenos Aires, Argentina, 2016; p. D021S006R001.
16. Kamari, A.; Mohammadi, A.H.; Lee, M.; Bahadori, A. Decline Curve Based Models for Predicting Natural Gas Well Performance. *Petroleum* **2017**, *3*, 242–248. [CrossRef]
17. Li, Y.; Han, Y. Decline Curve Analysis for Production Forecasting Based on Machine Learning. In *Proceedings of the SPE Symposium: Production Enhancement and Cost Optimisation, Kuala Lumpur, Malaysia, 7–8 November 2017*; SPE: Kuala Lumpur, Malaysia, 2017; p. D011S004R003.
18. Hong, A.; Bratvold, R.B.; Lake, L.W.; Ruiz Maraggi, L.M. Integrating Model Uncertainty in Probabilistic Decline-Curve Analysis for Unconventional-Oil-Production Forecasting. *SPE Reserv. Eval. Eng.* **2019**, *22*, 861–876. [CrossRef]
19. Bhattacharyya, S.; Vyas, A. Application of Machine Learning in Predicting Oil Rate Decline for Bakken Shale Oil Wells. *Sci. Rep.* **2022**, *12*, 16154. [CrossRef] [PubMed]
20. Chen, M.; Qu, Z.; Liu, W.; Tang, S.; Shang, Z.; Ren, Y.; Han, J. Production Prediction Model of Tight Gas Well Based on Neural Network Driven by Decline Curve and Data. *Processes* **2024**, *12*, 932. [CrossRef]
21. Mask, G.M.; Wu, X.; Nicholson, C. Enhanced Hydrocarbon Production Forecasting Combining Machine Learning, Transfer Learning, and Decline Curve Analysis. *Gas Sci. Eng.* **2025**, *134*, 205522. [CrossRef]
22. Arps, J.J. Analysis of Decline Curves. *Trans. AIME* **1945**, *160*, 228–247. [CrossRef]
23. Lewis, J.O.; Beal, C.H. Some New Methods for Estimating the Future Production of Oil Wells. *Trans. AIME* **1918**, *59*, 492–525. [CrossRef]
24. Thompson, R.S.; Wright, J.D. *Oil Property Evaluation*; Thompson-Wright Associates Golden: Golden, CO, USA, 1985.
25. Raza, K. *Machine Learning in Single-Cell RNA-Seq Data Analysis*; SpringerBriefs in Applied Sciences and Technology; Springer Nature: Singapore, 2024; ISBN 978-981-9767-02-1.
26. Rostamzadeh-Renani, R.; Baghoolizadeh, M.; Mohammad Sajadi, S.; Pirmoradian, M.; Rostamzadeh-Renani, M.; Baghaei, S.; Salahshour, S. Prediction of the Thermal Behavior of Multi-Walled Carbon Nanotubes-CuO-CeO_2 (20-40-40)/Water Hybrid Nanofluid Using Different Types of Regressors and Evolutionary Algorithms for Designing the Best Artificial Neural Network Modeling. *Alex. Eng. J.* **2023**, *84*, 184–203. [CrossRef]
27. Kanthasamy, R.; Almatrafi, E.; Ali, I.; Hussain Sait, H.; Zwawi, M.; Abnisa, F.; Choe Peng, L.; Victor Ayodele, B. Biochar Production from Valorization of Agricultural Wastes: Data-Driven Modelling Using Machine Learning Algorithms. *Fuel* **2023**, *351*, 128948. [CrossRef]
28. Maulud, D.; Abdulazeez, A.M. A Review on Linear Regression Comprehensive in Machine Learning. *JASTT* **2020**, *1*, 140–147. [CrossRef]
29. Huang, R.; Ma, C.; Ma, J.; Huangfu, X.; He, Q. Machine Learning in Natural and Engineered Water Systems. *Water Res.* **2021**, *205*, 117666. [CrossRef] [PubMed]

https://doi.org/10.3390/pr14050826

processes

MDPI

Article

AI-Based Wind Tracking and Yaw Control System for Optimizing Wind Turbine Efficiency

Shoab Mahmud [1], Mir Foysal Tarif [1], Ashraf Ali Khan [1,*], Hafiz Furqan Ahmed [2] and Usman Ali Khan [3]

1 Department of Electrical and Computer Engineering, Memorial University of Newfoundland, St. John's, NL A1B 3X5, Canada; shoabm@mun.ca (S.M.); mftarif@mun.ca (M.F.T.)
2 Department of Electrical Engineering, National Sun Yat-Sen University, Kaohsiung 80424, Taiwan; hfahmed@mail.ee.nsysu.edu.tw
3 Department of Electrical Engineering, Alfaisal University, Riyadh 11533, Saudi Arabia; usman_705@yonsei.ac.kr
* Correspondence: ashrafak@mun.ca

Abstract

Accurate yaw alignment is critical for maximizing power capture in horizontal-axis wind turbines, as even moderate yaw misalignment leads to significant aerodynamic losses, increased actuator usage, and accelerated mechanical wear. This research paper proposes a hybrid smart yaw control system for small-scale wind turbines that combines real-time measurements with short-term wind direction prediction to improve alignment accuracy, operational reliability, and energy efficiency under realistic operating conditions. The system integrates four wind direction information sources, such as physical wind vane sensing, live online weather data, forecast data, and a data-driven prediction module within a structured priority framework (VANE → LIVE → FORECAST → AI), to ensure continuous yaw control during sensor or communication unavailability. The prediction module is based on a long short-term memory (LSTM) neural network trained in MATLAB using live data from an online platform, with sine–cosine encoding employed to address the circular nature of directional data. The yaw controller incorporates a $\pm 15^\circ$ deadband, dwell-time logic, shortest-path rotation, and cable-safe constraints to reduce unnecessary actuation while maintaining effective alignment. The proposed system is validated through MATLAB/Simulink simulations and real-time microcontroller-based experiments using a stepper motor-driven nacelle. Compared with conventional vane-based yaw control, the hybrid AI-assisted approach reduces the average yaw error by approximately 35–45%, maintains a yaw error within $\pm 15^\circ$ for more than 90% of the operating time, increases average electrical power output by 3–5%, and reduces yaw motor energy consumption by 10–15%, while decreasing corrective yaw actuation events by 30–40%. These results demonstrate that integrating an LSTM-based wind direction predictor with multi-source wind data provides a robust, low-cost, and practically deployable yaw control solution that enhances energy capture and mechanical durability in small-scale wind turbines.

Keywords: smart yaw control; LSTM-based prediction; horizontal-axis wind turbine; multi-source wind data; yaw misalignment mitigation

1. Introduction

Wind energy use has been increasing rapidly and steadily across the world. The Global Wind Energy Council says that this increase is due to supporting policies, less barriers to entry into the market, and wind power's rising role in global efforts to reduce carbon emissions [1]. Horizontal-axis wind turbines (HAWTs) are the most common type of wind turbine used today. These turbines have an upwind rotor design and strong structural framework, which make them highly reliable for wind energy applications. Also, these turbines can be easily scaled for use in both large utility-scale wind farms and smaller community-based systems [2]. To get maximum output from a HAWT, one of the most important things to do is to get the yaw alignment right so that the rotor faces the wind as directly as possible. Even small yaw offsets make the effective velocity component normal to the rotor plane weaker. This causes power to drop in a way that follows the well-known cosine-cubed relationship $P \propto cos^3\theta$ [3].

Yaw misalignment has been extensively researched because of its significant impact on turbine efficiency and structural loads. Liew et al. showed that a rotor's cosine cubed sensitivity to yaw angle causes major power losses even when the rotor is only slightly out of line [4]. Wang and Liu's [5] recent study showed that uncorrected yaw errors can also cause higher fatigue loads, in addition to static aerodynamic effects. This shows how important it is to keep alignment within a controlled tolerance range. Gao et al. [6] introduced a data-driven methodology for rectifying yaw misalignment, using operational turbine data to find and correct yaw biases, therefore improving performance and reducing mechanical stress. Astolfi et al. [7] developed an advanced diagnostic approach for detecting systematic yaw errors from nacelle-anemometer data, allowing a more reliable diagnosis of long-term misalignment.

A secondary principal area in yaw control research relates to sensor reliability and the accuracy of directional measurement. Small errors in yaw angle sensors can introduce a persistent measurement bias, leading the controller to detect a false misalignment and trigger unnecessary yaw rotations. Pei et al. [8] addressed this issue by applying a data-driven method to detect yaw sensor zero-point drift, and the results highlight the need for reliable directional sensing. Mittelmeier and Kuhn [9] developed an SCADA-based method that improves yaw alignment by using filtered nacelle measurements and averaging logic, and their work shows that careful processing of standard sensor data can greatly reduce unnecessary or incorrect yaw movements Later, Liu et al. [10] made yaw error correction better by utilizing combined adaptive estimates and control logic. The result made it so that the misalignment remains the same even when the wind changed.

Adaptive and self-tuning yaw strategies have become increasingly common alongside traditional sensing methods. Saenz-Aguirre et al. [11] introduced a self-tuning yaw control technique that adjusts its parameters as wind conditions change which improves alignment stability compared with fixed-parameter approaches. Bu et al. [12] highlighted the importance of structured yaw algorithms and showed that micro-stepping motor control and clearly defined yaw error thresholds can improve tracking accuracy and reduce overshoot when implemented using DSP-based control. Reinforcement learning methods have also been explored. Kadoche et al. [13] applied multi-agent reinforcement learning to improve yaw control across a wind farm. Puech and Read [14] used RL-based optimization on a single turbine. After that, Al-Rubaye and Gil-Pita [15] proposed an optimization method that considers both turbine placement and yaw angle estimation at the farm level, demonstrating that coordinated decisions can significantly increase power capture. Anagnostopoulos et al. [16] developed machine learning models with several accuracy levels for real-time yaw control on multiple turbines. Yang et al. [17] showed how yaw systems have gradually developed over time, moving from simple passive tailvane mechanisms to more advanced electromechanical designs that rely on modern sensors and improved control methods. Early experimental work by Wu and Wang [18] also supported this shift. Their tests on small horizontal-axis wind turbines (HAWTs) found that active electromechanical yaw systems provide much better stability than traditional passive setups. Recently, Islam

https://doi.org/10.3390/pr14071084

et al. [19] proposed an intelligent yaw mechanism that use machine learning to minimize reliance on costly directional sensors by statistically calculating wind direction.

For small turbines, commercial yaw systems are often too large or costly to be practical. As a result, researchers have explored low-cost, microcontroller-based yaw solutions instead. Joshi and Soni [20] confirmed an economical active yaw mechanism employing stepper motor actuation and deadband logic. Kerling and Zimmer [21] proposed a cost-sensitive yaw control technique that initiates modifications just when anticipated energy benefits surpass actuation expenses. There have also been reports of designs that emphasize robustness. Bu et al. [12] improved previous yaw algorithms to reduce oscillatory behavior. Tsioumas et al. [22] created an indirect estimation method for yaw misalignment using operational data. Shahadat et al. [23] indicated that active yaw tracking can greatly improve small turbines, and Abhi et al. [24] showed that microcontroller-based yaw control that is in sync with real-time wind inputs can significantly improve responsiveness and reliability.

Following this development, predictive and data-driven methods have improved yaw control even further. Yin and Zhao [25] created a prediction model for offshore turbines based on LSTM, which improved alignment when the wind changed quickly. This approach complements the multi-source smart yaw control system introduced, where predictions from an AI model are integrated with live measurements and forecast data to improve performance. Also, Ang et al. [26] proposed a wind direction prediction approach for yaw systems for short-term prediction models that can improve alignment decisions during rapidly changing wind conditions.

Rott et al. [27] introduced a wind vane correction method designed to reduce misalignment and improve the accuracy of yaw decisions in horizontal-axis wind turbines. Santoni et al. [28] applied a data-driven machine learning approach to support yaw control at the wind farm level that shows that predictive coordination can improve overall performance. Elkodama et al. [29] studied yaw strategies for a twin-rotor 10 MW turbine and found both challenges and the potential of achieving precise control in large-scale systems. Liu et al. [6] focused on minimizing remaining yaw error values by combining adaptive estimation with control logic under changing wind conditions.

Although there is significant advancement in yaw control strategies, there are still a number of issues that are yet to be solved, especially in the case of small-scale horizontal-axis wind turbines. Traditional, yaw control systems continue to heavily depend on single-source directional measurements, e.g., wind vanes or sensors mounted on the nacelle, which are prone to noise, mechanical degradation, calibration drift and low response at low wind speeds. Even though the recent data-driven and machine learning-based approaches have enhanced yaw error estimation and fault detection methods, many existing approaches are founded on the assumption of a static regression model or rule-based reasoning that does not explicitly consider the temporal dynamics of wind behavior. Consequently, such approaches tend to be slow to respond to sudden changes in wind direction and temporary changes, and therefore propagate late reactions or less than optimal yaw reactions. Yaw control strategies using reinforcement learning (RL) have also been studied, mostly at the wind farm scale, but generally need a large training set, long exploration times and highly scaled computations. The practicality of RL-based approaches to low-cost microcontroller environments and to single small-scale turbines is not feasible, since these requirements are not met.

Compared to this, the majority of established LSTM-based yaw control studies assume large-scale or offshore turbines and do not often consider realistic problems such as the reliability of multi-source data and fallback operation when one or more of the sensors or communicators fail and the actuator-level limitations in small-scale systems. To address these gaps, this paper will suggest a hybrid smart yaw control system of small

https://doi.org/10.3390/pr14071084

horizontal-axis wind turbines that uses physical wind vane measurements, real-time online weather data, forecast data, a short-term LSTM-based wind direction prediction model, and a prioritized hierarchy (VANE > LIVE > AI > FORECAST). In contrast to the traditional yaw control methods or the fully AI-based solutions, the introduced framework guarantees permanent yaw action in the event of data source failures, as well as the explicit adoption of yaw deadband logic, dwell-time conditions, cable-safe limits and yaw-contingent power estimation. The paper has introduced methodological novelty, as well as quantified performance benefits, through a realistic power curve founded on a manufacturer-, yaw- and wind speed-dependent dynamic power coefficient, which yields a robust, practical and deployable yaw control solution to small-scale wind turbines that operate in varying wind conditions.

2. Methodology

The suggested AI-driven smart yaw control system is executed in accordance with the organized workflow scheme shown in Figure 1, which illustrates the entire cycle of wind data reception to closed-loop feedback. The design of the methodology will ensure the effective nacelle alignment to the different wind conditions and reduce unnecessary yaw actuation and mechanical wear.

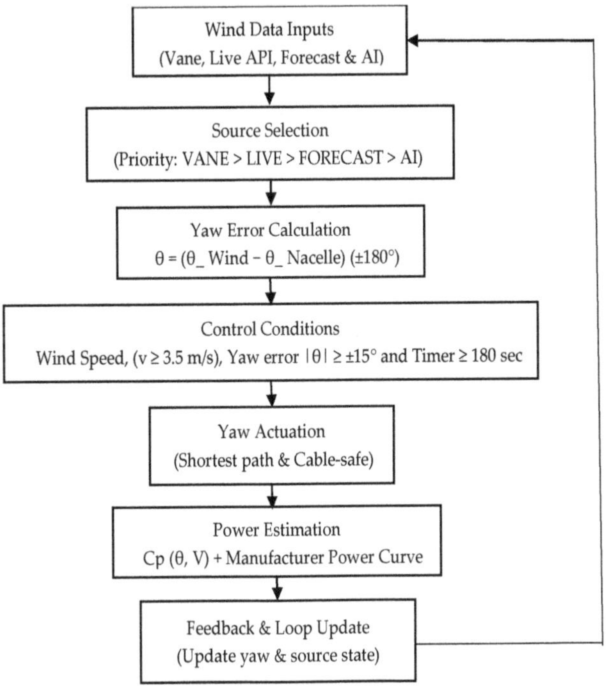

Figure 1. Workflow diagram of the proposed AI-based yaw control system.

This workflow starts with the wind data input, in which the wind direction and wind speed are obtained form in four sources, for example, a physical wind vane attached to the turbine, live online weather data through an application programming interface (API), forecast wind data, and a short-term prediction model based on AI. Multiple sources make it more robust and permit the system to continue functioning even in the presence of sensor failures or communication failures. Any signal of wind direction is standardized between the range of 0 and 360.

https://doi.org/10.3390/pr14071084

Next, source selection is performed using a priority-based logic defined as VANE > LIVE > FORECAST > AI. The wind vane is preferred because it provides direct on-site measurements with minimal latency. If the vane becomes unavailable or unreliable, the controller switches to live online data. Forecast data is used when both vane and live inputs are unavailable, while the AI-predicted wind direction serves strictly as a fallback when no physical or online measurements exist. This hierarchy ensures reliable control while limiting unnecessary reliance on prediction.

Once a valid wind reference is selected, the controller computes the yaw error, defined as the angular difference between the selected wind direction and the current nacelle heading measured by a magnetic encoder. The yaw error is wrapped within $\pm 180°$ to avoid discontinuities at angular boundaries and to ensure that the shortest rotational path is always selected.

The yaw error is then evaluated under control conditions. Yaw actuation is enabled only when the wind speed exceeds 3.5 m/s, the absolute yaw error is greater than or equal to $\pm 15°$, and the error persists for at least 180 s. These conditions prevent unnecessary yaw motion during low-wind operation, filter out short-term fluctuations, and balance aerodynamic efficiency against mechanical wear.

When the conditions are satisfied, yaw actuation is initiated. The nacelle is rotated using a stepper motor along the shortest path either clockwise or counterclockwise, while enforcing a cable-safe constraint to prevent excessive twisting beyond $\pm 180°$. Following alignment, power estimation is performed using a yaw- and wind speed-dependent power coefficient combined with the manufacturer's power curve. Finally, a feedback and loop update stage continuously updates the yaw position, power output, and source status, forming a closed-loop control system that adapts dynamically to changing wind conditions.

2.1. Energy Available in the Wind

Wind energy is a natural and renewable source that can be used to create power. A wind turbine does this by transferring the kinetic energy of moving air into electricity. Figure 2 illustrates the swept area of the wind turbine, which plays an important role in determining how much wind energy the turbine can capture. The amount of power a turbine makes depends on a few important things, such as the density of the air, the speed of the wind, and the size of the area that the blades sweep.

Figure 2. Real wind turbine used for experimental testing at Bonavista, NL.

https://doi.org/10.3390/pr14071084

The kinetic energy of a unit mass of moving air is given by the following formula:

$$E_k = \frac{1}{2} m v^2 \quad (1)$$

where m is the mass of the air and v is its velocity.

Considering the air density ρ, and the cross-sectional area A swept by the turbine blades, the total available power in the wind flowing through that area becomes

$$P = \frac{1}{2} \rho A v^3. \quad (2)$$

This equation shows that wind power increases with the cube of the wind speed, making wind velocity the most critical factor in power generation.

2.2. The Operating Principle of the Yaw Control Mechanism

In horizontal-axis wind turbines, maintaining the alignment of the nacelle with the wind direction is critical to ensure optimal power generation. The yaw control system in this paper is designed to actively monitor and correct the turbine's orientation based on real-time wind data.

The system includes a wind vane sensor and live wind data from weather station to detect wind direction. These sources send analog or pulse signals to a microcontroller, which processes the input to calculate the yaw error and the difference between the nacelle's current heading and the actual wind direction.

To determine the amount of extractable wind power, we use the equation derived from Betz's law [4]:

$$P(V,\theta) = 0.5 \rho A v^3 C_{p.max} \cdot f\ (\mathrm{V},\ \theta\ ) \quad (3)$$

where

$P(V,\theta)$ = Power extracted by the wind turbine;

ρ = Air density $(\mathrm{kg/m^3})$;

A = Rotor-swept area m^2;

v = Wind speed $(\mathrm{m/s})$;

$C_{p.max}$ = Maximum achievable power coefficient under ideal alignment $\theta = 0°$;

$f\ (\mathrm{V},\ \theta\ )$ = Yaw-dependent derating factor $(0 \leq f \leq 1)$ that captures the reduction in aerodynamic efficiency caused by yaw misalignment and its dependence on wind speed.

Aerodynamic theory and experimental studies have consistently shown that wind turbine power under yawed inflow conditions decreases according to a cosine-based yaw loss relationship, commonly expressed as $P(\theta) \propto \cos^n(\theta)$, where the exponent n typically ranges between 2 and 3 depending on rotor aerodynamics and operating conditions [3,9,30]. This formulation captures the dominant physical effect of yaw misalignment, namely the reduction in the axial wind velocity component normal to the rotor plane.

Based on this established principle, the yaw- and wind speed-dependent derating function illustrated in Figure 3 is defined as

$$f(V,\theta) = \cos^n(\theta)\, g(V,\theta) \quad (4)$$

where $g(V,\theta)$ is a bounded correction term introduced to account for secondary aerodynamic and operational effects that are not explicitly captured by the pure cosine law. These effects include changes in effective power coefficient C_p with wind speed, partial load versus near-rated operation, and rotor–wake interaction characteristics under yawed inflow, as discussed in previous experimental and SCADA-based studies [3,9,30].

https://doi.org/10.3390/pr14071084

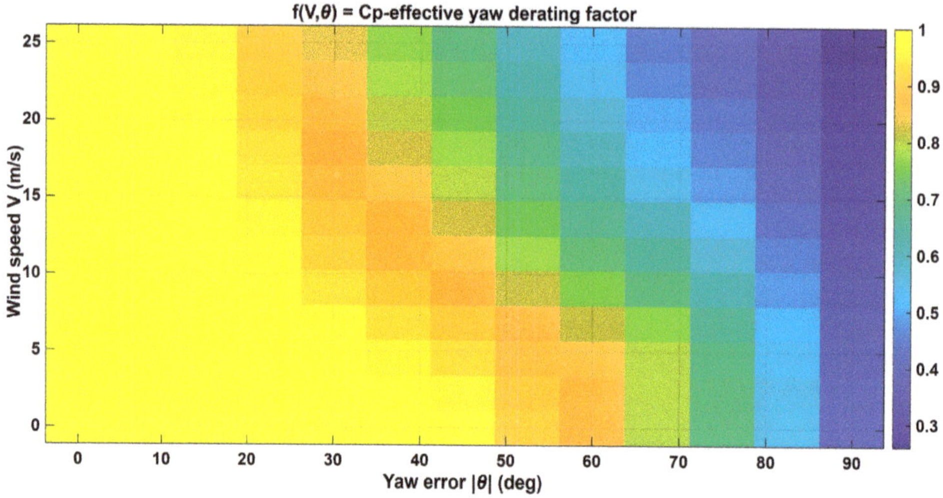

Figure 3. Yaw- and wind speed-dependent aerodynamic power derating factor.

In this paper, $g(V, \theta)$ is modeled in a physically consistent and bounded form as

$$g(V, \theta) = 1 - \alpha(V)\sin^2(\theta) \tag{5}$$

where $\alpha(V)$ is a wind speed-dependent coefficient constrained within $0 \leq \alpha(V) \leq 0.2$. This formulation ensures that the overall derating function remains smooth, bounded, and monotonic with increasing yaw misalignment, while preserving the dominant cosine-based yaw loss behavior. Similar corrective formulations have been reported in the literature to represent deviations from ideal cosine behavior observed in real turbines, particularly at higher wind speeds and near-rated operation [9].

The resulting derating surface shown in Figure 3 is an author-generated, physics-based two-dimensional lookup table implemented in MATLAB R2025b and applied within the Simulink environment. Although the efficiency values are not obtained from direct experimental measurements, the model is grounded in established aerodynamic yaw loss theory and validated at the system level. Specifically, the computed power output $P(V, \theta) = P_{\text{ref}}(V)\, f(V, \theta)$ remains bounded by the manufacturer's reference power curve, and system-level simulations confirm physically realistic power variations under changing yaw conditions. As such, Figure 3 provides a transferable and aerodynamically consistent representation of yaw-induced power loss suitable for yaw control analysis and performance assessment.

In order to obtain realistic and physically significant simulation outcomes, the wind turbine in this paper is specified in terms of the Falcon Silence 3.6 kW horizontal-axis wind turbine as the reference model to be used in calculating the aerodynamic and power values of the wind turbine in Figure 4. The turbine is up-wind and three-blade in nature and the rotor diameter is about 4.0 m, which determines the effective swept area utilized in the aerodynamic modeling. The turbine has a practical range of wind speed, cut-in wind speed of approximately 3 m/s, rated wind speed of approximately 11–12 m/s and a cut-out wind speed of 25 m/s, as indicated by the manufacturer [31].

https://doi.org/10.3390/pr14071084

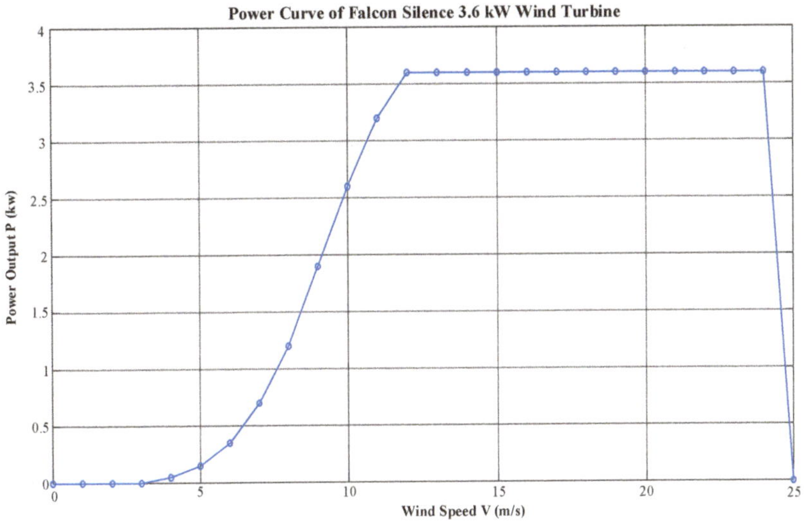

Figure 4. Power curve of the reference turbine.

In as much as the reference model wind turbine has blade pitch control in actual operation, a dedicated blade pitch control model is not directly applied in this research paper. Rather, the direct effect of the aerodynamic control of the pitch is indirectly described in the form of the manufacturer-based power curve of the Falcon Silence 3.6 kW wind turbine. In addition, the model also includes dependence on the effective power coefficient of the yaw and wind speed to demonstrate the power losses associated with a misalignment in the yaw. Consequently, the combined use of the manufacturer's power curve and the yaw-dependent power coefficient delivers realistically represented turbine power regulation without introducing any further pitch actuator dynamics. This is beneficial as the main purpose of this paper is to explore the effects of yaw misalignment.

Accordingly, the turbine power output is calculated as

$$P_{out}(V,\theta) = P_{ref}(V) \cdot f\ (\mathrm{V},\ \theta\ ) \tag{6}$$

where $P_{ref}(V)$ is obtained directly from the manufacturer's power curve. This represents the effective power coefficient reduction due to yaw misalignment and wind speed variation. By explicitly modeling this yaw-dependent power coefficient, the proposed approach captures yaw-induced aerodynamic losses while preserving realistic turbine operating limits, thereby ensuring the validity of the simulation results for practical yaw control applications.

When the wind turbine is misaligned with the wind, power is reduced due to the yaw angle θ. In such cases, the effective power is as follows:

$$P(V,\theta) = 0.5\rho A v^3 C_{p.max} \cdot f\ (\mathrm{V},\ \theta\ ) \tag{7}$$

where θ is the relative yaw misalignment angle. The power output is maximized when $\theta = 0°$, meaning that the nacelle is directly facing the wind.

The wind power depends on the cube of the wind speed, so the power that can be captured when the turbine is misaligned becomes the following:

$$\mathrm{P}\ (V,\theta)\ \propto\ (v\cos(\theta)^3) = \ v^3\cos^3\theta \tag{8}$$

https://doi.org/10.3390/pr14071084

The cosine-cubed relationship comes from the fact that only part of the wind that hits the rotor at a right angle can be used to produce power. When the turbine is turned away from the wind by an angle θ, the effective wind speed is reduced to $v\ cos\theta$. This is why the term $\cos^3\theta$ is widely used as a practical indicator of the fraction of power retained when the nacelle is not perfectly aligned with the wind. This relationship clearly explains why power loss increases rapidly as yaw misalignment grows. As the yaw angle increases, the cosine-cubed term decreases sharply, resulting in a significant reduction in captured power. This behavior highlights the importance of accurate yaw control in maintaining high turbine efficiency and maximizing energy production [32].

The reduction in power caused by yaw misalignment can also be expressed more explicitly using the standard cosine-cubed formulation, which estimates the fraction of power lost as the nacelle deviates from the wind direction. In this research, a 3.6 kW small-scale wind turbine is considered as the reference system for evaluating the impact of a yaw error.

$$P_{loss}(\theta) = P_{rated}\left(1 - cos^3\theta\right) \tag{9}$$

Similarly, the percentage loss in power due to yaw misalignment can be computed using the following [33]:

$$\text{Loss (\%)} = \left(1 - cos^3\theta\right) * 100 \tag{10}$$

Table 1 shows that a 15° yaw misalignment leads to a loss of about 0.356 kW or 356 W for the 3.6 kW turbine, which equals roughly 9.88% of the available power. This level of loss is meaningful, but it is still manageable without requiring the yaw motor to work constantly. Smaller errors such as 5° or 10° result in only small power losses, but keeping the turbine within such a narrow range would require the motor to move almost constantly, increasing mechanical wear and reducing the overall lifespan of the yaw system. On the other hand, allowing the misalignment to grow beyond 20° results in a steep increase in power loss, reaching more than 17–26%, which would noticeably reduce the turbine's overall energy output. For these reasons, a ±15° threshold offers a practical middle ground that keeps aerodynamic losses at an acceptable level to avoid unnecessary motor activity. This is the main reason that this paper adopts ±15° as the yaw error limit.

Table 1. Power output and gross power loss of a 3.6 kW wind turbine under different yaw error conditions.

Yaw Error (°)	Fraction of Retained Power	Power Output (kW)	Gross Power Loss (kW)	Loss (%)
5°	0.9886	3.56	0.041	1.14%
10°	0.9551	3.44	0.162	4.49%
15°	0.9012	3.24	0.356	9.88%
20°	0.8297	2.99	0.613	17.02%
25°	0.7444	2.68	0.920	25.56%

In this research work, the microcontroller continuously monitors the yaw angle. If the error $|\theta| \geq \pm 15°$ and wind speed exceeds around 3.5 m/s, it sends signals to the stepper motor driver, which then rotates the bipolar stepper motor. The motor precisely moves the nacelle in the shortest direction either clockwise or counterclockwise and maintains cable safety to reduce the yaw error and realign with the wind.

https://doi.org/10.3390/pr14071084

The required number of motor steps to correct the yaw angle θ

$$n = \frac{\theta NM}{360} \quad (11)$$

where

N = Number of pulses or steps.

θ = Required yaw correction angle $(^\circ)$.

N = Gear ratio of the nacelle.

M = Steps per revolution $\times$ micro-stepping factor.

This system makes sure that the nacelle is positioned very accurately while using the least amount of motor power and wear. The system keeps an ideal balance between mechanical reliability and energy savings by making corrections only when required. Thus, this real-time yaw control method guarantees greater use of available wind resources and greatly improves turbine performance.

2.3. Wind Data Acquisition and Preprocessing

In order to have realistic simulation conditions, the wind data were obtained in Bonaventure, Newfoundland and Labrador. The data are based on real-time measurements of wind speed and wind direction collected during the analysis of a microcontroller-based system, which retrieves live meteorological data from OpenWeatherMap [34] (City ID: 5905393) via an API. It has a 5 min interval between data points, which provides the high temporal resolution necessary to determine short-term variability in wind direction, which is of importance to active yaw control, and to establish a smooth yaw response and power output response without steps like artifacts. The chosen site is in an area in Newfoundland and Labrador that is a good coastal location for wind turbines and its representativeness is also supported independently with the publicly available records of Environment and Climate Change Canada [35], which are used to validate the wind regime that was present during the period of gathering the data.

The raw CSV data that were preprocessed in MATLAB to retrieve the most important parameters to be included in the analysis of the yaw control method included date and time, latitude, longitude, wind speed, and wind direction. The values of wind direction were brought to the range of 0–360 and the values of wind speed were kept in meters per second. To achieve numerical stability, invalid and missing entries were eliminated.

The table below (Table 2) shows representative samples of the historical data that show how the real-world wind conditions are integrated into the simulation framework. The entire dataset is about 80,000 s ($\approx$22.2 h) of continuous data, which is directly fed into the Wind Environment subsystem of Simulink and allows realistic analysis of the yaw dynamics and power response, taking into consideration realistic operating conditions.

The processed wind speed and direction time series were then imported into MATLAB and used directly as inputs to the Wind Environment subsystem of the Simulink model. Table 2 summarizes the wind dataset used for the active yaw control analysis. Although live meteorological data via an API are used as the primary input for yaw control performance and power analysis, a physical wind vane is retained as an integral component of the proposed system architecture. The wind vane provides local, instantaneous wind direction measurements and serves as a reference for control verification, sensor-level validation, and fallback operation when online data are unavailable or delayed. In the absence of a full-scale turbine installation, the wind vane is not used as the sole source for long-duration performance analysis; however, its inclusion supports the feasibility, robustness, and practical deployability of the yaw control system.

https://doi.org/10.3390/pr14071084

Table 2. Sample of wind data used for yaw control simulation.

Longitude (x)	Latitude (y)	Date/Time (UTC)	Wind Dir (Deg)	Wind Speed (m/s)
−53.11	48.67	5 December 2025 0:30	160	7.78
−53.11	48.67	5 December 2025 0:35	158	7.25
−53.11	48.67	5 December 2025 0:40	156	7.23
−53.11	48.67	5 December 2025 0:45	155	7.88
−53.11	48.67	5 December 2025 0:50	154	7.86
−53.11	48.67	5 December 2025 0:55	152	7.88
−53.11	48.67	5 December 2025 1:00	155	7.77
−53.11	48.67	5 December 2025 1:05	153	7.65

This data usage strategy ensures reliable yaw control behavior and consistent system validation within the scope of the present study.

The use of high-resolution, real-world wind speed and direction data, as described in Section 2.3, highlights the inherently time-varying and non-stationary nature of the wind environment relevant to active yaw control. While such data provide realistic inputs for system evaluation, effective yaw control performance further depends on the ability to anticipate short-term wind direction changes rather than responding purely in a reactive manner. This requirement motivates the adoption of data-driven predictive control strategies capable of learning temporal wind patterns, which form the basis for the LSTM-based yaw control approach introduced in the following section.

2.4. Motivation for LSTM-Based Predictive Yaw Control Strategy

Table 3 below provides a summary and compares popular data-driven, machine learning (ML) and reinforcement learning (RL) methods to wind turbine yaw control, and highlights their usefulness in small-scale systems especially. Traditional non-ML data-driven models, e.g., linear and polynomial regression, are computational simple and easy to implement, but they are not capable of learning in time, thereby restricting their application in quickly evolving wind settings.

Table 3. Comparison of data-driven, ML and RL approaches for wind turbine yaw control.

Category	Approach	Wind Trend Learning	Real-Time Use	Training Reliability	Computational Cost	Suitability for Small Turbines
Data-Driven (Non-ML)	Linear/Polynomial Regression	Low	High	High	Very Low	Limited (cannot track rapid direction changes)
Machine Learning (ML)	Random Forest/Tree-Based Models	Moderate	Moderate	High	Moderate–High	Suitable for static estimation, not short-term prediction
Reinforcement Learning (RL)	RL-based Yaw Control	High	Low–Moderate	Low	Very High	Impractical for small turbines
Machine Learning (ML)	LSTM	High	High	High	Moderate	Highly suitable and robust

ML-based approaches, such as support vector regression and tree-based models, achieve moderate improvements but are limited in their short-term predictive accuracy as well as real-time efficiency. Yaw control strategies developed by RL methods have excellent

https://doi.org/10.3390/pr14071084

adaptive potential but demand large amounts of training, large computational resources, and ongoing exploration, so are inapplicable to embedded microcontroller systems. The LSTM-based approach that is used in the research is able to address these shortcomings by explicitly learning temporal dynamics in wind data and maintaining stable training dynamics and moderate computational cost, providing a feasible and effective predictive solution to yaw control in small wind turbines.

2.5. Block Diagram

Figure 5 below is the functional layout of the proposed smart wind turbine system in the form of a block diagram, with real-time yaw control introduced to enhance the efficiency of energy collection. This system can be subdivided into two key subsystems, one of them is the power-generation unit and the other is the yaw control mechanism. To make an output compatible with the grid, the turbine initially transforms wind energy into electricity by rectifying the produced AC into DC and inverting it back into AC with the help of the inverter. The microcontroller also performs yaw correction by accessing data from the physical wind vane as well as the online weather station via the API. The yaw control section has a wind vane that is characterized by 12 V DC with a constant analog display of wind direction. Furthermore, the microcontroller is equipped with an OpenWeatherMap server, which is linked to it by the embedded Wi-Fi module and sends secure HTTP requests with an API key to retrieve LIVE and FORECAST wind direction information. This online platform has been selected because it provides globally tested, high-resolution weather data that are updated every few minutes and are aided by numerous meteorological models and a real-life network of observations [34]. After fusing physical and online data, the controller will be able to work even when one of the sources becomes unavailable. Once these inputs are received, the controller computes the error in yaw and passes accurate control signals to the stepper motor driver, which drives the stepper motor and causes the nacelle to spin. The feedback loop maintains the turbine in the correct orientation, as the wind changes to minimize unnecessary movement and prevents the mechanical parts of the turbine from over rotating. Subsequently, the system ensures consistent alignment, provides reliable functionality, and optimizes the total energy generation when the patterns of weather change. The modular's design also makes it easier to upgrade the individual components without redesigning the entire system, and the use of online data enables the turbine to operate successfully even in places where the installation of multiple sensors can be very expensive and unfeasible.

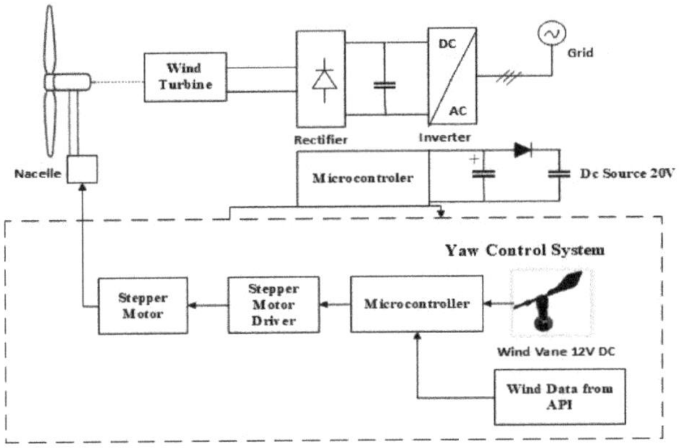

Figure 5. Block diagram of the proposed smart wind turbine system. Arrows indicate the direction of power and control signal flow. The dashed box represents the yaw control subsystem. Standard electrical symbols denote the rectifier, inverter, DC source, and associated components.

https://doi.org/10.3390/pr14071084

In addition, with the integration of local sensing and cloud-based information, the system attains durability in the case of unforeseen weather conditions. The fact that the microcontroller can cross-check the data on the vane and API enhances the accuracy and minimizes the possibility of errors due to short-lived changes. All in all, this combined architecture enhances the reliability and durability of the wind turbine.

2.6. System Algorithm

Figure 6 describes the operations of the yaw control system of the wind turbine in steps. It is initiated by measuring wind speed and wind direction using a wind vane sensor and live wind data via an API. The controller then verifies that the speed of the wind has surpassed the cut-in threshold, otherwise the system stops and waits until the appropriate conditions are reached. Once the wind speed reaches a certain level, the controller compares the yaw error, and this is the difference between the orientation of the turbine nacelle and the direction the wind is blowing. This error is then assessed by the system as an equal or larger error of $\pm 15^{\circ}$. When the deviation is out of this tolerance band, the yaw motor is turned on to move the nacelle in the correct direction. When the yaw error remains acceptable, the turbine maintains its current position and waits without taking any action, which can result in unwarranted movement of the motor. This minimizes wear of the mechanical components and does not allow the system to respond to short-term variations, which are not significant in influencing power production. The flowchart indicates the decision-making logic that makes the nacelle move whenever the energy gained is greater than the actuation cost. The controller ensures that the motor is not in operation when the wind is too slow to warrant movement by ensuring that the wind speed is checked before each correction. The delay functions incorporated in the loop are used to filter out the noise sensor readings and stabilize the response in windy conditions. This assessment is constantly repeated, creating a closed system that continuously provides an updated orientation with the change in the direction of the wind. Due to such a logical design, the turbine is at right angles most of the time and does not experience quick and sideways movements. The general design enhances power harvesting, shields hardware and secures similar reactions to various wind directions. Practically, this circulation assists the turbine to move effortlessly and also makes every correction during the yaw significantly add to performance.

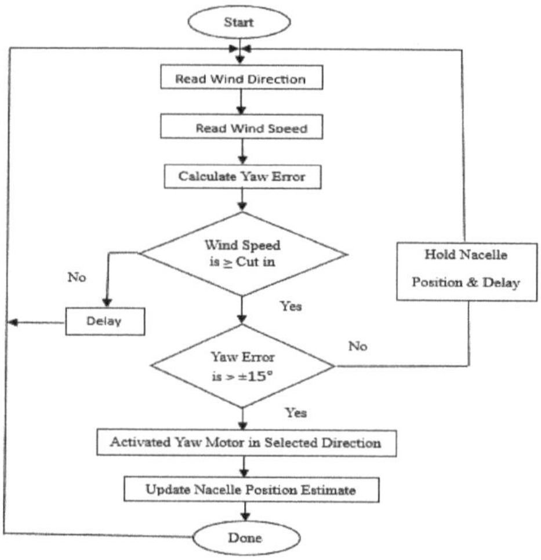

Figure 6. Flowchart of real-time yaw control algorithm.

https://doi.org/10.3390/pr14071084

2.7. *Yaw Correction Logic Based on Wind Direction and Rotor Alignment*

The Smart Wind-Tracking System in Figure 7 has a yaw mechanism that is driven by a step motor that is in turn used to check the direction that the wind is moving. The entire control logic is implemented by a microcontroller, and a magnetic encoder feeds precise information about the position of the nacelle. This closed-loop system guarantees accurate yaw positioning, minimizes yaw error, collects maximum energy and reduces mechanical forces on other parts of the system, including bearings, shafts and the gear box.

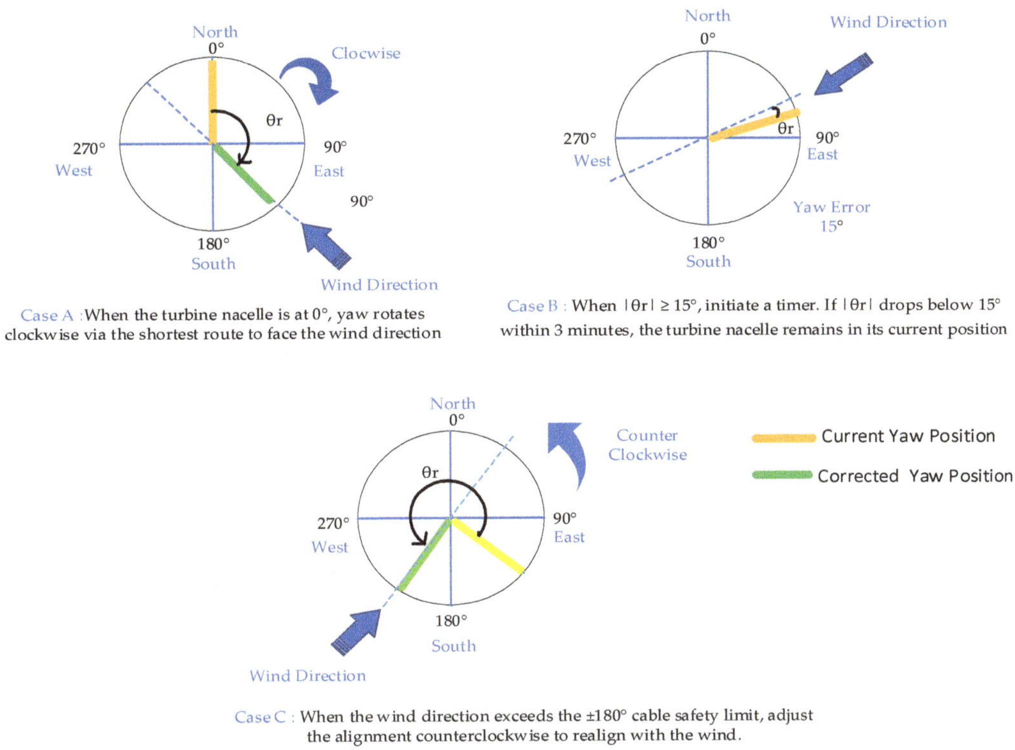

Figure 7. Yaw control strategy based on wind direction. The blue arrows and dashed lines indicate the incoming wind direction used to determine the nacelle yaw adjustment.

In case A, the minimum distance between the two points is calculated. Upon a change in the wind direction, the controller determines the yaw error (θ_r) and spins the nacelle around the shortest possible rotational track clockwise or counterclockwise, to bring it back into line with the wind. This will see that it completes the correction of the orientation fast and efficiently and reduces unnecessary motor travel.

Case B—Deadband $\pm 15°$ and 3 min Stability Timer. Here, yaw correction is only implemented in cases where the absolute yaw error is more than $\pm 15°$. When $|\theta r| \geq \pm 15°$, the controller initiates a 3 min timer to ensure that the misalignment is not temporary and that small variations do not cause misalignment. When the error equal or exceeds $15°$ in the entire 2 min, then the nacelle is turned back to the right position.

However, when the error decreases below the threshold within this interval, then the controller maintains the nacelle in the same position. The deadband-plus-timer plan eliminates unwarranted actuation, minimizes the wear of motors, and prevents constant oscillation caused by small changes in wind.

https://doi.org/10.3390/pr14071084

Case C—Cable Safety Limit ($\pm 180°$). At the point where the wind direction passes over the $\pm 180°$ axis, the system will enable the cable-safe logic. Rather than rotating the same way and over-twisting the cables, the controller reverses the direction of rotation and proceeds counterclockwise to arrive at the new alignment. This makes sure that the yaw cables do not exceed the range of twisting, but at the same time, the rotor is directed properly to the wind.

2.8. Baseline Yaw Control Strategy

A traditional rule-based yaw controller is used to determine the effectiveness of the proposed strategy of yaw control. This is the kind of yaw control logic that is currently popular in small and medium wind turbines. In this scheme, the nacelle yaw angle is only calculated based on the instantaneous measurements of the wind direction without prediction, optimization, and adaptative decision-making.

The baseline controller is based on a predefined deadband of the yaw to control the yaw motion. Under the circumstances that the absolute misalignment between the nacelle's orientation and the direction of the incoming wind is more than the given threshold, the yaw actuator is activated to reverse the alignment. On the other hand, in the event that the error in the yaw is still within the deadband, no corrective action is taken and the position of the nacelle is held constant. This method restricts unnecessary movement of the yaw axis and still allows the approximate direction of the wind to be maintained.

The baseline controller does not have any wind forecasting, artificial intelligence, multi-source data fusion, or learning-based adaptation. The current wind direction input is all that triggers all of the yaw actions. Mechanical constraints, yaw rate limits and turbine parameters are the same in order to guarantee a fair and unbiased comparison in the proposed control framework. Therefore, any difference in performance that is observable in the results can be solely accounted for due to the control strategy itself and not by the differences in system configuration or operating conditions.

The methodology described in the previous section establishes the data sources, preprocessing steps, and control strategies used to evaluate yaw control performance under realistic wind conditions. Building on this foundation, the following section presents the overall control system architecture and block diagram of the proposed yaw control framework. The control diagram illustrates how wind data, prediction modules, and control logic are integrated within the Simulink environment to generate yaw commands and evaluate system performance.

3. Control Design

Figure 8 depicts the logic of decision-making employed by the smart yaw control system in order to pick the most dependable wind direction source under different operation conditions. The flowchart is developed based on two condition priority structures aimed at maintaining continuous and fault-tolerant operation of the yaw.

This starts with reading the wind vane sensor, which is the highest priority in the source hierarchy. The controller checks the availability of the vane sensor and AI-predicted validity every 5 s. When the vane is in operation, yaw control is carried out directly with the measurements of the vane. In cases where the vane is not available because of sensor noise, hardware error, or temporary blockage, the system will jump to the live wind data source that is acquired via the OpenWeatherMap using the API key.

In the case of the live data pathway, the controller once more tests the availability of data and the validity of the related AI prediction. When the live source is on, the yaw controller will still use the real-time online measurements and watch to ensure that the vane is coming back. If the vane source and live source fail, for example, because of network

https://doi.org/10.3390/pr14071084

breakage, the system will switch to live AI-predicted data, which also offer short-term predictions based on the most recent online dataset. This also means that the yaw alignment can still be achieved without the direct sensor or network input.

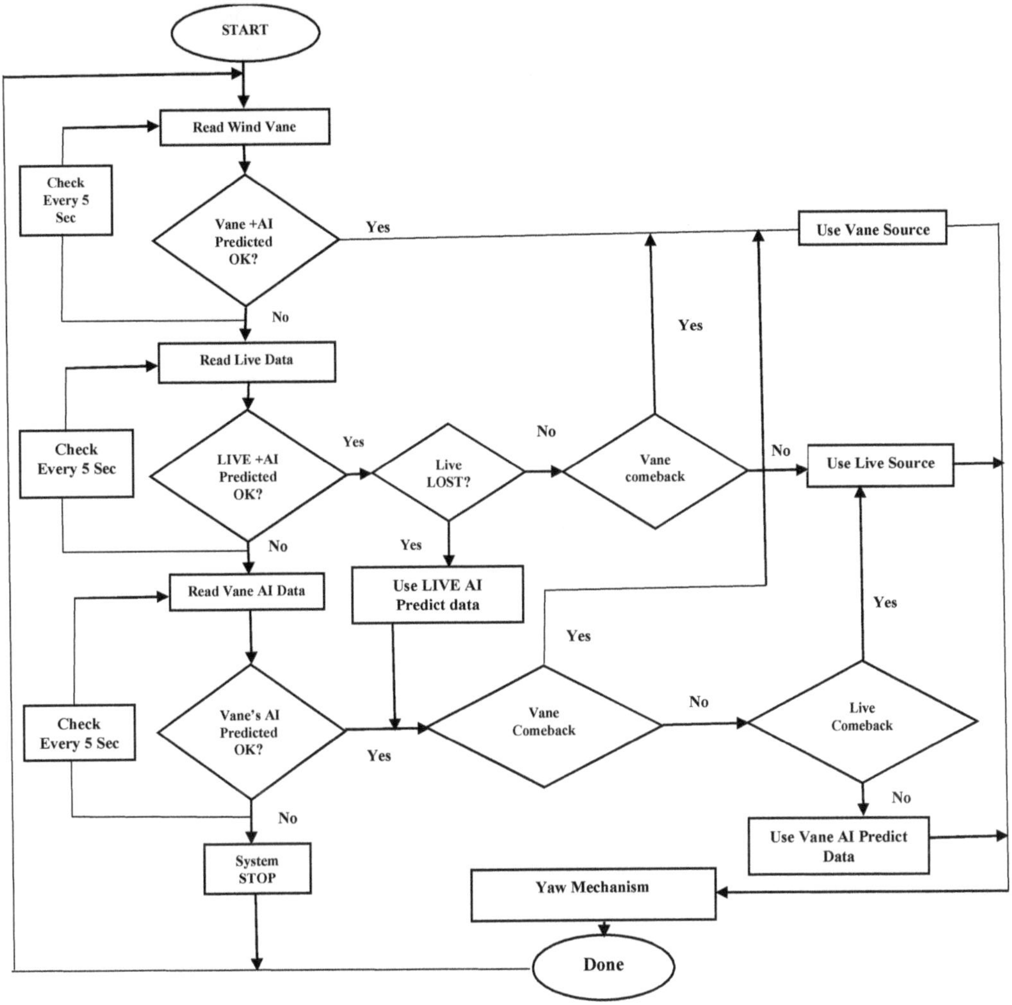

Figure 8. Priority-based yaw control logic for wind turbine alignment.

In the second scenario, where there is only one source (either vane, or live) available, the system makes use of it and falls back on its corresponding AI-predicted data in the event that the primary measurements are lost. In this fallback mode, the controller will continue checking every 5 s whether the vane or live sensor has resumed operation. Whenever a primary source becomes available again, the controller immediately switches back to it, placing AI prediction as the fallback option used only when no physical source is available.

This control diagram given in the above part outlines the general model and signal path of the proposed yaw control system. With this conceptual architecture in place, the next section describes the practical system design and implementation processes such as model development, controller setup and integration in the MATLAB Simulink R2025b environment.

https://doi.org/10.3390/pr14071084

4. System Design and Implementation

4.1. Hardware Implementation

All the hardware implementation details of the proposed smart yaw control system are provided in Figure 9 below, which provides the physical relationship of the major components of the system at the central microcontroller between the sensing subsystem and the actuation subsystem. The turbine structure has an analog wind vane that measures the direction of local wind, and is a continuous analog signal with a voltage range of 0–5 V, representing a full 0–360-degree wind direction. The wind vane itself operates on the 5 V of the microcontroller, and the wind direction analog output of the wind vane is hard-wired into one of the analog input pins of the controller, with the microcontroller converting the corresponding voltage into one of the wind direction angles with its internal analog-to-digital converter (ADC). On the actuation side, the microcontroller is connected to a stepper motor driver, and has three digital control signals, one being a step (PUL) signal that is used to step the driver in a direction, the other a direction (DIR) signal, which is used to turn the driver clockwise or anticlockwise, and the last being an enable (EN) signal, which is used to switch the driver off or on. These control signals are formed by the microcontroller, based on the calculated error of the yaw and the strategy that is adopted by the yaw control. This external 12 V DC power is then used to drive the motor driver, in which the required current is applied through a stepper motor that is mechanically connected to the wind turbine nacelle to make the wind turbine rotate on the yaw. It is also done through the distribution and grounding of power with the help of a breadboard that gives a common electrical reference point to all the components. Overall, it should be mentioned that the figure indicates the flow of information regarding wind sensing to control the decision-making process to future yaw actuation controlled by high-power motor circuitry with the necessary distance between low-power control electronics and high-power motor circuitry.

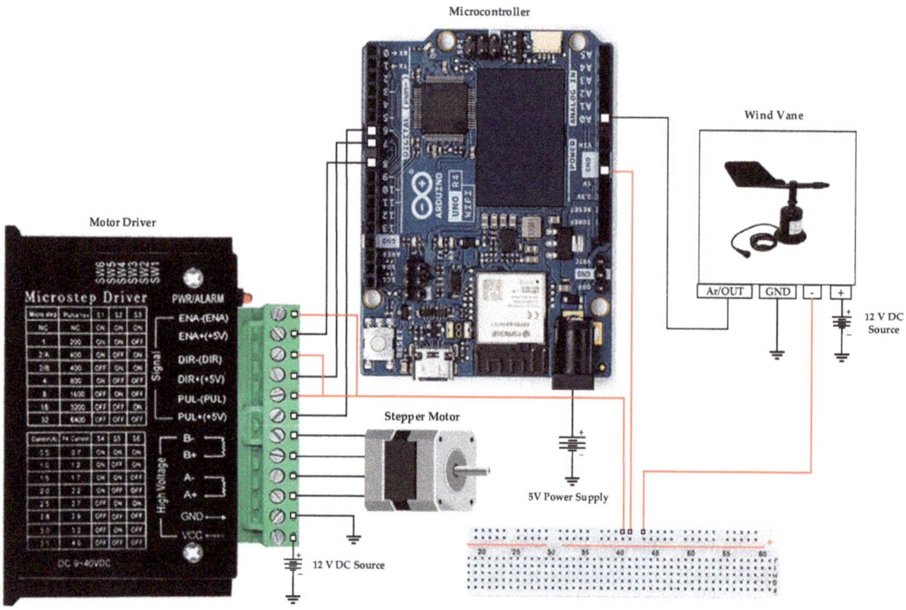

Figure 9. Connection diagram of hardware setup.

The components include the following:

https://doi.org/10.3390/pr14071084

- Microcontroller: Arduino Uno R4 WiFi (Arduino, Turin, Italy);
- Wind vane: RS-FXJT05 (Rika Sensors, Handan, China);
- Stepper motor: NEMA 17 Stepper Motor (Wantai Motor, Changzhou, China);
- Motor driver: TB6600 Stepper Motor Driver (Toshiba, Tokyo, Japan).

4.2. Software Implementation: MATLAB/Simulink Model

The offered smart yaw control system is produced in the MATLAB/Simulink environment, trying to model and establish the dynamic behavior of a horizontal-axis wind turbine under different wind conditions. Figure 10 depicts the entire Simulink architecture, which consists of four primary subsystems, namely the Wind Environment, Yaw Controller, Yaw Mode Selector and Power Output Calculation blocks. The Wind Environment subsystem is a system that takes time-varying wind speed and wind direction input based on vane, live and forecast data sources and therefore allows realistic modeling of the changing atmospheric conditions. The Yaw Controller subsystem is another closed-loop device, which reduces the yaw misalignment within a 15-degree deadband and maintains a 180-degree cable-safe rotation limit, which guarantees aerodynamic efficiency and mechanical safety. The reference model is a 3.6 kW horizontal-axis wind turbine, which is utilized to calculate the instantaneous electrical power output as a result of the existing yaw alignment and wind conditions. The Yaw Mode Selector enables the operator to change between yaw active operating mode and fixed-heading operation, so that the results of operating in controlled and uncontrolled yaw conditions may be directly compared. In the Power Output Calculation subsystem, the simulated wind speed is mapped to the generator's electrical output using a manufacturer-derived power curve, with an additional yaw-dependent efficiency factor incorporated to realistically model power losses caused by yaw misalignment. In general, this Simulink model can be viewed as a digital representation of the wind turbine system and a stable virtual environment to test the viability of the yaw aligning process, cable safety mechanism, and the general performance of power optimization during the pre-implementation phase.

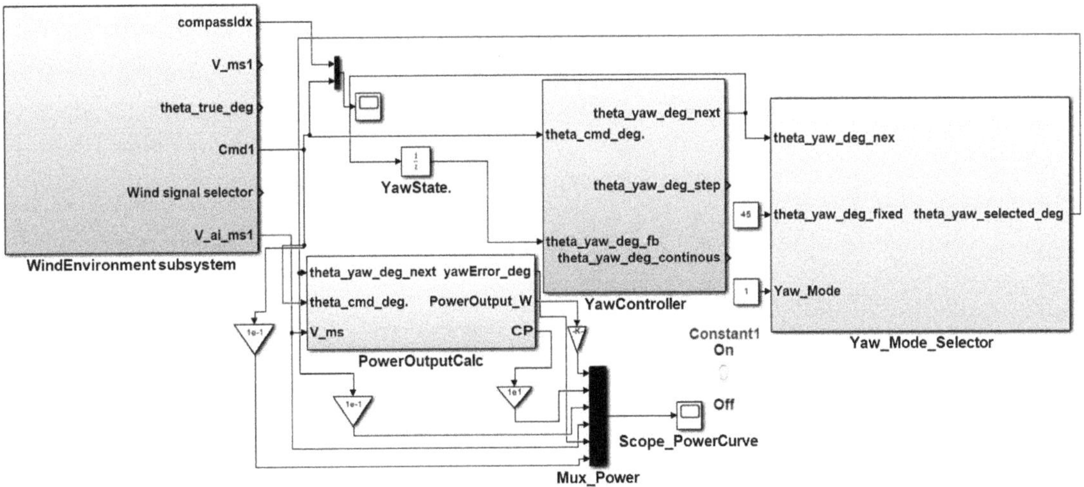

Figure 10. MATLAB/Simulink-based design for dynamic yaw alignment and power output.

4.2.1. Wind Environment Subsystem

The subsystem below in Figure 11 represents the wind direction acquisition, validation, and AI-assisted selection logic used by the smart yaw control system. It processes wind

https://doi.org/10.3390/pr14071084

direction inputs from three independent sources—the physical wind vane, live online data, and forecast data—each of which is conditioned using noise modeling, discrete filtering, delay blocks, and angle unwrapping to ensure continuity across the 0–360° range. Validity checks are applied to each source using threshold logic and enable switches so that unreliable or unavailable signals are automatically rejected. In parallel, an AI predictor block estimates the wind direction based on historical wind speed and direction trends when conventional sources are degraded or unavailable. A hierarchical switching mechanism then selects the most reliable wind direction signal based on predefined priority and availability conditions, while maintaining angular consistency through wrapping logic. The selected wind direction output is forwarded to the yaw controller, ensuring robust, continuous yaw reference generation under varying sensor availability and environmental conditions.

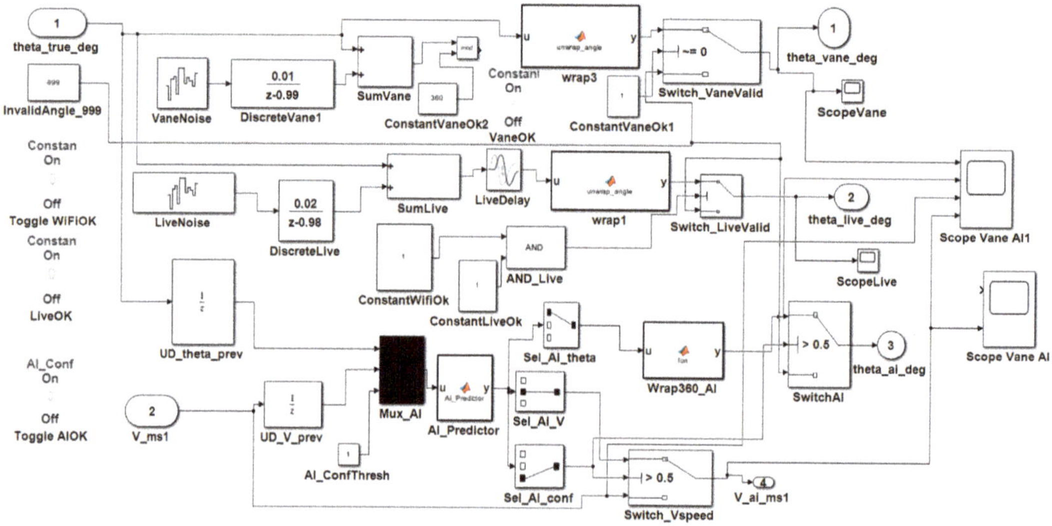

Figure 11. Wind Environment subsystem in MATLAB/Simulink showing the priority-based selection of vane, live, and forecast wind inputs. The circled numbers denote signal port identifiers used for internal routing in the Simulink model.

4.2.2. Yaw Controller Subsystem

Figure 12 below is the Yaw Controller subsystem, which enables the rotation of the nacelle in such a way that the turbine opens to the approaching wind and even shields the cables to ensure the smooth functioning of the motors. The system continuously takes the difference between the commanded wind direction and the feedback yaw angle and compares the yaw error at the safety limit of $\pm 180°$. A Cable-Safe Planner is utilized to ensure that the nacelle does not cross the $\pm 180°$ mechanical boundary and this eliminates the twisting of the cables when the nacelle is operated over long durations. Rate limiters and acceleration filters as well as dwell timers are also utilized by the controller and stabilize the stepper motor in such a way that yaw correction is accurate and efficient in varying wind conditions.

https://doi.org/10.3390/pr14071084

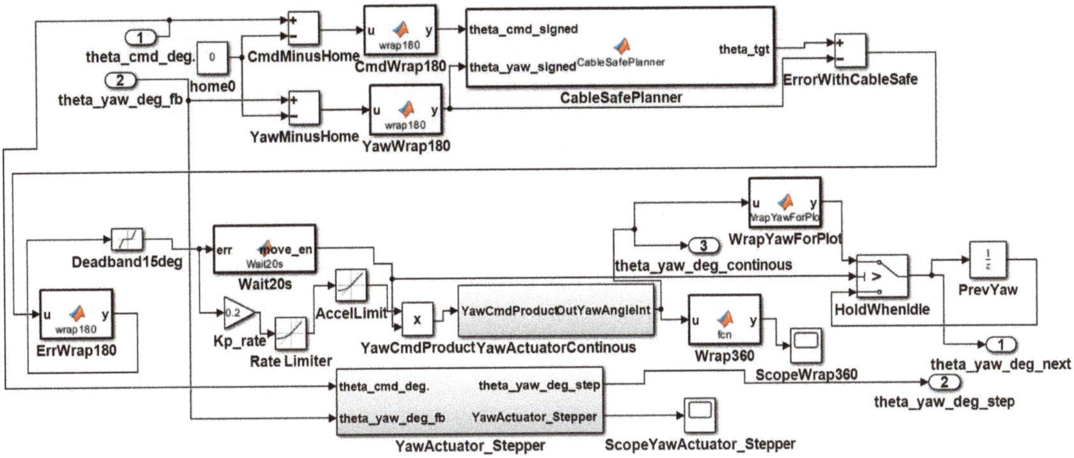

Figure 12. Yaw Controller subsystem for nacelle alignment and cable-safe operation.

4.2.3. Yaw Mode Selector Subsystem

This subsystem determines the active or passive yaw control that the turbine is to pursue or if the turbine is to remain in a fixed heading. Figure 13 shows that the Yaw Mode input signal selects and the switch between the controller's continually updated yaw command or the preset fixed position. An angle value is kept within the range of +180 to −180 and a wrap180 fixed block is utilized to allow the smoothness of the angles as they rotate and ensure that there are no sharp turns. The design makes this system flexible and enables easy comparison of fixed-yaw and active controlled operation when performing simulation and testing.

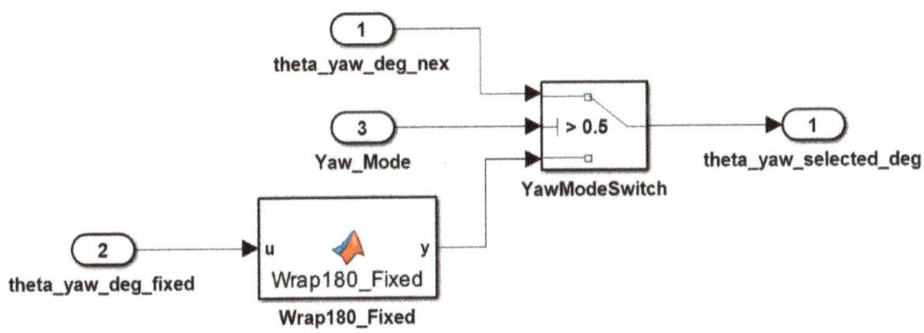

Figure 13. Mode selection logic for enabling or disabling active yaw control.

4.2.4. Power Output Calculation Subsystem

The subsystem of estimating the electrical output of the turbine in Figure 14 shows that the output depends on both the wind velocity and the orientation of the yaw. The first step it takes is to compute the error, which is the yaw, and wrap it within the range of −180 and +180. The second step is to take the error and multiply it by a cosine-cubed function, which is used to establish the proportion of the wind speed that can be effectively utilized to generate power. This fixed wind speed is entered into a look-up table and the system interpolates this with the 3.6 kW turbine power curve to obtain a real-time estimate of power.

https://doi.org/10.3390/pr14071084

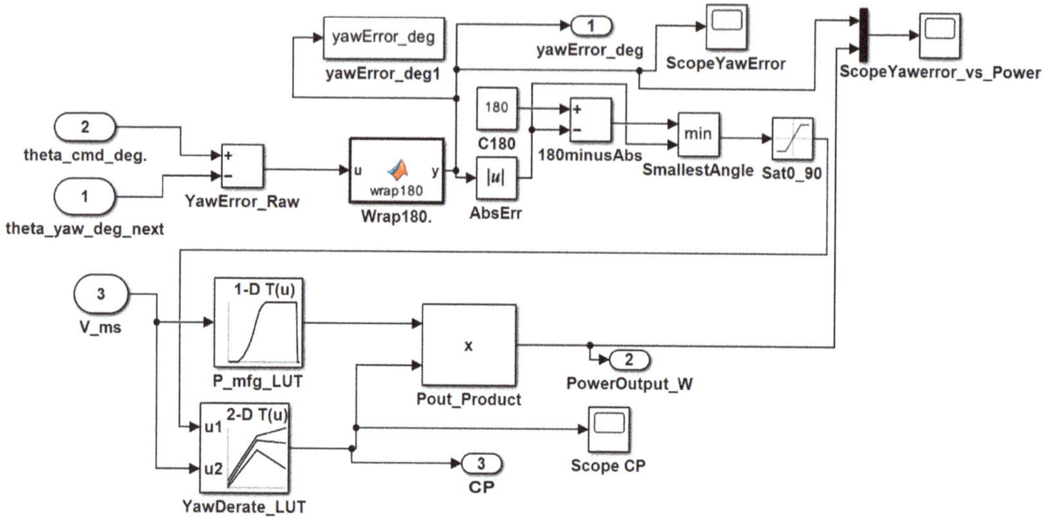

Figure 14. Power and yaw error calculation model.

Having the yaw control system developed and installed, the next part is to consider its behavior in the conditions of real wind. The discussion and findings are aimed at conducting a comparison of the suggested control strategy to the baseline approach, showing the improvements in the yaw alignment, power response, and the behavior of the whole system.

5. Result and Discussion

5.1. Real-Time Microcontroller Output Log

Figure 15 presents a block diagram representation derived from the Arduino serial monitor's output, illustrating the real-time operation of the smart yaw control system and its priority-based source selection logic (VANE > LIVE > FORECAST). At system startup, the controller initializes the hardware, updates its geolocation for Mount Pearl, Newfoundland and Labrador (Canada), and establishes a Wi-Fi connection to retrieve wind information from the OpenWeatherMap service.

Under normal conditions, when all sources are available, the nacelle is aligned using the physical vane measurement (e.g., 270.5° W), resulting in a very small yaw error, typically within ±0.5°. If the Wi-Fi connection is lost, the controller continues operating in vane mode, ensuring uninterrupted yaw alignment. When the vane signal becomes unavailable but live data is restored, the controller automatically transitions to live data mode and updates the yaw position accordingly. Once the vane signal is restored, the system immediately returns to vane mode, reflecting the highest priority assigned to the physical sensor.

In scenarios where both vane and live inputs are unavailable, the controller switches to forecast mode and uses predicted wind direction data (e.g., 250° WSW) to maintain turbine alignment. The arrows in the diagram explicitly show both fallback and recovery transitions between operating modes, demonstrating continuous monitoring and automatic failover. Overall, the figure confirms that the microcontroller autonomously manages source switching, maintains stable yaw alignment, and ensures reliable operation even during temporary sensor or network interruptions.

https://doi.org/10.3390/pr14071084

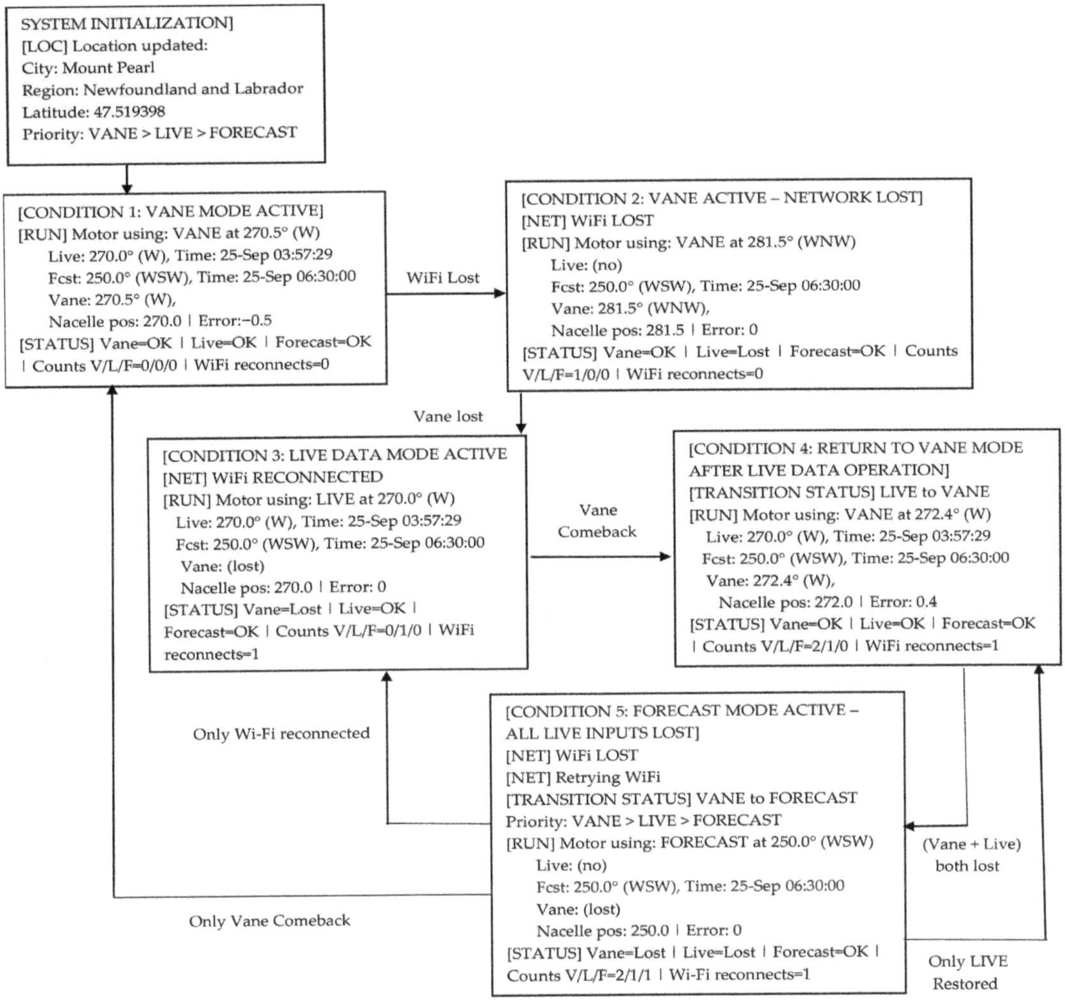

Figure 15. Microcontroller output of smart yaw control source selection and fallback logic.

The current arrangement of the microcontroller takes predicted wind direction data via the OpenWeatherMap API, and these predictions are usually in rough time increments (e.g., +3 h), which cannot be used in a small wind turbines to make real-time yaw adjustments. Since small turbines can have short-term variations in the wind direction, it is unwise to rely on just vane measurements or the use of forecasting information several hours old, which may result in short-term misalignment and low efficiency. In order to fill this gap, an AI-based model of short-term prediction, which is trained under MATLAB, is added to predict the wind direction within the next few minutes. This predictive capability allows the controller to respond faster and to have consistent control in sudden directional variations.

Moreover, there is a tendency of inconsistencies between the vane's measured output of wind direction at the turbine and the live weather station data. These variations are because the station and the turbine are situated at different points and the terrain, the elevation and other local obstacles also alter the air flow. The system should have its own GPS module to remove this spatial mismatch and so that the online data are able to reflect

https://doi.org/10.3390/pr14071084

the accurate position of the turbine. Geolocation using GPS would enable the real-time forecasting of the data to be requested at the exact coordinates of the turbine, which would enhance the directional accuracy and consistency of the yaw control process.

This research paper is discussed in two phases. First, the wind direction prediction model is trained and validated by measured wind datasets to measure the prediction accuracy. Second, the validated output of predicted values and the prediction errors is incorporated into the yaw control structure and its effect on the yaw misalignment and power generation of turbines is tested in the same wind conditions.

5.2. AI Training and Performance Evaluation Using MATLAB

This part shows the AI-based wind direction prediction model that was developed in this paper. The focus of the AI module is to help make active yaw control decisions by short-term wind direction prediction to aid long-term wind predictions. The AI model is designed as a predictive support and redundancy system in situations where there might be noise in direct wind measurements, latency in direct wind measurements or short-term unavailability. To be clear and transparent, the condition of the AI model is measured separately, and combining its results with the analysis of the yaw control and power-generation analysis will be carried out in the following sections.

Various regression-based learning methods in MATLAB were first considered when predicting the wind direction in the short term, including linear regression, support vector regression, regression tree models, and recurrent neural networks. The initial appeal of tree-based regression models is that they are easy to compute, can be interpreted, and implementation can be embedded or edge-assisted. Nevertheless, the direction of the wind is highly time dependent and circular in nature and thus constrained the usefulness of static regression models, especially during wind conditions that vary rapidly.

On these grounds, an LSTM recurrent neural network was chosen as the ultimate prediction model. The time series nature of the information required in active yaw control support suits LSTM networks well and can be used to learn short-term temporal patterns in the sequence of wind measurements, which are essential in support of active yaw control. The LSTM model is trained on inputs such as wind direction and speed that are coded in sine–cosine form so that the seam of the 0–360 boundary is continuous and so that numerical accuracy is enhanced during training.

It is necessary to underline the fact that the yaw actuator is not directly controlled by the AI model. Under the suggested system architecture, the decisions to actuate the yaw are always determined based on the real-time measurements of wind direction on the active data source, and AI prediction is also performed separately to introduce redundancy, validation, and situational awareness. This paper provides certainty and safety in yaw behavior and also provides a control framework with short-term predictive information as necessary. Consequently, the AI module improves robustness and reliability without interfering with trusted real-time measurements.

5.2.1. AI Training Workflow and Dataset Structure

The AI model was trained using real-time wind speed and wind direction data collected through a microcontroller-based system that retrieves live meteorological information from OpenWeatherMap. The dataset was sampled at 5 min intervals and contained 3665 time-synchronized samples recorded on 5 December 2025, enabling effective sequential learning for short-term wind direction prediction.

The raw angular measurements were pretrained by converting to sine–cosine representations in order to take into consideration the circular nature of wind direction. This transformation removes the artificially introduced discontinuity at the 0°/360° boundary

https://doi.org/10.3390/pr14071084

and turns the angular data into continuous numerical characteristics that can be trained by machine learning algorithms. An example is that a wind direction of 360° would be denoted as (cos = 1, sin = 0) and 20 as (cos = 0.94, sin = 0.34). This encoding also guarantees that the transition of directional changes and wrap-around transitions are learned properly by the model.

The training set is organized in the form of a sliding time window in which there are three sequential wind direction measurements, WindDirT3, WindDirT2, and WindDirT1, which denote the direction of the wind in the furthest past, the nearest past, and the nearest future, respectively. The values of each of these are represented in terms of the relevant sine–cosine pair. Another input feature of the model is wind speed, to enable the model to represent the interaction of the wind magnitude and directional change. The AI model generates the target output as WindDirT, which is the predicted direction of the wind in the coming time step, which is 5 min of the wind. During supervised training, the model learns the nonlinear temporal relationship between past wind direction states (T3 → T2 → T1), wind speed, and the future wind direction WindDirT. This formulation enables short-horizon wind direction prediction that is directly relevant for yaw control decision support.

A sample subset of the structured training data is illustrated in Table 4 below, demonstrating how the cosine–sine encoding of the consecutive wind directions, the wind speed of that direction and the forecasted direction of the target are encoded. Through exposure to patterns such as stable wind conditions (e.g., 360° → 360° → 360°), gradual transitions (e.g., 360° → 20°), and sudden directional shifts (e.g., 20° → 360°), the model learns both steady and rapidly changing wind behaviors. The model of prediction is performed with the help of the LSTM neural network in MATLAB. The MATLAB implementation used for training the LSTM prediction model and for the real-time yaw control system is provided in the Supplementary Material (Supplementary Material S1). The network takes a sequence of input features sequentially and fills in the predicted yaw angle as a scalar value using a formulation (atan2d) that is an inverse trigonometric, which is angular consistent and avoids discontinuities. This approach ensures that the predicted yaw direction remains physically meaningful across the full 0–360° range.

Table 4. Sample of AI training dataset.

Wind Dir	Cos T3	Sin T3	Wind Dir	Cos T2	Sin	Wind Dir T1	Cos T1	Sin T1	Wind Speed	Wind Dir	Cos T	Sin T
T3			T2		T2					T		
360	1	0	360	1	0	360	1	0	5.81	360	1	0
360	1	0	360	1	0	360	1	0	6.17	20	0.94	0.34
360	1	0	360	1	0	20	0.94	0.34	6.17	20	0.94	0.34
360	1	0	20	0.94	0.3	20	0.94	0.34	6.17	20	0.94	0.34
20	0.94	0.34	20	0.94	0.3	20	0.94	0.34	6.17	20	0.94	0.34
20	0.94	0.34	20	0.94	0.3	20	0.94	0.34	6.17	20	0.94	0.34
20	0.94	0.34	20	0.94	0.3	20	0.94	0.34	7.15	360	1	0
20	0.94	0.34	20	0.94	0.3	360	1	0	7.15	360	1	0
20	0.94	0.34	360	1	0	360	1	0	7.15	360	1	0
360	1	0	360	1	0	360	1	0	7.15	360	1	0

Figure 16 below illustrates the prediction performance of the trained LSTM-based wind direction model using the dataset summarized in Table 4. The first subplot in Figure 16 compares the actual wind direction with the corresponding LSTM-predicted

https://doi.org/10.3390/pr14071084

values. The close overlap between the blue solid curve (actual) and the red dashed curve (predicted) demonstrates that the model accurately tracks the temporal evolution of wind direction. Importantly, smooth transitions are observed across the 0°/360° boundary, confirming that the sine–cosine encoding and angular reconstruction effectively prevent artificial jumps in the prediction output. The second subplot shows the circular prediction error computed using an angular distance formulation. The error remains centered around 0° and is largely confined within ±10°, indicating stable and consistent short-term predictions throughout the dataset. This level of angular accuracy is well within acceptable limits for yaw control decision support, where small residual misalignment results in only marginal aerodynamic power loss. The third subplot presents the regression fit between predicted and actual yaw angles. The data points cluster closely around the fitted regression line, yielding a high correlation coefficient (R = 0.9789) and a mean squared error (MSE) of 113.32°. These results confirm a strong agreement between predicted and measured wind directions and validate the effectiveness of the LSTM-based approach for short-horizon wind direction prediction.

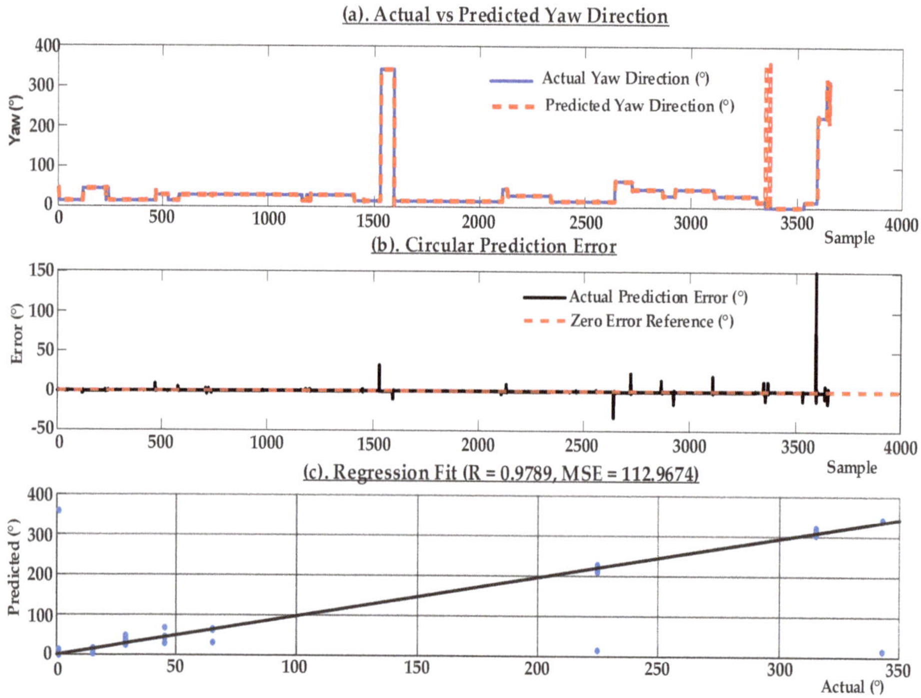

Figure 16. Performance of the trained LSTM-based wind direction prediction model showing (**a**) actual vs predicted yaw direction, (**b**) circular prediction error, and (**c**) regression fit between predicted and actual yaw angles (R = 0.9789, MSE = 113.32°).

Taken together, the structured input data summarized in Table 4 and the performance results shown in Figure 16 indicate that the AI module successfully learned the short-term temporal evolution of wind direction ($T_3 \rightarrow T_2 \rightarrow T_1 \rightarrow T$). The model's output, $WindDir_t$, represents the predicted wind direction at the next time step corresponding to a 5 min prediction horizon, which is well suited for yaw control decision support. The offline evaluation using historical sequential data confirms that the LSTM-based model can accurately capture short-term wind direction dynamics and provide reliable predictions. However, while these results validate the intrinsic predictive capability of the AI model, they do not by themselves demonstrate performance when the predictor is embedded within a complete yaw control framework under real-time operating conditions, which is addressed separately through system-level simulations.

To address this, the following section evaluates the trained AI model within a closed-loop yaw control system using live wind data acquired through a microcontroller-based interface. This system-level validation examines how AI prediction interacts with real wind measurements, motor actuation, yaw error, and power output in real time. The analysis is supported by MATLAB simulation results and representative samples of live operational data, as is illustrated in Figure 16 and summarized in Table 4. It is noted that, like all data-driven models, the AI predictor performs best within the range of conditions represented in the training data. The model is therefore designed as a supportive and fallback component, rather than a sole control authority, ensuring that real-time measurements always take precedence when available.

5.2.2. AI Prediction Result and MATLAB Validation

In the validation stage, wind measurements are obtained exclusively via a microcontroller-based system using live data from the OpenWeatherMap API. The trained LSTM-based wind direction prediction model is evaluated over approximately 14 h of real-time operation. Wind speed and wind direction are retrieved continuously through a Wi-Fi connection and transmitted directly to MATLAB for processing, visualization, and control analysis. This configuration enables system-level validation of the proposed yaw control framework without reliance on local wind sensors or a physical turbine installation.

For clarity of evaluation, only the live data source is activated during this stage. This deliberate choice avoids source switching and isolates the yaw control response to real-time wind inputs. The AI model operates in parallel and generates short-term wind direction predictions; however, it does not directly actuate the yaw motor. All yaw commands are computed exclusively from live wind measurements, ensuring deterministic, safe, and physically interpretable control behavior.

Figure 17 presents the MATLAB simulation results corresponding to the operating events summarized in Table 5. The subplots illustrate the interactions between wind measurements, AI prediction, yaw control, and turbine power response.

Table 5. Representative samples of live wind data acquired via a microcontroller-based OpenWeatherMap API, showing AI-predicted yaw direction, motor yaw response, and yaw error used for validating the proposed yaw control system.

Time (s)	Source	Wind Dir (°)	Wind Speed	AI Predicted Yaw (°)	Motor Yaw (°)	Yaw Error (°)	Yaw Movement	Wi-Fi
1581.2	LIVE	80	3.60	5	80.0	+0.0	—(initial)	ON
1762.06	LIVE	80	3.60	5	80.0	+0.0	0.0	ON
3525.78	LIVE	293	1.79	262	293.0	−0.0	−147	ON
3526.92	LIVE	293	1.79	315	293.0	−0.0	0.0	ON
4131.01	LIVE	60	4.12	355	59.9	+0.1	+126.9	ON
4132.13	LIVE	60	4.12	5	59.9	+0.1	0.0	ON

Figure 17a shows the live wind direction (wind source), the AI-predicted direction, and the motor yaw position. The wind source data are not missing; rather, they largely overlap with the motor yaw position during steady-state operation. This overlap occurs because the yaw controller successfully aligns the nacelle with the live wind direction to maximize aerodynamic efficiency. When alignment is achieved, both curves coincide, making them visually indistinguishable in certain intervals. During periods of stable wind, such as around t ≈ 1581 s and t ≈ 1762 s, the yaw motor closely tracks the live wind direction (≈80°), resulting in a yaw error near zero. Although the AI-predicted direction exhibits minor deviations, it does not directly actuate the motor. The controller prioritizes real-time live measurements for actuation while using AI prediction only for anticipatory support. When abrupt wind direction changes occur, the transient behavior becomes visible. For example, at t ≈ 3525.8 s, the wind direction shifts rapidly to approximately 293°, producing a temporary yaw error while the motor remains near its previous position. AI prediction indicates an approaching directional shift; however, the controller computes the yaw error exclusively from live measurements and commands the shortest rotation path. The motor then settles near 293°, restoring a near-zero yaw error. Similar behavior is observed at t ≈ 4131.0 s when the wind shifts to approximately 60°. These responses confirm that AI captures directional trends, while the control system ensures safe and accurate physical alignment. Figure 17b relates wind speed to both measured turbine power and predicted power output. During steady alignment, measured power follows the Falcon Silence 3.6 kW reference power curve, as shown in Figure 17c. Temporary power reductions occur during brief yaw misalignment following abrupt wind changes, which are corrected as the yaw error returns toward zero. A noticeable discrepancy between predicted and measured power is observed in certain intervals, reaching approximately 50% during rapid wind fluctuations. This deviation arises primarily from high-frequency wind speed variations present in the MATLAB simulation environment. In contrast to real-world atmospheric behavior, the simulated wind data include sharper and faster fluctuations, which amplify prediction sensitivity and result in larger instantaneous power differences. Importantly, these discrepancies occur during transitional periods rather than steady-state operation. Once wind speed stabilizes and yaw alignment is restored, predicted and measured power values converge toward the reference curve. Future implementation will incorporate improved wind filtering and adaptive prediction smoothing techniques to reduce transient prediction errors and enhance robustness under rapidly varying wind conditions. Figure 17c presents the manufacturer's reference power curve of the Falcon Silence 3.6 kW turbine. The simulated measured power remains bounded by this curve throughout operation, confirming that the implemented aerodynamic and derating models maintain physical realism.

Figure 17d illustrates the yaw error evolution. Error spikes occur only during abrupt wind direction transitions. These spikes are short-lived and rapidly decay as the controller re-establishes alignment. During steady wind conditions, the yaw error remains near zero, confirming effective tracking performance. Figure 17e shows the turbine efficiency variation. Efficiency remains high when yaw alignment is maintained and decreases temporarily during misalignment events. Once corrective yaw action is completed, efficiency returns to near-optimal levels, demonstrating the effectiveness of the LIVE-prioritized yaw control strategy.

Table 5 below presents representative samples of live wind data used to validate the proposed AI-assisted yaw control framework. Wind speed and wind direction are acquired in real time by a microcontroller through a Wi-Fi connection to the OpenWeatherMap API and transmitted directly to MATLAB for processing. The trained LSTM model generates short-term yaw predictions, which are logged alongside the measured wind direction, motor yaw position, yaw error, and yaw movement. The yaw controller computes motor actions exclusively from the live wind measurements, while AI prediction serves as a supporting reference and does not directly actuate the nacelle. All values are reported with appropriate numerical precision consistent with measurement resolution and control accuracy. This table supports Figure 17 by illustrating how the yaw control system responds to

https://doi.org/10.3390/pr14071084

real-time wind variations while maintaining stable alignment and minimizing unnecessary yaw movements.

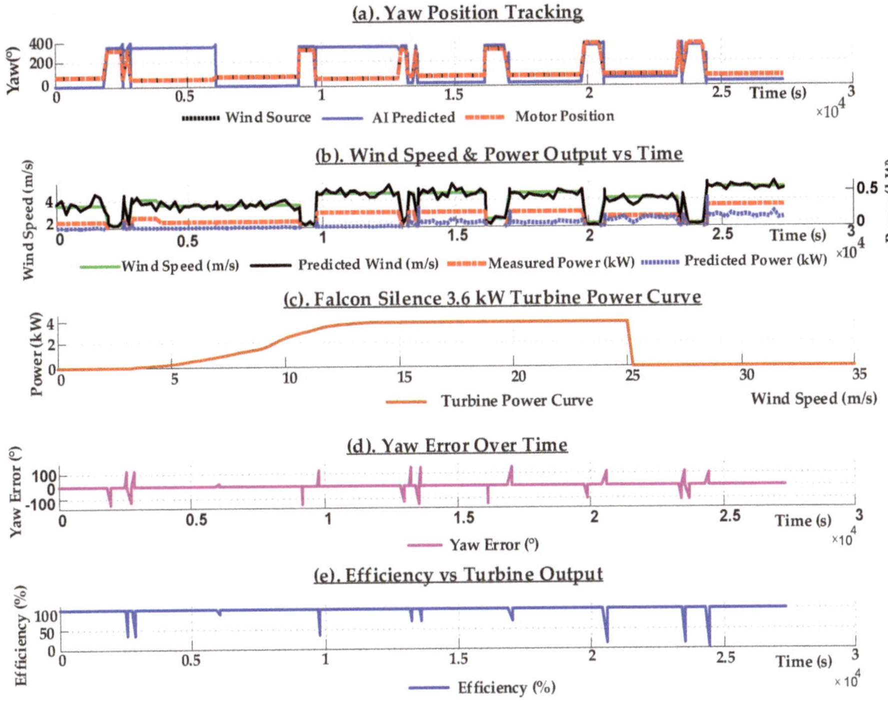

Figure 17. MATLAB simulation results of the AI-assisted yaw control system, showing (**a**) yaw position tracking with live wind direction, AI prediction, and motor position; (**b**) wind speed and corresponding measured and predicted turbine power versus time; (**c**) Falcon Silence 3.6 kW reference power curve; (**d**) yaw error dynamics; and (**e**) turbine efficiency variation during operation.

Taken together, Figure 17 and Table 5 confirm that the AI prediction model functions as a supportive, non-intrusive component within the proposed yaw control architecture. The microcontroller-based live wind data serve as the sole control input, while the AI provides additional situational awareness and redundancy without overriding real-time measurements. This structure allows the system to maintain accurate alignment during rapid wind direction changes, reduce mechanical stress through shortest-path rotation, and demonstrate the feasibility of AI-assisted yaw control using online meteorological data.

The results presented in Section 5.2 demonstrate that the proposed LSTM-based AI model can reliably capture short-term wind direction dynamics and generate accurate yaw predictions under real-time operating conditions. The validation confirms that the AI module functions as a stable and supportive component within the yaw control framework, without directly influencing actuator commands when trustworthy live measurements are available. Having established the predictive accuracy and robustness of the AI model in MATLAB, the next section focuses on evaluating how this AI-assisted yaw control strategy affects turbine-level performance metrics. Section 5.3 therefore investigates the resulting yaw alignment behavior, power output, efficiency, and overall performance improvements when the validated prediction model is integrated into the complete yaw control and power-generation analysis.

https://doi.org/10.3390/pr14071084

5.3. Output of MATLAB/SIMULINK

5.3.1. LIVE-Based Yaw Control Performance Analysis

Figure 18 present the turbine response under LIVE-Based yaw control using real wind direction and wind speed data obtained from the Government of Canada dataset [35]. The data are plotted at 10 min intervals to clearly show the operational behavior of the turbine over time.

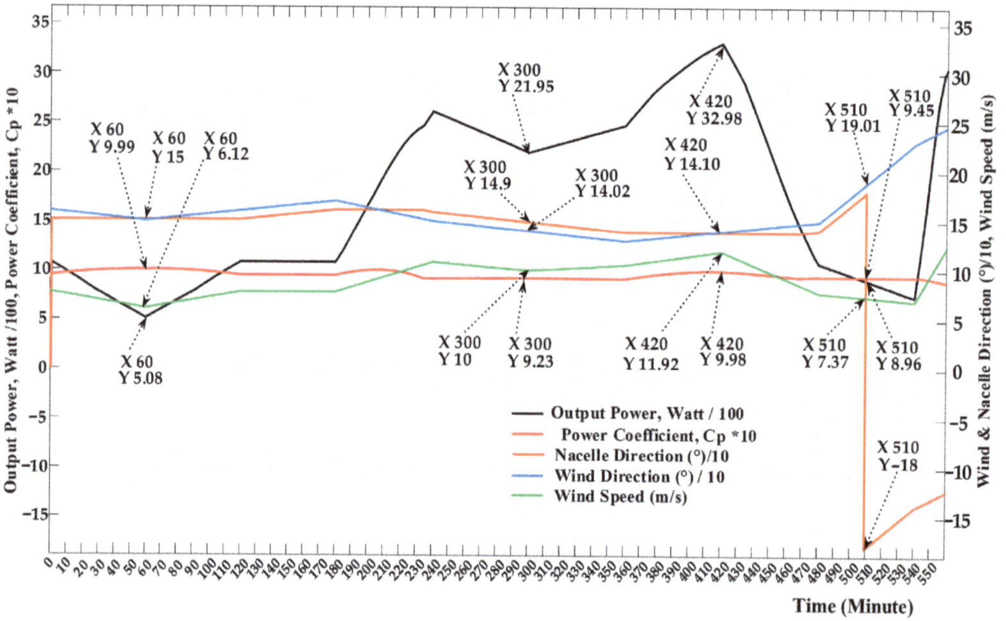

Figure 18. Turbine performance under live yaw control using 10-min interval data, showing output power, power coefficient (Cp), wind direction, nacelle direction, and wind speed. For visualization on a common axis, scaling factors are applied: power (W/100), Cp (×10), and wind/nacelle direction (°/10). Yaw angles are plotted in the −180° to +180° range for clarity.

At t = 60 min, the wind direction (blue) and nacelle direction (brown) are aligned at approximately 15 × 10 = 150°. This alignment indicates proper yaw tracking under LIVE control. As a result, the yaw error is nearly zero, allowing efficient aerodynamic energy conversion. The power coefficient Cp is 9.99/10 = 0.99, which is close to the theoretical maximum, confirming optimal rotor alignment. The wind speed is 6.12 m/s, and accordingly, the output power is 5.08 × 100 = 508 W (0.508 kW). The moderate power level is directly related to the moderate wind speed at this time.

At t = 300 min, the wind direction is 14.02 × 10 = 140.2°, while the nacelle direction is 14.9 × 10 = 149°. The slight difference between these values indicates a small transient yaw misalignment. Consequently, the power coefficient slightly decreases to 9.23/10 = 0.923. The wind speed increases to 10 m/s, which significantly increases the available aerodynamic power. As a result, the turbine output rises to 21.95 × 100 = 2195 W (2.195 kW). The small reduction in Cp compared to t = 60 min is due to the minor yaw deviation during directional adjustment.

At t = 420 min, the wind direction and nacelle direction fully overlap at 14.10 × 10 = 141°, indicating perfect alignment. The wind speed increases further to 11.92 m/s, providing higher available wind energy. Because of this complete alignment, the power coefficient reaches 9.98/10 = 0.98, which is close to maximum aerodynamic efficiency. The turbine's

https://doi.org/10.3390/pr14071084

output reaches its highest observed value of 32.98 × 100 = 3298 W (3.298 kW), which is consistent with the increased wind speed and optimal yaw alignment.

At t = 510 min, the wind direction is 19.01 × 10 = 190°, while the nacelle direction is shown at −180°. This apparent negative value occurs due to the yaw angle being represented within the normalized range of −180° to +180°. When the nacelle reaches the rotational limit in one direction, the control system performs cable management (cable untwisting logic), causing the angle to wrap to −180° instead of continuing beyond +180°. This prevents excessive cable twisting inside the nacelle and maintains mechanical safety. Although the representation changes sign, the physical yaw alignment remains correct in circular angle space. At this time, wind speed decreases to 7.37 m/s, leading to a lower available aerodynamic power. The power coefficient is 9.45/10 = 0.945, slightly reduced due to transient adjustment and lower wind speed. The turbine output becomes 8.96 × 100 = 896 W (0.896 kW), which corresponds to the reduced wind speed conditions.

Overall, Figure 18 demonstrates that under LIVE yaw control, the nacelle consistently aligns with the real wind direction, maintaining high Cp values during steady conditions and restoring alignment quickly during directional changes. Power output variations directly follow wind speed variations and yaw alignment quality.

5.3.2. AI-Based Yaw Control Performance

Figure 19 presents the turbine performance when the LIVE wind source becomes unavailable and yaw alignment is maintained using the AI-based yaw prediction system. The data are shown at 10 min intervals to clearly illustrate system behavior during AI-driven operation.

At t = 60 min, the wind direction and nacelle direction coincide at approximately 17 × 10 = 170°, indicating accurate AI-based yaw alignment. Because the rotor faces the wind properly, aerodynamic efficiency is high. The power coefficient Cp equals 9.99/10 = 0.99, which is close to its optimal value. With a wind speed of 7.12 m/s, the turbine produces 8.50 × 100 = 805 W (0.805 kW). The moderate wind speed explains the moderate output power, while proper alignment explains the high Cp value.

At t = 174 min, the wind direction is 18.95 × 10 = 189.5°, while the nacelle direction is shown at −180°. This negative value results from angle normalization within the −180° to +180° range and cable management logic. Instead of rotating beyond +180°, the controller wraps the angle to −180° to prevent excessive cable twisting inside the nacelle. Although numerically different, −180° is physically equivalent to +180°, meaning alignment is maintained within circular angle space. The Cp value slightly decreases to 0.937 due to minor transient misalignment during angle transition. With wind speed at 8.77 m/s, output power rises to 1.552 kW, consistent with increased wind energy availability.

At t = 237 min, the wind direction is 170.9°, while the nacelle direction appears near +179°. This again reflects angular normalization at the rotation boundary. The controller selects the shortest rotational path while respecting cable limits. Cp reduces slightly to 0.902 due to transient adjustment. However, wind speed increases to 11.71 m/s, resulting in a significantly higher output power of 2.907 kW. The power increase is primarily driven by wind speed rather than yaw misalignment.

At t = 300 min, wind direction is 159.9°, and nacelle direction is 168.9°, indicating a small yaw deviation of approximately 9°. This minor misalignment explains the slightly reduced Cp value of 0.911. Wind speed remains high at 11.01 m/s, leading to a strong output power of 2.687 kW. The small efficiency reduction is consistent with the observed yaw offset.

https://doi.org/10.3390/pr14071084

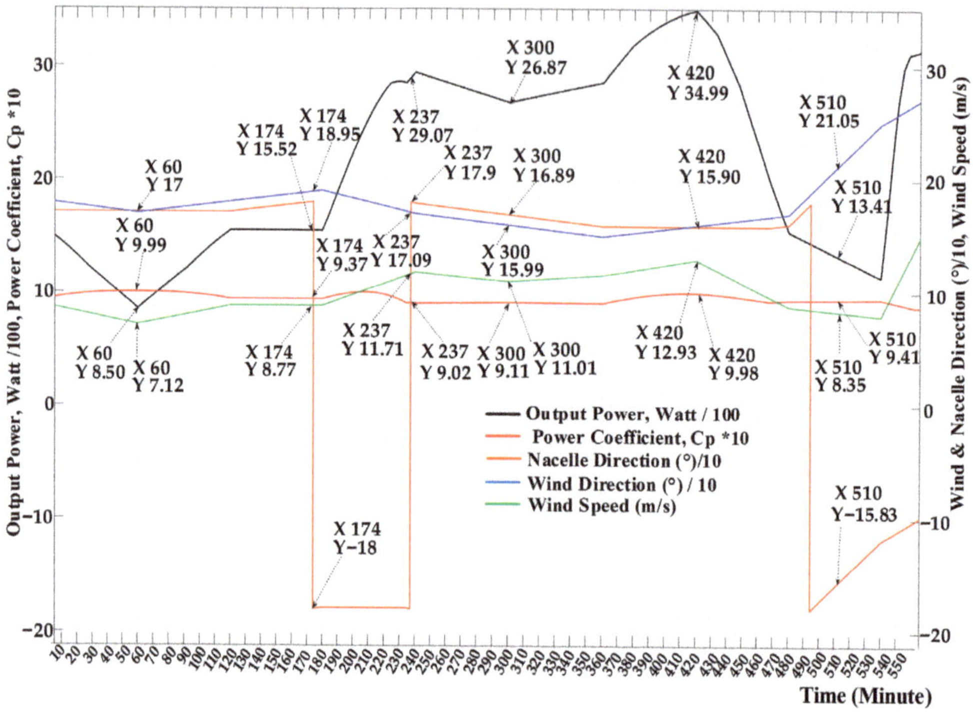

Figure 19. Turbine response under AI-based yaw control using 10-min interval data, showing turbine output power, power coefficient (Cp), wind direction, nacelle direction, and wind speed. For visualization on a common axis, scaling factors are applied: power (W/100), Cp (×10), and wind/nacelle direction (°/10). Yaw angles are represented in the normalized range of −180° to +180°.

At t = 420 min, wind direction and nacelle direction overlap at approximately 159°, indicating perfect alignment under AI control. Wind speed reaches 12.93 m/s, providing maximum available aerodynamic energy. Consequently, Cp increases to 0.98, and turbine output reaches its maximum observed value of 3.499 kW. This confirms that AI-based yaw control can maintain near-optimal alignment even without LIVE input.

At t = 510 min, wind direction is 210.5°, while the nacelle direction is shown as −158.3°. This representation results from angular wrapping and cable untwisting logic. In circular angle space, −158.3° is equivalent to 201.7°, which is close to the wind direction. The controller avoids excessive mechanical rotation by maintaining yaw within ±180°. Cp is 0.941, indicating good aerodynamic alignment. With wind speed reduced to 8.35 m/s, output power decreases to 1.341 kW, which is consistent with lower wind energy.

Overall, Figure 19 demonstrates that AI-based yaw control maintains stable alignment and efficient power capture even when LIVE wind measurements are unavailable. Temporary deviations are mainly due to cable management boundaries and angle normalization rather than control instability.

5.3.3. Overall Performance Comparison Between LIVE and AI-Based Yaw Control

Table 6 provides a detailed comparison of turbine performance under LIVEbased and AI-based yaw control at selected operating times (60, 300, 420, and 510 min). The results clearly indicate that the AI-based yaw control consistently achieves higher power output than the LIVEBased method under the same wind conditions.

https://doi.org/10.3390/pr14071084

Table 6. Comparison of LIVE-based and AI-based yaw control performance using MATLAB results.

Time (min)	Yaw Control Method	Power Output (kW)	Power Coefficient (C_p)
60	LIVE-based	0.508	0.99
60	AI-based	0.805	0.99
300	LIVE-based	2.195	0.923
300	AI-based	2.687	0.911
420	LIVE-based	3.298	0.98
420	AI-based	3.499	0.98
510	LIVE-based	0.896	0.945
510	AI-based	1.341	0.941

At 60 min, both control strategies maintain excellent aerodynamic efficiency with a power coefficient (Cp) of 0.99, indicating proper alignment with the wind. However, the AI-based system produces 0.805 kW compared to 0.508 kW under LIVE control. The higher power output under AI control suggests improved alignment timing and reduced transient misalignment losses during early stage wind variation.

At 300 min, wind speed increases and both systems operate at higher power levels. The LIVE-based method generates 2.195 kW with Cp = 0.923, while the AI-based method produces 2.687 kW with Cp = 0.911. Although the Cp value for AI control is slightly lower at this instant, the overall power output is significantly higher. This indicates that the AI model likely adjusted the nacelle earlier during wind transitions, reducing the duration of yaw error and allowing more effective energy capture over the interval.

At 420 min, both methods achieve near-maximum aerodynamic efficiency with Cp = 0.98, reflecting excellent yaw alignment. Even under these optimal conditions, AI control delivers a higher output of 3.499 kW compared to 3.298 kW under LIVE control. This demonstrates that anticipatory yaw adjustments under AI control contribute to slightly improved overall energy extraction, especially during preceding wind changes.

At 510 min, wind speed decreases and overall power output drops for both systems. However, AI-based control still outperforms LIVE control, producing 1.341 kW compared to 0.896 kW. The Cp values are very close for AI (0.941) and for LIVE (0.945), indicating that the main difference in output is not due to steady-state efficiency but rather improved yaw response timing during wind direction variation.

Overall, AI-based yaw control consistently provides higher power output across all evaluated time points. The key reason for this improvement is the AI model's ability to predict short-term wind direction trends and initiate earlier nacelle movement. This reduces the magnitude and duration of yaw misalignment, stabilizes aerodynamic performance, and enhances total energy capture compared to the purely reactive LIVE-based yaw control strategy.

5.4. Real-Time Microcontroller Output During AI Prediction Mode

Here below is the output from the microcontroller serial monitor, showing how the system behaves in AI prediction mode.

```
[PRIORITY]: VANE >  LIVE  >  AI FORECAST
    [SOURCES] Snapshot (angle, speed, time)
        VANE: 264.7° (W), Time: 09- Nov 07:08:27
         LIVE: 270.0° (W), 7.2 m/s, Time: 09- Nov 07:08:27
         FCST: 319.0° (NW), 7.8 m/s, Time: 09- Nov 10:05:00
```

https://doi.org/10.3390/pr14071084

[RUN] Motor using: VANE → 264.7° (W)
AI FORECAST Yaw (vs SRC= VANE), 294.3° (WNW) [Δ vs src = 29.6°]
Motor pos: 264.7° (W) | Error: 0.0°
[STATUS] VANE = OK | LIVE = OK | Forecast = OK | | Wi-Fi = OK | Counts V/L/F = 2/2/0 | Wi-Fi reconnects = 0 | Motor Pos = 264.7° | Err = 0.0°

[PRIORITY]: VANE > LIVE > FORECAST
[SOURCES] Snapshot (angle, speed, time)
VANE: (lost)
LIVE: (lost)
FCST: 319.0° (NW), 7.8 m/s, Time: 09- Nov 10:05:00
[RUN] Motor using: AI Predicted → 294.3° (WNW)
AI FORECAST Yaw (vs SRC= VANE), 294.3° (WNW) [Δ vs src = 0°]
Motor pos: 294.3° (W) | Error: 0.0°
[STATUS] VANE = LOST | LIVE = LOST | Forecast = OK | | Wi-Fi = LOST | Counts V/L/F = 2/2/0 | Wi-Fi reconnects = 0 | Motor Pos = 294.3° | Err = 0.0°

[PRIORITY]: VANE > LIVE > FORECAST
[SOURCES] Snapshot (angle, speed, time)
VANE: (lost)
LIVE: 270.0° (W), 7.2 m/s, Time: 09- Nov 07:16:35
FCST: 319.0° (NW), 7.8 m/s, Time: 09- Nov 10:05:00
[RUN] Motor using: LIVE → 270.0° (W)
AI Predicted Yaw (vs SRC LIVE, speed= 7.2 m/s from LIVE Time: 09- Nov 07:23:35): 299.5° (WNW) [Δ vs src = 29.5°]
Motor pos: 270.0° (W) | Error: 0.0
[STATUS] VANE = LOST | LIVE = OK | Forecast = OK | | Wi-Fi = OK | Counts V/L/F=2/2/0 | Wi-Fi reconnects= 0 | Motor Pos=270.0° | Err= 0.0°

[PRIORITY]: VANE > LIVE > FORECAST
[SOURCES] Snapshot (angle, speed, time)
VANE: (lost)
LIVE: (lost)
FCST: 319.0° (NW), 7.8 m/s, Time: 09- Nov 10:05:00
[RUN] Motor using: AI Predicted → 299.5° (WNW)
AI Predicted Yaw (vs SRC=LIVE, speed= 7.2 m/s from LIVE Time: 09- Nov 07:29:03): 299.5° (WNW) [Δ vs src = 0°]
Motor pos: 299.5° (W) | Error: 0.0°
[STATUS] VANE = LOST | LIVE = LOST | Forecast = OK | | Wi-Fi = OK | Counts V/L/F=2/2/0 | Wi-Fi reconnects= 0 | Motor Pos = 299.5° | Err= 0.0°

Although the findings provided in the last part indicate the technical efficiency of the offered yaw control approach and AI-based prediction model, practical implementation is also limited by economic considerations. Thus, the section below will assess cost implications of the proposed system, including hardware elements, implementation needs, and general cost-effectiveness of the proposed system in small-sized wind tur-bine applications.

5.5. Scalability of the Proposed Yaw Control System for Large-Scale Wind Turbines

Although the experimental validation was conducted on a low-power laboratory-scale wind turbine platform due to facility limitations, the proposed AI-based yaw control

https://doi.org/10.3390/pr14071084

strategy is inherently scalable to large-capacity wind turbines (e.g., 10 MW–15 MW systems). The developed control algorithm operates independently of turbine-rated power and primarily depends on wind direction sensing, yaw position feedback, and actuator control logic.

In large-scale commercial wind turbines, such as 10 MW or 15 MW units, the yaw mechanism typically employs high torque electric yaw drives and multiple geared motors instead of stepper motors used in laboratory validation. However, the control architecture remains fundamentally similar, consisting of wind direction sensors, nacelle position encoders, and a supervisory controller. Therefore, the proposed intelligent yaw alignment algorithm can be directly implemented in industrial-scale systems by integrating it into existing turbine control units (TCUs) without modification of the core control logic.

The primary difference between small-scale and large-scale implementation lies in actuator torque capacity and structural loading requirements, rather than in the control strategy itself. Since the algorithm processes the wind misalignment angle and generates corrective yaw commands independent of turbine size, the methodology remains applicable to multi-megawatt offshore and onshore wind turbines.

6. Cost–Benefit Analysis

The use of AI together with live online weather data gives the proposed yaw control system clear advantages in both cost and reliability compared with traditional sensor-based designs. Conventional systems depend on mechanical wind vanes and transmitter–receiver units to measure and send wind direction signals. These devices add to the overall hardware cost, need periodic calibration, and are often affected by wear, corrosion, or signal faults in harsh outdoor conditions. In this research work, the AI model and the live weather data replace the physical wind vane and the associated communication hardware. Real-time wind direction and speed are taken directly from reliable online weather sources, which reduces the amount of equipment required and simplifies the entire system. This lowers installation costs and reduces long-term maintenance demands. In addition, the control logic drives the yaw motor only when the deviation from the wind direction becomes larger than $\pm 15°$, so unnecessary movements are avoided and the motor experiences less mechanical stress, helping it last longer.

Comparative Cost–Benefit Summary

Table 7 below is a clear comparison between a traditional yaw control system and the proposed AI-based smart solution, which demonstrates improvements in functionality and total cost. The traditional design relies on physical wind vane sensors and transmitter–receiver units, which add to the cost of installation and require regular maintenance. These elements also impose an additional burden on the yaw mechanism, as the motor is forced to respond to frequent changes in the wind direction and, as a result, it runs almost continuously. Conversely, the AI-based system relies on real-time online wind information, short-term projection, and location selection through GPS to receive the correct directional information from the closest weather station. This saves the use of several hardware components, decreases the maintenance of sensors, and enhances accuracy in alignment, particularly in cases where the wind conditions vary rapidly. Two foundations are used to obtain the numerical values: the first one is the aerodynamic yaw misalignment theory of power retention and the second one is a relative index model of yaw actuation energy and maintenance cost. First, the effective power capture due to yaw alignment is evaluated using the standard cosine-cubed relationship $P(\theta) = P_0 cos^3\theta$, where P_0 is power at perfect alignment and θ is the yaw error. Conventional yaw systems typically allow persistent yaw errors in the range of 10–15° because reactive control and practical deadbands avoid constant actuation;

https://doi.org/10.3390/pr14071084

this corresponds to $cos^3(10°) = 0.955$ (95.5%) and $cos^3(15°) = 0.901$ (90.1%). Therefore, Table 6 reports 90–95% retained aerodynamic power. The proposed controller achieves a smaller yaw error during sustained wind direction trends because it combines multi-source data (vane + online data) with AI prediction as a fallback during data interruptions. For $\theta \leq 5°, cos^3(5°) = 0.989$ (98.9%); therefore, Table 6 shows 98–99% retained aerodynamic power. Importantly, the controller does not chase every small fluctuation, which would increase gearbox and bearing stress; yaw actuation is limited using the deadband (typically ±10–15°), dwell-time, filtered wind direction, and rate-limited yaw commands. Second, yaw actuation energy is estimated using yaw activity as an energy proxy because, for the same actuator and supply conditions, energy scales with total movement/actuation time; thus, yaw energy saving is computed as $\frac{A_{conv}-A_{AI}}{A_{AI}} * 100$, where A is cumulative yaw motion (sum of absolute step counts or yaw angle changes). With deadband and dwell-time, the proposed controller reduces yaw activity by approximately 10–15%, which is reported as the yaw energy-saving range. Finally, maintenance cost reduction is derived from a normalized maintenance index that separates maintenance into hardware servicing and yaw wear servicing, $C_{maint} = C_{HW} + C_{yaw}$, with the conventional system normalized to 1.0. Because exposed electrical/communication hardware and wiring represent a major portion of servicing effort in small yaw systems, conservative weighting is applied ($C_{HW,conv} = 0.60$, $C_{yaw,conv} = 0.40$). Since the transmitter–receiver modules are fully removed while the vane remains, hardware-related maintenance, Υ_{HW}, is conservatively reduced by 35–50%, and wear-related maintenance, Υ_{yaw}, is reduced by 15–20% due to fewer yaw actions; applying these reductions gives $C_{maint,AI} = 0.60\ (1 - \Upsilon_{HW}) + 0.40\ (1 - \Upsilon_{yaw})$, which results in an overall reduction of approximately 27–38% and is reported as a conservative rounded range of 30–40% to account for deployment variability.

Table 7. Comparative cost–benefit analysis of conventional and AI-based smart yaw control systems.

Aspect	Conventional System	AI-Based Smart System	Benefit/Savings
Wind Direction Input	Physical wind vane sensor	Online live wind data via AI	Eliminates sensor cost
Data Communication	Transmitter–receiver modules	Cloud-based online data	Removes hardware modules
Yaw Motor Operation	Continuous adjustments	Threshold-based activation	Lower Motor stress, longer lifespan
GPS-Assisted Location	Not available	GPS used to obtain nearest weather-station data	Higher directional accuracy, improved power capture
Power Capture Efficiency	90–95%	98–99%	3–5% higher power generation
Energy Use for Yaw	High	Low	10–15% energy saving
Maintenance Frequency	Regular servicing	Minimal maintenance	30–40% maintenance cost

7. Limitations and Future Directions

The present system uses a physical wind vane for real-time yaw measurement and an AI model trained on both on-site sensor data and live online wind direction information. This combined structure helps the turbine operate continuously, because the controller can shift between the vane, online weather data, and AI prediction whenever one source becomes unavailable. Since the AI model learns from both types of inputs, it can estimate short-term wind direction when the vane becomes unstable or when online updates are slow. Even with these strengths, the system still has some limitations. The online weather data comes from nearby meteorological stations rather than the exact turbine location, so small differences between the station readings and the actual on-site wind field may appear.

https://doi.org/10.3390/pr14071084

These differences become more noticeable when both the vane and the live online data are unavailable and the controller must rely only on AI prediction. In such cases, the AI has no fresh input to correct itself, and its estimates may be less accurate. Adding GPS would allow the system to automatically choose the nearest weather station and improve alignment accuracy and power capture.

The mechanical wind vane also introduces limitations. There are many factors such as exposure to dust, moisture, temperature changes, and corrosion that can affect its moving parts and reduce sensitivity, which may cause slow or uneven responses. At low wind speeds, the vane may not align properly because of inertia, and this can lead to unstable readings. Calibration drift, electrical noise, and the use of a single measurement point further reduce accuracy, especially in turbulent areas or complex terrain. Severe gusts or storms can even damage the vane or cause temporary data loss, which would require maintenance or recalibration.

The system also depends on a stable internet connection for live data updates, and the current AI model—although trained on both sensor and weather station data—uses a fixed training set and does not update itself automatically during operation. Future work will focus on GPS-assisted selection of weather stations, adaptive AI models that can learn from ongoing turbine performance, and a stronger hybrid-sensing approach that combines vane feedback with short-term prediction to keep the system reliable during network outages. Expanding the design for larger turbines, testing it at different locations, and adding faster onboard processing will help improve yaw response, mechanical protection, and long-term performance.

8. Concluding Remarks

The smart yaw control system developed in this study demonstrates that real-time sensing, intelligent control, and data-driven prediction can be effectively integrated to enhance wind turbine performance. The proposed framework combines a physical wind vane sensor with an AI-based predictive model trained on live meteorological data, enabling continuous nacelle alignment with changing wind direction. This coordinated approach significantly reduces yaw misalignment, minimizes unnecessary motor actuation, and decreases mechanical wear and associated power losses.

Both simulation and hardware validation results confirm that the AI-supported controller maintains yaw deviation within $\pm 15^{\circ}$, ensuring stable turbine operation and power output close to the rated capacity. The implementation of a multi-source priority control strategy further enhances system reliability by allowing uninterrupted operation in the event of sensor malfunction or communication failure. Additionally, the optimized control logic limits excessive yaw motor movement, thereby extending mechanical component lifespan and reducing maintenance requirements.

Although experimental validation was conducted on a laboratory-scale prototype due to facility limitations, the proposed yaw control architecture is inherently scalable to large-capacity wind turbines (e.g., 10–15 MW systems). The control algorithm operates independently of turbine-rated power and can be integrated into industrial yaw drive systems with appropriately sized actuators and torque mechanisms. Therefore, the methodology is applicable not only to small and medium wind turbines but also to utility-scale and offshore wind energy systems.

Future enhancements, including GPS-based weather station selection, adaptive AI retraining mechanisms, and advanced sensor fusion techniques, could further improve alignment accuracy, system robustness, and overall energy capture in next-generation wind energy installations.

https://doi.org/10.3390/pr14071084

Supplementary Materials: The following supporting information can be downloaded at: https://www.mdpi.com/article/10.3390/pr14071084/s1, Supplementary Material S1: MATLAB Code for AI-Based Yaw Prediction and Smart Yaw Control Implementation.

Author Contributions: Conceptualization, S.M., M.F.T., A.A.K. and U.A.K.; Methodology, S.M., M.F.T. and H.F.A.; Software, S.M., M.F.T., H.F.A. and U.A.K.; Validation, S.M. and M.F.T.; Formal Analysis, M.F.T.; Investigation, S.M., M.F.T. and A.A.K.; Resources, U.A.K.; Writing—Original Draft, S.M. and M.F.T.; Writing—Review and Editing, S.M., M.F.T., H.F.A. and U.A.K.; Visualization, A.A.K.; Supervision, A.A.K. and U.A.K.; Project Administration, A.A.K. and H.F.A.; Funding Acquisition, A.A.K. All authors have read and agreed to the published version of the manuscript.

Funding: This research received no external funding.

Data Availability Statement: The original contributions presented in this study are included in the article/Supplementary Material. Further inquiries can be directed to the corresponding author(s).

Conflicts of Interest: The authors declare no conflicts of interest.

References

1. Global Wind Energy Council. *Global Wind Report 2024*; GWEC: Brussels, Belgium, 2024. Available online: https://www.gwec.net/hubfs/Website-2023/documents/GWEC-2024.pdf?hsLang=en (accessed on 3 September 2025).
2. Burton, T.; Jenkins, N.; Sharpe, D.; Bossanyi, E. *Wind Energy Handbook*, 3rd ed.; John Wiley & Sons: Chichester, UK, 2021; pp. 17–21.
3. Burton, T.; Sharpe, D.; Jenkins, N.; Bossanyi, E. *Wind Energy Handbook*, 2nd ed.; Wiley: Chichester, UK, 2011; pp. 97–98.
4. Liew, J.Y.; Urbán, A.M.; Andersen, S.J. Analytical Model for the Power–Yaw Sensitivity of Wind Turbines Operating in Full Wake. *Wind Energy Sci.* **2020**, *5*, 427–437. [CrossRef]
5. Wang, Y.; Liu, Z. Fatigue Analysis of Wind Turbine and Load Reduction through Wind-Farm-Level Yaw Control. *Energy* **2025**, *326*, 136266. [CrossRef]
6. Gao, L.; Li, H.; Wang, J.; Yang, C. Data-Driven Yaw Misalignment Correction for Wind Turbines. *Energies* **2021**, *14*, 3481. Available online: https://arxiv.org/pdf/2109.08998 (accessed on 3 September 2025).
7. Astolfi, D.; Pandit, R.; Lombardi, A.; Terzi, L. Diagnosis of Wind Turbine Systematic Yaw Error through Nacelle Anemometer Measurement Analysis. *Sustain. Energy Grids Netw.* **2023**, *34*, 101071. [CrossRef]
8. Pei, Y.; Qian, Z.; Jing, B.; Kang, D.; Zhang, L. Data-Driven Method for Wind Turbine Yaw Angle Sensor Zero-Point Shifting Fault Detection. *Energies* **2018**, *11*, 553. [CrossRef]
9. Mittelmeier, N.; Kühn, M. Determination of optimal wind turbine alignment into the wind and detection of alignment changes with SCADA data. *Wind Energy Sci.* **2018**, *3*, 395–408. Available online: https://wes.copernicus.org/articles/3/395/2018/ (accessed on 4 September 2025). [CrossRef]
10. Liu, Y.; Liu, S.; Zhang, L.; Cao, F.; Wang, L. Optimization of the Yaw Control Error of Wind Turbine. *Front. Energy Res.* **2021**, *9*, 626681. Available online: https://www.frontiersin.org/articles/10.3389/fenrg.2021.626681/pdf (accessed on 7 September 2025). [CrossRef]
11. Saenz-Aguirre, A.; Zulueta, E.; Fernandez-Gamiz, U.; Ramos-Hernanz, J.A.; Lopez-Guede, J.M. Self-Tuning Yaw Control Strategy of a Horizontal-Axis Wind Turbine Based on Machine Learning. In *Numerical Methods for Energy Applications*; Mahdavi Tabatabaei, N., Bizon, N., Eds.; Springer: Cham, Switzerland, 2021; pp. 879–900. [CrossRef]
12. Bu, F.; Huang, W.; Hu, Y.; Xu, Y.; Shi, K.; Wang, Q. Study and Implementation of a Control Algorithm for Wind Turbine Yaw Control System. In Proceedings of the IEEE Canada Electrical Power Conference (EPC), Montreal, QC, Canada, 26–28 October 2009; pp. 1–7. [CrossRef]
13. Kadoche, E.; Gourvénec, S.; Pallud, M.; Levent, T. Multi-Agent Reinforcement Learning for Wind Farm Yaw Control. *Energies* **2023**, *16*, 7892. [CrossRef]
14. Puech, A.; Read, J. Reinforcement Learning for Optimal Single-Turbine Yaw Control. *Energies* **2023**, *16*, 4521. Available online: https://arxiv.org/pdf/2305.01299 (accessed on 12 September 2025).
15. Al-Rubaye, S.Z.; Gil-Pita, R. A Novel Optimization Method for Maximizing Wind Farm Performance through Turbine Positioning and Yaw Angle Estimation. *Energy Convers. Manag.* **2026**, *347*, 120546. [CrossRef]
16. Anagnostopoulos, S.; Bauer, J.; Clare, M.C.A.; Piggott, M.D. Multi-Fidelity ML Models for Real-Time Active Yaw Control in Wind Farms. *Energies* **2023**, *16*, 6124. Available online: https://arxiv.org/pdf/2303.16274 (accessed on 8 September 2025).
17. Yang, J.; Liu, Y.; Zhang, Z.; Ma, Z.; Wang, Y. Review of Control Strategy of Large Horizontal-Axis Wind Turbines Yaw System. *Wind Energy* **2021**, *24*, 97–115. [CrossRef]

https://doi.org/10.3390/pr14071084

18. Wu, Z.; Wang, H. Research on Active Yaw Mechanism of Small Wind Turbines. *Energy Procedia* **2012**, *16*, 53–57. [CrossRef]
19. Islam, M.; Biswas, P.; Rahman, M.; Rahman, A.; Hossain, M. An Intelligent Wind Turbine with Yaw Mechanism Using Machine Learning to Reduce High-Cost Sensors Quantity. ResearchGate Preprint. 2023. Available online: https://ijeecs.iaescore.com/index.php/IJEECS/article/view/29384 (accessed on 24 February 2026).
20. Joshi, A.Y.; Soni, S.G. Design of Active Yaw Control Mechanism for Small Horizontal Axis Wind Turbines. *Int. J. Res. Sci. Innov.* **2015**, *2*, 47–49. Available online: https://rsisinternational.org/Issue11/47-49.pdf (accessed on 9 September 2025).
21. Kerling, I.M.; Zimmer, D. A Yaw Control Approach Using Economic Aspects. In *Proceedings of the 5th International Conference on Contemporary Problems of Thermal Engineering (CPOTE 2018), Gliwice, Poland, 18–21 September 2018*; Institute of Thermal Technology: Gliwice, Poland, 2018. Available online: https://elib.dlr.de/123975/1/CPOTE2018_final_kerling.pdf (accessed on 18 September 2025).
22. Tsioumas, E.; Karakasis, N.; Jabbour, N.; Mademlis, C. Indirect Estimation of the Yaw-Angle Misalignment in a Horizontal Axis Wind Turbine. In Proceedings of the 2017 IEEE 26th International Symposium on Industrial Electronics (ISIE), Edmonton, AB, Canada, 21–23 June 2017; pp. 1239–1246. [CrossRef]
23. Shahadat, M.M.Z.; Chaity, T.T.; Ahmed, M.M.; Kader, M.I.; Snigdho, N.A. Enhancing the Wind Energy Harvesting Capacity of a Horizontal Axis Wind Turbine (HAWT) Using a Novel Wind Tracking System. In Proceedings of the IEEE International Conference on Power, Electrical, Electronics and Industrial Applications (PEEIACON 2024), Rajshahi, Bangladesh, 12–13 September 2024; pp. 1–6. [CrossRef]
24. Abhi, S.B.; Hossain, M.I.; Hoque Suny, R.; Zahura, F.T.; Hazari, M.R.; Jahan, E.; Mannan, M.A. Design and Implementation of a Smart Wind Turbine with Yaw Mechanism. In Proceedings of the 3rd International Conference on Robotics, Electrical and Signal Processing Techniques (ICREST 2023), Dhaka, Bangladesh, 7–8 January 2023; pp. 358–362. [CrossRef]
25. Yin, X.; Zhao, X. LSTM Predictive Yaw Control for Offshore Wind Turbines. *Energies* **2021**, *14*, 6301.
26. Yang, Y.; Solomin, E.V.; Shishkov, A.N. Wind Direction Prediction Based on Nonlinear Autoregression and Elman Neural Networks for the Wind Turbine Yaw System. Preprint, SSRN 4128942. 2022. Available online: http://dx.doi.org/10.2139/ssrn.4128942 (accessed on 1 February 2026).
27. Rott, A.; Höning, L.; Hulsman, P.; Lukassen, L.J.; Moldenhauer, C.; Kühn, M. Wind Vane Correction During Yaw Misalignment for Horizontal-Axis Wind Turbines. *Wind Energy Sci.* **2023**, *8*, 1755–1770. [CrossRef]
28. Santoni, C.; Zhang, Z.; Sotiropoulos, F.; Khosronejad, A. A Data-Driven Machine Learning Approach for Yaw Control Applications of Wind Farms. *Theor. Appl. Mech. Lett.* **2023**, *13*, 100471. [CrossRef]
29. Elkodama, A.; Abdellatif, A.; Shaaban, S.; Rushdi, M.A.; Yoshida, S.; Ismaiel, A. Investigation into the Yaw Control of a Twin-Rotor 10 MW Wind Turbine. *Appl. Sci.* **2024**, *14*, 9810. [CrossRef]
30. *IEC 61400-1:2019*; Wind Energy Generation Systems–Part 1: Design Requirements. International Electrotechnical Commission: Geneva, Switzerland, 2019.
31. Falcon Silence 3.6 kW Wind Turbine. Turbinawiatrowa. Available online: https://www.turbinawiatrowa.com/EN-H4/oferta/3/turbina-wiatrowa-36kw-falcon-silence.html (accessed on 5 January 2026).
32. Hansen, M.O.L. *Aerodynamics of Wind Turbines*, 3rd ed.; Routledge: New York, NY, USA, 2015; pp. 21–24.
33. Sereema. *Yaw Misalignment: (R)Evolution*; Technical White Paper; Sereema SAS: Montpellier, France, 2019; p. 11. Available online: https://www.sereema.com/documents/yaw-misalignment-white-paper (accessed on 17 November 2025).
34. Historical and Live Weather Data for Paradise, Newfoundland and Labrador (City ID: 6324733). Available online: https://openweathermap.org/city/5905393 (accessed on 12 January 2026).
35. Environment and Climate Change Canada. Historical Climate Data—Search and Download. Government of Canada. 2024. Available online: https://climate.weather.gc.ca/historical_data/search_historic_data_e.html (accessed on 10 January 2026).

https://doi.org/10.3390/pr14071084

MDPI AG
Grosspeteranlage 5
4052 Basel
Switzerland
Tel.: +41 61 683 77 34

Processes Editorial Office
E-mail: processes@mdpi.com
www.mdpi.com/journal/processes

www.ingramcontent.com/pod-product-compliance
Lightning Source LLC
Chambersburg PA
CBHW061601220926
51850CB00005B/394
9783725876792